Water and Life in Tonle Sap Lake

Chihiro Yoshimura • Rajendra Khanal •
Uk Sovannara
Editors

Water and Life in Tonle Sap Lake

Springer

Editors
Chihiro Yoshimura
Tokyo Institute of Technology
Tokyo, Japan

Rajendra Khanal
Tokyo Institute of Technology
Tokyo, Japan

Policy Research Institute
Kathmandu, Nepal

Uk Sovannara
Tokyo Institute of Technology
Tokyo, Japan

ISBN 978-981-16-6631-5 ISBN 978-981-16-6632-2 (eBook)
https://doi.org/10.1007/978-981-16-6632-2

This Springer imprint is published by the registered company Springer Nature Singapore Pte Ltd.
The registered company address is: 152 Beach Road, #21-01/04 Gateway East, Singapore 189721, Singapore

Foreword

Tonle Sap Lake (TSL) is situated in the central plains of Cambodia. The lake basin harbors some 32 % of the nearly 15 million national population. TSL is an integral part of the Mekong River which originates in the Himalayas and flows through five other riparian countries in Southeast and East Asia, namely China, Lao PDR, Myanmar, Thailand, and Vietnam. Through the interconnecting Tonle Sap River, it forms a unique "pulsing" hydrological system serving for an enormously diverse aquatic ecosystem with rich fish resources to supply protein to local livelihoods. The majority of protein for consumption by Cambodians come from inland fisheries in the TSL flood plain. TSL is ranked fourth globally in terms of its productivity among freshwater bodies, making a remarkable contribution to the economic growth of the country by contributing around 10% to the Cambodian national gross domestic product. In the meantime, the socioeconomic development activities as a whole in the TSL floodplain invariably accompanies exploitative transformations of watershed lands that contradictorily constitute important part of the UNESCO-accredited Biosphere Reserve being set for the whole of lake and its basin. In addition, the upstream development along the Mekong River has long been adversely affecting the soundness of lake ecosystem.

Although numerous attempts have been made to address the TSL ecosystem sustainability over the past decades at the national, regional, and global levels, the pace of TSL degradation has been accelerating. There could be many possible reasons why such attempts have been far less than satisfactory. The consequences and impacts of environmental degradation have implications to a wide range of academic disciplinary fields, including but not limited to socioeconomic, agriculture policy and management, manufacturing technology, legislation, natural and life sciences, and engineering. They have been very useful in increasing the depth of knowledge and for undertaking important policy interventions, but few were meant to focus on transdisciplinary research involving all of the major national institutions in Cambodia jointly working on TSL.

With the above in the background, a unique transdisciplinary research collaboration (TDRC) project, entitled "Establishment of Environmental Conservation Platform of Tonle Sap Lake" was agreed to be implemented during the period of 2016 and 2021, under the framework of Science and Technology Research Partnership for Sustainable Development (SATREPS) being funded by the Japan International Cooperation (JICA) and Japan Science and Technology Agency (JST). The institutions participated from Cambodia included the Institute of Technology of Cambodia (ITC), the Ministry of Environment (MOE), Tonle Sap Authority (TSA), the Ministry of Water Resource and Meteorology (MOWRM), and the Royal University of Phnom Penh (RUPP), while the counterpart institutions from Japan included Tokyo Institute of Technology (Tokyo Tech) in collaboration with the Institute for Global Environmental Strategies (IGES) and Yamagata University (YU). The International Lake Environment Committee Foundation (ILEC) had indirect but close collaboration with the project.

To address the complex environmental issues facing TSL, a TDRC platform was established for undertaking collaborative and transdisciplinary research in the broad fields of natural science, bioresource and environmental science, agriculture science and technology, and global environment studies. Throughout the process of TDRC, special emphasis was placed on technology transfer, knowledge sharing, resource sharing, and international research collaboration, etc. with a particular focus on capacity development. The work undertaken spans the general categories of (I) Socioeconomics and Governance, (II) Climate and Hydrology, (III) Hydrodynamics, (IV) Sediment Dynamics, (V) Physicochemical Water Quality, (VI) Microbial Community, (VII) Flora and Fauna, (VIII) Chemical Pollution, (IX) Sanitation and Health Risk, and (X) Environmental Shifts and Management. In all, there are 48 thematic chapters in this book covering the activities undertaken by the teams formed under the above ten categories. The five-year intensive field research activities were supported by the latest scientific and technology knowledge, tools, and methodologies, with particular emphasis placed on sharing of the acquired interim research outputs among the concerned research teams.

In this book, all of the work undertaken by these teams is presented in 48 chapters, with two synthesis chapters (Chaps. 49 and 50) categorized under (XI) Outlook for Sustainability. A set of bullet-point "Key points" is provided in the concluding part of each chapter, echoing the thematic outline provided in the respective Abstract section in the front part. The above two synthesis chapters are entitled "Transdisciplinary Research Collaboration for Environmental Conservation," and "Recommendations for Further Research and Environmental Management." Chapter 49 illustrates the outputs of TDRC on TSL produced by the team over the course of this project. In a nutshell, the chapter points out that environmental issues are multidimensional, that the consequence and impacts of environmental degradation are on a wide range of fields, and that TDRC must be conducted in such a way that the evidence generated through research is implemented in an effective way for policy implementation. Chapter 50 provides two sets of recommendations for the potential users of TDCR outputs as well as the readers of this book, i.e., (1) Recommendations for Future Research and (2) Recommendation for Environmental

Management in a well-organized format. As stated in the Final Remark part of this chapter, "the project team has succeeded in taking steps in the journey to fully place TSL in the context of limnology and environmental science," and the readers of this book would be able to grasp "the essence of why TSL is so fascinating in terms of limnology, biogeochemistry, and ecology."

In all, the book highlights the possibilities and limitations of our understanding of the TSL ecosystem and will be extremely useful for those working, or needing to work, on similar lake basin management situations in a transdisciplinary manner, particularly in the tropical climate region. Further, it will surely serve as a very useful reference source for researchers, scientists, policymakers, and stakeholders wishing to contribute to the development of strategies toward Integrated Lake Basin Management (ILBM) for sustainable resource use, an approach focusing on incremental, continuous, and long-term improvement of basin governance of lakes and their interlinked water systems. Advancement of TDRC as attempted for TSL will also prove to be of enormous benefit for integrated management of such systems globally.

Masahisa Nakamura

International Lake Environment
Committee Foundation (ILEC),
Kusatsu, Japan

Shiga University, Otsu, Japan
29.07.2021

Acknowledgments

This book is among the most significant outcomes of the international project "Establishment of Environmental Conservation Platform of Tonle Sap Lake (2016–2022)." The editors and authors are grateful to Japan Science and Technology Agency (JST) and Japan International Cooperation Agency (JICA) for the grant funding (grant-number JPMJSA1503). They have continuously supported our activities as one of the research projects under the Science and Technology Research Partnership for Sustainable Development (SATREPS) and enabled us to produce this technical book. We would also like to express our sincere appreciation to our Cambodian counterparts, namely the Tonle Sap Authority (TSA), Ministry of Environment (MOE), and Ministry of Water Resources and Meteorology (MOWRAM) for their substantial contribution to the project.

The editors deeply thank the Mekong River Commission (MRC) and the Water Utilisation Programme Finnish Component (WUP-FIN), especially Dr Matti Kummu, for the kind provision of the environmental data. These outstanding contributions were invaluable in assisting us to integrate the fragmented information on Tonle Sap Lake as presented in this book. Our sincere appreciations go also to the peer reviewers in the relevant fields, who generously provided us with excellent comments that help enrich all the major parts. They are listed on the following page.

Last but not least, we would also like to express our particular thanks of profound gratitude to Ms Aiko Yamashita (the project coordinator), Ms Kimiko Ando and Ms Manami Miyamoto (the project assistants), and Dr Mei Hann Lee (the editor at Springer) for their countless technical and administrative supports. The language (English) was proofread by Enago (Crimson Interactive Pvt. Ltd.). The accomplishment of this book would not have been possible if there were not for the financial support, dynamic efforts, and technical and administrative assistance of these many institutions and individuals.

Peer Reviewers

Part I Socioeconomics and Governance

Eric Zusman, Institute for Global Environmental Strategies (IGES), Japan
Matthew McCandless, International Institute for Sustainable Development (IISD), Canada

Part II Climate and Hydrology

Masayuki Fujihara, Kyoto University, Japan
Mean Sovanna, Kyoto University, Japan
Seng Theara, Kyoto University, Japan
Takeo Yoshida, National Agriculture and Food Research Organization, Japan

Part III. Hydrodynamics

Laurence Hawker, The University of Bristol, United Kingdom
So Kazama, Tohoku University, Japan

Part IV. Sediment Dynamics

Stephen E. Darby, University of Southampton, United Kingdom
Tsuyoshi Kinouchi, Tokyo Institute of Technology, Japan

Part V. Physicochemical Water Quality

Pham Quy Giang, Ha Long University, Vietnam
Binaya Kumar Mishra, Pokhara University, Nepal

Part VI. Microbial Community

Satoshi Tsuneda, Waseda University, Japan
Satoshi Wakai, Japan Agency for Marine-Earth Science and Technology (JAMSTEC), Japan

Part VII. Flora and Fauna

Bena Smith, Wildfowl & Wetlands Trust (WWT), United Kingdom
Satoshi Kameyama, National Institute for Environmental Studies (NIES), Japan
Tomos Avent, Wildfowl & Wetlands Trust (WWT), United Kingdom

Part VIII. Chemical Pollution

Carl Renan Estrellan, Eurofins Agroscience Services, United States of America
Charissa Ferrera, University of the Philippines, Philippines
Susan Gallardo, De La Salle University, Philippines

Part IX. Sanitation and Health Risk

Nguyen Thanh Gia, Hue University of Medicine and Pharmacy, Vietnam
Pham Duy Dong, National University of Civil Engineering, Vietnam

Part X. Environmental Shifts and Management

Makoto Umeda, Nihon University, Japan
Manoj Khanal, Tribhuvan University, Nepal
Tetsuro Kikuchi, Japan International Research Center for Agricultural Sciences (JIRCAS), Japan

Contents

Chapter 1
Introduction

Uk Sovannara, Rajendra Khanal, and Chihiro Yoshimura

1.1 Importance of Tropical Lakes

Freshwater ecosystems provide a myriad of benefits to humanity. It ranges from fundamental needs such as water, food, and energy supply, to tourist, recreation, and transportation needs. Among them, lakes hold only approximately 0.4% of global freshwater volume (Wetzel 1983). However, lakes (or *lentic* systems) are epic centers of biodiversity, harboring a variety of endangered and vulnerable faunal communities (e.g., fish and waterbirds) as well as the biodiversity of global significance (e.g., agricultural genetic diversity). Because of their immense value, the lake ecosystems have been enlisted as the "Biospherc reserves," meaning they are "learning places for sustainable development under diverse ecological, social and economic contexts" (UNESCO 2020). The Common International Classification of Ecosystem Services (CICES, v5.1; https://cices.eu/) has listed 90 "contributions that ecosystems make to human well-being," of which 58 are potentially applicable to lakes (Sterner et al. 2020).

Among the 1500 lakes that account for the most surface area of lakes globally, approximately 10% of them are located in the tropical region (Lewis 1996). Tropical aquatic environments, especially freshwater lakes and their associated floodplains and wetlands, play a pivotal role in supporting a wide array of ecological, socioeconomic, and cultural values, as well as remarkably high productivity and rich biodiversity (Dudgeon 2000; Gaston 2000; Kolding and Zwieten 2006; Uk et al. 2018; UNESCO 2020). The tropical freshwater ecosystems contribute to nearly 15%, or maybe even higher, of the world's reported total fisheries production (Kolding and

U. Sovannara (✉) · C. Yoshimura
Tokyo Institute of Technology, Tokyo, Japan

R. Khanal
Tokyo Institute of Technology, Tokyo, Japan

Policy Research Institute, Kathmandu, Nepal

C. Yoshimura et al. (eds.), *Water and Life in Tonle Sap Lake*,
https://doi.org/10.1007/978-981-16-6632-2_1

Zwieten 2006). Several studies (e.g., Lewis 1987; Irvine et al. 2016; Santoso and Toruan 2020; Ramírez et al. 2020) have also highlighted the uniqueness of tropical ecosystems, including high solar radiation intensity, rich precipitation, dominance by omnivorous fishes, adaptive radiation, and high species diversity, compared to lakes in the temperate zone.

In the face of unprecedented global change, however, tropical ecosystems have been put under serious threats. In the last half-century, anthropogenic pressures (e.g., growing resource consumption, rapid urbanization, and swelling population) have caused major ecological impacts, adversely affecting the tropical biodiversity and ecosystems in multiple and interacting ways (Vörösmarty et al. 2010; Irvine et al. 2016; Santoso and Toruan 2020). The stress from the fishery has already been one of the most serious threats to biodiversity and sustainable livelihoods across the tropics (Allan et al. 2005). Climate change and hydropower dam construction have also been considered or projected to affect the phenology and the inland fishery and aquacultures of the tropical systems (Barange and Perry 2009; NASA 2015; Lin and Qi 2017; Uk et al. 2018). To date, only less than 3% of the references found and indexed by Scopus as being within the category of "limnology" were grouped under tropical studies (Santoso and Toruan 2020). Moreover, despite numerous researches and the availability of quantitative information on various processes, structures, and functions, our overall understanding of the tropical lakes remains scarce and fragmented (Liljeström et al. 2012; Sarmento 2012; Uk et al. 2018). Santoso and Toruan (2020) also emphasized the need to further study the ecosystem services provided by tropical inland waters and their resilience to disturbance.

1.2 Tonle Sap Lake as a Unique Ecosystem

Tonle Sap Lake (TSL; "Tonle" means "river" and "Sap" means "not salty or fresh" in the Khmer language) is a unique flood-pulse and a complex hydroecological system. This shallow lake is an integral part of the Mekong River (MR) basin and the largest freshwater body in Southeast Asia, with a vast seasonally inundated floodplain surrounding the lake (Fig. 1.1). The expansive floodplain offers shelters and breeding colonies for numerous fauna and flora species, including the globally significant biodiversity (i.e., rare, near threatened, and globally critically endangered species; Cambodia National Mekong Committee 2004; Campbell et al. 2006; Sun and Mahood 2015). The intense interaction among hydrology, biogeochemistry, ecology, and the human domain forms a remarkable feature of the lake's ecosystem. We would recommend readers to watch the video clip on YouTube (URL. https://www.youtube.com/watch?v=8cvUyD6mpxk&t=5s) for obtaining a vivid overview of this lake prior to reading the following chapters.

Owing to the geographical and monsoonal setting, TSL has an extraordinary hydrological system with the largest natural freshwater flow reversal in the world. Its hydrology is intimately linked with MR, which was estimated to become connected during the mid-Holocene (Darby et al. 2020). MR flows southward over a distance of

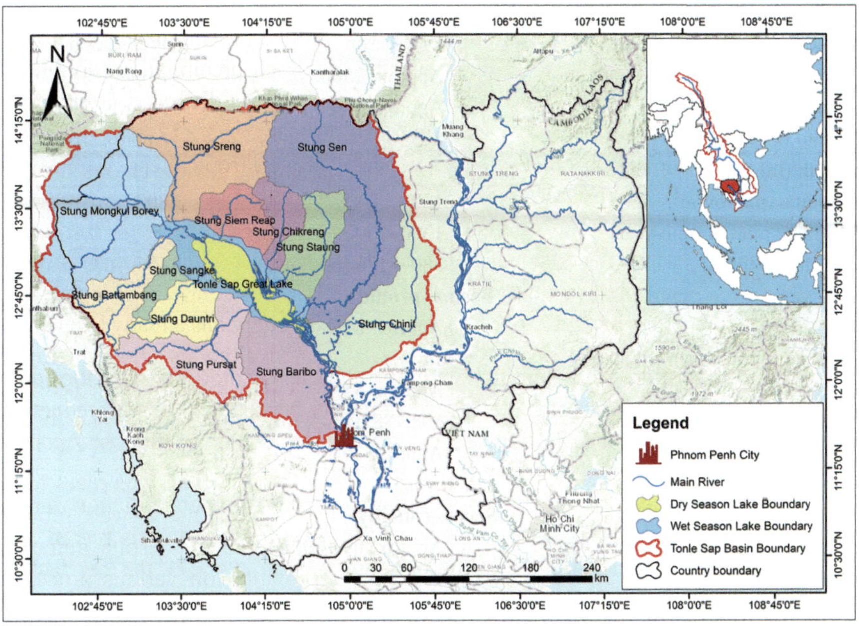

Fig. 1.1 Map of Tonle Sap Lake. Adopted from Binaya Shivakoti and Bao (2020)

4350 km across six countries, originating in the Tibetan highlands, viz., China, Myanmar, Laos, Thailand, Cambodia, and Vietnam, to empty into the South China Sea. In Cambodia, MR is connected to TSL via a 120-km-long Tonle Sap River (TSR) at Chaktomuk, the famous conjunction where four rivers converge. TSR is the only river in the world with a bidirectional flow. Thus, it plays a dual role as both the inlet and the outlet of TSL.

TSL has only a single river outlet TSR through which it discharges water to MR, usually from September/October to April/May. By May/June (the beginning of the wet monsoon season), MR's water level gets higher than TSL. It pushes the flow in TSR to flip its direction toward the upstream to fill the lake with an enormous volume of water, alluvial sediment, nutrients, and even fish until the lake peaks and receives no more water (September/October). The lake has no river outlet during the reversed flow period. It is due to such a unique phenomenon that TSL cyclically oscillates between deep and shallow conditions. Simultaneously, its surrounding floodplain perpetually shifts between terrestrial and aquatic phases (i.e., seasonal inundation). The seasonal water level fluctuation was relatively large, varying between 1 and 2 m in the low-water period (the dry season) to more than 8–10 m at the peak of the food (the wet season), resulting in the expansion of the lake's surface area of approximately sixfold, from roughly 2500 km^2 to more than 15,000 km^2, respectively (Table 1.1; Binaya and Bao 2020).

Such flood-pulse dynamics not only determine the hydrodynamic and sediment conditions (i.e., connectivity) but also drive high spatiotemporal variation in water

Table 1.1 Basics of Tonle Sap Lake basin

Lake profile	Dry season	Wet season
Water surface area (km^2)	Approx. 2500	Approx. 15,278
Length (km)	Approx. 120	Approx. 250
Width (km)	3–35	Approx. 100
Depth (m)	<2	<11
Volume (million m^3)	1300	8000
Average rainfall (mm/year)	1300–1500	
Discharge (m^3/s)	380–8176 (outflow)	104–7032 (inflow)
Basins' profile		
Area (km^2)	Approx. 85,786	
# of sub-basins/main tributaries	11	
Forest (%)	Approx. 55	
Flooded forests (ha)	608,188 ha (approx. 3%)	
Agriculture (%)	Approx. 45	
Urban (%)	< 0.5	
Socio-economy		
Population around the lake	Approx. five million	
Fishing families	135,000	
# Land-based villages	948	
# Land and water-based villages	36	
# of floating (water-based) villages	53	
Fish catch (1000 tons/year)	289–431	

Adopted from Shivakoti and Bao (2020)

quality and the biotic community (e.g., fish, plant, phytoplankton, and bacteria; Arias et al. 2012, Oyagi et al. 2017; Siev et al. 2018; Ung et al. 2019; Hiroki et al. 2020, Khanal et al. 2021, Halls & Hortle 2021). They concomitantly affect the region's socioeconomic setting (e.g., fishery and agriculture and water-related migration; Heinonen 2006; Hortle 2007). Recent evidence confirmed that TSL's productivity is supported by autochthonous primary production and macrophytes and allochthonous inputs from terrestrial plants and seasonally inundated floodplain (Hiroki et al. 2020). In the floodplain around TSL, the flooded forests—distribution, structure, and variation—are also primarily dependent on the flood-pulse dynamics (Arias et al. 2012). Fish in TSL has also strongly adapted to the unique habitats created by the seasonal flood-pulse dynamics, and the migratory fish species migrate in the extensive system of TSL, TSR, and MR.

For Cambodians, the water and life are inseparably intertwined. Such intimate interaction has been nurturing Cambodia's cultural identity and economy. For instance, the livelihood of at least five million people living a full-time life on the lake as floating communities (e.g., floating houses, schools, and hospitals) and in the vicinity of the lake as water-based communities (i.e., seasonally inundated communities) and land-based communities (i.e., permanently on land) has also adapted to the ebbs and flows and is also reliant upon the natural resources and services of the lake as the source of sustenance and income for generations (e.g., fish and water resources; Table 1.1, Bao et al. 2020). Anita (2015) referred TSL to as "a birthplace of civilizations" and also a major pillar of the emergence and thrift of the Khmer Empire (ninth–fifteenth century) owing to the great wealth the Great Lake generated (i.e., intense rice production and abundance of fish). She also mentioned the ancient Chinese archives that described the link between TSL and the Khmer civilization during the era of the kingdom of Chenla (sixth–eighth century). Until present, Cambodian people celebrate a 3-day-long Water Festival, known locally as "Bon Om Touk," on an annual basis, usually in November, marking the end of the wet season and the annual natural occurrence of the water outflow from TSL to MR.

Overall, the ecosystem of TSL plays an essential role in supporting the country's socioeconomic development, serving as the foundation of food security in Cambodia. Nonetheless, it is a large and complex data-deficient ecosystem in which there exist many interactions among various components, while our understanding of this ecosystem is fragmented and immature. With this backdrop, it is essential to integrate the relevant information regarding TSL to obtain a solid scientific understanding or develop management measures for water resources management, pollution control, wildlife management, and ecosystem management (Uk et al. 2018). Therefore, this lake ecosystem presents us with excellent frontiers in science, particularly in hydrology, sedimentology, wetland ecology, fisheries science, agriculture, and environmental science and engineering.

1.3 The Target and Structure of This Book

This book, Water and Life in Tonle Sap Lake, is the first comprehensive compilation of the environmental significance, research, and policy significance of TSL, highlighted by comparisons to other lakes in the world. The book covers multiple different topics and will be beneficial to readers in understanding the ecosystem and livelihood of TSL and taking actions to conserve the lake environment.

Management and conservation of tropical lake ecosystems are complicated and require multidimensional approaches. Most importantly, we need a fundamental understanding of tropical limnology. Equally important, various other key factors that link the connections between socio-ecological structures and scientific understanding, economics, and capacity development are also required to reverse the degradation trends and develop sustainable solutions to the effective management of tropical inland waters (Irvine et al. 2016; Santoso and Toruan 2020). Effective partnerships across regions and disciplines may also generate an environment that supports the development of tropical limnology (Santoso and Toruan 2020).

Moreover, the ecology of inland waters in the tropics should be of great concern to limnologists since the economic growth in countries in these areas is among the fastest in the world. The challenges and concerns, in turn, provide us with great opportunities for innovative science (Santoso and Toruan 2020). The pressing issues and challenges in managing the TSL ecosystem presented in this book are also relevant to other shallow lakes in other developing countries.

We attempted to concisely structure and organize the relevant information in both spatial and temporal dimensions together with the elaboration of the processes, underlying phenomena, and changes that govern the dynamics of TSL. This book may also be used as a textbook or a handy reference in environmental science and engineering at undergraduate and graduate levels for understanding and synthesizing new research directions relevant to the water environment. Each chapter focuses on one topic and outlines its basics and recent research outcomes which were synthesized from extensive field surveys, experimental and modeling works, and constructive discussion among scientists and stakeholders.

The book also provides an example of a transdisciplinary research collaboration, which established the Platform for Aquatic Ecosystem Research (PAER). Research and management scenarios for the future are discussed based on the anticipated change in the economic development, urbanization, climate change, and sociopolitical scenario in Southeast Asia. Thus, the book also accommodates general readers, academicians and researchers, the local community, environmental managers and policymakers, scientists, and relevant stakeholders interested in and relevant to shallow lake ecosystems. Finally, this book will also aid policymakers and researchers in developing effective practices in environmental management and relevant research plans regarding shallow tropical lakes.

This book is structured in 11 parts (Fig. 1.2), covering the lake ecosystem's key components and human life: socioeconomics, hydrodynamics, water chemistry, microbiology, ecology, sanitation, and environmental management. Each chapter is introduced by *Abstract*, followed by the main contents, and the important learnings are summarized *Key Points* section. *References and Further Readings* is at the end of each chapter. *Abstract* and *Key Points* provide readers with a concise summary of the chapter, helping the readers grasp the chapter's essence. The *References and Further Readings* section contains a compiled list of additional reading materials related to the chapter's topic.

Part I discusses the local community and livelihood, governance, and the impacts of recent environmental changes on them in and around the lake. Parts II and III focus on the hydrology and hydrodynamics of TSL and its tributaries. Parts IV and V cover the dynamics of sediment and physicochemical water quality, including sediment yield, transport, deposition, and resuspension. Part VI provides an overview of the microbial community in TSL, with particular attention on pathogens. Part VII describes the major biological components in relation to ecosystem functions and services. Part VIII presents the status of chemical pollution, focusing on metals and chemicals from agriculture in the lake basin. Part IX addresses sanitation and health risk assessment, extensively discussing its basic concept, methodology, and application. In Part X, the main focus is the decision-making tools and concepts to conserve the water environment. Lastly, Part XI summarizes the recommendations for strengthening science, engineering, and management regarding tropical lakes.

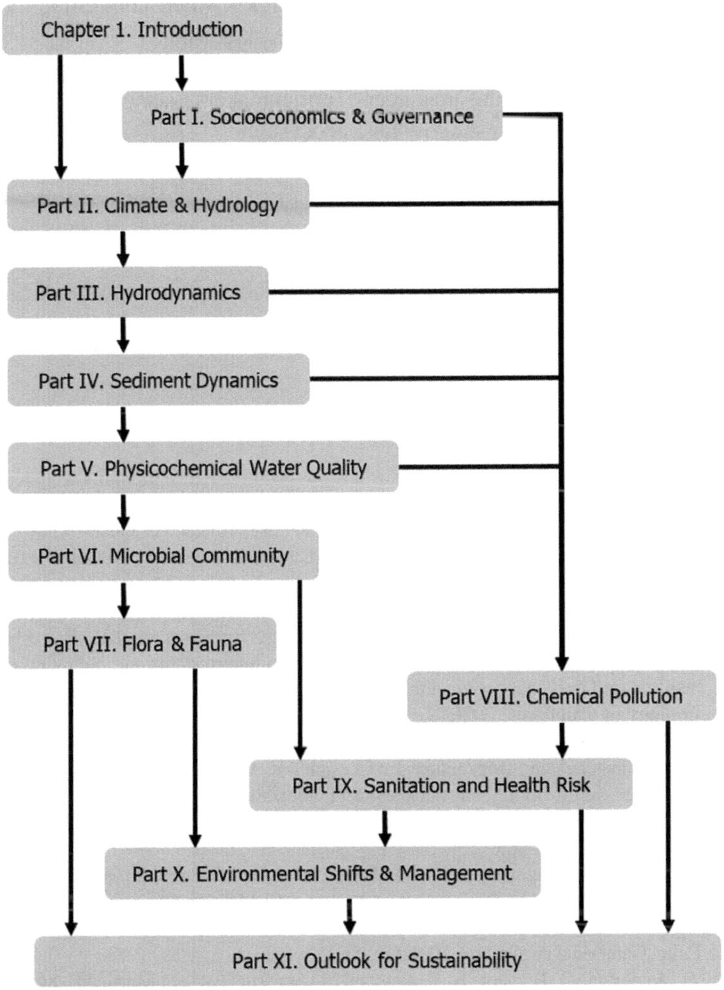

Fig. 1.2 Structure and flow of the content of this book

Key Points

- Tropical lake ecosystems are crucial in terms of their ecological, socioeconomic, and cultural virtues, being an epicenter for biodiversity and offering numerous ecosystem services for humanity.
- TSL is a unique ecosystem with the largest natural freshwater flow reversal in the world and the intense interactions among various domains that present us with excellent scientific challenges regarding the ecological and social importance of tropical lakes.
- This book aims to provide the latest scientific knowledge, modeling tools, and management practices on water and life in TSL in Cambodia, covering various

topics, including socioeconomics, hydrodynamics, water chemistry, microbiology, ecology, sanitation, and environmental management.

- The book is beneficial to a wide range of readers, viz., students, academicians, researchers, environmental managers, policymakers, scientists, and relevant stakeholders interested in shallow lake ecosystems.
- Each chapter contains an *Abstract*, main contents, and *Key Points*, followed by *Appendix* (if available) and *References and Further Readings*. *Abstract* and *Key points* provide readers with a concise summary of the chapter, helping the readers grasp the chapter's essence. *Appendix* covers additional useful resources (e.g., figure, movie, and model code) which are available on the website of Springer.

References

Allan JD, Abell R, Hogan Z, Revenga C, Taylor BW, Welcomme RL, Winemiller K. Overfishing of inland waters. Bioscience. 2005;55:1041.

Arias EM, Cochrane T, Piman T, Kummu M, Caruso SB, Killeen JT. Quantifying changes in flooding and habitats in the Tonle Sap Lake (Cambodia) caused by water infrastructure development and climate change in the Mekong Basin. J Environ Manag. 2012;112:53–66.

Bao NP, Uk S, Shivakoti RB, Chihiro Y, Khanal R, Siev S, Seingheng H, Aiko Y. Socio-economic dependency of local communities on Tonle Sap Lake. In: Shivakoti RB, Bao PN, editors. Environmental changes in Tonle Sap Lake and Its Floodplain: status and policy recommendations. Institute for Global Environmental Strategies, Tokyo Institute of Technology and Institute of Technology of Cambodia; 2020.

Barange M, Perry RI. Physical and ecological impacts of climate change relevant to marine and inland capture fisheries and aquaculture. In: Cochrane K, De Young C, Soto D, Bahri T, editors. Climate change implications for fisheries and aquaculture: overview of current scientific knowledge. FAO Fisheries and Aquaculture Technical Paper. No. 530. Rome: FAO; 2009. p. 7–106.

Cambodia National Mekong Committee. Sub-area analysis and development: the Tonle Sap Sub-Area. Phnom Penh, Cambodia: Cambodia National Mekong Committee; 2004.

Campbell CI, Poole C, Giesen W, Valbo-Jorgensen J. Species diversity and ecology of Tonle Sap Great Lake. Cambodia Aquat Sci. 2006;68:355–73.

Darby, E. S., Langdon, G. P., Best, L. J., Leyland, J., Hackney, R. C., Marti, M., Morgan, R. P., Ben, S., Aalto, R., Parsons, R. D., Nicholas, P. A., Leng, J. M. (2020). Drainage and erosion of Cambodia's great lake in the middle-late Holocene: the combined role of climatic drying, base-level fall and river capture. Quatern Sci Rev 236, 106265, ISSN 0277–3791.

Dudgeon D. The ecology of tropical Asian rivers and streams in relation to biodiversity conservation. Annu Rev Ecol Evol Syst. 2000;31:239–63.

Gaston KJ. Global patterns in biodiversity. Nature. 2000;405:220–7.

Halls AS, Hortle KG. Flooding is a key driver of the Tonle Sap dai fishery in Cambodia. Sci Rep. 2021;11:3806.

Heinonen U. Environmental impact on migration in Cambodia: Water-related migration from the Tonle Sap Lake Region. Int J Water Resources Dev. 2006;22(2)

Hiroki M, Tomioka N, Murata T, Imai A, Jutagate T, Preecha C, Avakul P, Fukushima M. Primary production estimated for large lakes and reservoirs in the Mekong River Basin. Sci Total Environ. 2020;747:141133. ISSN 0048-9697

Hortle KG. Consumption and the yield of fish and other aquatic animals from the Lower Mekong Basin. MRC Technical Paper No. 16, Mekong River Commission, Vientiane. 87 pp., 2007.

Irvine K, Castello L, Junquera A, Moulton T. Linking ecology with social development for tropical aquatic conservation. Aquatic Conserv: Mar Freshw Ecosyst. 2016;26:917–41.

Khanal R, Uk S, Kodikara D, Siev S, Yoshimura C. Impact of Water Level Fluctuation on Sediment and Phosphorous Dynamics in Tonle Sap Lake, Cambodia. Water Air Soil Pollut. 2021;232: 139.

Kolding J, van Zwieten PAM (2006). Improving productivity in tropical lakes and reservoirs. Challenge Program on Water and Food-Aquatic Ecosystems and Fisheries Review Series 1. Theme 3 of CPWF, C/o WorldFish Center, Cairo, Egypt. 139 pp. ISBN: 977-17-3087-8.

Lau-Bignon AW. Le Tonlé Sap, bassin historique de civilisations. L'Espace géographique 1. 2015;44:81–8. (in French)

Lewis WM Jr. Tropical limnology. Ann Rev Ecol Syst. 1987;18:159–84.

Lewis WM Jr. Tropical lakes: how latitude makes a difference. In: Schiemer F, Boland KT, editors. Perspectives in tropical limnology. Amsterdam: SPB Academic Publishers; 1996. p. 43–64.

Liljeström I, Kummu M, Varis O. Nutrient balance assessment in the Mekong Basin: nitrogen and phosphorus dynamics in a catchment scale. Int J Water Resources Dev. 2012;28(2):373–91.

Lin Z, Qi J. Hydro-dam—a nature-based solution or an ecological problem: the fate of the Tonlé Sap Lake. Environ Res. 2017;158:24–32.

NASA. (2015). Study: Climate change rapidly warming world's lakes. Retrieved from https://climate.nasa.gov/news/2378/study-climate-change-rapidly-warming-worlds-lakes/. Accessed on 3/4/2021

Oyagi H, Endoh S, Ishikawa T, Okura Y, Tsukawaki S. Seasonal changes in water quality as affected by water level fluctuations in Lake Tonle Sap, Cambodia. Geograph Rev Japan Ser B. 2017;90(2):53–65.

Ramírez A, Caballero M, Vázquez G, et al. Preface: recent advances in tropical lake research. Hydrobiologia. 2020;847:4143–4.

Santoso AB, Toruan RL. IOP Conf Ser: Earth Environ Sci. 2020;535:012007.

Sarmento H. New paradigms in tropical limnology: The importance of the microbial food web. Hydrobiologia. 2012;686:1–14.

Shivakoti RB, Bao PN, editors. Environmental changes in Tonle Sap Lake and its floodplain: status and policy recommendations, Institute for Global Environmental Strategies, Tokyo Institute of Technology and Institute of Technology of Cambodia, 2020.

Siev S, Yang H, Sok T, Uk S, Song L, Kodikara D, Oeurng C, Hul S, Yoshimura C. Sediment dynamics in a large shallow lake characterized by seasonal flood pulse in Southeast Asia. Sci Total Environ. 2018;631–632:597–607.

Sterner WR, Keeler B, Polasky S, Poudel R, Rhude K, Rogers M. Ecosystem services of Earth's largest freshwater lakes. Ecosyst Serv. 2020;41:101046. ISSN 2212-0416

Sun V, Mahood S. Wildlife Monitoring at Prek Toal Ramsar Site, Tonle Sap Great Lake, 2013 and 2014. Wildlife Conservation Society, Cambodia Program, Phnom Penh, 2015.

Uk S, Yoshimura C, Siev S, Try S, Yang H, Oeurng C, Li S, Hul S. Tonle Sap Lake: current status and important research directions for environmental management. Lakes Reserv Res Manag. 2018;23(3):177–89.

UNESCO (2020) Biosphere reserves. Retrieved from: https://en.unesco.org/biosphere/about. Accessed 28 Jan 2021

Ung P, Peng C, Yuk S, Tan R, Ann V, Miyanaga K, Tanji Y. Dynamics of bacterial community in Tonle Sap Lake, a large tropical flood-pulse system in Southeast Asia. Sci Total Environ. 2019;664:414–23.

Vörösmarty CJ, McIntyre PB, Gessner MO, Dudgeon D, Prusevich A, Green P, Glidden S, Bunn SE, Sullivan CA, Liermann CR, et al. Global threats to human water security and river biodiversity. Nature. 2010;467:555–61.

Wetzel RG. Limnology. 2nd ed. Philadelphia: Saunders College Publishing; 1983. 767 pp.

Part I
Socioeconomics and Governance

Chapter 2
Lake and Livelihoods: Threats to Their Sustainability

Pham Ngoc Bao, Khoeurn Kimleang, Boeut Sophea, Nobue Amanuma, Binaya Raj Shivakoti, and Rajendra Khanal

2.1 Lake Environment

The overall geometry of TSL varies from 1.3 km^3 to a maximum of 80 km^3 in volume and 0.5–9 m in depth in the dry and wet seasons, respectively (Kummu and Koponen 2003). TSL is located in a tropical region governed by the rainy season (May–October) and dry season (December–April) (Lamberts 2001; see also Chap. 6). TSL is known for its extraordinary flood pulse system (see Chap. 1 for details). The average air temperature of the lake varies between 20 °C and 36 °C (Fig. 2.1) in January and April, respectively (WMO 2016), and the water temperature is between 28 °C and 33 °C, with the mean monthly rainfall at Siem Reap exceeding 100 mm in the rainy season (WMO 2016). Three main sources of water fill the great lake basin: 53.5% from the Mekong River (MR) mainstream, 34% from the lake's tributaries, and 12.5% from precipitation. However, TSL discharges around 88.5% of its total outflow to MR, and 11.5% evaporates directly from the lake surface (Uk et al. 2018).

Supplementary Information The online version of this chapter (https://doi.org/10.1007/978-981-16-6632-2_2) contains supplementary material, which is available to authorized users.

P. N. Bao (✉) · N. Amanuma · B. R. Shivakoti
Institute for Global Environmental Strategies, Hayama, Japan
e-mail: ngoc-bao@iges.or.jp

K. Kimleang · B. Sophea
Institute of Technology of Cambodia, Phnom Penh, Cambodia

R. Khanal
Tokyo Institute of Technology, Tokyo, Japan

Policy Research Institute, Kathmandu, Nepal

C. Yoshimura et al. (eds.), *Water and Life in Tonle Sap Lake*,
https://doi.org/10.1007/978-981-16-6632-2_2

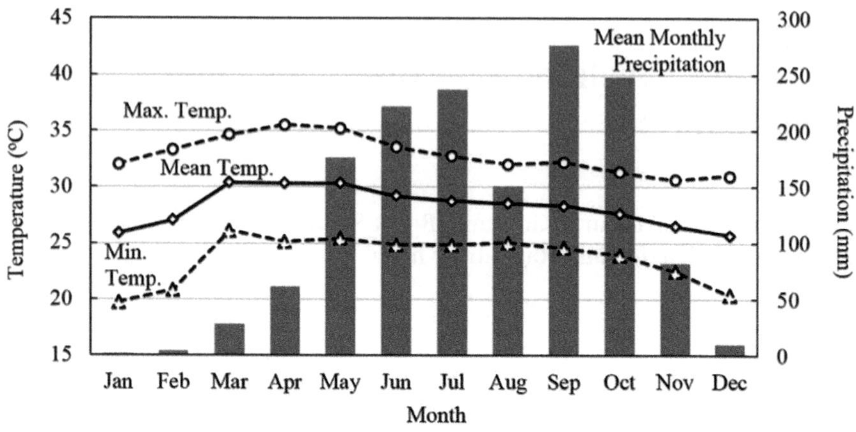

Fig. 2.1 Mean and maximum monthly temperature (°C) and precipitation (mm) in the TSL region observed at Siem Reap (Uk et al. 2018)

The rainy season brings the floodwaters from MR to TSL, forming an exceptional and highly productive floodplain ecosystem, and TSL is believed to be among the most productive freshwater ecosystems in the world (Keskinen et al. 2007). The flood pulse drives the expansion of the lake area and extensive inundation of the floodplain for several months, shaping a mosaic of natural and agricultural habitats and nutrient replenishment from MR (Arias et al. 2013) and helps in sustaining primary productivity and breeding of fish and aquatic species (see Chaps. 31–34 for details). The TSL ecosystem has eight subsystems (see Chaps. 31 and 32), providing TSL with exceptionally high biodiversity. Overall, TSL is home to more than 370 plant species, 197 phytoplankton species, 46 zooplankton species, 57 reptile species, 225 bird species, 46 mammal species, and 42 reptile species. At least 44 species are classified as threatened or endangered species (MRC 2018; Uk et al. 2018). Fish production from the lake area supplies 60% of Cambodian total inland fisheries (So 2010), or equivalent to 200,000 to 218,000 tons in weight and US$150–250 million in cash as the average from 2001 to 2010 (MAFF-CNMC 2003). Apart from the fisheries, the floodplain area of TSL has potential for agriculture as it is 1.4 million ha in total (Song et al. 2011). Other agricultural activities range from herding cattle, pigs, or ducks to the growing of corn and other cash crops such as tobacco, watermelon, and mung bean. Rice and a variety of vegetables are grown in the floodplain and on the riverbanks after the water has receded (Lamberts 2001).

2.2 Livelihood and Socioeconomics

To understand better the livelihoods in and around TSL, both qualitative and quantitative methods were used. A primary data survey was carried out in 2017 and 2018, previous pieces of literature were also collected, and an analysis of the impact on the environment and socioeconomic situations of 170 floating villages, particularly the fishing communities, was carried out (Nuorteva et al. 2010). The different types of households in TSL refer to the different degrees of fishery dependency between villages such as water-based, water–land-based, and land-based villages (Hap et al. 2006). Water-based villages depend on the fishery as their main source of income. Water–land-based villages earn from fishing-cum-farming. Land-based villages refer to those whose main income is from farming.

2.2.1 *Socioeconomic Survey*

For data collection, the socioeconomic survey was conducted using the following qualitative approaches in land-based, land–water-based, and water-based (floating village, Supplementary Video 2.1) communities: (a) ethnographic-type fieldwork, in particular, participant observations in the selected study sites; (b) semi-structured interviews with local residents; and (c) focus group discussion (FGD) with local NGOs and communities, and village chiefs. A FGD targeting local residents, fishers, local NGOs, and village chiefs were conducted to get a general understanding of the village and its livelihood situations. Three FGDs were conducted, with 253 interviewing samples; the interviewing households were from four villages in four major provinces located around TSL. They are Rohal Suong (land-based village) in Battambang Province, Muk Wat (water–land-based village) in Siem Reap, Kampong Luong (water-based village) in Pursat, and Chhnok Trou (water-based village) in Kampong Chhnang Province (Fig. 2.2). The semi-structured questionnaire was used, and the main issues of the household interviews were as follows: characteristics, geographical landscape of the community, linkages between TSL and their livelihoods, household income, farming and fisheries and their impacts on the livelihoods, access to and availability and uses of water; changes in socioeconomic conditions, living standards, and the surrounding environment during the past decade. The secondary data relied on published literature and statistical data (e.g., Cambodia Socio-Economic Survey 2014, 2015; Cambodia Census Population Survey 2013).

Fig. 2.2 Study areas for data collection (encircled in red)

2.2.2 Livelihood Conditions and Activities

About 1.7 million people are living in 1037 villages of TSL and surrounding floodplains (Sithirith 2011), where they have learned to adapt to hydrological changes and live closely with the local environment. The fishing villages have been classified into three groups based on their ecological zones: land-based villages (948), water-based villages (53), and water–land-based villages (36) (Sithirith 2011). People in land-based villages are primarily engaged in farming with occasional fishing. In water-based villages or floating villages, people depend on fishing because of their very limited access to land. Meanwhile, in water–land-based villages, people often spend approximately 6 months in a year on the land and the other 6 months on the water. The houses are built within 6–8 m above the ground to avoid flooding in the wet season. For most households, their main income comes from fishing, whereas their extra income comes from small-scale farming. The common practice for small-scale fishing is using small gear, such as gill nets and bamboo fence traps. Because majority of people in all three types of villages depend mainly on available resources (e.g., catched fishes, household farming products) and floodplains of TSL, they were the first group to become the most vulnerable to any negative impacts on the lake resources and ecosystem.

2.2.3 Declining Fisheries and Their Socioeconomic Impacts

According to our survey, the fisheries are the major source of their primary or secondary income. Results from our survey indicated that 74% of households in Chhnok Trou, which is a water-based village, rely largely on small-scale fishing. Their income is directly dependent on the daily fish catch, ranging from US$3 to 20 per day (Hap et al. 2006); thus, the fishery is the only source of revenue to support the family of most households. Small-scale fishing is the secondary source of income for most land-based villages, whereas rice cultivation is their main source of income. Thus, the rapid fall of fish catch in both quantity and quality is a critical issue identified from the surveys (see also Chap. 48 regarding the management of fish resource). Since fishing is the major source of income for households living in floating villages such as Chnok Trou, the decline of fishing capacity causes difficulty in supporting daily lives. Despite the intensification of fishing efforts at the household level, fish catch has not improved. Intensification in fishing and low fish catch also means waste in added time, financial and energy expenditures, and opportunity costs.

The decline of fish catch is related to the use of illegal fishing methods, lack of control of fishing gears, unequal access to fish catchments, an increase in the number of fishers, deterioration of fish habitats, and hydrological changes in the flood pulse (i.e., mainly inadequate flood water entering the lake during the wet season). Although the communities are not fully aware of the root causes of the fluctuation of water level in the wet season, the construction of a cascade of hydropower dams on the upstream countries as well as the impact of climate change is believed to alter the hydrological cycle affecting the seasonal inundation of the lake, sedimentation, and nutrient transport (Pokhrel, et al. 2018). It is quite evident that the continuation of disruption in the hydrological cycle and the declining fish catch is increasing the vulnerability of fishing communities, who have low assets base and rely on income from the daily fish catch.

2.2.4 Impact of Climate Change and the Depletion of Fisheries

The findings from the interviews showed that the most recent environmental change is the deterioration of water quality in both TSL and TSR. This resulted from the discharge of untreated wastewater from the floating villages and disposal of solid wastes into the lake's tributaries, run-off of agrochemicals, inadequate water, sanitation and hygiene. People who live in the floating village use this water directly for their daily usages such as washing raw food, hand washing, and bathing.

The local livelihoods and people's lives on and around TSL could be directly affected by the impacts caused by climate change, especially through disruption on the hydrological cycle of TSL. Extreme or unexpected weather events such as heavy storms causing high waves, severe floods (e.g., the flood events in 2000 and 2011), and droughts seem to occur more frequently and have huge economic impacts, particularly to the poor households. It has been indicated that there will be a significant change and shift in the pattern of rainfall in the overall MR region, including the TSL area, which will experience a drier and longer dry season, especially lower MR countries such as Cambodia and Vietnam. On the other hand, the rainy season will be shorter, but with higher precipitation, which may lead to higher flood risks (Snidvongs 2006; see also Chap. 43 regarding the impact of climate changes on hydrological regime). Consequently, this could influence fishery migration and affect the ecosystem productivity of TSL.

This is highly critical for a large population dependent on the resources and services of the TSL basin (Grundy-Warr and Sithirith 2016), agriculture and food security, terrestrial and freshwater ecosystems, and human health. Among the three types of surveyed villages under this project, the water-based communities (Kampong Luong and Muk Wat villages) as such are physically less affected than those living on land. However, coping with high waves caused by heavy storms is becoming a challenge for small-scale fishing activities. Meanwhile, severe water pollution can happen during low water level periods, affecting the fish population. As a result, the communities have to travel afar in search of better fishing ground. It means additional fuel cost of the fishing boats, extra physical labor, and risk of accidents.

2.3 Environmental Stress

2.3.1 Rapid Population Growth and Agricultural Activities

Cambodia's population has significantly increased after the collapse of the Khmer Rouge, from around 6.7 million in 1979 to 15 million in 2015 and is predicted to reach 22.5 million by 2050 (Uk et al. 2018). It has been reported that about five million people are living on and around TSL and approximately 62% of the population is estimated to live in rural areas (Ministry of Planning 2013). The population growth has resulted in increased food demands and pressures on the natural resources of the lake (Keskinen 2006). The rapid growth of every livelihood sector surrounding TSL and the increase in siltation and floods reduce the water quality and quantity, affecting strongly on the fish catchment in the region (Chadwick et al., 2008). The pollution from agrochemicals such as unregulated pesticides and fertilizers applied to increase agro-productivity has also significantly affected the lake ecosystem and indirectly affected public health (Uk et al. 2018).

A study in six provinces surrounding TSL indicated that prohibited pesticides are still traded at a high level despite the 1998 Sub-Decree on Standards and Management of Agricultural Materials (Yang Saing 2001), with 59% of the pesticides widely used in Cambodia being categorized by the World Health Organization as Class I + II, which means highly to extremely hazardous (Jensen et al. 2011). It has been found that 67% of farmers in the TSL region used pesticides for agricultural activities (Cambodia Center for Study and Development in Agriculture 2001), with about 3570 tons used in 2007 (Preap and Sareth 2015). Domestic wastes, untreated industrial effluent, the consumption of agrochemicals, and discharges and spillage of oil and fuel (WEPA 2020) are the sources of ecosystem pollution (see also Chap. 22 regarding basin-wide distribution of water quality and its affecting factors). Water pollution has threatened the supply of drinking water, thus urging an explosive development of harmful invasive plant species such as water hyacinth (Kuenzer 2013; Uk et al. 2018), for people living in 170 floating villages (Nuorteva et al. 2010) who directly consume the freshwater fishes and water from the lake.

2.3.2 *Deforestation and Geographical Changes*

Deforestation and land-use changes in the TSL basin and inundation area are substantial and detrimental to the breeding of aquatic species (see also Chap. 44 regarding the projection of land use change in TSL basin). Once there used to be approximately one million ha of flooded forest around the lake. The flooded forest areas had subsequently been reduced to about 614,000 ha by the 1960s, 362,000 ha by 1991, and 350,000 ha by 1997 (ADB 2005). Land-use changes are equally significant in the upland area of the TSL basin. Annual deforestation in this TSL basin is worse than in any other areas in Cambodia (Senevirathne et al. 2010). In less than 20 years, forest cover in the lake basin decreased by 43%, from 20,170 km^2 in 1990 to 11,436 km^2 in 2009, whereas the total agricultural land area in the basin increased by 30%, from 14,076 km^2 in 1990 to 18,858 km^2 in 2009 (Senevirathne et al. 2010) (Fig. 2.3). The area affected by deforestation in the TSL basin is equal to 15% of the basin's total land area. Higher sediment loads into TSL and its floodplain are anticipated from soil erosion and nutrient runoffs from the deforested and agricultural areas in the basin (Uk et al. 2018).

2.3.3 *Hydropower Dam Constructions Along Mekong River*

The number of hydropower dam stations along MR increased by 183% from 2000 to 2010, and the total water-holding capacity has increased fourfold compared to that of the 1990s (Lin and Qi 2017). Although MR was largely undeveloped prior to 1990, it is currently undergoing a rapid dam construction period, with 11 upstream operational dams in China (Lin and Qi 2017), and 120 dams are planned (ODM 2017) for the Lower MR and tributaries (Lin and Qi 2017). Hydrological structures (e.g., dams) would block seasonal fish migration between MR and TSL. Fish migrate down the river to breed in MR during the low-flow period, with their larvae being washed back up the river in the following wet season to grow in the shallow waters around the lake (Campbell et al. 2009). At peak migration periods, more than 50,000

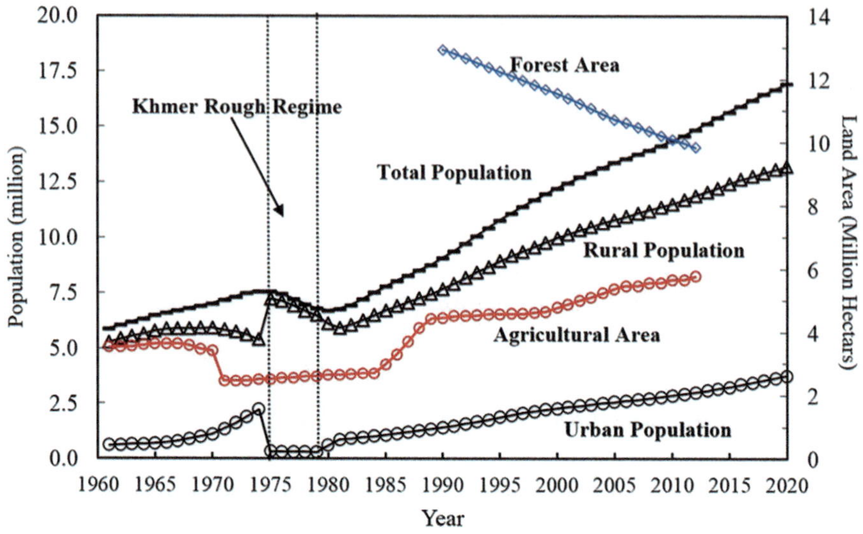

Fig. 2.3 Population and land use trends in Cambodia (Uk et al. 2018)

fish per minute swim pass a given point from TSL to MR. Consequently, any disturbance to the TSL ecosystem implies negative impacts on fisheries in the entire MR system. The construction of hydropower dams by 2040 is suggested, as analyzed, to reduce the losses of fisheries by 11% (MRC 2018). The lake could also be considered a pollution hotspot (Chea et al. 2016), with TSL water quality being influenced by both internal (e.g., lake sediments) and external nutrient sources (e.g., tributary point and non-point sources; groundwater (Burnett et al. 2017)). The load of sediments transported to the lower delta is estimated to reduce according to Development 2040. For instance, under the M1 early development scenario of Baseline 2007, 143 Mt./annum of sediment reaches Kratie, whereas under the M3 planned development scenario of 2040, only 4 Mt./annum of sediment reaches Kratie, a reduction in sediment loads of 97% (MRC 2018).

This threatens the ecosystem and livelihoods of the TSL's floodplain, causing the rise of dry season water level (Keskinen et al. 2007). This has been the major concern on the threats to and degradation of biodiversity and ecosystem production in TSL as well as on human beings dependent on the lake (Lin and Qi 2017).

Key Points

- TSL plays a major role in supporting the income-generating and livelihood activities of local communities, including fishing (representing one-third of people's primary occupation), farming, and collection of firewood as the secondary source of revenues for households.
- Recently, TSL's ecosystem has experienced a number of threats such as rapid population growth, water pollution, over-exploitation and illegal fishing, deforestation, and an increasing number of dams constructed along MR. This situation

has significantly affected the lake and livelihoods' sustainability. In addition, the negative impacts of climate change have further worsened the situation.

- To improve the community resilience and to cope with unexpected environmental and climate change, it is important to strengthen their capacity, by equipping them with sufficient knowledge and a good understanding of the effects of environmental and climate change impact and appropriate actions to enhance adaptive livelihood at the local level.

References

ADB. The Tonle Sap Basin strategy. Manila, Philippines: ADB; 2005.

Arias ME, Cochrane TA, Norton D, Killeen TJ, Khon P. The flood pulse as the underlying driver of vegetation in the largest wetland and fishery of the Mekong Basin. Ambio. 2013;42(7):864–76. https://doi.org/10.1007/s13280-013-0424-4.

Burnett WC, Wattayakorn G, Supcharoen R, Sioudom K, Kum V, Chanyotha S, Kritsananuwat R. Groundwater discharge and phosphorus dynamics in a flood-pulse system: TSL, Cambodia. J Hydrol. 2017;549:79–91. https://doi.org/10.1016/j.jhydrol.2017.03.049.

Cambodia Center for Study and Development in Agriculture. The situation of pesticide use in the Tonle Sap Catchment. Phnom Penh, Cambodia: Cambodia Center for Study and Development in Agriculture; 2001.

Campbell IC, Say S, Beardall J. TSL, the heart of the Lower Mekong. In: The Mekong. USA: Elsevier Inc.; 2009. p. 251–72. https://doi.org/10.1016/B978-0-12-374026-7.00010-3.

Chadwick MT, Juntopas M, Sithirith M, editors. Sustaining Tonle Sap: an assessment of development challenges facing the Great Lake. Bangkok: Sustainable Mekong Research Network (Sumernet); 2008. 144pp. ISBN 978–91–86125-06-6.

Chea R, Grenouillet G, Lek S. Evidence of water quality degradation in Lower Mekong Basin revealed by self-organizing map. PLoS One. 2016;11(1):1–19. https://doi.org/10.1371/journal.pone.0145527.

Grundy-Warr C, Sithirith M. Threats and challenges to the "floating lives" of the Tonle Sap. In: Tidwell AC, Zellen BS, editors. Land, indigenous peoples and conflict. London and New York: Routledge Taylor &Francis Group; 2016.

Hap N, Seng L, Chuenpagdee R. Socioeconomics and livelihood values of Tonle Sap Lake fisheries. IFReDI; 2006.

Jensen HK, Konradsen F, Jørs E, Petersen JH, Dalsgaard A. Pesticide use and self-reported symptoms of acute pesticide poisoning among aquatic farmers in Phnom Penh. Cambodia J Toxicol. 2011;2011:8. https://doi.org/10.1155/2011/639814.

Keskinen M. The Lake with floating villages: socio-economic analysis of the Tonle Sap Lake. Int J Water Resources Dev. 2006;22(3):463–80.

Keskinen M, Kakonen M, Tola P, Varis O. The Tonle Sap Lake, Cambodia: water-related conflicts with abundance of water. Econ Peace Security J. 2007;2(2)

Kuenzer C. Threatening Tonle Sap: challenges for Southeast Asia's largest Freshwater Lake. Pacific Geogr. 2013;40(August):29–31.

Kummu M, Koponen J. Modelling flood, sedimentation processes, and water quality for environmental impact assessment and management support in the Tonle. Paper to ICGRHWE 2003 conference in Three Gorges Dam (September), 2003. pp. 1–14.

Lamberts D. Tonle Sap fisheries: a case study on floodplain gillnet fisheries. Bangkok, Thailand: RAP Publication; 2001.

Lin Z, Qi J. Hydro-dam – A nature-based solution or an ecological problem: The fate of the Tonlé Sap Lake. Environ Res. 2017;158:24–32. https://doi.org/10.1016/j.envres.2017.05.016.

MAFF-CNMC. National Sector Review 2003: Fisheries Management Ministry of Agriculture, Forestry and Fisheries (MAFF) in association with Cambodia National Mekong Committee (CNMC) 15 p.; 2003

Mekong River Commission (MRC). The Council Study: The study on the sustainable management and development of the Mekong River Basin, including impacts of mainstream hydropower projects; 2018.

Ministry of Planning. Cambodia Inter-censual Population Survey 2013: Spatial distribution and growth of population. Phnom Penh: Cambodia; 2013.

National Institute of Statistics. Cambodia Inter-Censal Population Survey 2013 – Final Report. Phnom Penh, Cambodia: National Institute of Statistics, Ministry of Planning; 2013.

National Institute of Statistics (2014, 2015). Cambodia Socio-Economic Survey 2014 (2015). National Institute of Statistics, Ministry of Planning, Phnom Penh, Cambodia.

Nuorteva P, Keskinen M, Varis O. Water, livelihoods and climate change adaptation in the Tonle Sap lake area, Cambodia: learning from the past to understand the future. J Water Clim Change. 2010;1(1):87–101. https://doi.org/10.2166/wcc.2010.010.

Open Development Mekong (ODM). Hydropower dams. opendevelopmentmekong.net; 2017.

Pokhrel Y, Burbano M, Roush J, Kang H, Sridhar V, Hyndman DW. A review of the integrated effects of changing climate, land use, and Dams on Mekong River Hydrology. Water. 2018;10 (3):266. https://doi.org/10.3390/w10030266.

Preap V, Sareth K. Current use of pesticides in the agricultural; products of Cambodia. Submitted for the FFTC-KU International Workshop on Risk Management on Agrochemicals through Novel Technologies for Food Safety in Asia, November 10–14. Nakorn Pathom, Thailand, 2015.

Senevirathne, N., Mony, K., Samarakoon, L., & Hazarika, M. K. (2010). Land use/land cover change detection of Tonle Sap Watershed, Cambodia. In: 31st Asian Conference on Remote Sensing 2010, ACRS 2010. pp. 852–857.

Sithirith M. Political geography of the Tonle Sap: power, space and resources. PhD dissertation, National University of Singapore, 2011.

Snidvongs, A. Vulnerability to climate change related water resource changes and extreme hydrological events in Southeast Asia. A Final Report Submitted to Assessments of Impacts and Adaptations to Climate Change (AIACC), Project No. AS 07, 2006.

So N. Fisheries Resources in Cambodia: An Overview. Phnom Penh: Inland Fisheries Research and Development Institute, Fisheries Administration; 2010.

Song S, Lim P, Meas O, Mao N. The agricultural land use situation on the periphery of the Tonle Sap Lake. International Journal of Environment and Rural Development. 2011;2:2.

WEPA. State of water environmental issues, water-related issues and policies. Cambodia: wepa-db.net; 2020.

World Meteorological Organization (WMO) (2016). World Meteorological Organization. Retrieved from http://www.worldweather.org/en/city.html?cityId=347. Accessed 21 September 2016.

Yang Saing K. Pesticide use in Tonle Sap region. Phnom Penh, Cambodia; 2001.

Uk, et al. Tonle Sap Lake: Current status and important research directions for environmental management. Lakes Reserv. 2018:1–13. https://doi.org/10.1111/lre.12222.

Chapter 3
Fish Resources: Its Importance and Challenges

Hul Seingheng

3.1 Importance of Fish Production in Khmer Culture

Fish is the source of protein for more than five million people in Cambodia. Historically, Tonle Sap Lake (TSL) has provided multiple services as seen in Fig. 3.1, and it accommodates particularly an enormous amount of fish stock as stated in the record of a Chinese delegate Zhou Daguan (c. 1270 to c. 1350; Zhou and Harris 2007). The records mention there are plenty of fishes in the lake, called freshwater sea by the delegate. The missionaries even exaggerated that the fish quantity was like a mountain that could be counted as half out of the water. This natural existence has played an important role in sustaining society and economy since ancient times. Evidence of TSL's richness could be seen through sculptures of different kinds of aquatic species including fishes on the wall of Angkor Wat temple complex in Siem Reap province.

This great lake has been recognized as a food source that supported the army during the Khmer–Siem war during the post-Angkor regime (Rungswasdisab 1995). The abundance of fishes in Cambodia shares a major portion with TSL, which is used as a breeding ground by most of the species of fishes available in the lower Mekong River Basin (MRB) system (see also Chaps. 31–33). It is also commonly understood among Cambodians and has been dictated from generation to generation that "wherever there is water, there are fishes." This common perception has been embedded in Cambodian society. It is true, somehow, since some record has shown impression of fish presence in the lake that had been historically recorded by a Chinese missionary during the Angkor dynasty (Zhou and Harris 2007).

H. Seingheng (✉)
Institute of Technology of Cambodia, Phnom penh, Cambodia

Ministry of Industry Science, Technology and Innovation, Phnom Penh, Cambodia
e-mail: hul.seingheng@misti.gov.kh

C. Yoshimura et al. (eds.), *Water and Life in Tonle Sap Lake*,
https://doi.org/10.1007/978-981-16-6632-2_3

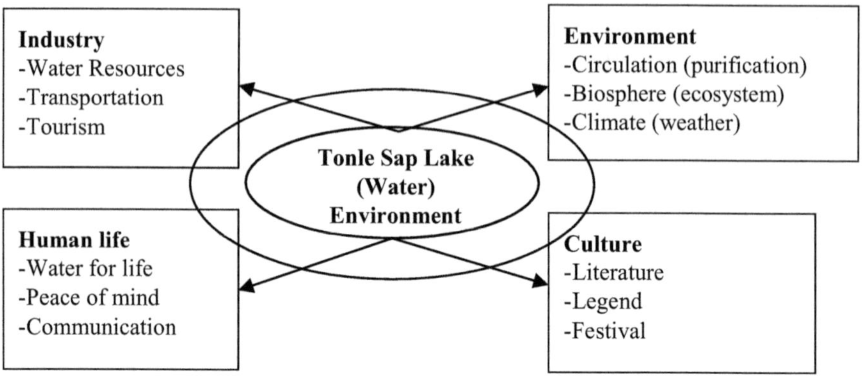

Fig. 3.1 Overview on the importance of Tonle Sap Lake

A study by Lim et al. (1999) showed that there are 120 species of fishes in the lake, which implies that 47% of species have been lost when compared to data from studies from 1936 to 1976. Some species are unique to this freshwater environment (see also Chap. 33 for fish taxonomy and ecology). Habitat characteristics such as flooded forest, open water, and/or agricultural fields show a positive correlation with the distribution of fish species in the lake (Chan et al. 2020). Some fish species such as *Pangasianodon gigas* and *Catlocarpio siamensis* reach more than 100 kg per head. TSL, as a fish resource, has been ranked fourth globally in terms of its productivity among freshwater bodies (Bonheur and Lane 2002). The lake has made a remarkable contribution to the economic growth of Cambodia. The total fish catch was found to contribute around 10% to the national gross domestic product of Cambodia. The lake itself plays an important role in supplying protein to local livelihoods (Baran et al. 2014). The majority of protein for consumption among Cambodians is from the inland fish catchment, especially from the TSL floodplain. The Statistic by Inland Fisheries Research and Development Institute (IFReDI) shows that inland fish capture shared a larger percentage than marine fish as the total inland fish capture was estimated to be around 570,000 tons per year (IFReDI 2012) (Table 3.1).

Because of the importance of maintaining the ecological landscape of the lake, Cambodia issued a Royal Decree in November 1992 to conserve the lake environment. Moreover, UNESCO has further enlisted this natural body as a biosphere reserve in 1997 since the lake environment must be recognized as a humankind resource. Governance efforts have been put in place to ensure that all stakeholders are properly performing their job. For instance, the Tonle Sap Authority (TSA) was established in 2009 as a coordination body to manage, preserve, and conserve the biosphere reserve (see Chap. 48 for details). In this context, the authority works with line ministries and other authoritarian institutions such as the Fishery Administration, National Mekong Committee, the Mekong River Commission, and other development partners. The current strategic plan till 2020 of TSA pays remarkable

Table 3.1 Inland fish capture by type of fish habitat (IFReDI 2012)

Sources	Type of fish sources		Annual consumption of fish and fish products (kg/person/year)
Inland capture resource	Inland fish	Floodplain residents	18.8
		Long-distance migrants	15.5
		Short-distance migrants	6.0
	Other inland aquatic animals		3.9
Marine capture resource	Marine fish		16.2
	Other marine aquatic animal		1.1
Aquaculture			1.3

attention to climate change. However, the strategy incorporates less information about the management of the lake such as the accountability of changes of flooding in the reservoir. The strategic direction of the lake management should include a stronger focus on the consequences of climate change.

Many studies have shown the decline in the number of species as well as the fish stock of the lake (see also Chap. 48). For instance, the Food Agriculture Organization report in 2001 documented the concern about overfishing on the fish stock ecosystem and fish productivity. The pressure is attributed to the overpopulation in the basin and the change in the hydrological regime caused by climate and artificial development (Campbell et al. 2006). In addition, the anthropological pressure through development around the lake basin has mounted the decline of fish resources as well as other aquatic biodiversity. The hydropower dam construction upstream could affect the uniquely balanced ecosystem in TSL (Lin and Qi 2017). Recent findings suggest that anthropogenic activities in the lake basin, as well as in MRB, in combination with climate changes, are key contributing factors in the degradation of the TSL ecosystem (Uk et al. 2018).

3.2 Environment Pressures on the Lake Ecosystem

Recent developments have shown the importance of water resource management in an integrated manner. It has been recognized that there are interconnections between human activities and water quality. Environmental problems attributable to increasing industrial and agricultural activities and population are matters of serious concern to human society. Therefore, environmental laws have been becoming more stringent. On the other hand, other factors for consideration are other development pressure, discharge of domestic waste to river channel, climate change, and unplanned population growth. These factors threaten the water sources of

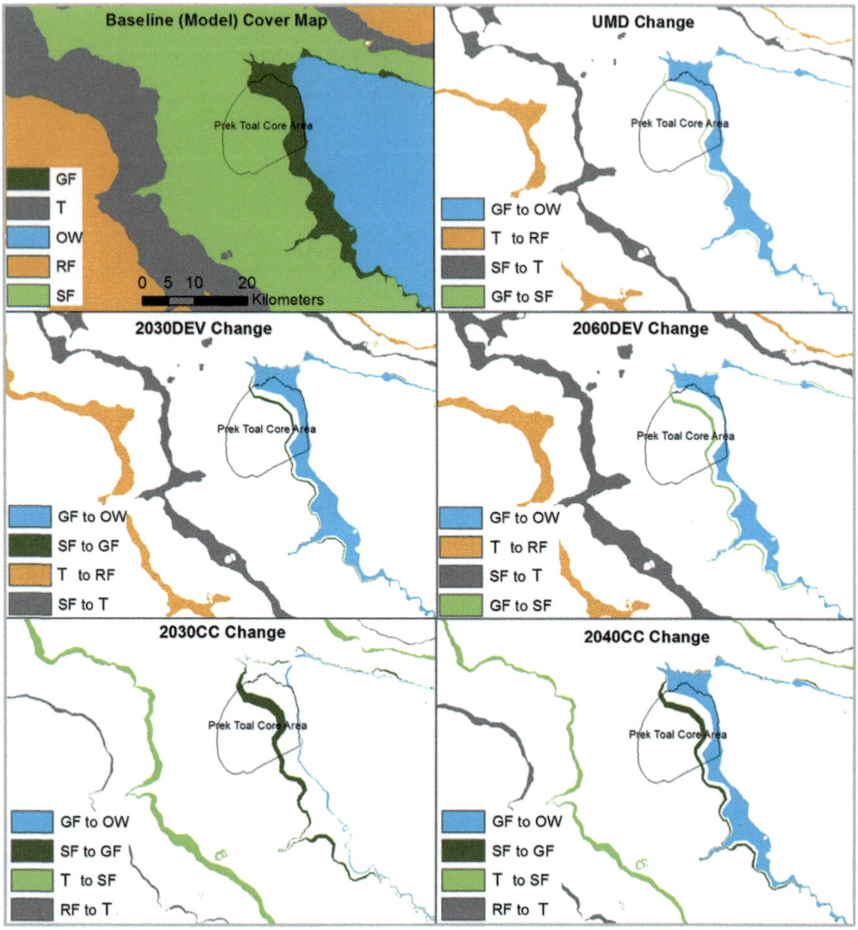

Fig. 3.2 The analysis of climate change prediction scenarios concerning the change of environmental habitats. *OW* open water habitats, *GF* gallery forest habitats, *SF* seasonally flooded habitats, *T* transitional habitats, *RF* rainfed habitats (Arias et al. 2012)

conventional treatment plants (Vanny et al. 2015) and the likelihood of people living in this floodplain. Some pollutants including nutrients from both natural and human factors could be a source of the undesirable water quality of TSL (Michaud 1991; Burnett et al. 2013).

Climate change is possibly the source of those hydrologic shifts in TSL (Eastham et al. 2008; Kummu and Sarkkula 2008; Kummu et al. 2014; see also Chap. 43 regarding the impact of climate change on hydrological regime). Consequently, the habitat covers have changed accordingly as shown in the climate change prediction scenarios of Arias et al. (2012) (Fig. 3.2). These predictions highlighted the effects of climate change on the surface area of the lake and pointed out the possible consequences of habitat changes for 2030 and 2040 for three different flooded zones of the lake. For instance, a significant gallery forest habitat could be lost to become open water by 2040. The hydrological change remains a persistent threat to the lake productivity due to habitat change. Oeurng et al. (2019) have made predictions for three time horizons (the 2030s, 2060s, and 2090s) and estimated the decrease of flow in general for 11 sub-basins of TSL. These prediction scenarios could serve as a foundation for the long development planning and management of this UNESCO biosphere reserve.

Aside from climate pressure, anthropological development is another challenge to maintaining the water environment. Socioeconomic development is a critical condition to be well understood from an environmental management perspective since these factors are interdependent with each other. Thus, countermeasures and adaptation approaches are always the best options for the long-term conservation of the lake ecosystem.

3.3 Countermeasures and Adaptation

To protect and conserve the lake ecosystem, an option that could be considered would be creating a more concise policy regarding the management and reduction of the changes of the water surface of the lake by proper maintenance of habitat areas or restoration of the habitat through expansion of the gallery forest. As studied by Arias et al. (2012), climate change contributes to altering flooding, resulting in changes in natural habitat surfaces. The analysis of scenarios taking into account the climate changes between 2030 and 2040, as seen in Fig. 3.2, shows that the area of the open water for the habitat cover of the lake could increase spatially from 2% to 21% and the rainfed habitats could decrease from 2% to 5% and habitats could decrease from 5% to 11%. Hydrological regime changes could be one of the reasons for this habitat change. This knowledge of the change of habitats for aquatic biodiversity including fishes could help reduce the abrupt decline of productivity of the lake.

Other or additional approaches to tackling the reduction of lake productivity could be the adaptation to the changes through raising awareness among key stakeholders and the public or readiness for the changes and choosing other effective management practices including enforcement of lake management frameworks and stronger coordination among the players to maintain the natural ecosystem in the lake (see also Chap. 48 regarding the management of flooded forests and fish resources). To make these options possible, a practical approach in terms of monitoring in time series the water regime changes and the change of habitat conditions in the lake is recommended. Once a better understanding of the changes of flooding and their relative consequences is gained, it will be a good time for proper policy to be established.

Change in fish productivity is seen to have a strong relation between anthropological development and unpredictable climate conditions in MRB since water quality and quantity primarily determine the richness of aquatic biota. It is necessarily important to have more scientific evidence before making any decision, policy development, and/or action for management in the basin. It is therefore a requirement to have a better understanding of the natural and social science perspectives through research. External factors including the need for collaboration/dialogue about common problems and solutions among countries along MR and having an open science-based foundation in the regional dialogues between the upper and lower Mekong region should be considered to ensure a geopolitical-free environment.

Key Points

- The natural fish resource in TSL is in a critical condition because of climate change and anthropological development, a food security concern that requires immediate attention from relevant stakeholders.
- Socioeconomic development in the TSL floodplain is strongly related to the watershed development in the biosphere reserve, whereas the upstream development along the MR is also bounded with the vitality of the lake ecosystem including the availability of fish resources.
- The impact of climate change on the TSL environment and fish abundance is unquestionable, but the extent of this impact must be correctly addressed through collective actions among actors of multidisciplinary fields including science, social science, politician, environmentalist, relevant governmental agencies, and private sectors.
- Fair and open dialogues based on scientific evidence among local and international stakeholders must be in place to deal with the endangerment of the fish resource in the lake since this transboundary issue requires a solution for water discharge regulation along MR due to development, understanding of common problems and benefits, sharing of interest from the economic point of view, and biodiversity conservation.

References

Arias ME, Cochrane TA, Piman T, Kummu M, Caruso BS, Killenn TJ. Quantifying changes in flooding and habitats in the Tonle Sap Lake (Cambodia) caused by water infrastructure development and climate change in the Mekong Basin. J Environ Manag. 2012;112:53–66.

Baran E, Chheng P, Ly V, Nasielski J, Sary S, Touch B, Tress J, Kaing K, Tan S. Fish resources in Cambodia (2001–2011). Phnom Penh, Cambodia: Atlas of Cambodia Maps on Socio-Economic Development and Environment, Save Cambodia's Wildlife; 2014. p. 178.

Burnett WC, Peterson RN, Chanyotha S, Wattayakorn G, Ryan B. Using High-resolution in situ radon measurements to determine groundwater discharge at a remote location: Tonle Sap Lake, Cambodia. J Radioanal Nucl Chem. 2013;296:97–103.

Bonheur N, Lane BD. Natural resources management for human security in Cambodia's Tonle Sap Biosphere Reserve. Environ Sci Pol. 2002;5:33–41.

Campbell IC, Poole C, Giesen W, Valbo-Jorgensen J. Species diversity and ecology of Tonle Sap Great Lake, Cambodia. Aquatic Sci. 2006;68:355–73.

Eastham J, Mpelasoka F, Mainuddin M, Ticehurst C, Dyce P, Hodgson G, Ali R, Kirby M.. Mekong River Basin water resources assessment: impacts of climate change, 2008

Chan B, Brosse S, Hogan ZS, Ngor PB, Lek S. Influence of local habitat and climatic factors on distribution of fish species in the Tonle Sap Lake. Water. 2020;12(3):1–16.

IFReDI. Food and nutrition security vulnerability to mainstream hydropower dam development in Cambodia, Synthesis report, 2012, pp. 44.

Kummu M, Sarkkula J. Impact of the mekong river flow alteration on the Tonle Sap Flood Pulse. Ambio. 2008;37(3):185–92.

Kummu M, Tes S, Yin S, Adamson P, Józsa J, Koponen J, Richey J, Sarkkula J. Water balance analysis for the Tonle Sap Lake–floodplain system. Hydrol Process. 2014;28:1722–33.

Lim P, Lek S, Touch ST, Mao SO, Chhouk B. Diversity and spatial distribution of freshwater fish in Great Lake and Tonle Sap River (Cambodia, Southeast Asia). Aquat Living Resource. 1999;12 (6):379–86.

Lin Z, Qi J. Hydro-dam- A nature-based solution or an ecological problem: the fate of the Tonlé Sap Lake. Environ Res. 2017;158:24–32.

Michaud JP. A citizen's guide to understanding and monitoring lakes and streams. Washington State Department of Ecology; 1991.

Oeurng C, Cochrane TA, Chung S, Kondolf MG, Piman T, Arias M. Assessing climate change impact on river flows in the Tonle Sap Lake Basin Cambodia. Water. 2019;11(3):618.

Rungswasdisab P. War and Trade: Siamese intervention in Cambodia, 1967–1851, Doctor of Philosophy Thesis, Department of History and Politics, University of Wollongong, Australia; 1995.

Uk, S., Yoshimura, C., Siev, So., Try, S., Yang, H., Oeurng, C., Li, S. and Hul, S. (2018). Tonle Sap Lake: Current status and important research directions for environmental management, Lakes Reserv: Res Manage 23 (3), 177–189.

Vanny L, Jiwen G, Seingheng H. Phnom Penh's municipal drinking water supply: water quality assessment. Sustain Water Resource Manage. 2015;1(1):27–39.

Zhou D, Harris P. A record of Cambodia: the land and its people, eBook. Silkworm Books; 2007.

Chapter 4
Governance and Human–Nature Relation

Mak Sithirith

4.1 Introduction

Tonle Sap Lake (TSL) is a unique freshwater lake in the world (see Chap. 1 for details). The management of the lake's resources has been challenging, changing over time from the management based on commercial fishing to the community, conservation, and open access. In early 2012, the Royal Government of Cambodia (RGC) abolished the fishing lots in TSL and returned large fishing areas for community uses. Although we welcome this intervention, we also raise the concern over how the returned fishing lot areas will be managed.

4.2 Governance of the Lake

TSL has been perceived as having global, regional, and national dimensions of governance for biodiversity conservation, water governance, and fishery management. Thus, the governance of TSL is shaped by the geography of the lake in the globe, the Mekong Region, and Cambodia's territory.

To enable and facilitate the discussion, a key question to be posed at the outset is, what is governance? Conceptually, governance is the process by which authority is conferred on rulers, by which they make the rules, and by which those rules are enforced and modified. In other words, governance is "the process of decision-making and the process by which decisions are implemented or not implemented" (UNESCO 2013; World Bank 2013). Eight major characteristics define "good governance," namely, participatory, consensus-oriented, accountable, transparent,

M. Sithirith (✉)
Royal University of Phnom Penh, Phnom Penh, Cambodia
e-mail: mak.sithirith@rupp.edu.kh

C. Yoshimura et al. (eds.), *Water and Life in Tonle Sap Lake*,
https://doi.org/10.1007/978-981-16-6632-2_4

Table 4.1 The current governance of Tonle Sap Lake

Scale	Governance principles		
	Stakeholders	Distribution of authority	Mechanism of accountability
Global	UNESCO, UNDP, IUCN, WWF, WCS, FFI, CI, Donors, MoE, TSBR	• Tonle Sap is dominated by fishery management. In 1993, TSL was designated as a biosphere reserve for biodiversity conservation. • Biosphere reserve classifies the lake into transitional, buffer, and core zones. • International convention on biodiversity conservation and Ramsar Convention. • MoE is responsible for biodiversity conservation. • Ecotourism in Prek Toal in Battambang.	• Accountable to the protection of biodiversity, less on humans. • MoE is accountable to the high-level government. • NGOs and UNESCO are accountable to donors. • Weak/poor accountability to resource users, and often there is a conflict between users. • There is weak coordination between government agencies, and this leads to institutional conflicts. • Weak resource access rights.
Regional	MRC, ADB/GMS, CNMC, ASEAN	• MRC Agreement 1995. • GMS Platforms. • ASEAN Agreement. • Intergovernmental agreements. • CNMC is coordinating 10 ministries in Cambodia that work on the Mekong.	• MRC is accountable to four lower Mekong countries, whereas the GMS to six countries including China and Myanmar. • CNMC is accountable to the national government.
National	FiA, MoE, MAFF, MoWRAM, CNMC, TSA, communities	• FiA manages fisheries: – RGC abolished commercial fishing lots. – Community fishery management. – Fishery sanctuary. – Public fishing areas. • Water management. • Agriculture.	• FiA is accountable to RGC and the RGC abolished the fishing lots and returned them to communities. • Community fishery is established to support the community members. • Public fishing is open access. • Management of water and agriculture conflicts.

responsive, effective and efficient, equitable and inclusive, and following the rule of law (UNESCO 2013).

The governance in TSL takes into account the three dimensions of governance plus the laws and policies, institutions, and process (see Table 4.1). In addition, Sokhem and Sunada (2006) discussed the governance of TSL concerning the governance of the Mekong, and thus, regional agreements such as the 1995 Mekong River Commission (MRC) Agreement and regional organizations such as the MRC and Asian Development Bank (ADB)/ Greater Mekong Subregion (GMS) have to be taken into account. Bonheur et al. (2001) analyzed the wetland governance in Cambodia, emphasizing the sectoral policies and institutions that influence wetland governance.

Management of TSL comes under the authority of many sector agencies with policies and laws specific to their mandates, but coordination among them remains weak. At least three agencies have major authority and legal frameworks for the management of the lake, and they are the Ministry of Agriculture, Forestry and Fisheries (MAFF); the Ministry of Environment (MoE); and the Ministry of Water Resources and Meteorology (MoWRAM).

The complexity of the TSL basin governance and the wide range of goods and services they support results in an equally complex set of government and resource user "stakeholders" that have legitimate interests in their sustainable management. Coordinated management of these resources is, therefore, critical for their conservation and sustainable utilization. Stakeholder coordination is required both at the national level, to ensure that the policy framework, laws, and institutional responsibilities, and budgetary resources are appropriate, and at the local level, where effective implementation and enforcement are required.

4.2.1 Global Dimension

TSL is perceived as a "global space" for biodiversity conservation, given that it is home to many biodiversity species including rare and endangered species: about 215 fish species, 42–46 reptile species, 225 bird species, and 15 mammal species (Bonheur and Lane 2002; Campbell et al. 2006). Thus, the governance of the global space is influenced by international, national, and local actors. Hence, in 1992, King Norodom Sihanouk requested the inclusion of TSL on the list of World Heritage sites. In addition, in November 1993, The Royal Decree officially designated TSL as a multipurpose protected area. In November 1995, the Technical Coordination Unit (TCU) was established within the MoE to provide coordination and technical support to the works on TSL, and the first inter-ministerial forum on the preservation of TSL, aimed at addressing the concerns about the lake, was held in March 1995 (Matsui et al. 2005).

At the international level, many conventions and treaties were signed and ratified by the RGC, including the Convention on Biological Diversity, known informally as the Biodiversity Convention in February 1995; the CITES Convention and the UNESCO agreement on Man and Biosphere Reserves in 1997; and the Ramsar Convention in 1999, to guide and support the works and activities of national and international organizations in TSL. TSL is one of the 610 biosphere reserves located in 117 countries in all regions of the world. They reinforce the significance of TSL as a "biosphere reserve" for biodiversity conservation.

4.2.2 Regional Dimension

The Tonle Sap River (TSR) acts as a key valve or artery connecting Mekong River (MR) to the lake, and thus, we cannot consider the TSL's biophysical characteristics without reference to the MR hydrological regime (see also Chap. 1). TSL takes in a lot of water and helps reduce flooding in MR during the peak flood season, and it releases water from the lake to MR in the dry season, helping reduce salt intrusion in the Mekong Delta. This is what the scientist called a "flood pulse" (Kummu and Sarkkula 2008).

The flood pulse makes TSL the "regional space." The flood pulse to TSL has been compared to that of a "heartbeat," and the flood pulse keeps the heart beating. If the heart stops, the system dies (Nikula 2005), and the entire MR ecosystem would be transformed adversely: the fisheries would collapse, indigenous knowledge would be subverted, the poor would go hungry, livelihoods would be disrupted, the communities would become dispossessed of basic means of survival, and the national economy would be severely affected. Thus, if we consider the trans-border hydrological and biophysical linkages of the lake with the Mekong region, we obtain a strong regional dimension concerning the lake's future ecological and environmental security (Kammu and Sarkkula 2008).

The 1995 MRC Agreement provides the basis for ensuring the functionality of the flood pulse and protecting TSL. The 1995 MRC Agreement protects the flood pulse in the Tonle Sap in two ways: first, the Agreement provides a general direction to safeguard TSL, and, second, it provides a specific guarantee to protect TSL (MRC Agreement 1995). The general direction in the Agreement includes the efforts by all parties "to protect the environment, natural resources, aquatic life, and conditions, and ecological balance of the Mekong River Basin (MRB) from the pollution and other harmful effects resulting from any development plans and uses of water and related resources in the Basin" (Article 3).

Article 5 of the Agreement guarantees the equitable and reasonable utilization of the MR water. It spells out a specific clause to protect the Tonle Sap as a "tributary of the MR and any intra-basin uses or inter-basin diversions shall be subject to notification to the Joint Committee" (MRC 1995). Article 6 of the MRC Agreement also raises two concrete points related to flood pulse and maintaining the flows in the mainstream:

(a) Of not less than the acceptable minimum monthly natural flow during each month of the dry season; and.
(b) To enable the acceptable natural reverse flow of the Tonle Sap to take place during the wet season. (Article 6, MRC Agreement in 1995).

The flood pulse influences the governance system of TSL, which in return ensures the acceptable minimum monthly natural flow in the dry season and the acceptable natural reverse flow during the wet season that contribute to enough water flow, ensuring that the flood pulse will still function in TSL. The MRC is the main regional body responsible for ensuring the flood pulse in MR and acceptable flow in the wet and dry seasons in TSL. Under the regional agreement, each country in the lower MR region forms a National Mekong Committee. The Cambodia National Mekong Committee (CNMC) was formed under the Mekong Agreement as the primary government agency coordinating natural resources management in the entire MRB in Cambodia as well as in the TSL basin (Sokhmem and Sunada 2006).

4.2.3 National Dimension

Last, but not least, TSL is perceived as a national space. Thus, the state has an "absolute power" to manage that space. This is at large based on how TSL could contribute to the economy of the country. Fishery and agriculture have been considered as the main engine for the national economy in TSL, but agriculture is small-scale, mainly for the household economy, and thus, the potential for the national economy is small, unlike with fishery, where the economic potential is significant. However, without water, agriculture and fishery will not be flourished, and thus, water management becomes an important part of the management of TSL. At the same time, without efforts to protect and conserve TSL, the abundance and richness of fishery, natural resources, and water will not be sustained, and therefore, the governance of TSL centers around the management of the fishery, water management, and agriculture and conservation. The following section will discuss the governance of TSL in great detail as an important national space.

Furthermore, the Tonle Sap Biosphere Reserve (TSBR) was then formally designated by Royal Decree on April 10, 2001. The TSBR Secretariat was established within the CNMC in 2001, and this raised the prerequisite for the environmental governance of the lake. After the establishment of the TSBR, the TCU was later upgraded to be the Secretariat for the TSBR and moved to the CNMC to respond to the growing need for coordination and planning. The CNMC coordinates 10 ministries, including the MoE as a member, responsible for biodiversity conservation, particularly in the TSL basin. The TSBR Secretariat is under the CNMC, but it also coordinates with the MoE for conservation. The establishment of the Secretariat for the TSBR in CNMC affirms its legitimacy as the coordinating body for the Tonle Sap and, thereby, its influence in all sectors of government (RGC 2001, TSBR Sub-decree, Article 7). The TSBR of Cambodia was established by the Royal Decree of the government of Cambodia to fulfill three key functions: (1) conservation function, (2) development function, and (3) logistic function (RGC 2001, TSBR Sub-decree, Article 7).

The TSBR territorializes the lake into three zones—(1) transitional zone, (2) buffer zone, and (3) core zone—for biodiversity conservation (see also Chap. 32). The core zone is territorialized into three main zones: (1) Prek Toal for bird colonies (21,342 ha), (2) Boeung Chmar as bird breeding areas (14,560 ha), and (3) Stung Sen as a unique gallery forest (6355 ha) (Sithirith 2011). This has led to the new form of structure of the governance of biodiversity conservation.

4.3 Human—Nature Relation: Influencing the Governance

Humans in TSL have organized themselves into three different groups situated in three different ecological zones: land-based village, water-based village, and water–land-based village. The land-based village is a village where villagers are engaged more in farming and less in fishing depending on the distance between the lake and land. The water-based village refers to a floating village, where fishing is a primary occupation for villagers. The third group is the water–land-based villages, which are physically located 6 months within the water and 6 months on land (see Table 4.2). These villages are in the ecological zone mostly affected by seasonal water levels (Sithirith 2011).

Of the total villages of 1037 around TSL, only 361 villages have organized into 175 community fisheries. This is about one-third of the total villages in TSL with a total household of 61,613 families. About one-third of the total households are dependent on fishing. Members of community fisheries are allowed to fish subsistence or small scale (Fishery Law 2006). However, people from about two-third of villages, or 676 villages, with a total household of 115,375 families, have access to fishing areas through open-access conditions. Fishing is not limited to community fisheries or open access. Non-members of the community could also fish inside the community fishery and in the open-access areas as long as they fish small scale. Members of community fisheries could also fish inside the open-access areas (Sithirith 2011).

The zoning in TSL does not take into full account the human settlements located in various natural zones. Various zones tend to conflict with fishing communities who use the lake since time immemorial, and the customary practices of various communities using the lake have been affected by the official zoning. Communities have been excluded and externalized by the zoning process. This has affected the way of life of fishing and farming communities around the lake.

The different communities use resources in the lake differently. Management of TSL has to take into account the differences of communities and the mode of resource uses in each type of community. In fisheries management, a fishing scale

Table 4.2 Typology of fishing villages by the province in the Tonle Sap (Sithirith 2011)

Province	Water-based village	Water–land-based village	Land-based village	Total
Battambang	10	2	117	129
Siem Reap	12	14	269	295
Kampong Thom	10	0	109	119
Kampong Chhnang	6	16	63	85
Pursat	15	1	238	254
Banteay Meanchey	0	3	152	155
Total	53	36	948	1037

is used to manage how different fishing communities fish in the lake. Fishing communities are allowed to engage in small-scale or subsistence fishing. On the one hand, different fishing communities practice small-scale fishing differently. On the other hand, small-scale fishing is no longer sufficient for survival, particularly for the water-based and land–water-based communities, where fishing is the primary occupation. As a consequence, small-scale fishing is not applicable in TSL (CDRI 2007).

4.4 Institutional Arrangements

The governance of TSL is complex, and its management is influenced by the global, regional, and national dimensions of TSL. The global conventions, regional agreements, and national sectoral frameworks have all weights influencing the institutional arrangement and governance system of TSL. However, despite the diverse factor, the management of TSL is still centered on three sectors: biodiversity, fishery, and water management for agriculture. Thus, diverse institutions and actors nationally, regionally, and globally working on these sectors shape the governance of TSL, each with the mandate, role, and responsibility in the management of each sector. The coordination among different agencies is inadequate, and this causes the weak governance of TSL. In addressing these, Tonle Sap Authority was established to facilitate the coordination and improve the governance of TSL.

The TSA was established in 2007, and it is chaired by the Minister of MoWRAM. The Minister of MoWRAM has also chaired the CNMC. This brings the MoWRAM, TSA, and CNMC under one umbrella at the national level. However, the Minister of MoWRAM is also a council member of the MRC at the regional level. These bring improved coordination concerning water governance in the MR and TSL regions.

The CNMC is the coordinating body for all types of works related to the Mekong development. It was established in line with the membership of Cambodia in the MRC, and therefore, the CNMC was established as a national arm of the MRC, coordinating MRC works at the national level. However, the ADB has expanded its works in the Mekong region through the GMS, and in Cambodia, under the GMS framework, the TSL Initiative was introduced to TSL, aiming at increasing the development in the TSL region. To address its needs, the ADB proposed to set up the TSL Basin Management Organization, parallel with the CNMC, to improve the coordination and the work in TSL. However, in 2007, the RGC did not take the ADB proposal forwarded, but instead, it established the TSA through a Royal Decree in September 2007 (RGC 2007), aiming at improving the coordination, conservation, and development in TSL and, more importantly, to address the governance issues outlined above.

The TSA was established and chaired by MoWRAM to coordinate government and nongovernment agency activities concerning TSL. Many representatives of various government agencies are assigned as members of the TSA. About 31 high-level representatives from different government ministries and institutions are appointed as members of the TSA (July 1, 2009). This brings the TSA closer to having the stakeholder representations. The stakeholder representation contributes to the goals of inclusive and deliberative decision-making. However, the stakeholder representations do not take into account the community and civil society representation in the TSA, and thus, the representation of the TSA only comes from government agencies, and it lacks representation from civil society. Thus, the decision-making is made within the line of government agencies.

The government has taken taught action following the political devotion to combat illegal fishing and the improved governance in TSL. The lower-level government institution has taken upward accountability following the pressure from the higher-level government. This form of accountability takes serious momentum when the Head of State declares the commitment, but it was doubtful how this momentum is going to be sustained. With regard to the mechanism of accountability, the government has used the court system to combat illegal fishing, and some illegal fishers were taken to the courts for their crime against the Fishery Law, and those found guilty were imprisoned, but only small fishing operators are arrested under this operation (WorldFish 2013).

The compliance and enforcement of laws in TSL by the Inter-Ministerial and Inter-Provincial Committee is not the same, is varied from province to province, and is biased by the interests of government agencies as it involves many institutions in the crackdown mechanism. The interpretation of the legal framework to combat illegal fishing is not the same as some of the institutions involved do not understand the Fishery Law and the legal framework (WorldFish 2013).

The zoning requires different modes of governance for each zone. The floating community needs to catch fish to feed their families, and therefore, they need to pump the water out. However, the farming community needs to keep water to irrigate their rice field. Any effort leading to flowing the water out of the field would affect rice productivity. Conflicts have occurred between different communities over the use of resources. The management of water is an essential element in the governance system at the local level.

At the provincial level, the Provincial Department of Water Resources and Meteorology provides the secretariat roles to the TSA. The commands from the TSA are channelled through MoWRAM, and the Provincial Department of MoWRAM acts to coordinate the works of the TSA at the provincial level and below that. However, at the TSL level, the TSA employed nine rangers across the lake to watchdog all activities inside TSL, and they are equipped with nine boats and other facilities. However, financially, they still face limited financial support, and therefore, their works have not been satisfactory.

Key Points

- The current governance system is even still complex and cannot guarantee sound environmental management and good governance. The current governance raises more concerns than prospective. The return of the lake to open access questions the current governance system. There is no governance system in this world to manage open access effectively.

- Governance of TSL has been changing over time from the management of TSL based on fishery management, focusing on the commercial exploitation of fishery resources, to community-based fishery management and conservation of biodiversity.
- The changes in the governance have happened rapidly without any clear plan or strategy; however, the policy changes at the national level are not necessarily being translated into action at the local level.
- The community is still struggling to define their governing system to manage its water and fisheries. Given the weak governance, the whole system is still suffering between too much water in the wet season and too little water in the dry season. The conflict over the use of water for irrigating rice and catching fish is still widespread. The implementation of the government intervention is weak because of the lack of resources and skills.

References

Bonheur, Nou, Lane, Benjamin D. Biodiversity conservation and social justice in the Tonle Sap watershed: the Tonle Sap biosphere reserve. International Conference on Biodiversity and Society, Columbia University Earth Institute, UNESCO, May 22–25, 2001; 2001.

Bonheur N, Lane BD. Natural resource management for human security in Cambodia's Tonle Sap biosphere reserve. Environ Sci Policy. 2002;5:33–41.

Cambodia Development Resource Institute (CDRI). We are living with worry all the time. A participatory poverty assessment of the Tonle Sap. Phnom Penh: CDRI; 2007.

Campbell I, Poole C, Giesen W, Valbo-Jorgensen J. Species diversity and ecology of Tonle Sap Great Lake, Cambodia. J Aquat Sci. 2006;68(3):355–73.

Fisheries Law. Fisheries law. Fisheries Administration, Ministry of Agriculture, Forestry and Fisheries, Royal Government of Cambodia, 2006.

Kummu M, Sarkkula J. Impact of the Mekong river flow alteration on the Tonle Sap Flood Pulse. J Human Environ. 2008;37(3):185–92.

Matsui, Satoru, Keskinen, Marko, Sokhem, Pech, Nakamura, Masahisa. Tonle Sap: experience and lesson learned brief, 2005.

Mekong River Commission (MRC). Agreement on the cooperation for the sustainable development of the Mekong River Basin, 5 April 1995. Bangkok: MRCs; 1995.

Nikula J. The lake and its people. MSc Thesis. Helsinki University of Technology; 2005.

Royal Government of Cambodia. Royal Decree on the establishment and management of the Tonle Sap Biosphere Reserve, Translation by Neou Bonheur, 2001. Available online at: http://www.tsbred.org/docs/law_and_regulation/Royal_Decree_On_creation_and_managemet_of_tsbr_Eng.Pdf (Accessed 22 July 2008).

Royal Government of Cambodia. Royal Decree on the establishment of tonle sap authority, Draft, Unofficial Translation, 2007. Available online at: http://www.foodsecurity.gov.kh/Otherdocs/Royal-Decree-tsba-Eng.pdfroyal. (Accessed Finnish Environment Institute, Helsinki 12 June 2008).

Sithirith M. Political Geography of the Tonle Sap: Power, space and resources. PhD Dissertation. Singapore: National University of Singapore; 2011.

Sokhem P, Sunada K. The Governance of the Tonle Sap Lake, Cambodia: Integration of Local, National and International Levels. Int J Water Resources Dev. 2006;22(3):399–416.

UNESCO. Concept of Governance, 2013. http://www.unesco.org/new/en/education/themes/strengthening-education-systems/quality-framework/technical-notes/concept-of-governance/. Accessed on 10 July 2013.

World Bank. The concept of Governance, 2013.. http://web.worldbank.org/WBSITE/EXTERNAL/COUNTRIES/MENAEXT/EXTMNAREGTOPGOVERNANCE/0,,contentMDK:20513159 accessed on 23 June 2012.

WorldFish. The Tonle Sap scoping report: Aquatic Agricultural System (AAS). Phnom Penh: WorldFish, Cambodia; 2013.

Chapter 5
Zoning and Its Impacts on Governance

Mak Sithirith

5.1 Zoning

Naturally, Tonle Sap Lake (TSL) and its floodplain consist of three ecological functioning zones: the terrestrial ecological zone, the floodplain, and the aquatic ecological zone. The terrestrial zone is covered by large land areas and flooded forest, homing to a land-based community that is not flooded by the peak flood of TSL. The floodplain is an area where it is flooded for 6 months and dry for another 6 months. The aquatic zone is an area where it is covered year-round by water. The natural ecological system influences how people make a living in TSL by using the lake's resources.

The economic and political zonings of TSL have been taking place since the French Protectorate Administration, and yet, it has not ended, but it has been zoned and zoning. The zoning of TSL is related to its valorization, commercialization, and the vital role that zoning plays in processes of capital accumulation. Thus, zoning becomes a process, serving various productive roles that help generate revenues and profits for different agents (Lefebvre 1991; Sithirith 2011). TSL has been subdivided into many zones, each with functional specializations, which sometimes overlap and contradict one another. Thus, TSL zoning is constantly in the process of being made, (re)produced, conceived in different ways, planned, mapped, and territorialized, as well as contested, resisted, and (re)imagined. This dynamic notion of zoning helps us to understand real resource management issues and problems that frequently relate to differing zoning conceptions of distinct agencies involved in resource governance (Sithirith 2011).

The zoning is rationalized on the basis of technical and scientific capacity and economic and political interests (Sithirith 2011; Lefebvre 1991). In rationalizing the

M. Sithirith (✉)
Royal University of Phnom Penh, Phnom Penh, Cambodia
e-mail: mak.sithirith@rupp.edu.kh

C. Yoshimura et al. (eds.), *Water and Life in Tonle Sap Lake*,
https://doi.org/10.1007/978-981-16-6632-2_5

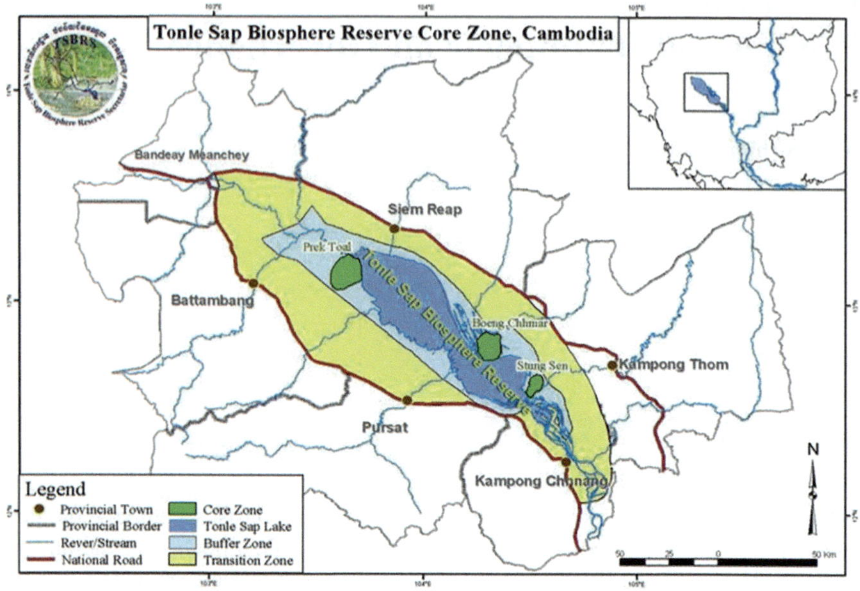

Fig. 5.1 Map of the Biosphere Reserve areas in Tonle Sap Lake (Source: TSBRS 2007)

zoning, agencies begin by identifying "the problems" affecting the area and then apply spatial and non-spatial strategies to address those problems. This is clearly illustrated through the TSL Biosphere Reserve (TSBR), which was established following the rationalization of zoning concerning the global significance of biodiversity of TSL under threat of environmental degradation. As a consequence, TSL was zoned into transitional, buffer, and core zones (Fig. 5.1, Royal Decree 2001; UNDP/GEF 2004).

Fisheries specialists construct their technical zones on the basis of their technical–professional specialization. We call this process creating zones of specialization. Lefebvre (1991) highlights that:

Specializations divide space among them and act upon its truncated parts, setting up mental barriers and practice-social frontiers. Thus, architects are assigned architectural space as their private property, economists come into possession of economic space, and geographers get their place in the sun, and so on. The ideologically dominant tendency divides space up into parts and parcels following the social division of labor (pp. 89–90).

Thus, the specialized zonings are dominated by the knowledge and skills of scientists, technocrats, experts, and bureaucrats. Since the French Protectorate Administration, the lake has been zoned into commercial fishing, conservation, and public fishing areas or open access. In 1997, the new zoning was created on the existing zoning system under the name of Biosphere Reserves, including the transitional, buffer, and core zones. The new and old zonings overlapped and conflicted, particularly the commercial fishing lots and the core zone of the

Biosphere Reserves. In 2012, the Royal Government of Cambodia (RGC) abolished the commercial fishing areas and has designated TSL into three zones for the flooded forest conservation, namely, Zone 1, Zone 2, and Zone 3 (see also Chap. 48). The flooded forest zoning is different from the Biosphere Reserve zoning, but it is likely overlapped, and these complicate the management of the lake resources.

5.2 Fisheries Management

Fishery management dominates the core elements of TSL governance, and it remains one of the key elements at the present day, although the mode of the fishery management system has slightly changed recently. Historically, a large part of TSL was managed through a commercial fishing system, some fishing areas were managed through a public fishing area, or "open access," and small fishing areas were managed via "fish sanctuary." This management system was practiced until 2012. Fishing conflicts between commercial and subsistence fishing intensified in TSL. In 2000, the RGC cancelled 56% of commercial fishing lots and introduced the community fishery (CFi) as a new model of management to TSL, but it does not address the increased fishing conflicts and the resource degradation in the lake. However, in 2012, the 100-year fishing lot system in TSL was abolished, and the whole lake was turned into open access and conservation.

Hence, on the basis of these facts, fishery management focuses on four key areas: (i) commercial fishing, (ii) the CFi, (iii) and open access and (iv) conservation or sanctuary. These management systems influence the governance of TSL, and therefore, studying the governance of TSL requires us to look back into these systems and see any areas that can be improved for the sake of lake management.

5.2.1 Commercial Fishing

Commercial fishing, established in 1908, was the main governance system of TSL. Thus, the commercial fishing business in TSL was managed on the basis of the private property system in which the private businessman received an exclusive right from the state to manage a fishing ground in TSL for commercial fishing. By 2000, the fishery sector contributes 10–11% of gross domestic product (Thouk 2009).

TSL was territorialized into small fishing areas known as the fishing lots, and the usufruct rights were assigned to the private individual who managed the fishing lots. Prior to 2000, about 507,730 ha in TSL was allocated for the commercial fishing areas, whereas 638,189 ha was for open access or the public fishing areas. The 507,730 ha of the commercial fishing areas were classified into 56 fishing lots.

The fishing lots were granted and managed, which was very controversial and untransparent. To get the fishing lots, the fishing lot owners spent tremendous amounts of money under the table. Thus, in return, fishing lot owners maximized fishing to generate enough returns and to ensure sustainable business in fishing. They caused conflict within the fishing business, particularly fishing conflicts between the fishing lots and fishing communities and between fishing and farming in the lake.

In 2000, the RGC introduced the first reform in fisheries management and released 56.46% of commercial fishing lots areas or 538,522 ha. In TSL, about 47% of the total commercial fishing lot areas were cancelled from commercial fishing lot areas in 2001 for local community uses. The number of fishing lots has been reduced from 56 in 2000 to 38 in 2001 (18 fishing lots have been cancelled). In 2012, the RGC abolished the fishing lot system and returned large fishing lot areas into open access, whereas some areas returned to conservation. The question is raised whether the open access system is an option for the management of fishery in TSL, compared with the commercial fishing lot system.

5.2.2 Community Fishery Management System

After the abolishment of the commercial fishing lot system, CFi was introduced to manage the cancelled fishing lot areas. About 412,205 ha of open access areas in TSL was transformed into community fisheries. CFi was established as a new emerging mode of management to manage fisheries in TSL. The new form of governance involving local fishing communities emerged, whereby about 175 community fisheries with a total household of 61,613 families in 361 villages across TSL were established in 2007 (Sithirith 2011).

The CFi was established by the Fisheries Administration (FiA). It has been suffering because of a lack of skills and experiences of FiA officers in managing CFi. It has been suffering also because of the push by community members to have full ownership over the CFi and the pull by FiA to have full control over CFi. The legal system for CFi management was developed by FiA, but in favor of FiA. This discourages the CFi members from constant involvement in managing the CFi owing to the unclear stream of benefits for the leadership as well as its members. At the same time, being members of CFi, they could only fish for subsistence while the non-members possibly could fish large scale. As a result, more illegal fishing is taking place within the CFi areas with no members daring to stop the illegal fishing. In return, members of CFi are reluctant to sacrifice for CFi. Thus, the governance of CFi is still weak.

5.2.3 Public Fishing or the Open-Access System

The management system of TSL since the French Protectorate Administration designates some areas in the lake for the mass fishing population, defining them as a “public fishing area.” The public fishing area is opened up to all people, and it is accessible by everybody as a citizen of Cambodia (Fishery Law 2006). However, the accessibility to the public fishing areas for fishing is regulated by the provincial fisheries cantonments through the use of fishing gear and other regulations. Everybody could access and fish in the public fishing area using small-scale fishing gear, and fishing with this scale in this area is only for household consumption, not for trade (Fiat Law 1987; Fisheries Law 2006). For small-scale fisheries or subsistence fisheries, access can be free and open.

The cancellation of a fishing lot and the return of it into the open-access raise more concerns about the governance of TSL. The removal of the private property resource regime and the return of the lake to the open-access regime raises the question of whether this is the right decision. Gareth Hardin (1968) calls open access an absence of property rights, and if this continues, he predicts that open access will lead to the "tragedy of the commons." The question is raised whether the open access in TSL could lead to the "tragedy."

5.3 Management of Conservation Area

Large conservation areas in TSL fall within the wetland areas. The management of conservation in TSL is classified into two categories: the "fish sanctuaries" and the "biosphere reserves." Some of them are overlapped, and it is managed by more than two institutions. However, its management is not guided by wetland principles, but by sectoral policies and institutions, which makes the wetland a cross-cutting issue.

5.3.1 Fishery Conservation

The fish sanctuary was established in the 1940s under the French Protectorate Regime in TSL, and it continues to exist since then in the lake. It has been created within the deep pools with permanent water in the lake, to provide refuge for fish broodstock to escape intensive fishing inside the lake during the dry season so as to improve replenishment of the fish stocks during the breeding season. The serious decline in some fish stocks and the threatened status of some fish species make the protection of broodstock a high priority (ADB, FAO & DoF 2003).

In TSL, there are eight fish sanctuaries, covering an area of 24,680 ha, situated all over TSL. These fish sanctuaries were established in the 1940s as part of the fishing lot systems. In 2012, the RGC cancelled the fishing lots (270,217 ha), and about 35% of the cancelled fishing lot areas, or 93,246 ha, were transformed into a fishery conservation area. The whole fishery conservation area is now classified into 23 conservation areas similar to fish sanctuaries. FiA is the sole agent responsible for the management of the fish sanctuary.

Large cancelled fishing lot areas in Battambang, Pursat, and Kampong Thom Provinces were converted into the fishery conservation area, accounting for 51%, 44%, and 26%, respectively. Former fishing lot no. 2 (50,134 ha) in Battambang is now entirely returned into the conservation. Former fishing lot no. 2 (7476 ha) is turned entirely into a conservation area in Pursat Province. Large areas of former fishing lot no. 6 in Kampong Thom are now turned into the conservation area.

5.3.2 Biosphere Reserves: Biodiversity Conservation

The increasing environmental concern since 1993 resulted in the formation of TSBR, initiated under the UNESCO Biosphere Man and the Biosphere Program, for the conservation of biodiversity. The Biosphere Reserve was classified into three zones: transitional, buffer, and core (UNESCO 2013).

The *transitional zone* covers an area of 899,600 ha, which is limited between the outer boundary of the buffer zone and National Road No. 5 and National Road No. 6 (RGC 2001). This is an area encircling the lake, which includes most of the floodplain and some areas of upland wet season rice production known as a *sreleu* (Sithirith 2011; Keskinen and Sithirith 2010). The transitional zone is set to promote the management of resources and human activities in the transitional areas of TSL in order to reduce the adverse effects on the buffer and core zones of TSL (Sithirith 2011).

The *buffer zone* covers the core area of TSL, extending outward up to the inner boundary of the transition zone with an estimated coverage of 541,482 ha (Sithirith 2011). The buffer zone is used for research; the management of flooded forest, fishery, agriculture, housing settlement, land use, water resources, navigation, and tourism; and preserving the environment and fish (RGC 2001). Activities in this area are managed to be consistent with the protection and conservation of the core areas. There is no clear control mechanism set for the buffer zone.

The buffer zone is classified into three core zones, namely, *Prek Toal*, *Boeng Tonle Chhmar*, and *Stung Sen*, covering an estimated area of 42,257 ha (Sithirith 2011). The design of the core zone is to securely protect the sites for conserving biodiversity, monitoring minimally disturbed ecosystems, and undertaking non-destructive research and other low-impact uses such as education. Thus, the core areas are defined likewise as national parks or wildlife sanctuary, which are devoted to long-term protection and conservation of natural resources and ecosystems. Management activities that would cause degradation and destruction of biodiversity are not permitted.

The control and management of the core zone and buffer zone become more problematic as about half of the buffer zone overlaps the fishery conservation areas (RGC 2007). The Prek Toal core zone overlaps former fishing lot no. 2 in Battambang Province, the Boeng Chhmar core area overlaps former fishing lot no. 6, and the Stung Sen core area overlaps former fishing lot no. 3 in Kompong Thom Province. Apart from fishing, TSBR is also home to 2218 people.

5.3.3 Flooded Forest Conservation

Since 2012, the RGC has made an effort to conserve the flooded forest around the lake. This happens following the cancellation of fishing lots in TSL. The conservation takes place through zoning the lake into three zones known as flooded forest zoning, namely, Zone 1, Zone 2, and Zone 3, targeting the conservation of flooded forest around TSL (Fig. 5.2, Sub-decree 2012).

Zone 1 covers an area of 395,578 ha, extending from National Road No. 5 and No. 6 down into the floodplain around the lake. This zone is largely covered by rice field and human settlements. There are diverse socioeconomic activities taking place in this zone, mostly farming and fishing. Zone 2 covers an area of 369,865 ha, located between Zone 1 and Zone 3. This zone has a low density of human settlement and few rice field but has more bushes and natural ponds. Socioeconomic activities in this zone are closely monitored, and some human activities are limited, subjecting to impact assessment study.

Ownership over land is subject to approval by the local administration. Some human activities such as clearance of flooded forest are not allowed, contributing to the protection of the flooded forest. Zone 3 covers the flooded forest areas, estimated at 647,406 ha. This zone is protected by the Sub-decree signed by Prime Minister Hun Sen that officially created 647,406-ha protected zones covering the flooded

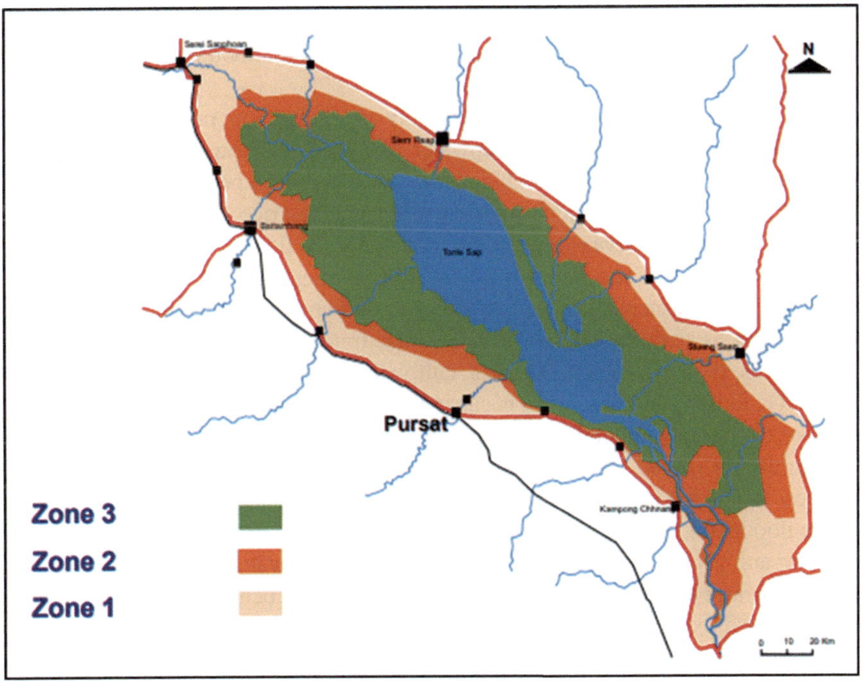

Fig. 5.2 Map of flooded forest zoning in Tonle Sap Lake (Source: ADB 2014)

forest and floodplains of TSL. Human activities, human settlement, agricultural activities, and ownership of land are not allowed in this zone. Any act opposing this will be penalized.

The flooded forest zoning suggests an overlap with the Biosphere Reserve zoning while the flooded forest zoning falls in the mandates of Tonle Sap Authority and the Biosphere Reserves falls in the authority of Tonle Sap Biosphere Reserve Secretariat and the Ministry of Environment. The difference is that the Biosphere Reserve zoning zones the whole lake including the terrestrial areas, the floodplain, and the aquatic areas, whereas the flooded forest zoning zones only the flood terrestrial areas and floodplain. Conservation in TSL has taken the sectoral approach in which fish sanctuary functions for fishery conservation, the Biosphere Reserves function for biodiversity conservation, and the flooded forest zoning acts for the conservation of flooded forest. These suggest a distinct governance approach toward conservation, and the weak coordination causes the conservation governance to overlap and complex.

5.4 Water Management and Farming/Agriculture

Communities around TSL organize the farming in the TSL floodplain into *sreleu* (rainfed lowland rice field), *srekandal* (medium deepwater rice field), and *srekrom* (deepwater rice/floating rice field). Areas named *sreleu* are called as such because they are located in the upper geographical area of the TSL floodplain, between 8 and 10 m a.s.l. (Sithirith 2011).

The rice cultivation in *sreleu* is entirely dependent on rainfall, not on lake hydrology. *Srekandal* (medium deepwater rice field) is located deeper inside the floodplain area, lower than *sreleu* but above *srekrom*, where people cultivate floating rice in the past, but at present, farmers cultivate receding rice, following the receding water in the lake. *Srekrom* (deepwater rice/floating rice field) is located deeper inside the TSL floodplain. The rising water in TSL influences the rice productivity of *srekrom*. Thus, farmers cultivate rice varieties that grow and rise according to water levels, such as that named *srove leung teuk*, translated as "rising water rice" in English. In the dry season, farmers cultivate in this area a "dry season rice." However, farming in these areas is carried out in a traditional way, small-scale and subsistence, largely dependent on rainfall, and sometimes on lake water.

Farming in *sreleu*, *srekandal*, and *srekrom* is affected by too much water in the wet season and too little in the dry season. In the wet season, there is too much water in the floodplain, and too much water sometimes causes flooding and damages rice farming and people's properties. In the dry season, the water from TSL recedes the floodplain and flows into the Mekong River without being used, as no storage system could extract the excess water and keep it in a reservoir or basin for use in the season of too little water. Indeed, too much and too little water makes farmers vulnerable in both the wet and dry seasons, and only 1% of the total water in Cambodia, or 750 million m^3, is used, and 95% of the use is for agriculture. The annual average outflow of TSL is estimated at 78,600 million m^3, and about 69,000 m^3 (88% of outflow) from TSL returns to the Mekong River via the TSL River (Kummu and Sarkkula 2008; CNMC 2004). Thus, the water management to retain water from too much water in the wet season and store it for use in the situation of too little water in the dry season is the ideal situation in TSL that farmers dream of.

However, where the water from TSL reaches, there is a fishery domain, and FiA has authority over it. These conflict with the use of public and private lands (Articles 8 and 9, Fishery Law 2006). The ownership over the land that FiA has authority can be granted only there is a study and assessment conducted and approved by the Ministry of Agriculture, Forestry and Fisheries (MAFF) (Article 13, Fishery Law 2006). Dams, dikes, reservoirs, canal, ponds, and reclaiming floodplain areas are not allowed inside the fishery domain, but if there is a need to do so, MAFF is the sole agency responsible for providing the approval, based on the proper assessment and study (Articles 20 and 25, Fishery Law 2006). In addition, the Fishery Law also prohibits the planting of a lot, although the Fishery Law does not refer to other crops. The expansion of farming land inside the fishery domain is not permitted under the Fishery Law, and it does not take into consideration the customary practice of farming inside TSL (Article 23, Fishery Law 2006).

Key Points

- TSL has been zoned and zoning for different purposes, including fisheries management, flooded forest protection, and biodiversity conservation on technical and scientific capacity and economic and political interests. These zonings have created overlapping spaces and conflicted with natural zoning and a live space of communities.
- There are overlapping claims over different spaces in TSL by different agencies with legal mandates, technical capacity, and legitimate roles and responsibilities. These lead to strong competition and weak coordination. Farming in the TSL floodplain often conflicts with fishing, and water management does not well coordinate with farming and fishing.
- The official zonings and spaces often conflict with community areas. These have affected the livelihoods of fishing-dependent communities. The cancellation of the fishing lot system has resulted in over-fishing and anarchies in the lake, raising concerns about the future and the sustainability of the lake.

References

Asian Development Bank (ADB). Integrated urban environmental management in the Tonle Sap Basin. Phnom Penh: Ministry of Public Works and Transport; 2014.

Asian Development Bank (ADB), Food and Agriculture Organization (FAO), Fisheries Administration (DoF). General fisheries plan for the management and protection of TSL, Tonle Sap environmental management project. Phnom Penh: Fisheries Administration, Cambodia; 2003.

Cambodian National Mekong Committee (CNMC). Sub-area analysis and development: the Tonle sap sub-area sa–9c. Phnom Penh: CNMC; 2004.

Fiat Law. Fisheries law. Fisheries Administration, Fisheries Administration, Ministry of Agriculture, Forestry and Fisheries, Royal Government of Cambodia, 1987.
Fisheries Law. Fisheries law. Fisheries Administration, Ministry of Agriculture, Forestry and Fisheries, Royal Government of Cambodia; 2006.
Hardin G. Tragedy of the common. Science. 1968;162(3859):1243–8. https://doi.org/10.1126/science.162.3859.1243.
Keskinen M, Sithirith M. Tonle Sap Lake and its management: the diversity of perspectives & institution, Working Paper for Improving Mekong Water Allocation Projects (PN-67). Chiang Mai: M-Power; 2010.
Kummu M, Sarkkula J. Impact of the Mekong River flow alteration on TSL flood pulse. J Human Environ. 2008;37(3):185–92.
Lefebvre H. The production of space, Translated by Donald Nicholson-Smith. Malden, Massachusetts: Blackwell; 1991.
Sithirith M. Political geography of TSL: power, space and resources. PhD Dissertation. Singapore: National University of Singapore; 2011.
Royal Government of Cambodia (RGC). Royal decree on the establishment of Tonle Sap authority, draft, unofficial translation, 2007. Available Online at: http://www.foodsecurity.gov.kh/Otherdocs/Royal-Decree-tsba-Eng.pdfroyal. (Accessed Finnish Environment Institute, Helsinki 12 June 2008).
Royal Government of Cambodia. Royal decree on the establishment and management of TSL biosphere reserve, translation by Neou Bonheur, 2001. Available online at: Http://www.tsbred.org/docs/law_and_regulation/Royal_Decree_On_creation_and_managemet_of_tsbr_Eng.Pdf (Accessed 22 July 2008).
Thouk N. Community fish refuge husbandry in lowland agriculture ecosystem. Doctoral Thesis. Phnom Penh:. Cambodia: Build Bright University, Institute of Post Graduate Studies; 2009.
Tonle Sap Biosphere Reserve Secretariat (TSBR). Policy and strategy for the Tonle sap biosphere reserve, revised version – January 2007. In: Component one: ADB loan no 1939-Cam (SF), Tonle sap environmental management project, Cambodia national mekong committee, Tonle sap biosphere reserve secretariat; 2007. Available online at: http://www.Tsbred.Org/Docs/Misc/Final_Policy_Working_Paper.Pdf. Accessed 09 October 2008.
UNESCO. Concept of Governance, 2013. http://www.unesco.org/new/en/education/themes/strengthening-education-systems/quality-framework/technical-notes/concept-of-governance/. Accessed on 10 July 2013.
United Nations Development Program/Global Environmental Facility (UNDP/GEF). The Tonle sap conservation project. Cambodia: Phnom Penh: UNDP; 2004.

Part II
Climate and Hydrology

Chapter 6
Climate and Rainfall

Kumiko Tsujimoto

6.1 Monsoon and Rainy Season

Cambodia is influenced by the Asian summer monsoon (ASM) and the Asian winter monsoon (AWM). The ASM is established in early to middle May, associated with a strong convection and change of direction of the prevailing wind over the Bay of Bengal, the Indochina Peninsula, and the South China Sea (Zhang et al. 2002). In the boreal winter season, the entire Indochina is subject to the dry season under the influence of the AWM from mid-December until mid-April (Yen et al. 2011). Consequently, Cambodia has distinct rainy and dry seasons. Although the onset and withdrawal of the ASM and AWM are generally defined by the synoptic-scale wind field, the rainy season is defined based on the rainfall amount.

Figure 6.1 shows the monthly mean synoptic-scale atmospheric fields at 1000 hPa according to the Japanese 55-year reanalysis data (JRA-55) (Kobayashi et al. 2015; Harada et al. 2016). From March to April (Fig. 6.1a, b), the monthly mean wind is easterly over the South China Sea and southeasterly over the Gulf of Thailand. From May to September (Fig. 6.1c–e), the entire Cambodia is subject to a large-scale southwesterly wind under the establishment of the ASM. From October to November (Fig. 6.1f, g), the northeasterly wind by the AWM prevails over the Gulf of Tonkin and the northern South China Sea. From December to February (Fig. 6.1h, i), the large-scale northeasterly wind by the AWM becomes stronger with the intrusion of drier air. Hence, in a simple definition on a monthly basis, the summer monsoon season is from May to September, with the pre-monsoon season in March and April and the post-monsoon season in October and November. The dry season is in December, January, and February (Tsujimoto et al. 2018).

K. Tsujimoto (✉)
Okayama University, Okayama, Japan
e-mail: tsujimoto@okayama-u.ac.jp

C. Yoshimura et al. (eds.), *Water and Life in Tonle Sap Lake*,
https://doi.org/10.1007/978-981-16-6632-2_6

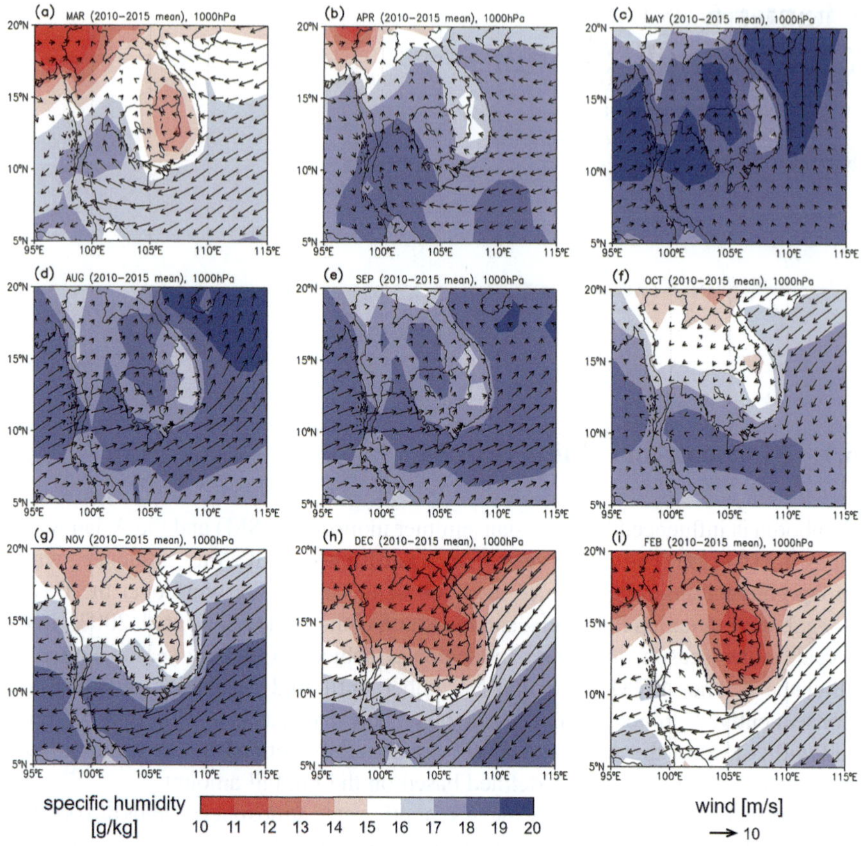

Fig. 6.1 Wind and specific humidity at 1000 hPa over the Indochina Peninsula. Monthly mean data for March (**a**), April (**b**), May (**c**), August (**d**), September (**e**), October (**f**), November (**g**), December (**h**), and February (**i**) as 6 year (2010–2015) averages are shown. Data are from JRA-55 historical reanalysis. Reprinted from Tsujimoto et al. (2018) with modifications

With abundant rainfall brought by the ASM, the rainy season roughly corresponds to the ASM season but is often associated with the pre- and post-monsoon rainfalls. In fact, it is well known by local people in Cambodia that it rains not only during the summer monsoon season but also during the pre- and post-monsoon seasons. These rainfalls are called “Phleang Koker” in the local parlance, and this phrase is associated with happiness.

The onset and withdrawal dates of the rainy season and their spatial distribution over and around the Indochina Peninsula, including Cambodia, were analyzed by Matsumoto (1997) using rain-gauge data in 1975–1987. He found that the rainy season starts earlier in the inland region of Thailand from late April to early May and withdraws later in the central eastern coast of Vietnam in late November. Based on his results, the onset date over the southern Indochina Peninsula (around Cambodia) is around April 28 to May 8, and the withdrawal date is around October 31–November 10. In his results, the area over and around Cambodia experiences a longer rainy season than other areas in Indochina. However, the actual distribution of rainfall within Cambodia had remained unknown until very recently due to the scarce rain-gauge data. Recent data shows diurnal, intra-seasonal, and year-to-year variations with regional differences within Cambodia (refer to Sect. 6.2).

6.2 Rainfall Amount and Its Distribution

Based on the rain-gauge observation at 30 stations in Cambodia in 2010–2015 (Tsujimoto et al. 2018), the annual rainfall amount in inland Cambodia is approximately 1087–1528 mm on station average and is much larger on the coast at nearly 4000 mm (Fig. 6.2). Over inland Cambodia, approximately 5–20% of the annual rainfall (63–278 mm) occurs in March and April (pre-monsoon season), 50–78% (519–1350 mm) occurs from May to September (summer monsoon season, under the fully established ASM effect), and 12–36% (181–457 mm) occurs in October and November (post-monsoon season).

During the pre-monsoon season, rainfall is dominant on the coast and over the Cardamom Mountains. During the summer monsoon season, the rainfall amount is distinctively large on the coast facing the Gulf of Thailand. In the inland, a larger amount falls in the northern region (around Kampong Thom, Preah Vihear, and Siem Reap provinces), and a smaller amount falls in the western region (around Battambang province). By contrast, the rainfall amount during the post-monsoon season is larger on the southwestern side of the lake (around Battambang and Pursat provinces), i.e., around the northeastern ridges of the Cardamom Mountains and the western lakeshore plain, than on the other regions in inland Cambodia. These predominant rainfalls in western Cambodia during the post-monsoon season mainly occur during nighttime through early morning and are brought by the land–lake–atmosphere interaction (refer to Sects. 6.3 and 6.5).

6.3 Diurnal Cycle of Rainfall

The diurnal cycle of rainfall presents the distribution of rainfall within a day. Generally, the diurnal cycle of convection tends to be stronger over land during summer, with maxima in the late afternoon or early evening under a dominant influence of daytime boundary layer heating (Chaboureau et al. 2004). As Wallace (1975) has summarized, daytime rainfall peaks are mainly controlled by thermodynamic processes that affect atmospheric stability.

Using the automatic weather station data at three locations in Cambodia observed by Masumoto et al. (2007), Tsujimoto et al. (2008) analyzed the diurnal cycle of rainfall and its regional difference. The result showed clear afternoon peaks in the

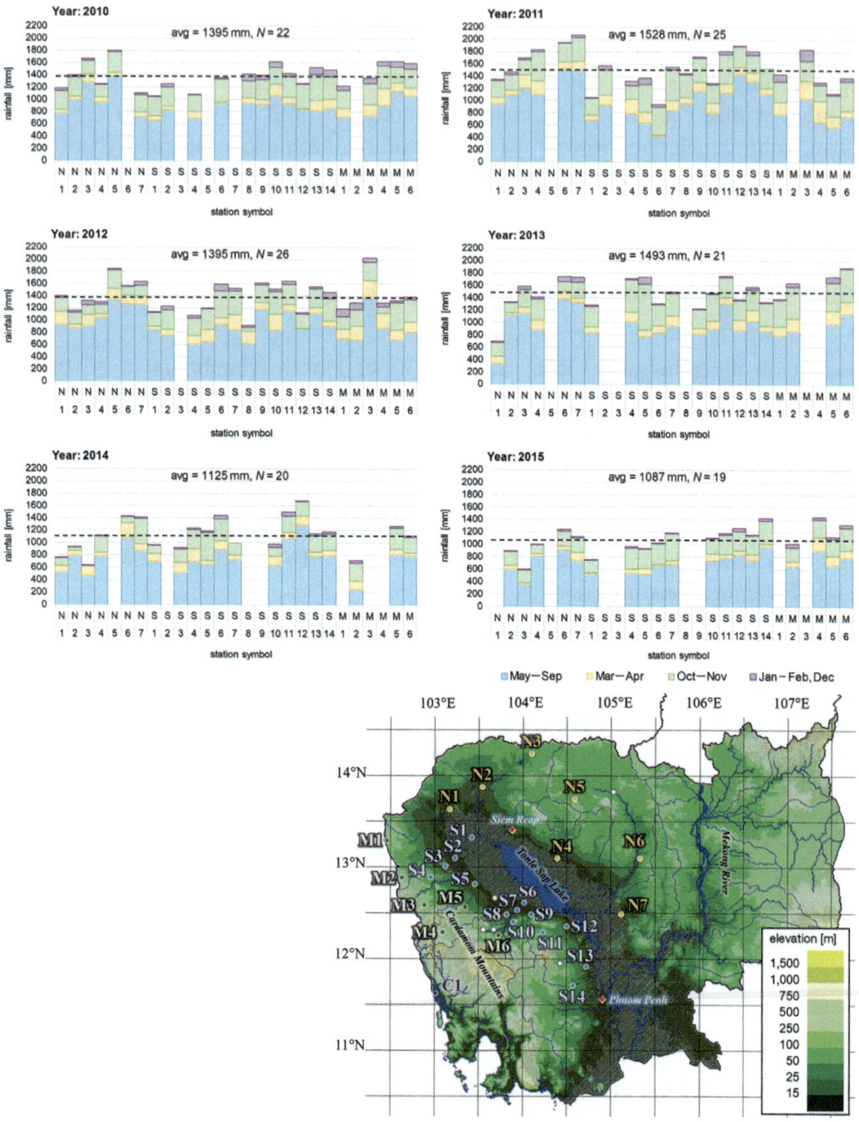

Fig. 6.2 Gauge-observed rainfall amount at each station. The number of available rain gauges, *N*, and annual rainfall amount averaged over available inland rain gauges (except for station C1) are shown in each panel with a broken black line. Reprinted from Tsujimoto et al. (2018)

city of and over a paddy field near Phnom Penh and a weaker afternoon peak over Tonle Sap Lake (TSL) near the city of Siem Reap. The observed tendency can be well explained by the thermodynamic stability, following the warmer surface over land during daytime.

The regional and seasonal differences in the diurnal cycle of rainfall are presented in more detail with potential mechanisms in Tsujimoto et al. (2018) as follows.

During the pre-monsoon season, rainfall maxima occur in the afternoon. Orographic lifting, as well as atmospheric instability by larger sensible heat from the drier ground, is likely the driving mechanism.

During the summer monsoon season, clear regional characteristics in the diurnal precipitation patterns are recognized, and they suggest the significant effects of local features even during the ASM season. The diurnal rainfall maximum occurs (i) in the early afternoon in the Cardamom Mountains, (ii) in the afternoon on the plain at the southwestern side of TSL, (iii) in the evening on the wide area at the northeastern side of the lake, and (iv) in the early morning on the coast. The potential mechanisms for each rainfall are (a) the daytime low-level convergence of the southwesterly ASM and the northeasterly lake breeze at the southwestern side of the lake as well as increased atmospheric instability by sensible heat from land (for (i) and (ii)); (b) the daytime low-level convergence of the ASM with the anabatic wind, as well as the orographic lifting of the moist lake air, over the northeastern slopes of the Cardamom Mountains (for (i)); (c) the low-level convergence of the ASM with the evening northeasterly land breeze at the northeastern side of the lake (for (iii)); and the (d) low-level convergence of the ASM with the nocturnal northeasterly land breeze enhanced by the katabatic wind from the southwestern slope of the Cardamom Mountains at the coastal area (for (iv)).

During the post-monsoon season, a diurnal maximum occurs at night and in the early morning. It probably results from the low-level convergence of the northeastern AWM with the southwestern land breeze enhanced by the katabatic wind from the northeastern slope of the Cardamom Mountains. Since the rainfall amount is smaller at the northeastern side of the lake where the AWM converges with the daytime lake breeze, the importance of the role of mountains on rainfall is proposed.

Such local atmospheric circulations as land–lake breezes and mountain–valley breezes are presented in more details in Sect. 6.5. The diurnal pattern of the rainfall in each season and in each region is, however, not clear on some days, and the analysis of the synoptic-scale atmospheric condition suggests the effect of the large-scale low-pressure system and disturbances on the appearance of the clear diurnal rainfall pattern.

6.4 Radiation and Evaporation

Tsujimoto et al. (2008) compared solar radiation amounts for the dry and rainy seasons with ground-based data in Cambodia and showed that the rainy season had nearly as much solar radiation as the dry season. Three reasons are considered: (1) the rainy season had larger amounts of extraterrestrial radiation; (2) nearly half the days in the rainy season were non-rainy days; and (3) the amount of solar radiation on rainy days reached 88% of that on non-rainy days. The third factor was attributed to the high frequency of evening rainfall events that is presented in

Sect. 6.3. Furthermore, this rainfall–radiation relationship indicated that the rainy season had larger amounts of net all-wave radiation because of the low reduction ratio of solar radiation and an increase in long-wave incoming radiation. Accordingly, the rainy season had high evaporation potential.

For the rainfed rice paddies that prevail in this region, sufficient radiation during the rainy season would be a great advantage for rice growing. In this regard, the diurnal cycle of rainfall presented in the previous section is one of the key factors required to evaluate solar radiation and to estimate evapotranspiration, which regulates rice production itself.

6.5 Land–Lake–Atmosphere Interaction and Its Effect on Rainfall

With the unique and dynamic hydrological characteristics around TSL (see also Chaps. 1, 8, 11, and 12), the rainfall pattern is highly affected by the land surface heterogeneity, in both time and space, resulting in rainfall characteristics within Cambodia having regional characteristics. As mentioned by Takahashi et al. (2010), the observed diurnal cycle of rainfall (refer to Sect. 6.3) is a "visualized" result of the interaction of several mechanisms under both large- and local-scale effects and is useful for the understanding of the local circulation, especially in regions where the observational network is sparse. The clear diurnal cycle and its regional differences found in the rain-gauge observation in Cambodia suggest the importance of not only the monsoon but also the effect of local features in the formation and development of rainfall in Cambodia. One of the noteworthy characteristics of Cambodia is the existence of TSL. The western side of the lake is at the foot of the Cardamom Mountains, and the Annam Range lies south and north along the eastern shore of the Indochina Peninsula to 400–500 km east of the lake.

Land–lake breezes, similar to land–sea breezes, are among the local circulations initiated by the thermal contrast of the land surface. Clear land–lake circulation, with a daytime lake breeze (onshore flow) and nighttime land breeze (offshore flow) in the presence of a linear convective system at its front, has been recognized in Cambodia by both the analysis of observations and numerical simulations (Tsujimoto and Koike 2008, 2013). Although the effect of the mountain–valley circulation interacting with land–lake circulation is often discussed in studies on the diurnal cycle of rainfall around the lake, the surrounding area of TSL is almost flat in comparison with other lakes in the world; thus, the single effect of land–lake circulation would clearly appear, although the effects of mountains also exist.

Especially during the post-monsoon season, the existence of the nocturnal clear land breeze circulation accompanied by a small linear cloud system has been suggested (Tsujimoto and Koike 2013). Generally, the development of the nocturnal land breeze tends to be weaker under the smaller land–lake thermal contrast compared with that during daytime, and it is suggested to be especially weak at low latitudes where the Coriolis force is weak (Yan and Anthes 1987; Mapes et al. 2003). However, over TSL, the existence of a linear updraft system over the lake, forming along the southwest lakeshore around 22:00 and moving northeast to the middle of the lake, is suggested by the numerical simulation of a regional atmospheric model, and the corresponding linear clouds are recognized by the satellite images in the early morning. The heavier air mass from the land meets the extraordinarily warm and humid air mass over the lake, triggering updrafts under the conditionally convective instability. The characteristic high surface water temperature is favorable for the generation of the land breeze and updraft systems. The high surface water temperature of TSL is produced by the tropical climate along with efficient energy absorption because of the shallowness of the water body. This unique feature can generate a clear nocturnal land breeze circulation accompanying a migrating updraft system over the lake despite its low latitude.

Besides the land breeze, the clear predominant post-monsoon rainfalls that occur after the withdrawal of the ASM have been observed in western Cambodia from midnight to early morning (Tsujimoto et al. 2014, 2018). Most of these events occurred under the condition of the strong nocturnal northeasterly wind, which is considered as the katabatic wind from the Annam Range (Tsujimoto and Koike 2012). This region is one of the most important grain production regions in Cambodia, and thus, water resource management there is essential for the nation's food production and economy.

A detailed study on the post-monsoon rainfall and its mechanism was conducted for the year 2009 (Tsujimoto and Koike 2012). In 2009, rainfall kept occurring for approximately 1 month after the withdrawal of the ASM, mostly over the western side of TSL. The rainfall events in this period were divided into two categories by referring to the global reanalysis data; one is the "large-scale rainfall" that is induced by the inflow and convergence of water vapor from outside Cambodia, and the other is the "local rainfall" that falls even under the non-convergent condition and is due to the redistribution of water vapor within Cambodia. In Fig. 6.3, the lower panel shows the rainfall amount under the non-convergent condition (i.e., local rainfall) comparing with the total (local and large-scale) rainfall amount, and the upper panel shows the ratio of the local rainfall amount to the total. The values in Fig. 6.3 are shown with a 3 h interval in Cambodian local time. As shown in Fig. 6.3, the contribution of the local rainfalls that fall even under the non-convergent condition was larger from midnight to early morning, accounting for more than 50%.

From late November in 2009, alternatively, there was almost no rainfall. During this period, the water vapor field was continuously divergent with the seasonal march, which decreased the precipitable water in Cambodia and made an unfavorable condition for precipitation. In this period, the nocturnal precipitable water with local precipitation was 45 mm, whereas that without local precipitation was 28 mm. Hence, for the initiation of the local precipitation, the local wind convergence and the abundance of water vapor are the prerequisites, although this area is located in the tropical Asian monsoon region and generally has a large amount of water vapor. Even before mid-November with sufficient precipitable water, there were days without local rainfall. The wind field revealed that when the land breeze from the southwest lakeshore was very weak (<0.15 m/s), there was no rainfall. Moreover,

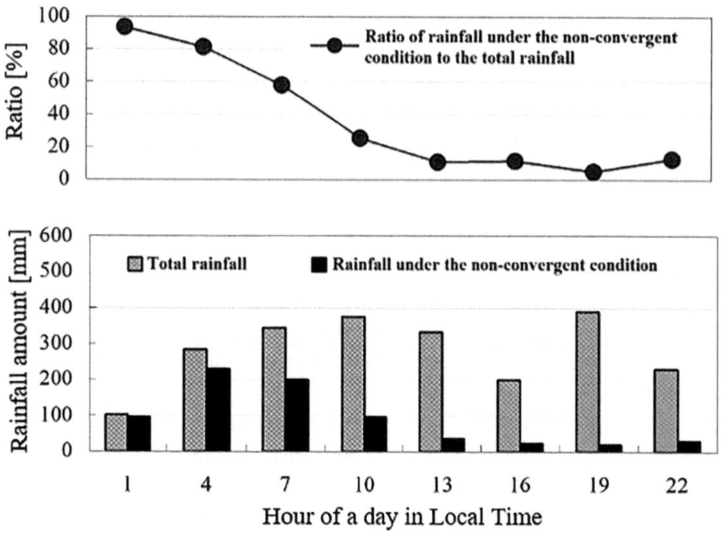

Fig. 6.3 Three-hour rainfall amount (gray bar in the bottom panel) and its breakdown under the non-convergent condition (i.e., local rainfall, black bar in the bottom panel) shown with the ratio of rainfall amount under the non-convergent condition to the total rainfall amount (upper panel). The amount of precipitation is the total accumulated amount for the given hour ±90 min recorded by the 30 rain gauges. Reprinted from Tsujimoto and Koike (2012) with modifications

even when the southwestern land breeze was strong enough, if the northeasterly from the northeastern side of the lake was weak (<0.6 m/s), there would be no rainfall. By contrast, when the nocturnal northeasterly was very strong (>4.0 m/s), it prevented the development of the southwestern land breeze, and thus, there was no rainfall.

Key Points

- Seasonal distribution of rainfall in the vicinities of TSL is under the strong influence of the ASM, with pre- and post-monsoon rainfalls as well.
- Over inland Cambodia, approximately 5–20% of the annual rainfall (63–278 mm) occurs in March and April (pre-monsoon season), 50–78% (519–1350 mm) occurs from May to September (summer monsoon season, under the fully established ASM effect), and 12–36% (181–457 mm) occurs in October and November (post-monsoon season).
- The characteristic diurnal cycle of precipitation and its regional and seasonal dependencies have been recognized over this region.
- The land–lake breezes brought by the interaction between TSL and the mountain–valley circulations are considered affecting the rainfall amount and its spatiotemporal distributions in pre- and post-monsoon seasons.
- The enhanced nocturnal land breeze circulation observed over TSL in the post-monsoon season is unique compared with the earlier studies suggesting weak land breeze circulation over the low latitudes. The characteristic high lake water temperature of TSL is considered to affect the strong land breeze circulation.

References

Chaboureau JP, Guichard F, Redelsperger JL, Lafore JP. The role of stability and moisture in the diurnal cycle of convection over land. Q J R Meteorol Soc. 2004;130:3105–17. https://doi.org/10.1256/qj.03.132.

Harada Y, Kamahori H, Kobayashi C, Endo H, Kobayashi S, Ota Y, Onoda H, Onogi K, Miyaoka K, Takahashi K. The JRA-55 reanalysis: representation of atmospheric circulation and climate variability. J Meteorol Soc Japan Ser II. 2016;94:269–302. https://doi.org/10.2151/jmsj.2016-015.

Kobayashi S, Ota Y, Harada Y, Ebita A, Moriya M, Onoda H, Onogi K, Kamahori H, Kobayashi C, Endo H, Miyaoka K, Takahashi K. The JRA-55 reanalysis: general specifications and basic characteristics. J Meteorol Soc Japan Ser II. 2015;93:5–48. https://doi.org/10.2151/jmsj.2015-00.

Mapes BE, Warner TT, Xu M. Diurnal patterns of rainfall in northwestern south America. Part III: Diurnal gravity waves and nocturnal convection offshore. Mon Weather Rev. 2003;131:830–44. https://doi.org/10.1175/1520-0493131<0830:DPORIN>2.0.CO;2.

Masumoto T, Tsujimoto K, Somura H. Hydro-meteorological observation and analysis of observed data at Tonle Sap Lake and its environs, urban and paddy areas. Tech Rep NIRE. 2007;206:219–36.

Matsumoto J. Seasonal transition of summer rainy season over Indochina and adjacent monsoon region. Adv Atmos Sci. 1997;14:231–45. https://doi.org/10.1007/s00376-997-0022-0.

Takahashi HG, Fujinami H, Yasunari T, Matsumoto J. Diurnal rainfall pattern observed by Tropical Rainfall Measuring Mission Precipitation Radar (TRMM-PR) around the Indochina Peninsula. J Geophys Res Atmos. 2010;115:D07109. https://doi.org/10.1029/2009jd012155.

Tsujimoto K, Koike T. Mechanism of locally-induced convection development and its effects on vapor transportation over Tonle Sap Lake Area. J Hydraul Eng. 2008;52:247–52. https://doi.org/10.2208/prohe.52.247.

Tsujimoto K, Koike T. Requisite conditions for post-monsoon rainfall in Cambodia by looking through 2009 rainfall data. J Hydrosci Hydraul Eng. 2012;30(1):1–14.

Tsujimoto K, Koike T. Land-lake breezes at low latitudes: the case of Tonle Sap Lake in Cambodia. J Geophys Res Atmos. 2013;118(13):6970–80. https://doi.org/10.1002/jgrd.50547.

Tsujimoto K, Masumoto T, Mitsuno T. Seasonal changes in radiation and evaporation implied from the diurnal distribution of rainfall in the Lower Mekong. Hydrol Process. 2008;22(9):1257–66. https://onlinelibrary.wiley.com/doi/abs/10.1002/hyp.6935

Tsujimoto K, Koike T, So Im M, Aida K, Tamagawa K, Nukui T, Sobue S. Validation of Satellite Precipitation Products over Cambodia. Trans Jpn Soc Aeronaut Space Sci. 2014;12:Tn_41–6. https://doi.org/10.2322/tastj.12.Tn_41.

Tsujimoto K, Ohta T, Aida K, Tamakawa K, So Im M. Diurnal pattern of rainfall in Cambodia: its regional characteristics and local circulation. Prog Earth Planet Sci. 2018;5(1):39. https://doi.org/10.1186/s40645-018-0192-7.

Wallace JM. Diurnal variations in precipitation and thunderstorm frequency over the conterminous United States. Mon Weather Rev. 1975;103:406–19. https://doi.org/10.1175/1520-0493(1975)103<0406:dvipat>2.0.co;2.

Yan H, Anthes RA. The effect of latitude on the sea breeze. Mon Weather Rev. 1987;115:936–56. https://doi.org/10.1175/1520-0493115<0936:TEOLOT>2.0.CO;2.

Yen MC, Chen TC, Hu HL, Tzeng RY, Dinh DT, Nguyen TTT, Wong CJ. Interannual variation of the fall rainfall in Central Vietnam. J Meteorol Soc Japan Ser II. 2011;89A:259–70. https://doi.org/10.2151/jmsj.2011-A16.

Zhang Y, Li T, Wang B, Wu G. Onset of the summer monsoon over the Indochina Peninsula: Climatology and interannual variations. J Clim. 2002;15:3206–21. https://doi.org/10.1175/1520-442(2002)015<3206:ootsmo>2.0.co;2.

Chapter 7
Inundation and Water Surface Temperature: Satellite-Based Observation

Yoichi Fujihara, Keisuke Hoshikawa, Hideto Fujii, Takashi Nakamura, and Sokly Siev

7.1 Importance of Hydrological Data

Tonle Sap Lake (TSL) has a unique hydrological environment in which large-scale pulsations occur annually (see details in Chap. 1). This unique hydrological cycle creates vast floodplains and flooded forests around the lake (see also Chap. 32), sustaining fishery and aquaculture that provide up to 80% of the protein consumed in Cambodia (Hortle 2007). Hence, monitoring the hydrology and water quality of the lake is crucial for environmental conservation and sustainable use of the lake.

To understand fish habitats and identify areas that need protection, monitoring of the inundation characteristics (e.g., number of inundation days and starting time of floods) under the forest canopy is necessary. Nevertheless, the presence of the canopy makes it difficult to discern the inundation status using satellite data that are conventionally used to monitor flooding. Although several studies have been conducted to estimate the floodplain area, monitoring inundation under the canopy of forests using conventional methods is not possible. Moreover, monitoring of the

Y. Fujihara (✉)
Ishikawa Prefectural University, Nonoichi, Japan
e-mail: yfuji@ishikawa-pu.ac.jp

K. Hoshikawa
Toyama Prefectural University, Imizu, Japan

H. Fujii
Yamagata University, Tsuruoka, Japan

T. Nakamura
Tokyo Institute of Technology, Tokyo, Japan

S. Siev
Institute of Technology of Cambodia, Phnom Penh, Cambodia

Ministry of Industry, Science, Technology and Innovation, Phnom Penh, Cambodia

C. Yoshimura et al. (eds.), *Water and Life in Tonle Sap Lake*,
https://doi.org/10.1007/978-981-16-6632-2_7

inundation and water level changes over large floodplains using ground-based observations is costly. Another essential parameter for lake management is water temperature since it has a major impact on biogeochemical processes and ecosystems. However, there are no long-term data on water temperature. Although intensive observations of water temperature have been reported previously (Oyagi et al. 2017), little is known regarding the changes in the lake's water environment.

Given the deficiency of hydrological and water quality data for TSL, we present the following three techniques in this chapter. First, we developed a method to monitor inundation under the canopies of flooded forests using vegetation and water indices obtained from satellite data. Second, by combining inundation maps and a digital elevation model (DEM), we developed a method for estimating changes in water level without ground-based observations. Third, we estimated surface water temperature from satellite data, investigated its long-term trend, and analyzed the relationship between water temperature and lake water level.

7.2 Inundation Area

In our study, TSL and its surrounding areas were categorized as (a) open water; (b) flooded forest, where inundation expands during the rainy season, causing flooding beneath the forest canopy; or (c) non-flooded land or area that is free from inundation, regardless of the land use type.

We used satellite imagery acquired from the Terra satellite, which uses a Moderate Resolution Imaging Spectroradiometer (MODIS) sensor. These images had a spatial resolution of 250–500 m, and the satellite revisit time was 1 day. MOD09A1 and MOD09Q1 products are 8-day composite images adjusted for atmospheric conditions, eliminating cloud effects to the highest possible extent. From the MOD09A1 and MOD09Q1, we used the bands RED (621–670 nm), NIR (Near Infrared, 841–876 nm), BLUE (459–479 nm), GREEN (545–565 nm), and SWIR (shortwave infrared, 1628–1652 nm). We also computed the normalized difference vegetation index (NDVI), enhanced vegetation index (EVI), and normalized difference water index (NDWI). These indices are frequently used to analyze flood conditions (e.g., Sakamoto et al. 2007; Fujihara et al. 2016). We also used NDWI–EVI, that is, the difference between the water index and the vegetation index, which was developed to consider both forests and inundation conditions.

We defined training areas for (a) open water, (b) flooded forest, and (c) non-flooded land, and extracted data from each area to develop an inundation estimation method. We retrieved training data for periods during which the inundation reached the forests. Hence, we selected 4 days (DOY 305 in 2000, DOY 257 in 2001, DOY 281 in 2002, and DOY 289 in 2011) when the lake water level was high and cloud impact on MODIS data was minimal. Training areas for these days were set up for the MODIS data, and training data were extracted for open water (2517 pixels), flooded forest (4984 pixels), and non-flooded land (5173 pixels). We additionally plotted these training data on a two-dimensional (2D) scatterplot to determine the thresholds and discrimination equations that could successfully classify open water, flooded forest, and non-flooded conditions.

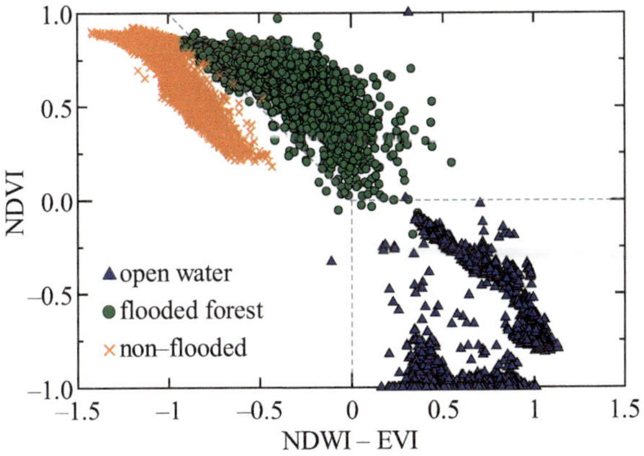

Fig. 7.1 Scatterplot of the training dataset x-axis: NDWI–EVI, y-axis: NDVI. Open water for NDVI < 0, points for flooded water for points plotted above $y = -x$, and points for non-flooded plotted below $y = -x$

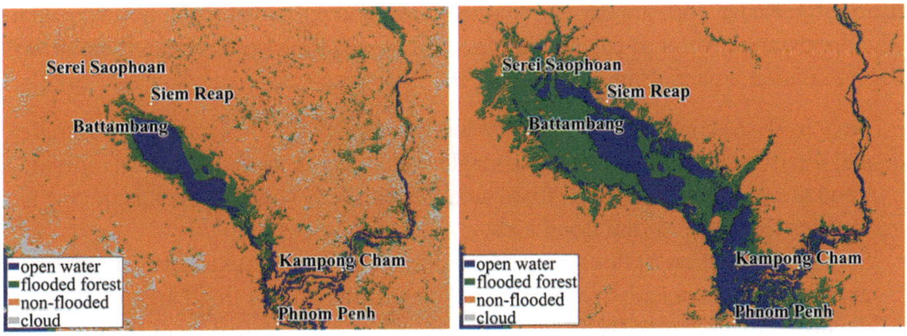

Fig. 7.2 Inundation maps. Left: DOY 209 in 2002; right: DOY 281 in 2002

Four indices (NDVI, EVI, NDWI, and NDWI–EVI) were then combined to create 2D scatterplots to verify the accuracy of the training data classification for (a) open water, (b) flooded forest, and (c) non-flooded land. We determined these 2D scatterplots, with NDWI–EVI on the horizontal axis and NDVI on the vertical axis, to be accurate (Fig. 7.1). The overall classification accuracy using this discrimination method was 0.975. The classification accuracy was 0.999, 0.947, and 0.999 for open water, flooded forests, and non-flooded land, respectively, which indicates that the classification accuracy is reasonably good.

Figure 7.2 shows an example of the inundation map created using the above method. During the rainy season, flooding started on DOY 209 in 2002 and reached its maximum level on DOY 281 in 2002. The flooded forest area is not too extensive at the beginning of the rainy season but occupies a much wider area around the lake at the peak of the flooding. Open water spreads widely in the northeastern part of the lake and around the city of Phnom Penh. Conventional satellite methods (e.g., Sakamoto et al. 2007) can only distinguish two classes (inundated and non-inundated) and cannot capture the water conditions under the canopies of flooded forests. As demonstrated, our method can be used to successfully classify three categories: (a) open water, (b) flooded forest, and (c) non-flooded land and can confirm the presence of water under the canopy. At the peak of the flooding, the total inundation area was approximately 16,600 km^2, of which 5100 km^2 (30.5%) was open water, and 11,500 km^2 (69.5%) was flooded forest, including grassland and shrubland. These results can be used to validate and improve two-dimensional hydraulic models (Tanaka et al. 2018) and provide basic information for fish and ecosystem conservation.

7.3 Water Level

Combining the developed inundation observation method with a DEM, we developed a new method for estimating inundation levels that does not require in situ observation data (Tanaka et al. 2019). The DEM used in this study was sourced from the Shuttle Radar Topography Mission (SRTM), and its spatial resolution was 90 m. The floodwater level was calculated based on the relationship between the extent of the inundation area and the DEM using the following steps (Tanaka et al. 2019):

- Step 1—Extraction of flood areas: Floods in the study area are highly complex, and the overall flood area is not a single area but comprises numerous individual flood areas. In this step, each flood area is identified. This approach is unique and clarifies the edges of the flood area.
- Step 2—Extraction of flood area edges: First, the original (extracted) flood area is expanded to the outside by two grids. Then, the expanded flood area is shrunk to the inside by one grid. By conducting this process, the noise data of the flood area are removed, and the outermost buffer of the flood area becomes a simple shape. Finally, the DEM values at these outermost buffers can be obtained.
- Step 3—Spatial interpolation: The extracted elevation data were transformed into the floodwater level. We adopted the inverse distance weighting interpolation method, wherein the floodwater level of each grid point is calculated as the weighted average value at m closest points.

These three steps were repeated from large to small flood areas progressively until the nth flood area, where n is the total number of flood areas. Considering that there were 48 scenes of MODIS data per year, these steps can be applied to all scenes to estimate the water level as the inundated area expands and shrinks (Fig. 7.3).

We verified the accuracy of the flood level estimation method by comparing the results with the actual water levels obtained from the TSL floodplains in 2002. Although some plots based on MODIS data for the river points were overestimated, there were linear relationships between the observed and estimated water levels. The root-mean-square error (RMSE) value for the floodplains was 1.96 m. Although the estimated water levels were larger than the actual water levels for sites upstream of MR, the observed and estimated water levels for the downstream sites tended to be consistent. Figure. 7.3 shows the results of the comparison between the observed and estimated water levels around TSL. In this figure, the circles represent estimated water levels derived from a combination of inundation maps and a DEM without

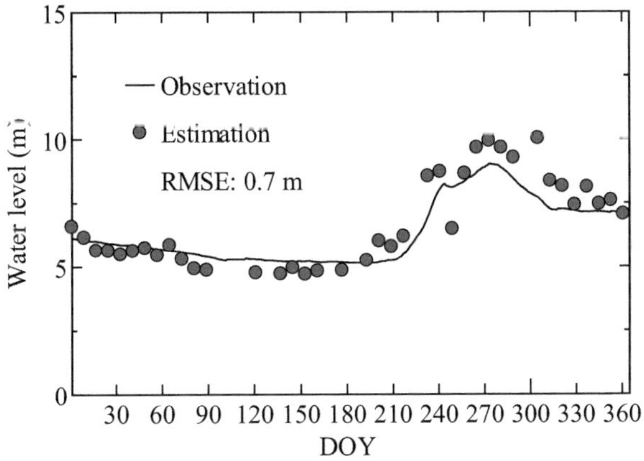

Fig. 7.3 Time series of water level in the floodplain. Observations were recorded at the floodplain station (Ou Khasch, lon: 104.835, lat: 11.664)

using any ground-based observation data. The results show that the method accurately estimated the water levels ranging from low during the dry season to high levels during the rainy season. The RMSE at this station was 0.7 m. These RMSE values may not be sufficient for floodwater management, risk management, and decision-making because we used SRTM data. The SRTM DEM has a relative vertical error of 3–5 m (e.g., Sun et al. 2003). Considering this error, the RMSE values for the floodplains were less than 2 m, indicating that most of the water level estimation errors could be attributed to the DEM error. A LiDAR DEM is an accurate DEM, but it cannot be used for the Mekong River (MR) floodplain. If we were to use one of the LiDAR datasets, the estimation accuracy would be less than 0.5 m, in which case, the estimated floodwater level could be used for floodwater management and decision-making. Such an attempt to estimate floodwater levels using LiDAR data is one of the most interesting studies in the future.

7.4 Surface Water Temperature

Various regions have witnessed a rise in lake water temperature because of the rise in air temperature caused by recent climate change. As mentioned in the previous section, water temperature is an essential parameter for lake management. However, the information regarding long-term water temperature trends in TSL is scarce, while the changes in the water environment in TSL are not well understood although Oyagi et al. (2017) reported no thermocline in the lake.

We investigated the trend in surface water temperature of TSL from 2000 to 2018 using 8-day composite data MOD11A2 that is a MODIS data product and has 46 images per year. The R-square between the observed water temperature and MODIS-derived water temperature was approximately 0.54 in TSL. We used the Google Earth Engine, a cloud-based platform for global-scale geospatial analysis that uses Google's computational capabilities for various high-impact societal

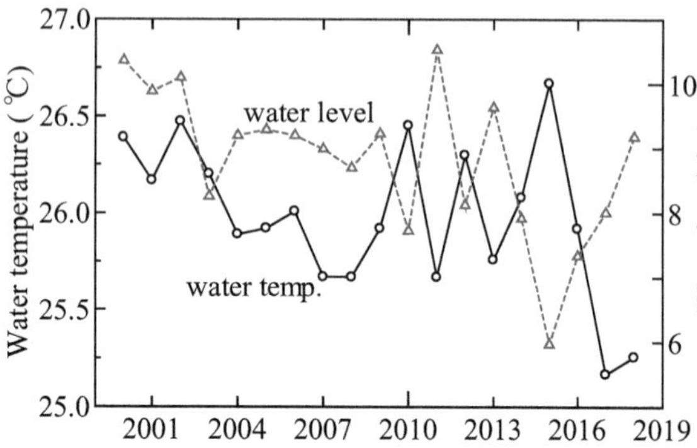

Fig. 7.4 Annual water temperature and maximum water level. Water level was observed at Kampong Luong

applications, including water management, climate monitoring, and environmental protection. It is an integrated platform designed to empower traditional remote sensing scientists and a much wider audience lacking the technical capacity required to use conventional supercomputers or large-scale cloud computing (Gorelick et al. 2017).

We investigated the lake water temperature fluctuations for open water areas throughout the year. Even with 8-day composite data from MODIS, the data tended to be influenced by clouds. The median image for the year was obtained for each pixel from 46 land surface temperature values to remove this influence. Additionally, the median surface temperature within the lake boundary was obtained to calculate the annual and lake average water temperatures. The trend of this median surface temperature was analyzed for data spanning 19 years. We also analyzed the yearly median air temperature of Phnom Penh City. Additionally, as the maximum water level of the lake has been shown to affect water quality (e.g., Hoshikawa et al. 2019), its relationship with water temperature was also analyzed.

Consequently, the northwestern part of the lake was found to have lower water temperatures than those of the southeastern part, with a difference of 1.5 °C. Figure 7.4 shows the annual water temperature data for the 2000–2018 period and the annual maximum water level variation of the lake. We determined that the annual mean water temperature for the 2000–2009 period was 26.0 °C (maximum 26.5 °C and minimum 25.7 °C), whereas for the 2010–2018 period, it was 25.9 °C (maximum 26.7 °C and minimum 25.2 °C). We found that the annual fluctuation of water temperature has been increasing since 2009. Moreover, the surface temperature showed a decreasing trend of 0.30 °C per 10 years, and there was no trend for the air temperature of Phnom Penh City. Considering that the water temperature rise due to global warming has shown an upward trend of 0.34 °C/10 years worldwide (O'Reilly et al. 2015), the decrease in water temperature of TSL is an interesting phenomenon.

For the 2000–2009 period, there was a positive correlation between annual water temperature and annual maximum water level. However, after 2009, it changed to a negative correlation, and the variability between annual water temperature and annual maximum water level increased. It appears that both annual water temperature and annual maximum water level have fluctuated more since 2009. The construction large dams in the upstream area are presently underway, and it has been suggested that this will change the hydrological environment in MR (e.g., Hecht et al. 2019). Although the relationship between water temperature fluctuation and dam construction cannot be accurately established in our study, such an investigation is necessary in the future. Because of the lack of hydrological data on TSL, little long-term water quality information has been available, including water temperature. However, according to our results it is essential to note that water temperatures in TSL have definitely begun to change since 2009.

Key Points

- The combination of NDVI, EVI, and NDWI allowed us to assess the inundation area even under the canopies of flooded forests.
- The estimated water level based on inundation maps and a DEM accurately reproduced the observed water level during low- and high-water seasons. The RMSE value for the floodplains was 1.96 m.
- The annual mean water temperature for the 2000–2009 period was 26.0 °C (maximum 26.5 °C and minimum 25.7 °C), whereas for 2010–2018, it was 25.9 °C (maximum 26.7 °C and minimum 25.2 °C), and the variation has been increasing since 2009.

References

Fujihara Y, Hoshikawa K, Fujii H, Yokoyama S, Kotera A, Nagano T. Analysis and attribution of trends in water levels in the Vietnamese Mekong Delta. Hydrol Processes. 2016;30(6):835–45. https://doi.org/10.1002/hyp.v30.6.

Gorelick N, Hancher M, Dixon M, Ilyushchenko S, Thau D, Moore R. Google Earth engine: planetary-scale geospatial analysis for everyone. Remote Sensing Environ. 2017;202:18–27. https://doi.org/10.1016/j.rse.2017.06.031.

Hecht JS, Lacombe G, Arias ME, Dang TD, Pimanh T. Hydropower dams of the Mekong River basin: a review of their hydrological impacts. J Hydrol. 2019;568:285–300. https://doi.org/10.1016/j.jhydrol.2018.10.045.

Hortle KG (2007) Consumption and the yield of fish and other aquatic animals from the Lower Mekong Basin. MRC Technical Paper No. 16, Mekong River Commission, Vientiane, Lao PDR

Hoshikawa K, Fujihara Y, Siev S, Arai S, Nakamura T, Fujii H, Sok T, Yoshimura C. Characterization of total suspended solid dynamics in a large shallow lake using long-term daily satellite images. Hydrol Processes. 2019;33(21):2745–58. https://doi.org/10.1002/hyp.13525.

O'Reilly CM, Sharma S, Gray DK, Hampton SE, Read JS, Rowley RJ, et al. Rapid and highly variable warming of lake surface waters around the globe. Geophys Res Lett. 2015;42: 2015GL066235. https://doi.org/10.1002/2015GL066235.

Oyagi H, Endoh S, Ishikawa T, Okumura Y, Tsukawaki S. Seasonal changes in water quality as affected by water level fluctuations in Lake Tonle Sap, Cambodia. Geograph Rev Japan Ser B. 2017;90(2):53–65. https://doi.org/10.4157/geogrevjapanb.90.53.

Sakamoto T, Nguyen NV, Kotera A, Ohno H, Ishitsuka N, Yokozawa M. Detecting temporal changes in the extent of annual flooding within the Cambodia and the Vietnamese Mekong Delta from MODIS time-series imagery. Remote Sensing Environ. 2007;109:295–313. https://doi.org/10.1016/j.rse.2007.01.011.

Sun G, Ranson KJ, Kharuk VI, Kovacs K. Validation of surface height from shuttle radar topography mission using shuttle laser altimeter. Remote Sensing Environ. 2003;88:401–11. https://doi.org/10.1016/j.rse.2003.09.001.

Tanaka T, Yoshioka H, Siev S, Fujii H, Fujihara Y, Hoshikawa K, Sarran L, Yoshimura C. An integrated hydrological-hydraulic model for simulating surface water flows of a shallow lake surrounded by large floodplains. Water. 2018;10(9):1213. https://doi.org/10.3390/w10091213.

Tanaka K, Fujihara Y, Hoshikawa K, Fujii H. Development of a flood water level estimation method using satellite images and a digital elevation model for the Mekong floodplain. Hydrol Sci J. 2019;64(2):241–53. https://doi.org/10.1080/02626667.2019.1578463.

Chapter 8
Hydrology of the Inflow River Basins

Rattana Chhin, Sokly Siev, Ichiro Yoneda, Takashi Nakamura, Chihiro Yoshimura, and Hideto Fujii

8.1 Overview of the Lake Basin

The river basins in Cambodia are classified into five major groups, namely, Upper Mekong, 3S (Sekong, Se San, and Srepok Rivers), Tonle Sap, Mekong Delta, and Coastal (ADB 2014). The Tonle Sap group is located in the center of the country and covers the entire catchment of Tonle Sap Lake (TSL) including TSL, Tone Sap River (TSR), and each of its tributary catchment. The TSL basin covers an area of approximately 85,850 km^2 (45.0% of the total land area of Cambodia; ADB 2014) including 11 major tributaries (Fig. 8.1; Table 8.2), making the largest river basin group of all.

TSL is a shallow lake in a gentle depression, lying in the northeast to the southeast direction in the central area of Cambodia (Uk et al. 2018). The water depth in TSL varied greatly between wet and dry seasons (see Chap. 1). With essentially flat bottom, the water depth in the lake is uniform in most of the permanently flooded areas (Day et al. 2011). The lake basin is physiographically characterized by a few distinct topographical features (Tanaka et al. 2018). The northern part is formed by an escarpment of the sandstone Dangrek Mountains. The southwest is dominated by the granite Cardamom Mountains, which forms the basin boundary between the

R. Chhin
Institute of Technology of Cambodia, Phnom Penh, Cambodia

S. Siev (✉)
Institute of Technology of Cambodia, Phnom Penh, Cambodia

Ministry of Industry, Science, Technology and Innovation, Phnom Penh, Cambodia
e-mail: siev.sokly@misti.gov.kh

I. Yoneda · H. Fujii
Yamagata University, Tsuruoka, Japan

T. Nakamura · C. Yoshimura
Tokyo Institute of Technology, Tokyo, Japan

C. Yoshimura et al. (eds.), *Water and Life in Tonle Sap Lake*,
https://doi.org/10.1007/978-981-16-6632-2_8

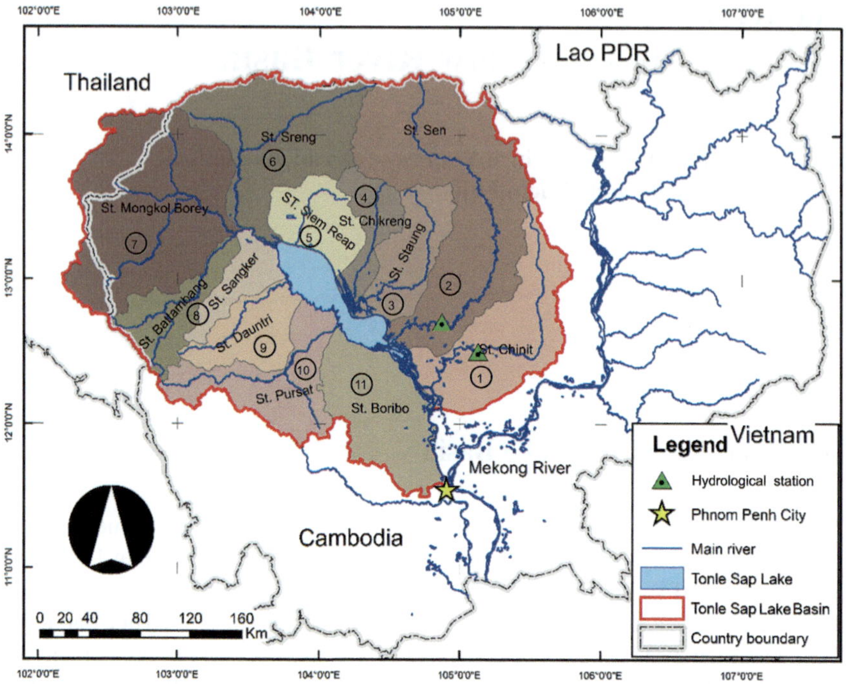

Fig. 8.1 The Tonle Sap Lake basin and its subbasins with the river network. The number on each tributary is the same as ID number in Table 8.2. The two hydrological stations were used for hydrology model calibration. Note: "St." is the abbreviation of "Steoung" in Khmer language which means "River"

rivers flowing down in the lake and the coastal area. The central flat lowland of TSL is interrupted by isolated hills, which form a basin boundary between the tributaries of the lake and the Mekong River. Most of the basin comprises lowlands with elevations of less than 100 m above the mean sea level and gentle slopes (Oeurng et al. 2019). There are 10 main soil types within the basin, namely, acrisol, arenosol, cambisol, ferralsol, fluvisol, gleysol, lepthosol, luvisol, planosol, and plinthosol. The land cover over the basin is dominated by 55.0% forest land with 3.0% flooded forest, and 45.0% agricultural land at the time of 2018 (Oeurng et al. 2019).

8.2 Model Application to the Lake Basin

8.2.1 Distributed Hydrological Model

Hydrological modeling is the state of the art of applying physics theory and mathematical solution in the field of hydrology to represent the hydrological process of a real-world hydrological system (e.g., surface water, soil water, wetland, groundwater, and estuary). Hydrological models are among the available tools used to forecast and predict the quantity and quality of water for decision-makers (Chow et al. 1988). The application of those models is attractive and useful for solving several practical problems of environmental engineering, flood protection, water resource management, and the field of applied hydrology in general (Montanari 2011). The models could be also used for the prediction of the impacts of natural and anthropogenic changes on water resources and to quantify the spatial and temporal availability of the resources.

Several hydrological models developed by different institutions could be applied to TSL, such as Agricultural Non-Point Source (Young et al. 1989), Hydrological Simulation Program-FORTRAN (Bicknell et al. 1996), MIKE SHE (DHI 1993), Soil and Water Assessment Tool (Arnold et al. 1998), Agricultural Policy/Environmental Extender (Williams and Izaurralde 2008), Rainfall–Runoff–Inundation model (Sayama et al. 2012), and geomorphology-based hydrological model (GBHM, Yang et al. 2002). Among them, GBHM has been recently applied to TSL (e.g., Tanaka et al. 2018). GBHM solves the continuity, momentum, and energy equations using the two modules, namely, the hillslope module and the river routing module. The advantage of using GBHM is to achieve fast computation in representing spatial and temporal variability. It can represent the heterogeneity of the land use and soil type especially in the large river basin (e.g., Mekong river basin, the TSL basin).

In GBHM, the target basin is divided into grids, and a digital elevation model is used to determine the flow direction and accumulation pattern that creates the river network. Each subbasin is divided into several flow intervals that are defined as a function of distance from the subbasin's outlet. Lateral flow to the mainstream is estimated by accumulating runoff at each grid in one hillslope unit. This means that all hillslopes of a flow interval drain into the mainstream in this model. The flow interval–hillslope system enables GBHM to realize a fast flow computation even in a large basin. The hillslope unit is viewed as a rectangular inclined plane with a defined length and unit width. The inclination angle is given by the corresponding surface slope.

In the hillslope model, each grid is divided into four layers: canopy, soil surface, unsaturated zone, and groundwater. Vegetation covers the surface soil and prevents direct rainfall onto the land. The deficit of canopy interception is calculated using vegetation coverage and leaf area index. The evapotranspiration module simulates the water volume that is evaporated from the surface soil and transpirated from the canopy, where pan observation could also be used. In the module, Priestley–Taylor's method is applied for the canopy water storage, root zone, surface storage, and soil surface. To describe the unsaturated zone water flow, a vertical one-dimensional Richards equation is used with soil infiltration rate and soil moisture contents in the root zone. Saturated water flow and exchange with the river is described using basic mass balance equations and Darcy's Law. The simulation module of surface water flow estimates the infiltration excess and saturation excess discharging into the river system as lateral flow.

In the river routing system, the Pfafstetter numbering system is applied to track water flow efficiently from upstream to downstream (Yang et al. 2002). The water routing on the river network is determined along the river stream using one-dimensional kinematic wave equations.

8.2.2 Model Performance

The input data for the model include weather, topography, soil property, and land cover (e.g., Tanaka et al. 2018). The parameters to be calibrated in the model include soil anisotropy ratio, surface roughness, saturated soil moisture, residual soil moisture, saturated hydraulic conductivity, and hydraulic conductivity of groundwater (Table 8.1). A detailed description of these parameters can be found in Yang et al. (2002).

Daily discharges were simulated for 10 years for calibration and compared with the observations at the evaluation sites for the Chinit River and the Sen River (Fig. 8.2). The model well captured the seasonal variation of river discharge (black dashed line versus light-blue solid line), and this seasonal variation is also consistent with the rainy season in TSL. In the Chinit River, the Nash–Sutcliffe efficiency (NSE) coefficient of the water discharge of 0.66 and 0.71 were obtained for the calibration and the validation, respectively (Fig. 8.2a). Similarly, in the Sen River, the NSE coefficient of water discharge accounted for 0.68 and 0.65 for the calibration and validation period, respectively (Fig. 8.2b). Hence, the accuracy of the discharge simulated by the GBHM model was in the acceptable ranges. The calibrated parameters can be applied to the ungauged rivers surrounding TSL (small rivers besides the main tributaries described in Table 8.2). Further details can be referred to in Tanaka et al. (2018).

Table 8.1 Land use and soil characteristic parameters calibrated for the study catchments

Parameter	Code in GBHM	Chinit River	Sen River	Calibrated value
Soil anisotropy ratio, r_a	Anik	1–16.5	1–16.5	0.5×
Surface roughness, n	Surfn	0.05–0.5	0.05–0.5	10.36×
Saturated soil moisture (%)	Wsat	0.45–0.47	0.45–0.47	Default value
Residual soil moisture (%)	Wrsd	0.15–0.2	0.071–0.083	+0.1
Saturated hydraulic conductivity, K_s (mm/h)	Ksat1	8.7–32.5	8.7–32.5	0.02×
Hydraulic conductivity of groundwater, K_g (mm/h)	Kg	0.003–0.074	0.3–1.6	Ksat1 × 0.02

×/+ mean to multiply by/add the number with the default value of each land use and soil parameters

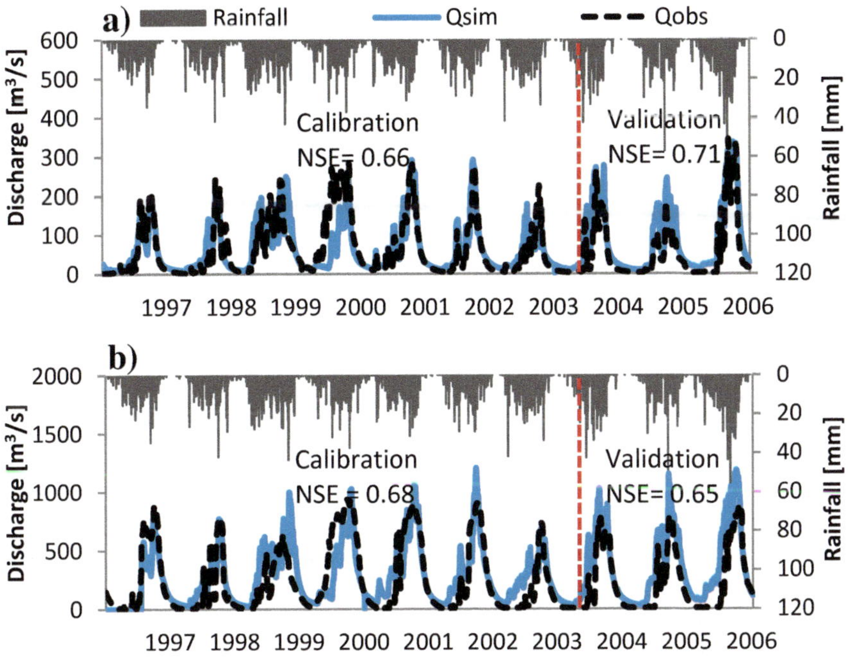

Fig. 8.2 Observed (black dot) and simulated (blue line) discharges at the gauging sites with rainfall (Tanaka et al. 2018). (**a**) The Chinit River, (**b**) the Sen River. Qobs means the observed discharge, whereas Qsim means the simulated discharge. The rainfall data (gray line) are the average over the whole catchment of each tributary, which were interpolated from observed rain gauges

8.3 Hydrology of the River Basins

The inflow to TSL originates from various sources (see Fig. 1.1 in Chap. 1 for basic information of the lake). The annual average total runoff from all the tributaries of TSL accounts for 29.1 km^3/year corresponding to 34.1% of annual total runoff to TSL, of which the Sen River contributes to the largest portion at 32.0% of total runoff from all tributaries (Kummu et al. 2014). The Chinit, Sangke, Pursat, and Baribo rivers each provide 10.0% or more. Elsewhere, the individual tributary flow contributes a minor proportion of the whole, with mean annual flow volumes below 1 km^3. A large portion of the total annual discharge of approximately 81.0% occurs during the rainy season (May–October). The peak discharge takes place in October when, on average, 26.6% of the total annual discharge occurs. Moreover, the mean total rainfall is 1222 mm during the rainy season, which is approximately 75.0% of the whole annual rainfall, whereas it is approximately 411.3 mm during the dry season. A detailed description of the seasonal rainfall and climate in Cambodia was provided in Chap. 6. The monthly average daily evaporation ranges from 3.5 mm/day in October to 5.8 mm/day in March, and the total evaporation was approximately 1627 mm/year (Kummu et al. 2014).

Table 8.2 Basic statistics of the observed and GBHM simulated river flow, annual rainfall, and runoff ratio for each tributary of TSL (observed flow and annual rainfall were obtained from Oeurng et al. (2019))

No.	Tributary	Area (km^2)	Observed flow (m^3/s)			GBHM simulated flow (m^3/s)			Annual rainfall (mm)			Mean runoff ratio (obs)	Mean runoff ratio (GBHM)
			Min	Mean	Max	Min	Mean	Max	Min	Mean	Max		
1	Chinit	8237	0.06	65	601	5	73	315	1058	1453	1839	0.17	0.19
2	Sen	16,360	0.1	249	1476	26	311	1240	1104	1385	1839	0.35	0.43
3	Staung	4357	0.01	28	277	2	36	171	1125	1470	1924	0.14	0.18
4	Chikreng	2714	0.01	11	395	3	42	212	964	1271	1646	0.10	0.38
5	Siem Reap	3619	0.04	6	132	5	32	135	1028	1286	1553	0.04	0.22
6	Sreng	9986	0.01	45	340	24	210	803	1138	1336	1560	0.11	0.50
7	Mongkol Borey	14,966	0.3	18	303	17	126	409	1261	1487	1749	0.03	0.18
8	Sangke	6053	0.67	62	1020	19	141	472	1109	1390	1664	0.23	0.53
9	Dauntri	3696	0.05	4	260	27	153	524	832	1151	1652	0.03	1.14
10	Pursat	5965	0.01	83	1264	6	106	427	1103	1493	2031	0.29	0.38
11	Baribo	7154	0.02	27	287	4	44	145	1038	1303	1568	0.09	0.15
Total		**83,107**	**1**	**598**	**6355**	**139**	**1274**	**4855**	**11,760**	**15,025**	**19,025**		

Note that the runoff ratio was estimated using mean flow and mean annual rainfall

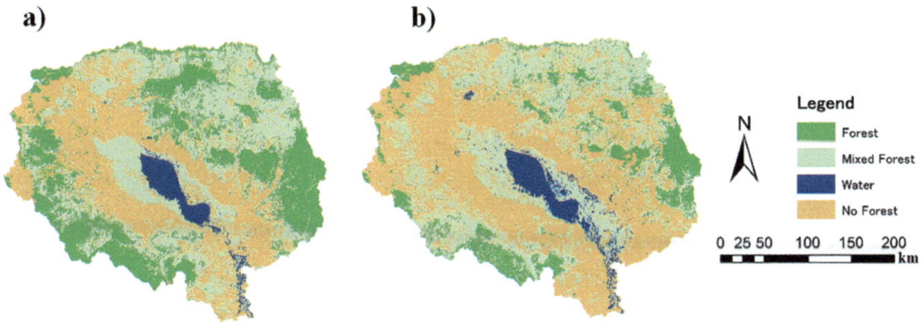

Fig. 8.3 Land cover change over TSL basin between 2000 (**a**) and 2014 (**b**). The land cover data were obtained from the Open Development Cambodia

The statistics of the observed flow in each tributary show a comparable minimum flow value among all tributaries, which is less than 0.7 m^3/s (Table 8.2). Meanwhile, the mean and maximum flow values are highly variable among the tributaries. The Sen River has the highest mean flow (249 m^3/s) and maximum flow (1476 m^3/s). The Dauntri River has the lowest mean flow (4 m^3/s), and the Siem Reap River has the lowest maximum flow (132 m^3/s). However, the annual rainfall over each tributary catchment is comparable across all tributaries. Based on the model simulation, the lowest daily minimum flow is seen at the Staung River (2 m^3/s), and the largest daily maximum flow is seen at the Sen River (1240 m^3/s), for which the data were obtained at the outlet of each tributary and the simulation period of 1997–2006. The runoff ratio of the Sen River that has the largest catchment size, mean, and maximum flow is also the highest. Conversely, the Mongkol Borey River that has the second largest catchment size shows the lowest runoff ratio. This indicates the large catchment-to-catchment variability of water loss characteristics in the TSL basin. The runoff ratio estimated from the model shows similar values to those of the observation, except for the Sangke, Dauntri, and Baribo Rivers.

The change in land use/land cover is considered one of the most important factors affecting the intensity and frequency of overland flow and surface wash erosion. Especially, hillslope processes are easily altered by land use changes. An increase in vegetation cover leads generally to a decrease in runoff generation at the catchment scale. In the TSL basin, the area of forest cover shows a dramatic decrease from 25.0% in 2000 to 11.0% in 2014 based on forest cover change data of the Open Development Cambodia (Fig. 8.3). Through forest cover change between 2000 and 2014 scenarios and the application of GBHM, the seasonal average of discharge of the Sen and Chinit Rivers has increased to approximately 23.0–87.0% and 38.0–166.0%, respectively, because the characteristic of land cover in the river basin has changed and so have the parameters (Table 8.1). Especially, the surface roughness of forest cover is relatively larger than other land covers, resulting in the slow velocity of the water flowing in the river channel. Additionally, the soil anisotropy ratio of forest cover type permits more lateral flow under the soil layer rather than surface runoff. This implies the importance of forest cover in maintaining the river flow while losing the forest cover (changed to other land covers) could increase the surface runoff.

Key Points

- The seasonal variation of water discharge and water depth of TSL is largely fluctuated because of the distinct seasonality of regional rainfall and the reverse flow in the TSR.
- A distributed hydrological model (GBHM) was applied to simulate the hydrological processes in the TSL basin. The model showed sufficient accuracy in terms of river discharge.
- A distributed hydrological model (GBHM) was applied to simulate the hydrological processes in the TSL basin. The model showed sufficient accuracy in terms of river discharge.
- TSL basin showed a large variability of water loss among tributary basins.
- The change in land use/land cover was considered as one of the most important factors affecting the intensity and frequency of overland flow.

References

Arnold JG, Raghavan S, Ranjan SM, Jimmy RW. Large area hydrologic modeling and assessment part I: Model development. Jawra J Am Water Resour Assoc. 1998;34:91–101.

Asian Development Bank (ADB). Cambodian National Water Status Report 2014. Prepared by Egis-eau. Phnom Penh. December 2014; 2014.

Bicknell BR, Imhoff JC, Kittle JL Jr, Donigian AS Jr, Johanson RC. Hydrological simulation program-FORTRAN. User's manual for release 11. Athens, Greece: US EPA; 1996.

Chow VT, Maidment DR, Mays LW. Applied hydrology. New York: McGraw-Hill; 1988. p. 572.

Danish Hydraulic Institute (DHI). MIKE SHE WM—Water movement module, a short description. Danish Hydraulic Institute: Hørsholm, Denmark; 1993.

Day MB, Hodell DA, Brenner M, Curtis JH, Kamenov GD, Guilderson TP, Peterson LC, Kenney WF, Kolata AL. Middle to late Holocene initiation of the annual flood pulse in Tonle Sap Lake. Cambodia J Paleolimnol. 2011;45(1):85–99.

Kummu M, Tes S, Yin S, Adamson P, Józsa J, Koponen J, Richey J, Sarkkula J. Water balance analysis for the Tonle Sap Lake-floodplain system. Hydrol Process. 2014;28(4):1722–33.

Montanari A. 2.17—Uncertainty of hydrological predictions. In: Peter W, editor. Treatise on water science. Oxford: Elsevier; 2011. p. 459–78.

Oeurng C, Cochrane TA, Chung S, Kondolf MG, Piman T, Arias ME. Assessing climate change impacts on river flows in the Tonle Sap Lake Basin, Cambodia. Water. 2019;11(3):618.

Sayama T, Ozawa G, Kawakami T, Nabesaka S. Rainfall–Runoff–Inundation Analysis of the 2010 Pakistan flood in the Kabul River Basin. Hydrol Sci J. 2012;57(2):298–312.

Tanaka T, Yoshioka H, Siev S, Fujii H, Fujihara Y, Hoshikawa K, Ly S, Yoshimura C. An integrated hydrological-hydraulic model for simulating surface water flows of a shallow lake surrounded by large floodplains. Water. 2018;10(9):1213.

Uk S, Yoshimura C, Siev S, Try S, Yang H, Oeurng C, Li S, Hul S. Tonle Sap Lake: current status and important research directions for environmental management. Lakes Reserv: Res Manage. 2018;23(3):177–89.

Williams J, Izaurralde R. Agricultural policy/environmental extender model theoretical documentation. Temple, TX: Blackland Research and Extension Center; 2008.

Yang D, Herath S, Musiake K. A hillslope-based hydrological model using catchment area and width functions. Hydrol Sci J. 2002;47(1):49–65.

Young R, Onstad C, Bosch D, Anderson W. AGNPS: a nonpoint-source pollution model for evaluating agricultural watersheds. J Soil Water Conserv. 1989;44:168–73.

Chapter 9
Groundwater and Surface Water Exchange in the Lake Basin

Sith Ratino, Heng Seangmeng, Doung Ratha, Chhuon Kong, Eng Khy Eam, Sokly Siev, Sive Thea, Rajendra Khanal, and Chihiro Yoshimura

9.1 Interactions Between Groundwater and Surface Water

Groundwater and surface water have strong interactions in a wide range of spatio-temporal scale, which often involve the exchanges of the water masses between the surface and the soil. These interactions and exchanges depend on the condition of the force, landscape, soil type, hydraulic characteristic, and climate (Cavazza and Pagliara 2000). The interactions commonly occur in the form of loss of the surface water to the groundwater aquifer, seepage of the groundwater to the surface water, or a combination of both. The interactions affect not only the water exchanges but also the contamination between the groundwater and the surface water (Winter et al. 1998). Hence, understanding the interactions between the groundwater and the surface water in the catchment scale plays an essential role in better water resource management, which is the important key for relevant policymakers and stakeholders.

S. Ratino (✉) · H. Seangmeng · D. Ratha · C. Kong · E. K. Eam
Institute of Technology of Cambodia, Phnom Penh, Cambodia
e-mail: ratino.sith@itc.edu.kh

S. Siev
Institute of Technology of Cambodia, Phnom Penh, Cambodia

Ministry of Industry, Science, Technology and Innovation, Phnom Penh, Cambodia

S. Thea
Ministry of Industry Science Technology and Innovation, Phnom Penh, Cambodia

R. Khanal
Tokyo Institute of Technology, Tokyo, Japan

Policy Research Institute, Kathmandu, Nepal

C. Yoshimura
Tokyo Institute of Technology, Tokyo, Japan

C. Yoshimura et al. (eds.), *Water and Life in Tonle Sap Lake*,
https://doi.org/10.1007/978-981-16-6632-2_9

Winter et al. (1998) indicated that the interactions between the groundwater and the surface water is a crucial key to address water resource-related issues as follows: (1) the conjunctive utilities of both groundwater and surface, (2) the environmental flow to sustain the aquatic life (fauna) that maintains the diverse ecology and the status of water quality, (3) the consideration of groundwater resources into watershed planning and management, and (4) the assessment and control of the impacts of groundwater contamination to the surface water or vice versa. These water resource-related issues are commonly found in the Cambodian context. However, the interactions between groundwater and surface water can be essential to address other water resource-related issues such as wetland, floods along valleys, and water right-related issues (Winter et al. 1998; Winter 1999; Woessner 2000; Sophocleous; 2002; Safeeq and Fares 2016).

The interactions between the groundwater and rivers are controlled by (1) the hydraulic conductivity and its distribution, (2) the river water level or stage concerning the nearby or adjacent groundwater heads, and (3) the position and geometry of the river within the fluvial plain. The position and geometry of the river can produce various conditions including (a) the effluent or fully gaining river, (b) the influent or fully losing river (hydraulically connected to the groundwater), (c) the influent or fully losing river (hydraulically disconnected to the groundwater), (d) the flow-through river, and (e) the parallel-flow river (Woessner 1998; Winter et al. 1998; Hoehn 1998; Khan and Khan 2019).

Cambodia recently has faced the challenges of shortage of water resources for different kinds of utilities (domestic, irrigation, etc.) although there are affluent river basins across the country. People mainly rely on both river water and groundwater sources specifically during both wet and dry seasons (Sreymom and Sokhen 2015). However, the study of groundwater and surface water interaction is scarce in the entire Cambodian catchment. Only a few previous studies report on the groundwater and surface water interaction in the manner of contamination and flood (May et al. 2010; Rechards et al. 2015, 2019). Located in the north of Cambodia, the Stung Sen River, the largest tributary of Tonle Sap Lake (TSL) and the largest river basin in Cambodia, originates in the mountainous area of Preah Vihear Province, meandering around the plain in Kampong Thom Province before it flows into TSL (see also Chap. 8 for the overview of the lake basin). The catchment has conserved water resources of both groundwater and surface water for thousands of populations, industries, irrigation, etc. However, the shortage of water remains the main challenge in the catchment, particularly during the dry season. The river is relatively deep (approximately 20 m) with less water during the dry season, which makes river water extraction limited and almost impossible. Hence, understanding the groundwater and surface water interaction in the catchment is essential for the better management of water resources.

9.2 Groundwater and Surface Water in the Catchment of the Stung Sen River

The groundwater level has strong spatiotemporal variability throughout the Stung Sen catchment area. At the downstream part of the catchment, the groundwater level clearly corresponds to the river water level, particularly during the rainy season (Fig. 9.1). This can be explained by the fact that the groundwater monitoring station

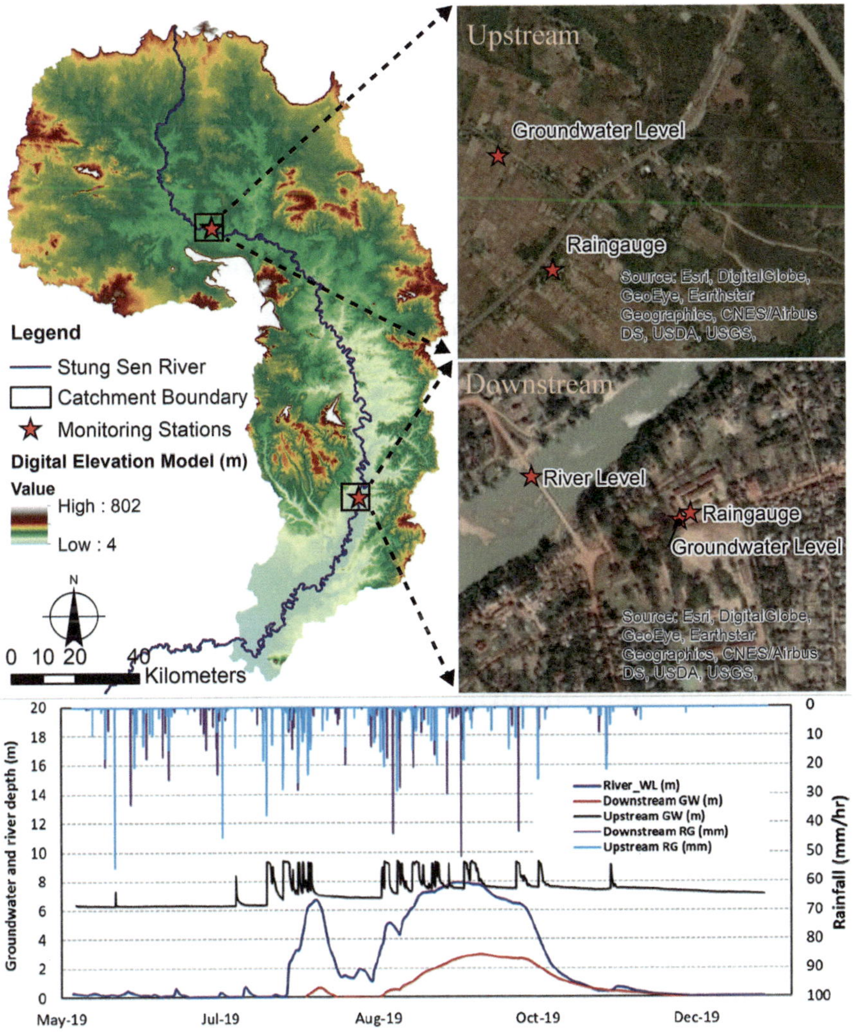

Fig. 9.1 Monitoring stations located downstream and upstream of the catchment of the Stung Sen River in 2019 (rain gauge, river water level, and groundwater water level). **Note:** the larger depth means the higher level

is located adjacent to the river at approximately 160 m apart, resulting in strong interactions between groundwater and surface water. However, this does not always mean that during the rainy season (high river water level), the groundwater flows into the river. Obviously, the river water pushes the water to adjacent groundwater aquifer, resulting in the reverse or negative hydraulic gradient, which occurs during the rising limb of the river hydrograph. By contrast, during the recession limb, the groundwater discharges back to the river, resulting in a decreasing trend of groundwater level.

During the dry season, the groundwater depth appears to be zero and negative, resulting from the relatively low groundwater table, which is below the recordable well depth (approximately 20 m) (Fig. 9.1). This clearly shows that the water in the river might flow into the groundwater aquifer because the river depth at the monitoring point is obviously less than 20 m. This process always happens during the dry season, during which the baseflow is relatively low and might result in drought, the so-called losing stream, which is either hydraulically connected to or disconnected from groundwater.

At the upstream part of the catchment, the trend of groundwater level follows that of rainfall patterns (Fig. 9.1). Although there is no available data of river water level, concerning the coverable catchment area (headwater), the river water level upstream must be controlled by the seasonal variability. This can be explained by the fact that the groundwater level upstream does not significantly interact with the river, although the location of groundwater level monitoring is adjacent to the river with a distance of approximately 820 m and a recordable depth of 20 m. From the monitoring water level, it can be observed that the groundwater and surface water interaction in the catchment has strong spatial and temporal variabilities. May et al. (2010) have reported that in the Tonle Sap River, the Upper Mekong River, and the Lower Bassac and Mekong Rivers, the groundwater and surface water interaction is confirmed at 15, 20, and 70 km away from the rivers, respectively.

9.3 Groundwater and Surface Water Exchange

Monitoring of the interactions between the groundwater and the surface water is commonly costly and arduous. Recently, the modeling approach is considered to be cost-effective to assess the exchanges between the groundwater and the surface water and has been developed and applied worldwide. Among the integrated groundwater and surface water models, SWAT-MODFLOW has been popularly applied for the study of groundwater and surface water interaction from a wide range of scale (Bailey et al. 2017, 2020; Ochoa et al. 2019; Mosase et al. 2019; Sith et al. 2019). Groundwater recharge, groundwater head, and groundwater and surface water exchanges are the common results obtained from this model.

The SWAT-MODFLOW was applied in the Stung Sen catchment to understand groundwater and surface water exchange. Figure 9.2 illustrates the mean monthly groundwater recharge in the Stung Sen catchment. The groundwater recharge

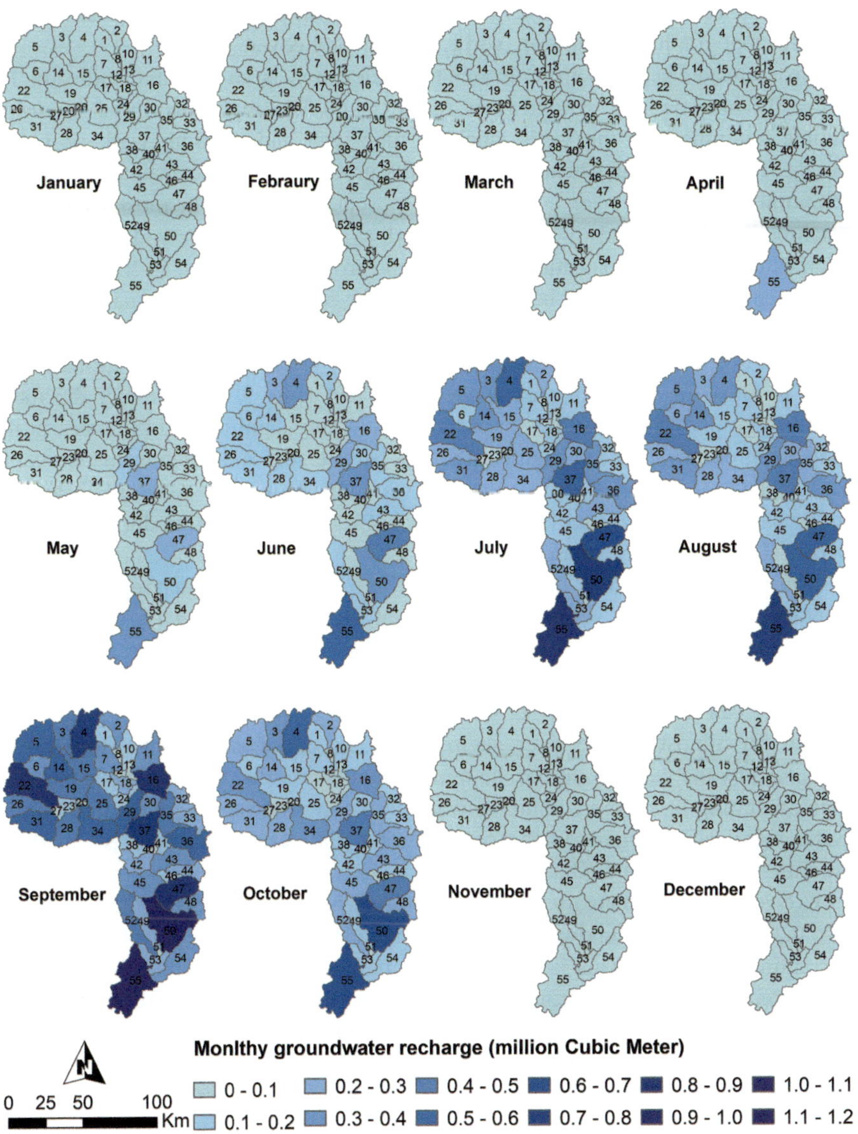

Fig. 9.2 Monthly mean groundwater recharge in the Stung Sen catchment

obviously has seasonal variability, for which the high recharge can be observed starting from May to October during the rainy season. However, the limited groundwater recharge can be observed starting from November to April during the dry season. It also shows strong spatial variability throughout the catchment area, in which high recharge values are found at the plain area of catchment adjacent to TSL. At the headwater area (upstream), the recharge values are medium to high, depending on the characteristics of each sub-catchment including the amount of rainfall, topographical condition, soil type, and land cover.

Groundwater head or groundwater table elevation is an essential indicator for primarily investigating groundwater resources. It is simply measured by using simple equipment in the wells, such as tube well and opening well, or in the ponds. The groundwater head represents the groundwater table, which has spatio-temporal fluctuations over a wide range of scale. Recently, in most Cambodian catchments, the groundwater is considered as both the primary and secondary sources of water for domestic, industrial, and irrigation utilities. In the Stung Sen catchment, the groundwater head shows a strong spatial variability, mainly following the topographical condition. The groundwater head is found to be high at the upper part or mountainous areas of the catchment, whereas the lowest groundwater head is found in the plain area at the lower part (Fig. 9.3). Although the mean monthly groundwater head does not likely show any significant differences in the map, the groundwater head slightly increases between 0.5 and 1.0 m from October to December at the upper part or mountainous areas. However, at the plain area or lower part, the groundwater head increases between 2.0 and 5.0 m during the same period (October to December).

The exchange between the groundwater and the surface water represents their interactions that has a strong spatiotemporal variability (Fig. 9.4). The recharge means that the water flows from the river to the groundwater aquifer, whereas the discharge means that the water flows from the groundwater aquifer to the river. The discharge can be significantly observed from June to November during the rainy season for almost all the river segments in the entire catchment. This can be explained by the fact that during the rainy season, groundwater recharge is predominant, resulting in a huge amount of water in the groundwater aquifer, from which water significantly moves to the river. During this period, the river is called the gaining river (Khan and Khan 2019).

Conversely, from January to April (the dry season), the water in the river flows into the underlying groundwater aquifer for almost all the river segments throughout the entire catchment. This is called the losing river, which is either hydraulically connected or disconnected from the groundwater (Khan and Khan 2019). During our field investigation at the end of December, this losing river was found in the lower part of the catchment (at the water level monitoring station) during the dry season, when the river was almost dry. However, in December, it is considered the transition period between the rainy and dry seasons, for which the recharge and discharge may occur simultaneously over the entire catchment from the upper part to plain areas. Previous studies report that the gaining river occurs during the rainy season in the catchment scale (Ochoa et al. 2019; Mosase et al. 2019), whereas some studies report that the gaining river occurs during the dry season (May et al. 2010; Ochoa et al. 2019). Thus, the gaining and losing rivers occur in different conditions not only due to the seasonal patterns but also various other factors including the position and geometry of the river, the position of the river stage and adjacent groundwater head, and the hydraulic conductivity and its distribution (Woessner 2000).

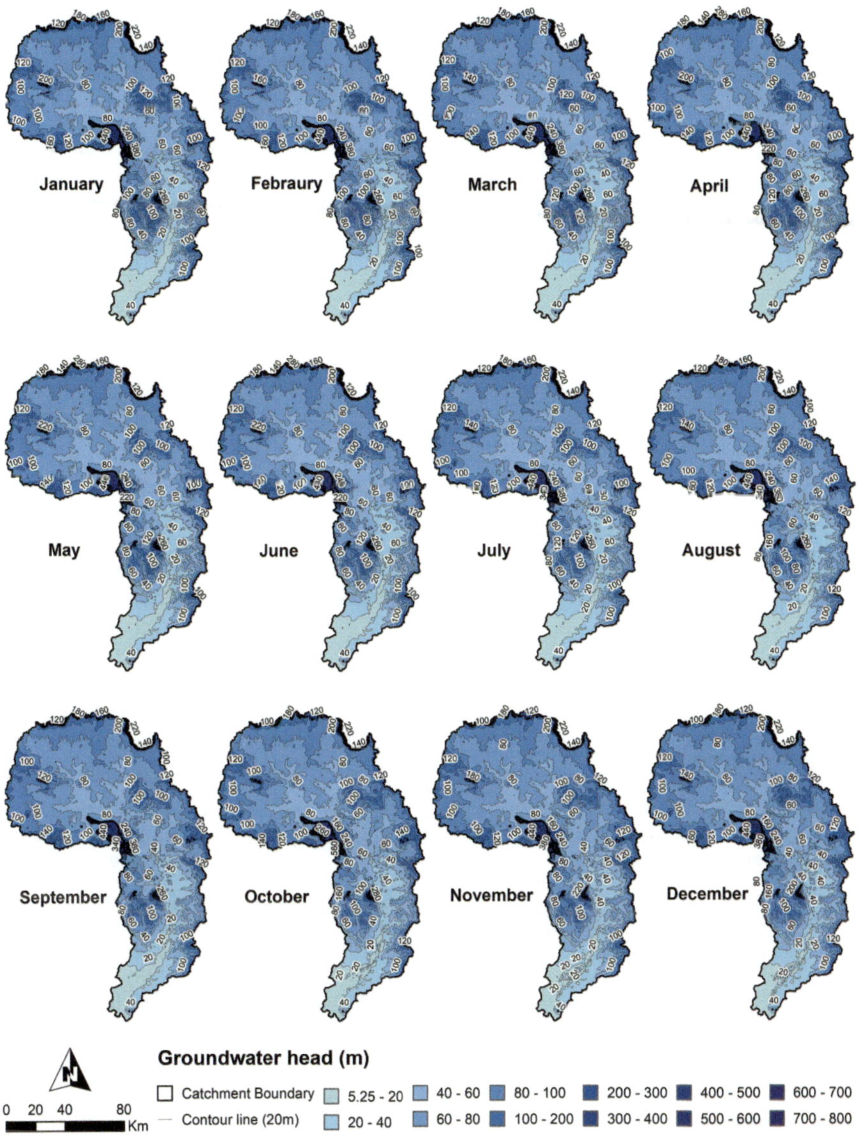

Fig. 9.3 Monthly mean groundwater head (groundwater table elevation) in the Stung Sen catchment

Although the recharge and discharge are equally separated by seasonality and well distributed over the entire catchment, the discharge is relatively lower in terms of water volume (88.5×10^6 m^3/year) as compared to the recharge of 97.8×10^6 m^3/year. The recharge into the underlying groundwater aquifer may cause significant

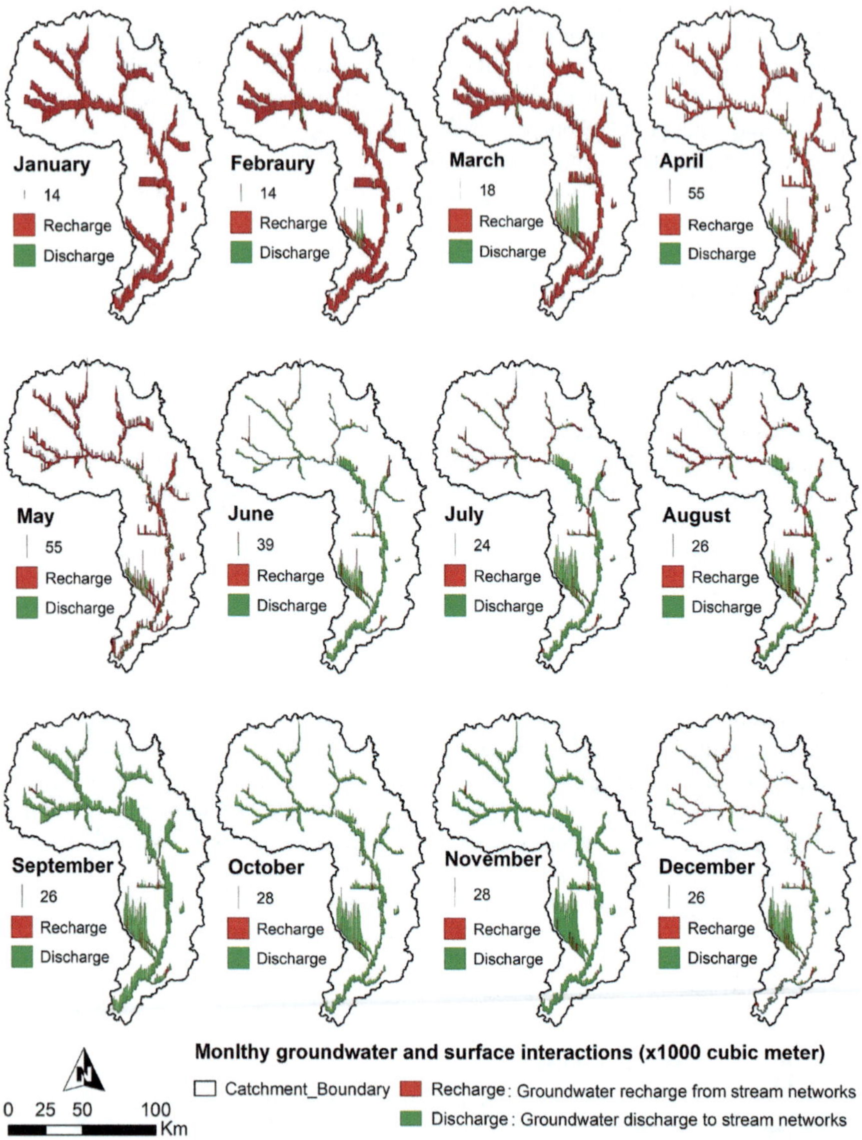

Fig. 9.4 Monthly mean groundwater and surface water exchanges (recharge and discharge) in the Stung Sen catchment

drought during the dry season, which particularly affects those who rely on the river water or live adjacent to the river itself.

This study does not consider any influences from TSL. However, at the lower part of the catchment, the water exchanges might be significantly affected by the hydraulic condition of TSL. Thus, the impacts of the hydraulic condition of the TSL on the water exchange in the Stung Sen catchment require further study.

Key Points

- The groundwater recharge has seasonal variability, for which the high recharge can be observed during the rainy season and the limited groundwater recharge can be observed during the dry season.
- The groundwater head has strong spatial variability, which mainly follows the topographical condition. Groundwater head is found to be high at the upper part or mountainous part of the catchment, whereas the lowest groundwater head is found in the plain area at the lower part of the catchment.
- The annual volume of water flows from the river to the groundwater aquifer is relatively higher than that from the groundwater aquifer to the river.

References

Bailey RT, Hendrik R, Katrin B, Indrajeet C, Jeffrey A. SWATMOD-Prep: graphical user interface for preparing coupled SWAT-MODFLOW simulations. J Am Water Res Assoc (JAWRA). 2017;53(2):400–10.

Baily RT, Park SP, Bieger K, Arnold JG, Allen PM. Enhancing SWAT+ simulation of groundwater flow and groundwater-surface water interactions using MODFLOW routines. Environ Modell. Softw. 2020;126 https://doi.org/10.1016/j.envsoft.2020.104660.

Cavazza S, Pagliara S. Groundwater and surfacewater interactions. Groundwater. 2000;I:96–110.

Hoehn E. (1998). Solute cxchange between river water and groundwater in headwater environments. In: Hydrology, water resources and ecology in headwaters. Proceedings of the HeadWater' 98 Conference Held at Meran/Merano, Italy, April 1998, pp. 165–171. IAHS Publ. No. 248.

Khan HH, Khan A. Chapter 14–Groundwater and surface water interaction. In: GIS and geostatistical techniquies for groundwater science. Elsevier; 2019. p. 197–207.

May R, Jinno K, Tsutsumi A. Influence of flooding on groundwater flow in central Cambodia. Environ Earth Sci. 2010;63:151–61.

Mosase E, Ahiablame L, Park S, Bailey R. Modelling potential groundwater recharge in the Limpopo river basin with SWAT-MODFLOW. Groundwater Sustain Dev. 2019;9:100260.

Ochoa CG, Sierra AM, Vives L, Zimmermann E, Bailey R. Spatio-temporal patterns of the interaction between groundwater and surface water in plains. Hydrol Process. 2019;34(6): 1371–92.

Richards LA, Magnone D, Van Dongen B, Bryant C, Boyce A, Ballentine CJ, Polya DA. Seasonal influences on ground-surface water interactions in an arsenic-affected aquifer in Cambodia. In: AGU Fall Meeting Abstracts, vol. 2015; 2015.

Richards LA, Magnone D, Sültenfuß J, Chambers L, Bryant C, Boyce AJ, van Dongen BE, Ballentine CJ, Sovann C, Uhlemann S, Kuras O, Gooddy DC, Polya DA. Dual in-aquifer and near surface processes drive arsenic mobilization in Cambodian groundwaters. Sci Total Environ. 2019;659:699–714.

Safeeq M, Fares A. Groundwater and surface water interactions in relation to natural and anthropogenic environmental changes. In: Fares A, editor. Emerging issues in groundwater resources, advances in water security. Prairie View, TX: Springer; 2016.

Sith R, Watanabe A, Nakamura T, Yamamoto T, Nadaoka K. Assessment of water quality and evaluation of best management practices in a small agricultural watershed adjacent to Coral Reef area in Japan. Agric Water Manage. 2019;213:659–73.

Sophocleous M. Interactions between groundwater and surface water: the state of the science. Hydrogeol J. 2002;10:52–67.

Sreymom S, Sokhem P. Challenge and perspectives for water security and climate change in selected catchment, Cambodia. Climate Change and Water Governance in Cambodia, CDRI, 2015.

Winter TC. Relation of streams, lakes, and wetlands to groundwater flow systems. Hydrogeol J. 1999;7:28–45.

Winter TC, Harvey JW, Franke OL, Alley WM. Groundwater and surface water: a single resources. U.S. Geol Survey Circ. 1998;1139:1–53.

Woessner WW. Changing views of stream-groundwater interaction. In: Van Brahana J, Eckstein Y, Ongley LW, Schneider R, Moore JE, editors. Proceedings of the Joint Meeting of the XXVIII Congress of the International Association of Hydrogeoloists and the Annual Meeting of the American Institute of Hydrology. St. Paul, MN: American Institute of Hydrology; 1998. p. 1–6.

Woessner WW. Stream and fluvial plain groundwater interactions: rescaling hydrogeological thought. Groundwater. 2000;38(3):423–9.

Chapter 10
Improvement of a Hydrological Model Performance by Satellite Rainfall Product

Hideto Fujii, Ichiro Yoneda, Yoichi Fujihara, Keisuke Hoshikawa, and Takashi Nakamura

10.1 Rainfall Observation Density in the Tonle Sap Lake Basin

The World Meteorological Organization recommends the installation of one rainfall station every 600–900 km^2 on flat terrain and 100–250 km^2 on mountainous areas (WMO 2008). There are approximately 1300 rainfall observation stations called Automated Meteorological Data Acquisition System in Japan, in which the rainfall observation density is approximately one in 289 km^2 (17 km × 17 km). In the case of the Tonle Sap Lake (TSL) basin (83,000 km^2), most of the area is flat with less than 200 m above sea level. According to the WMO criteria, at least 90 rainfall stations are required in its basin. The weather observation network in Cambodia comprises 20 synoptic stations and approximately 200 manual rainfall stations. However, some of the manual rainfall stations are not working or not maintained well. Rainfall data sent to the meteorological department in Phnom Penh are only 34 stations (Phalla 2010). Of these, the number of rainfall stations located in the TSL basin is estimated to be 20–25. Because of insufficient number of available rainfall stations, the performance of runoff discharge by a hydrological model is insufficient.

H. Fujii (✉) · I. Yoneda
Yamagata University, Tsuruoka, Japan
e-mail: fhideto@tds1.tr.yamagata-u.ac.jp

Y. Fujihara
Ishikawa Prefectural University, Nonoichi, Japan

K. Hoshikawa
Toyama Prefectural University, Imizu, Japan

T. Nakamura
Tokyo Institute of Technology, Tokyo, Japan

C. Yoshimura et al. (eds.), *Water and Life in Tonle Sap Lake*,
https://doi.org/10.1007/978-981-16-6632-2_10

Recently, rainfall from satellite observations (hereinafter, satellite rainfall) is often used for hydrological models (Biemans et al. 2009; Getirana et al. 2011; Fujihara et al. 2014). Satellite rainfall is reported to be effective in improving the performance of hydrological models in areas with few rainfall observations points and be used in hydrological models in the MR basin (Lauri et al. 2014; Wang et al. 2016; Mohammed et al. 2018; Nguyen et al. 2018). Kakizawa et al. (2011) compared the gauged and satellite rainfall and reported the effectiveness of the Global Precipitation Climatology Project (GPCP) version 1.1 for the flood calculations in the MR basin. Also, Kotsuki and Tanaka (2013) applied the gauged rainfalls and several satellite rainfalls to a hydrological model in the whole Mekong River (MR) basin and reported the efficiency of GPCP.

In the present study, we applied GPCP satellite rainfall to a hydrological model in the TSL basin and analyzed the effect of satellite rainfall on reproductivity of runoff discharge under a calibration period from 1998 to 2002 (5 years) and a validation period from 2010 to 2015 (6 years). The catchment area of TSL varies greatly in dry and wet seasons depending on the inundation area. It is approximately 83,030 km^2 in the dry season and 48,700 km^2 in the peak flood season (Yoneda et al. 2019).

10.2 Gauge-Based and Satellite Rainfall Data

Figure 10.1 shows the gauged rainfall stations collected from the Ministry of Water Resources and Meteorology of Cambodia and Mekong River Commission. The number of collected gauged rainfall was 16 in 1998 and 1999, 14 in 2000, 38 in 2001, and 28 in 2002. The number was 22 on average. Conversely, the collected rainfall data were only five stations for the validation period from 2010 to 2015; these stations are Kampong Thom in the Sen River basin, Siem Reap in the Siem Reap River basin, Battambang in the Sangker River basin, Pursat in the Pursat River basin, and Kampong Chhnang in the Boribo River basin (see also Chap. 8 for the overview of these river basins).

The satellite rainfall products GPCP (Huffman et al. 1997, 2001) and Global Satellite Mapping of Precipitation (Okamoto et al. 2005; Kubota et al. 2007; Aonashi et al. 2009; Ushio et al. 2009) are applied to the hydrological models in the MR basin thus far. We applied the GPCP version1.2, whose spatial and temporal resolutions are 1° and 1 day, respectively (Huffman et al. 1997; Huffman and Bolvin 2013). GPCP is produced by combining microwave, infrared, and acoustic data measured through several sensors in low-earth-orbit satellite and geostationary satellite and gauged rainfall of Global Precipitation Climatology Center (Huffman et al. 1997; Huffman and Bolvin 2013).

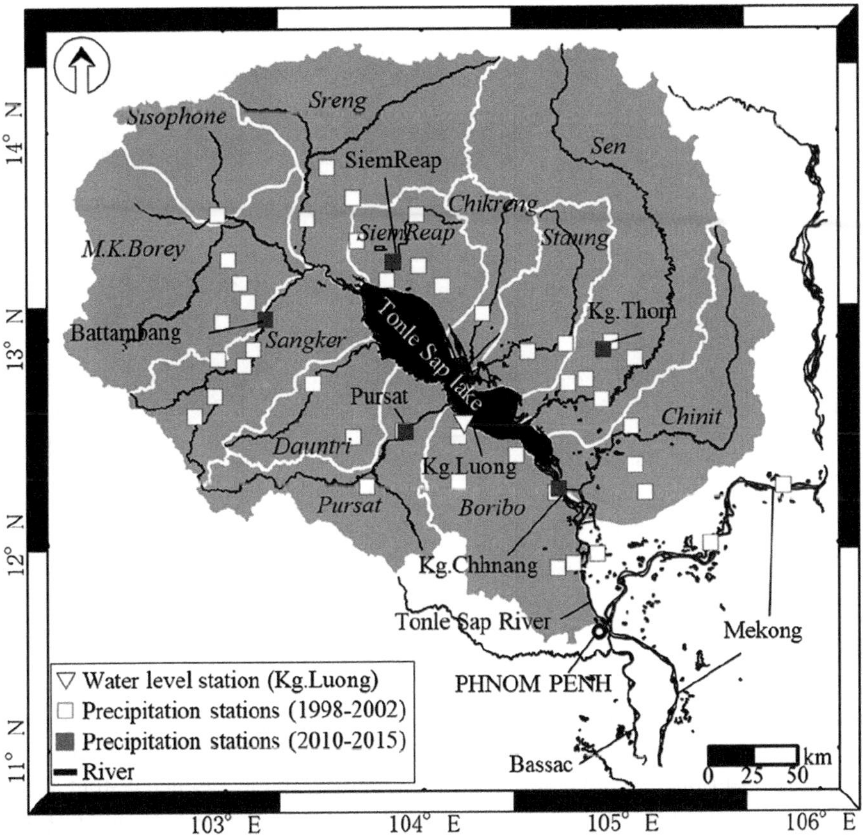

Fig. 10.1 Rainfall stations in the Tonle Sap Lake basin

10.3 Adjustment and Evaluation of Satellite Rainfall

Figure 10.2 shows the spatial distribution of the 5-year average of annual rainfalls of gauged, original, and adjusted GPCP. The spatial distribution of gauged and GPCP rainfalls showed a similar pattern that is lower in the northwestern part of the basin. However, the original GPCP was much larger than gauged rainfall. Hence, the minimum and maximum were 980 and 1750 mm in gauged rainfall, but they are 1630 and 2330 mm in the original GPCP. It is 45% of the overestimation compared with that of gauged rainfall.

The adjustment of the original GPCP was done to modify the overestimation based on the ratio of gauged rainfall to the original GPCP using the 5-year average in the rainy season from May to October in every 0.01° of longitude and latitude. Gauged rainfall was interpolated into 0.01° via inverse distance weighting method, which is suitable for rainfall interpolation (Ashouri et al. 2015; Lauri et al. 2014; Getirana et al. 2011).

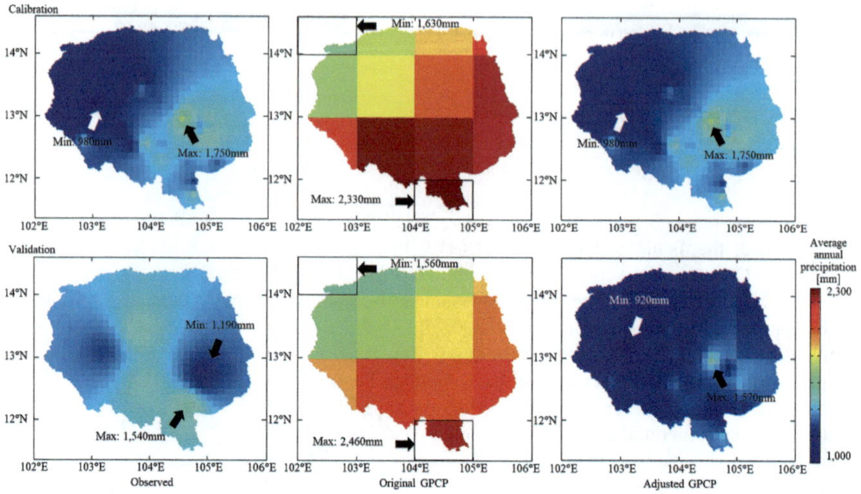

Fig. 10.2 Distribution of average of annual rainfall (upper: calibration period, lower: validation period, left: gauged rainfall, center: original GPCP, right: adjusted GPCP

Figure 10.3 shows the comparison of monthly rainfall between the observed, original, and adjusted GPCP. As mentioned above, the original GPCP shows higher values in almost all stations in most of the months. The mean annual rainfall of the adjusted GPCP and gauged rainfalls is modified almost the same (Fig. 10.2, right).

The evaluation of satellite rainfall was conducted through two steps: (1) the effect of GPCP adjustment and (2) the effect on simulated runoff discharge from seven tributaries in the TSL basin. The Nash–Sutcliffe efficiency (NSE) was used for the evaluation in both steps (Nash and Sutcliffe 1970).

Table 10.1 shows the comparison of NSEs between the gauged and original GPCP monthly rainfalls and also the gauged and adjusted GPCP rainfalls. The mean NSE is improved from 0.37 to 0.73 in the calibration period and from 0.32 to 0.66 in the validation period, although there are variations among the stations. Significant improvement in NSE was shown at three stations except for Siem Reap.

10.4 Improved Performance of a Hydrological Model by Satellite Rainfall

The hydrological model used in this study is the Nedbor–Afstromnings Model (NAM) (DHI 2016; Fujii et al. 2003; Yoneda et al. 2019). NAM is a lumped model that comprised three kinds of tanks, ground surface tank, seepage flow tank, and base flow tank, with nine parameters in total. The model is driven by rainfall, potential evapotranspiration, and the catchment area. The values of the NAM parameters are determined using multi-objective optimization for each parameter so that each of them takes the minimum value, with the two objective functions: the

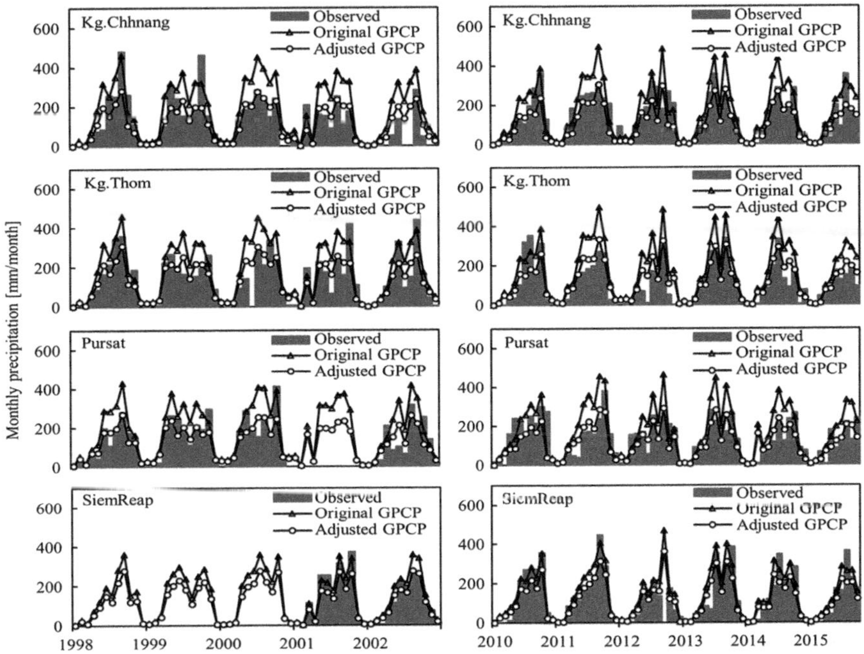

Fig. 10.3 Comparison of monthly rainfall among observed and satellite rainfall (original GPCP and adjusted GPCP)

Table 10.1 Comparison of NSEs between gauged and original GPCP monthly rainfall and gauged and adjusted GPCP at four stations

	Calibration period		Validation period	
Rainfall station	Gauged and original GPCP	Gauged and adjusted GPCP	Gauged and original GPCP	Gauged and adjusted GPCP
Kg. Chhnang	0.07	0.63	0.43	0.66
Kg. Thom	0.33	0.69	0.04	0.73
Pursat	0.18	0.72	−0.05	0.54
Siem Reap	0.90	0.88	0.81	0.78
Mean	0.37	0.73	0.32	0.66

Note: Battambang rainfall station was not used for the analysis because the data for the calibration period could not be collected

average flow rate error and root-mean-square error of the calculation period (Madsen 2000). The shuffled complex evolution algorithm is used for multi-objective optimization (Duan et al. 1992). The parameters were determined for the adjusted GPCP rainfall.

The observed and simulated runoff discharge by hydrological model from the gauged and adjusted GPCP rainfalls in each basin is compared (Table 10.2). In the calibration period, the mean NSE of seven basins was slightly improved from 0.53 to

Table 10.2 Comparison of NSEs of simulated runoff discharge from gauged and satellite rainfall in seven basins

	Calibration period		Validation period	
Name of basin	Gauged rainfall	Adjusted GPCP	Gauged rainfall	Adjusted GPCP
Chinit	0.71	0.83	0.52	0.70
Sen	0.62	0.78	0.71	0.75
Staung	0.37	0.44	0.48	0.55
Chikreng	0.21	0.30	0.13	0.24
Siem Reap	0.59	0.54	0.45	0.40
Pursat	0.62	0.52	0.12	0.41
Boribo	0.63	0.53	0.31	0.38
Mean	0.53	0.56	0.39	0.49

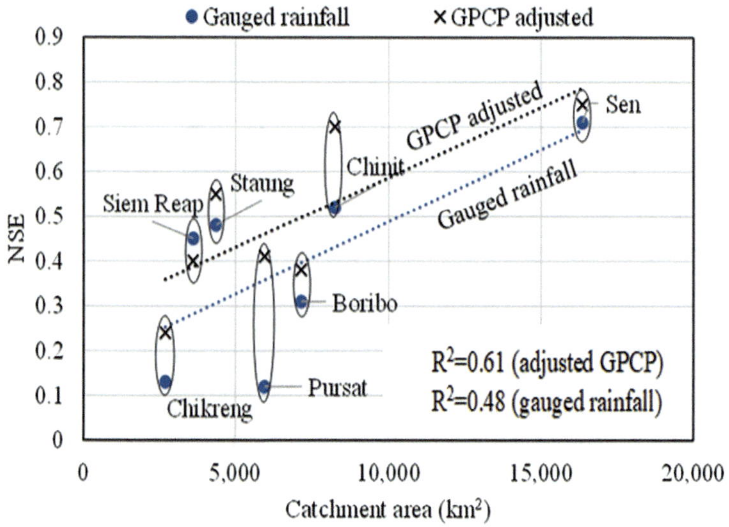

Fig. 10.4 Relationship between catchment area and NSEs from simulated runoff discharge using gauged and GPCP rainfall in validation period

0.56. Although the NSE was improved in Chinit, Sen, Staung, and Chikreng, it was decreased in the other three basins, namely, Siem Reap, Pursat, and Boribo. Conversely, the mean NSE of the seven basins in the validation period was improved from 0.39 to 0.49. Looking at each basin, the NSEs in six basins, excluding Siem Reap basin, were better in GPCP rainfall than in gauged one. Thus, the application of GPCP rainfall improved the reproduction performance of the hydrological model, compared with the simple application of ground rainfall alone, in most basins in the TSL basin. However, further analysis on the reason why the performance in some basins such as Siem Reap has not improved is necessary.

Figure 10.4 shows the relationship between the catchment area of seven basins and NSEs of discharge by hydrological model from gauged rainfall and GPCP rainfall. The NSE tends to increase as the basin area increases. Furthermore, comparing the NSE based on linear regression, the NSE from the adjusted GPCP rainfall indicated improvement by approximately 0.1 compared with gauged rainfall in most of the basins.

Key Points

- Through the bias adjustment based on the gauged rainfall, the mean NSE of the adjusted GPCP rainfall at four rainfall stations was improved from 0.37 to 0.73 in the calibration period and from 0.32 to 0.66 in the validation period.
- The GPCP satellite rainfall improved the performance of the hydrological model in six basins out of seven basins in the TSL basin.
- From the comparison of NSE between observed and simulated discharge from the gauged and adjusted GPCP rainfalls, the mean NSE of the seven basins was slightly improved from 0.53 to 0.56 in the calibration period and improved from 0.39 to 0.49 in the validation period.
- The application of the satellite rainfall in areas with few rainfall stations is recommended.

References

Aonashi K, Awaka J, Hirose M, Kozu T, Kubota T, Liu G, Shige S, Kida S, Seto S, Takahashi N, Takayabu YN. GSMaP passive microwave precipitation retrieval algorithm: Algorithm description and validation. J Meteorol Soc Jpn. 2009;87A:119–36.

Ashouri H, Hsu KL, Sorooshian S, Braithwaite DK, Kna KR, Cecil LD, Nelson BR, Prat OP. Persian-CDR Daily precipitation climate data record from multisatellite observations for hydrological and climate studies. Am Meteorol Soc. 2015;96:69–83.

Biemans H, Hutjes RWA, Kabat P, Strengers BJ, Gerten D, Rost S. Effects of precipitation uncertainty on discharge calculations for main river basins. J Hydrometeorol. 2009;10:1011–25.

DHI. MIKE 11 A modeling system for rivers and channels reference manual. Denmark: DHI Water & Environment; 2016.

Duan Q, Soroshian S, Gupta V. Effective and efficient global optimization for conceptual rainfall-runoff models. Water Resour Res. 1992;28(4):1015–31.

Fujii H, Garsdal H, Ward P, Ishii M, Morishita K, Boivin T. Hydrological roles of the Cambodian floodplain of the Mekong River. Int J River Basin Manage. 2003;1(3):253–66.

Fujihara Y, Yamamoto Y, Tsujimoto Y, Sakagami J. Discharge simulation in a data-scarce basin using reanalysis and global precipitation data, A case study of the White Volta basin. J Water Resource Protect. 2014;6:1316–25.

Getirana ACV, Espinoza JCV, Ronchail J, Filho OCR. Assessment of different precipitation datasets and their impacts on the water balance of the Negro river basin. J Hydrol. 2011;404: 304–22.

Huffman GJ, et al. The global precipitation climatology project (GPCP) combined precipitation dataset. Bull Am Meteor Soc. 1997;78:5–20.

Huffman GJ, Adler RF, Morrissey MM, Bolvin DT, Curtis S, Joyce R, McGavock B, Susskind J. Global precipitation at one-degree daily resolution from multi-satellite observations. J Hydrometeorol. 2001;2:36–50.

Huffman GJ, Bolvin DT. 2013. Version 1.2 GPCP one-degree daily precipitation data set documentation. http://precip.gsfc.nasa.gov. Last access March 20, 2013.

Kakizawa K, Sunada K, Suetsugi T. Verification on the use of global precipitation data in the Mekong River basin. Trans Japan Soc Civil Eng B1 (Hydro Eng). 2011;67(4):289–94.

Kotsuki S, Tanaka K. Uncertainties of precipitation products and their impacts on runoff estimates through hydrological land surface simulation in Southeast Asia. Hydrol Res Lett. 2013;7(4): 79–84.

Kubota T, Shige S, Hashizume H, Aonashi K, Takahashi N, Seto S, Hirose M, Takayabu YN, Nakagawa K, Iwanami K, Ushio T, Kachi M, Okamoto K. Global precipitation map using satellite borne microwave radiometers by the GSMap project, production and validation. IEEE Trans Geosci Remote Sens. 2007;45(7):2259–75.

Lauri H, Räsänen TA, Kummu M. Using reanalysis and remotely sensed temperature and precipitation data for hydrological modeling in monsoon climate. Mekong River case study. J Hydrometeorol. 2014;15:1532–45.

Madsen H. Automatic calibration of a conceptual rainfall-runoff model using multiple objectives. J Hydrol. 2000;235:276–88.

Mohammed IN, Bolten JD, Srinivasan R, Lakshmi V. Satellite observations and modeling to understand the Lower Mekong River Basin stream flow variability. J Hydrol. 2018;564:559–73.

Nash JE, Sutcliffe JV. River flow forecasting through conceptual models, part I A discussion of principles. J Hydrol. 1970;10:282–90.

Nguyen TH, Masih I, Mohamed YA, Zaag PVD. Validating rainfall-runoff modeling using satellite-based and reanalysis precipitation products in the Sre Pok Catchment, the Mekong River Basin. Geosciences. 2018;8:164.

Okamoto K, Iguchi T, Takahashi N, Iwanami K, Ushio T (2005) The global satellite mapping of precipitation (GSMaP) project. In: 25th IGARSS Proceedings, pp. 3414–3416.

Phalla P. Report on the status of weather observation in Cambodia. In: JMA/WMO workshop on quality management in surface, climate and upper-air observations in RAII(Asia), 2010.

Ushio T, Kubota T, Shige S, Okamoto K, Aonashi K, Inoue T, Takahashi N, Iguchi T, Kachi M, Oki R, Morimoto T, Kawasaki Z. A Kalman filter approach to the global satellite mapping of precipitation (GSMaP) from combined passive microwave and infrared radiometric data. J Meteorol Soc Japan. 2009;87A:137–51.

Wang W, Lu H, Yang D, Sothea K, Jiao Y, Gao B, Peng X, Pang Z. Modeling hydrologic processes in the Mekong River Basin using a distributed model driven by satellite precipitation and rain gauge observations. PLoS One. 2016;11(3):e0152229.

World Meteorological Organization Guide to hydrological practices. Vol. 1 Hydrology from measurement to hydrological information: WMO-No.168, 2008.

Yoneda I, Fujii H, Fujihara Y. Improvement of hydrological and hydraulic model applying GPCP and its characteristics in the Tonle Sap Lake. J Rainwater Catchment Syst. 2019;25(1):23–31. (in Japanese with English abstract)

Part III
Hydrodynamics

Chapter 11
Flood Pulse and Water Level

Heejun Yang, Sokly Siev, Uk Sovannara, and Chihiro Yoshimura

11.1 Flood Pulse in the Floodplain

Evaluation of water level changes on Tonle Sap Lake (TSL) and River (TSR) is crucial to understand the large flood pulse system. Changes in the water level are often generated by complex hydrological factors, which are dependent on the physiographical features of the river basin and the spatiotemporal distribution of the rainfall such as catchment retention, diurnal and seasonal evaporation variations, transpiration from plants, and infiltration (Singh et al. 2010). In particular, the water levels are related to river floods and can be used to manage water resources, so the accurate analysis of the flood pulse based on measuring water levels becomes crucial (Kerh and Lee 2006). Thus, this chapter focuses on changes in the water level in TSL and TSR to precisely evaluate short-term, periodical, and long-term changes in the water levels in relation to the flood pulse and the relationship among the water levels at different locations.

The water levels were recorded at three stations located in the TSL ecosystem, namely, Phnom Penh Port (PPP), Prek Kdam (PK), and Kampong Luong (KL), for 15 years from July 1, 1998, to December 31, 2013 (Fig. 11.1). Those stations at PPP, PK, and KL are located in the intersection between Mekong River (MR) and TSR, in TSR, and in TSL, respectively (see Fig. 12.1a in Chap. 12). All water level and bed

H. Yang (✉)
Kyoto University, Kyoto, Japan
e-mail: heejun@bep.vgs.kyoto-u.ac.jp

S. Siev
Institute of Technology of Cambodia, Phnom Penh, Cambodia

Ministry of Industry, Science, Technology and Innovation, Phnom Penh, Cambodia

U. Sovannara · C. Yoshimura
Tokyo Institute of Technology, Tokyo, Japan

C. Yoshimura et al. (eds.), *Water and Life in Tonle Sap Lake*,
https://doi.org/10.1007/978-981-16-6632-2_11

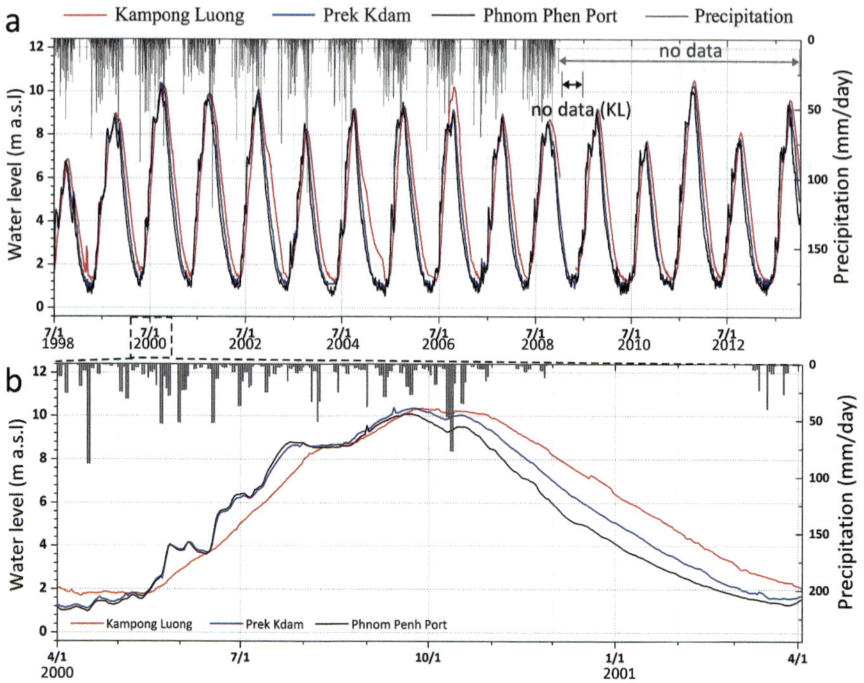

Fig. 11.1 Time series data of (**a**) the water level and precipitation from July 1, 1998, to December 31, 2013, and (**b**) those from April 1, 2000, to April 3, 2001. The no data in the panel corresponds with the missing values of the water level and precipitation

elevation data were converted into the Ha Tien mean sea level (MSL) in this chapter. Precipitation data were obtained from a rainfall gauge station at Pursat.

The water levels periodically fluctuated: the highest water level in September or October and the lowest water level from April to June. The maximum water levels were 10.09 m at PPP, 10.38 m at PK, and 10.36 m at KL, whereas the minimum water levels were 0.64 m at PPP, 0.71 m at PK, and 1.19 m at KL. The water level was in the order PPP > PK > KL during the rising water level (i.e., flow reversal), but those were changed during falling water level to the order KL > PK > PPP. The reverse flow occurred mainly from May to October. The duration of the reverse flow differed year by year from 80 to 120 days at PPP and 110 to 150 days at PK. The results correspond well to those of the previous studies explaining the flow reversal in the wet season and the normal flow in the dry season (Kummu and Sarkkula 2008; Siev et al. 2018; Uk et al. 2018).

11.2 Periodical Cycle in the Water Level

A time series analysis was performed to identify characteristics of and interactions between water levels at three stations. First, we applied auto-correlation, which is one of the methods for confirming the periodicity of the time series (Kim et al. 2005). As a result, the auto-correlation coefficients of daily precipitation quickly decreased to zero value, representing no periodic characteristic (Kim et al. 2005, Fig. 11.2a). By contrast, the auto-correlation coefficients of water levels at PPP, PK, and KL oscillate with a decreasing trend as the time lag grows, indicating strong periodicity. The maximum auto-correlation coefficient showed a lag of 365 days, reflecting a 1-year cycle.

Fourier transform was also applied for identifying significant frequency bands of the water level at each station. A spectral analysis can detect characteristics of periodic wave motion in a frequency domain (Yang et al. 2015; Yang and Shibata 2020). Significant amplitude spectra of all the water levels at frequencies of 0.0027 and 0.0057 cycles per day (CPD), which corresponds to periodicity of 365 and 174 days, respectively (Fig. 11.2b). The amplitude was 2.0 m at 365-day periodicity (annual component) and approximately 0.5 m at 174-day periodicity (semiannual component). This indicates that there were two periodicities, which controlled the periodic changes of the water levels, but the contribution of the semiannual component was weaker for the change in the water level than that of the annual component.

Significant amplitude spectra were also presented at the frequency of 0.00027 CPD (periodicity of 10 years) (Fig. 11.2b). Hydrological parameters are related to climate variabilities such as El Niño Southern Oscillation (ENSO) and Pacific Decadal Oscillation (PDO) (Hanson et al. 2006). For instance, the water level in Lake Urmia, Iran, was correlated to the Southern Oscillation Index (SOI) and the North Atlantic Oscillation (NAO) with periodicities from 11.4 to 0.2 cycles per year (Jalili et al. 2016). The component of the 10-year periodicity might be derived from the climate variability, which is probably the ENSO, because it is related to the Asian

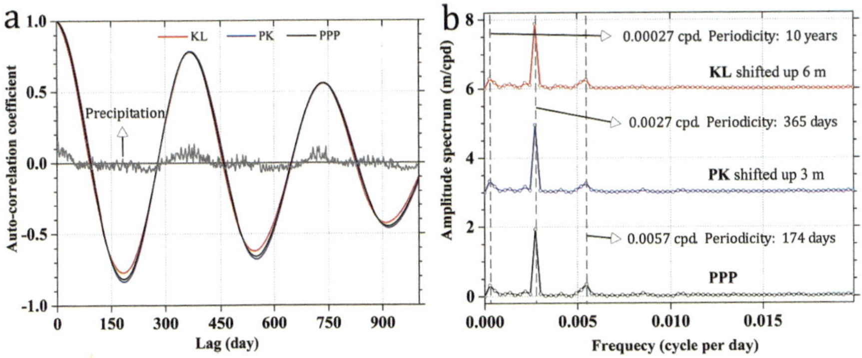

Fig. 11.2 (**a**) Auto-correlation analysis of water levels at Kampong Luong (KL), Prek Kdam (PK), and Phnom Penh Port (PPP), and precipitation. (**b**) ier transform of the water levels at the stations

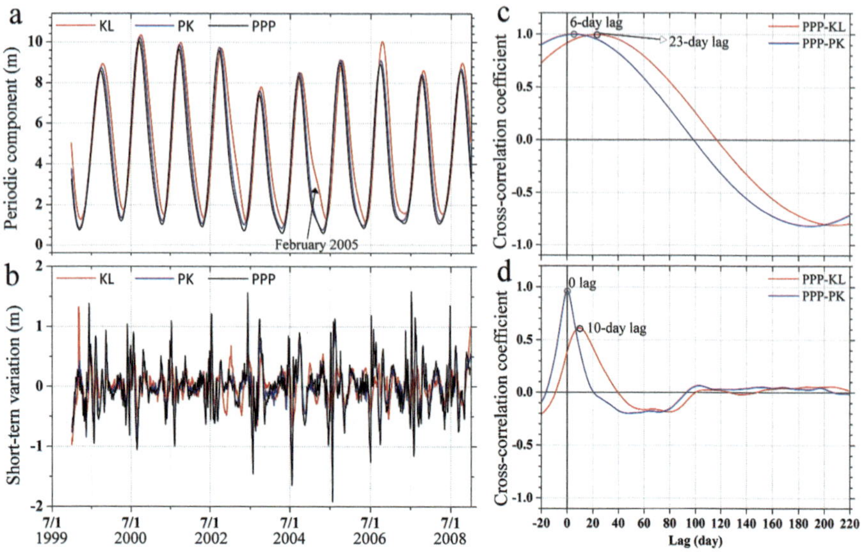

Fig. 11.3 (**a**) Periodic components (i.e., semi-annual and annual) of the water levels and (**b**) short-term variation of the water levels. Decomposition was performed by a fast Fourier transform filter. The cut-off frequency was 0.0057 cycles per day (i.e., 174 days). (**c**) Cross-correlation analysis of periodic components of the water level and (**d**) short-term variation of the water level as outputs against the Phnom Penh Port (PPP) station

summer monsoon. As mentioned in Chap. 6, Cambodia has distinct wet and dry seasons (the Asian monsoon in summer and in winter) that affect rainfall patterns. However, solid evidence such as coherence analysis between the ENSO and the water level of TSL and TSR is required to understand the relationship between climate variability and long-term periodic change in the water level.

To confirm the correlation of the periodic component, as well as the short-term variations of the water levels at the three stations, the water levels were decomposed to periodic components and short-term variations using the fast Fourier transform (FFT) filter (Yang et al. 2020b). The cut-off frequency was 0.0057 CPD (i.e., 174 days). Figure 11.3a and b depicts the periodic component (i.e., FFT low-pass filtered water levels) and the short-term variations (i.e., FFT high-pass filtered water levels), respectively. The periodic component of the water levels mainly oscillated as a 1-year cycle with high amplitudes (approximately 2.0 m) (Fig. 11.3a), but the water levels were distorted during the dry seasons (e.g., February 2005 in Fig. 11.3a). This may indicate that the semiannual component is superposed on the annual component, and the short-term variations indicated non-periodic fluctuations.

11.3 Cross-Correlation of Water Levels

A cross-correlation analysis was conducted on the periodic components (Fig. 11.3c) and the short-term variations (Fig. 11.3d) using the water level at PPP as an input and those at PK and KL as outputs. The maximum cross-correlation coefficient of the periodic components in the water level at PK was 0.99, attained after 6 days, and that at KL was 0.99, attained after 23 days (Fig. 11.3c). The short-term variations were also observed to be highly correlated (Fig. 11.3d), with a coefficient of 0.97 at PK with 0-lag and 0.61 at KL, attained after 10 days. These results reveal that flood pulse exerts controls over the changes in water levels at PK and KL. Cross-correlation coefficients of the water levels at stations of PK and KL were highly correlated with the water levels of PPP on the periodic components and short-term variations. This indicates that the flood pulse controls change in the water levels at PK and KL. However, the cross-correlation coefficient of the short-term variations at KL was lower, approximately 0.61, indicating that the flood signal was weakened.

Assuming that water flows along the river path (one-dimensional flow) and water depth is sufficiently smaller than the horizontal flow area (shallow surface water flow), the water level at PPP propagates to PK and KL stations with a high cross-correlation coefficient. The cross-correlation coefficient of the short-term variations at PK was high, implying that the river water between PPP and PK stations mainly flows along the river path. However, the cross-correlation coefficient of the short-term variations at KL was low, indicating the shifting width of the water body in TSL. Figure 11.4 depicts water levels at three stations in the high-water level period and the normal and reverse flows with the riverbed elevation. The bed elevation in TSR is approximately −8 m and that in TSL is approximately 0.5 m, but the water levels in the high-water level period at three stations are almost 9 m (Fig. 11.4), although the bed elevations at PK and KL are different, with approximately 8 m. This means that during the reverse flow, the river water flows toward the floodplain at the entrance of TSL, which causes the expansion of the TSL size.

Besides, according to Chadwick et al. (2008), of the incoming water source of TSL, 51% is from MR via TSR; 5%, from the floodplain of TSR; 31%, from TSL's

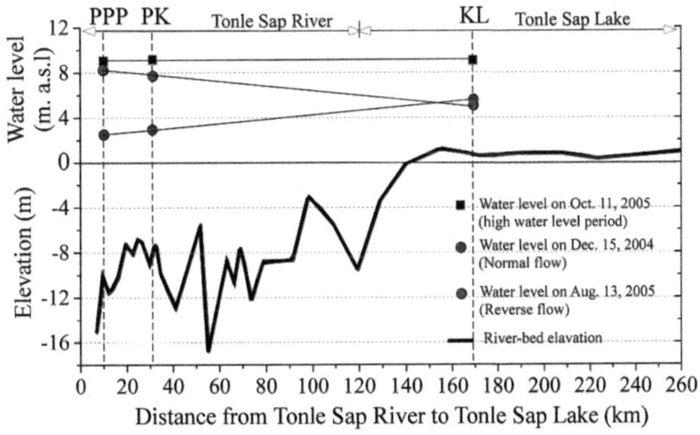

Fig. 11.4 Water levels at Phnom Penh Port (PPP), Prek Kdam (PK), and Kampong Luong (KL) and the riverbed elevation from Tonle Sap River to Tonle Sap Lake referred to Xu et al. (2020). a.s.l stands for above sea level

tributaries; and 13%, from precipitation. Kummu et al. (2014) also mentioned that 34% of the water inflow into TSL comes from its basin; 12.5%, from direct rainfall on the lake surface; and 53.5%, from the reverse flow of the MR (see also Chaps. 1 and 8). Therefore, the contribution of different water sources cannot be ignored to understand the water level at KL. The other incoming sources of TSL and the flow reversal toward the floodplains of TSL are perhaps the reasons why the cross-correlation coefficient at KL is lower than that of PK (Fig. 11.3d).

The short-term variations can be considered to be the rapid water level rise temporarily due to local heavy rainfall in the MR basin. Propagation velocity of the temporary water level rise from PPP to PK and KL was calculated based on the results of time lags of the short-term variations (Fig. 11.3d) and the shortest river distances (approximately 31 km between PPP and PK and approximately 169 km between PPP and KL, estimated using ArcGIS 10.6.1). The average velocity is 0.36 m/s for the section between PPP and PK or above and 0.20 m/s for the section between KL and PPP. The difference in the propagation velocity can be derived from the shifting width of the water body, implying that the flood pulse signal weakens in TSL. The observed mean flow velocity at a single point (at PK) was 0.88 m/s (refer to Chap. 12), which is different from the aforementioned estimates but in the same order of magnitude.

11.4 Long-Term Shift of Water Levels

To understand the long-term change in water levels, the Mann–Kendall test was applied at a confidence level at 0.05 (i.e., p-value of 0.05) to the monthly water levels from July 1998 to December 2013 at the PPP station and from July 1998 to December 2018 at PK and KL stations. The Mann–Kendall test is suitable for non-normal distribution data (e.g., Gocic and Trajkovic 2013; Yang et al. 2020a). The water level was subtracted by the mean value (i.e., $x_i - x_m$; x_i is a water level and x_m is a mean water level) because our purpose is to detect the trend of the water levels.

The water levels of the KL and PK stations had a negative z-value, indicating the decreasing trend with significant p-values (Table 11.1). However, p-values are over 0.3 at the PPP station, indicating that there is no significant long-term change in the water level. The reduction of the water level for 20 years from 1999 to 2018 at KL and PK stations was −0.96 m and −0.74 m, respectively. For further understanding of the long-term change in water levels, the trend test was conducted using the annual maximum and minimum water levels at KL and PK stations. The annual maximum water levels had a decreasing trend, whereas the annual minimum water levels exhibited no changes.

Those results imply that the reduction of the annual water levels at KL and PK stations for 20 years is influenced by the decrease of the maximum water levels at the KL and PK stations during the wet seasons. According to Räsänen et al. (2012), the cascade dams in the upper MR affected the downstream flow regimes significantly

Table 11.1 The Mann–Kendall trend test (S) for three water level stations with the monthly water level and the water levels at KL and PK with the annual water level. The subscript max and min stand for maximum and minimum, respectively. Var stands for the variance. The decreasing water level for 20 years was calculated by multiplying the number of data with the slope

Mann-Kendall test							
	Monthly water level			Annual water level			
	KL	PK	PPP	KLmax	KLmin	PKmax	PKmin
Number of data	246	246	186	20	20	20	20
Confidence level	0.05	0.05	0.05	0.05	0.05	0.05	0.05
S	−2796	−2509	−791	−70	−26	−64	−2
$Var(S)^{1/2}$	1290.0	1290.0	848.9	30.8	30.8	30.8	30.8
z-value	−2.17	−1.94	−0.93	−2.24	−0.81	−2.04	−0.03
p-value	0.03	0.05	0.35	0.02	0.42	0.04	0.97
Trend	**Yes**	**Yes**	No	**Yes**	No	**Yes**	No
Slope	−0.004	−0.003	–	−0.098	–	−0.115	–
Decreasing water level for 20-y (m)	−0.96	−0.74	–	−1.96	–	−2.42	–

by dampening the wet season flows and increasing the dry season flows, and the altered flow regimes may have significant implications for the mainstream and floodplain ecosystems of TSL. Pokhrel et al. (2018) also reported a reduction of more than 20% in the peak of flood pulse near Stung Treng station in a Cambodian part of MR. Thus, the decrease of maximum water levels at TSL and TSR in this study could be derived from the reduction of the water level in MR during the wet seasons. This causes a decrease in the annual water level at TSL and TSR. Various reasons for the alteration of the flow regimes in MR have been provided by researchers such as dam constructions, flow regulations of dams, climate variability, and floodplain operations (e.g., Pokhrel et al. 2018) and the precipitation decrease in the upstream MR (Wang et al. 2020). In any case, the alteration of the flow regimes certainly affects the water level of TSL and TSR; thus, continuous water level monitoring is required to further understand effects of the flood pulse system on TSL and TSR relative to the alteration of the flow regimes in MR, which supports water resource management in these basins and studies on water, sediment, and nutrient balances in TSL.

Key Points

- Time series analysis and Fourier transformation were applied to the water levels at Kampong Luong (KL) located in TSL, Prek Kdam (PK) in TSR, and Phnom Penh Port (PPP) at the intersection between MR and TSR.
- In both the long-term (i.e., annual cycle) and local rainfall-derived short-term variations, the water levels at PK and KL had correlations with that at PPP with time lags.
- Those time lags in the long-term and short-term variations were found to be 6 and 0 days for PK and PPP and 23 and 10 days for KL and PPP, respectively.

- During the reverse flow, which occurs from May to October, the short-term variation of the water level is weakened in TSL by the widened water surface in the lake.
- The long-term trend also revealed that the average water levels at TSL and TSR decreased for 20 years from 1999 to 2018; this is caused by the reduction of the discharge in MR and the reverse flow in TSR.

References

Chadwick M, Juntopas M, Sithirth M. Synthesis. In: Chadwick M, Juntopas M, Sithirth M, editors. Sustaining Tonle Sap: an assessment of development challenges facing the great lake. Bangkok, Thailand: The Sustainable Mekong Research Network; 2008. p. 6–9.

Gocic M, Trajkovic S. Analysis of changes in meteorological variables using Mann-Kendall and Sen's slope estimator statistical tests in Serbia. Glob Planet Chang. 2013;100:172–82. https://doi.org/10.1016/j.gloplacha.2012.10.014.

Hanson RT, Dettinger MD, Newhouse MW. Relations between climatic variability and hydrologic time series from four alluvial basins across the Southwestern United States. Hydrogeol J. 2006;14(7):1122–46. https://doi.org/10.1007/s10040-006-0067-7.

Jalili S, Hamidi SA, Ghanbari RN. Climate variability and anthropogenic effects on Lake Urmia water level fluctuations, northwestern Iran. Hydrol Sci J. 2016;61:1759–69. https://doi.org/10.1080/02626667.2015.1036757.

Kerh T, Lee CS. Neural networks forecasting of flood discharge at an unmeasured station using river upstream information. Adv Eng Softw. 2006;37(8):533–43.

Kim JH, Lee J, Cheong TJ, Kim RH, Koh DC, Ryu JS, Chang HW. Use of time series analysis for the identification of tidal effect on groundwater in the coastal area of Kimje. Korea, J Hydrol. 2005;300(1):188–98. https://doi.org/10.1016/j.jhydrol.2004.06.004.

Kummu M, Sarkkula J. Impact of the Mekong River flow alteration on the Tonle Sap flood pulse. Ambio. 2008;37:185–92.

Kummu M, Tes S, Yin S, Adamson P, Józsa J, Koponen J, Richey J, Sarkkula J. Water balance analysis for the Tonle Sap Lake–floodplain system. Hydrol Process. 2014;28:1722–33.

Pokhrel Y, Shin S, Lin Z, Yamazaki D, Qi J. Potential disruption of flood dynamics in the lower Mekong River basin due to upstream flow regulation. Sci Rep. 2018;8:17767. https://doi.org/10.1038/s41598-018-35823-4.

Räsänen TA, Koponen J, Lauri H, Kummu M. Downstream hydrological impacts of hydropower development in the upper Mekong Basin. Water Resour Manag. 2012;26:3495–513. https://doi.org/10.1007/s11269-012-0087-0.

Siev S, Yang H, Sok T, Uk S, Song L, Kodikara D, Oeurng C, Hul S, Yoshimura C. Sediment dynamics in a large shallow lake characterized by seasonal flood pulse in Southeast Asia. Sci Total Environ. 2018;631–632:597–607. https://doi.org/10.1016/j.scitotenv.2018.03.066.

Singh V, Jain SK, Shukla S. Response of hydrological factors and relationships between runoff and sediment yield in the Sub Basin of Satluj River, Western Himalaya, India. Int J Civ Struct Eng. 2010;2(1):205–21.

Uk S, Yoshimura C, Siev S, Try S, Yang H, Oeurng C, Li S, Hul S. Tonle Sap Lake: current status and important research directions for environmental management. Lakes Reserv Res Manag. 2018;23(3):177–89.

Wang Y, Feng L, Liu J, Hou X, Chen D. Changes of inundation area and water turbidity of Tonle Sap Lake: responses to climate changes or upstream dam construction? Environ Res Lett. 2020;15(9):0940a1.

Xu Z, Li C, Li A, You Z, Yao W, Chen Y, Huang L. Morphological characteristics of Cambodia Mekong Delta and Tonle Sap Lake and its response to river-lake water exchange pattern. J Water Resour Protect. 2020;12:275–302.

Yang H, Shibata T. Aquifer classification and pneumatic diffusivity estimation using periodic groundwater level changes induced by barometric pressure. Hydrol Res Lett. 2020;14(3): 111–6. https://doi.org/10.3178/hrl.14.111.

Yang H, Shimada J, Matsuda H, Kagabu M, Dong L. Evaluation of a freshwater lens configuration using a time series analysis of a groundwater level and an electric conductivity in Minami-Daito Island, Okinawa Prefecture, Japan. J Groundw Hydrol. 2015;57:187–205. https://doi.org/10.5917/jagh.57.187.

Yang H, Kagabu M, Okumura A, Shimada J, Shibata T, Pinti LD. Hydrogeochemical processes and long-term effects of sea-level rise in an uplifted atoll island of Minami-Daito, Japan. J Hydrol Reg Stud. 2020a;31:100716. https://doi.org/10.1016/j.ejrh.2020.100716.

Yang H, Shimada J, Okumura A, Shibata T, Pinti LD. Freshwater lens oscillation induced by sea tides and variable rainfall at the uplifted atoll island of Minami-Daito, Japan. Hydrogeol J. 2020b;28:2105–14. https://doi.org/10.1007/s10040-020-02185-z.

Chapter 12
Discharge Measurement and Hydraulic Characteristics in the Tonle Sap River

Hideto Fujii, Takashi Nakamura, Lun Sambo, Ly Sarann, Keisuke Hoshikawa, and Yoichi Fujihara

12.1 Discharge Measurement by ADCP with RTK GPS

An Acoustic Doppler Current Profiler (ADCP) is a multilayered flowmeter that uses the Doppler effect in ultrasonic waves to measure water flow discharge (e.g., Gordon 1996; Oberg and Mueller 2007). Advances in ocean, lake, and river discharge observation technology using ADCP have been remarkable, and there are increasing opportunities to use ADCP instead of the traditional current meter in river discharge observation. ADCP can significantly improve observation accuracy and time for observation. In the USA, an observation standard of ADCP measurement from moving boats has been established by the US Geological Survey (USGS), with river discharge observation by ADCP now common (Mueller et al. 2013; Parsons et al. 2013). The technology and methods for recording ADCP-based discharge measurements differ from conventional measurement methods using mechanical meters. ADCP can also observe the riverbed shape, moving bed and complicated flow in a river mouth or estuary (Hackney et al. 2018; Sassi et al. 2011; Hidayat et al. 2011). The needs of ADCP observation technology extend not only to the measurement of the cross-sectional profile of velocity but also to the estimation of suspended

H. Fujii (✉)
Yamagata University, Tsuruoka, Japan
e-mail: fhideto@tds1.tr.yamagata-u.ac.jp

T. Nakamura
Tokyo Institute of Technology, Tokyo, Japan

L. Sambo · L. Sarann
Institute of Technology of Cambodia, Phnom Penh, Cambodia

K. Hoshikawa
Toyama Prefectural University, Imizu, Japan

Y. Fujihara
Ishikawa Prefectural University, Nonoichi, Japan

C. Yoshimura et al. (eds.), *Water and Life in Tonle Sap Lake*,
https://doi.org/10.1007/978-981-16-6632-2_12

sediment and the simultaneous measurement of riverbed fluctuations (Darby et al. 2016; Binh et al. 2020).

Tonle Sap River (TSR) that connects Mekong River (MR) to Tonle Sap Lake (TSL) is characterized by a unique flow regime showing a seasonally repeated reversal flow, which is caused by the shifting difference in water level between the lake and the MR. This flow reversal pattern normally occurred from mid-May to early September, creating a unique ecosystem on the vast floodplain of the lake (see Chaps. 1 and 8 for the hydrology of TSL and inflow river and Chap. 11 for flood pulse and water level). To capture such unique flow characteristics, extensive streamflow observations have been implemented in the MR basin in several surveys (e.g., MRC 2004; Nguyen et al. 2015; Hackney et al. 2015). Furthermore, flow velocity distributions and cross-sectional flow rates obtained through ADCP are often used to validate hydraulic models. For example, ADCP observations from Kampong Cham and Prek Kdam to Chak Tomuk and Neak Luong carried out in the MRC WUP-JICA project (MRC 2004) were used to calibrate and validate the one-dimensional (1D) hydraulic model MIKE 11 (Fujii et al. 2003).

ADCP is a powerful tool for measuring streamflow; however, it is only accurate when used in conjunction with appropriate techniques (Mueller et al. 2013). For example, GPS is often employed with ADCP to complement the accuracy of bottom tracking especially when the riverbed is moving due to high-flow-velocity. The positioning accuracy of GPS is several meters for a single GPS, several tens of cm for a differential GPS, and several cm for a real-time kinematic (RTK) GPS. Therefore, for high-precision observation, although it is necessary to install a base station, an RTK GPS is desirable. So far, there are several observations of the flow in the Tonle Sap River with ADCP, but few of them applied RTK GPS. One of the objectives in this chapter is to demonstrate the effectiveness of RTK GPS, even if bottom tracking is not available because of the riverbed movement under high-flow-velocity measurement. In our research, a monitoring method of the flow discharge was developed by combining ADCP and RTK GPS, and velocity profile and discharge were periodically observed from March 2017 to March 2018 using ADCP to obtain a better understanding of the dynamic flow of TSR.

12.2 Application of ADCP with RTK GPS

The ADCP transducer emits an ultrasonic signal from transducers, which is reflected by suspended particles in the water; the change in the frequency of the returning ultrasonic wave is proportional to the flow velocity. A River Surveyor M9 (SonTek-Xylem Co.) ADCP unit was used in our survey. The M9 has eight transducers for flow velocity profiling and one vertical transducer for water depth observation. It takes shallow flow measurements up to 5 m in four 3.0 MHz and deep-field measurements up to 40 m in four 1.0 MHz bands. Water depth is measured by irradiating a 0.5 MHz ultrasonic wave vertically downward and recording the reflections. RTK GPS with an error of 3 cm is used to obtain high-accuracy

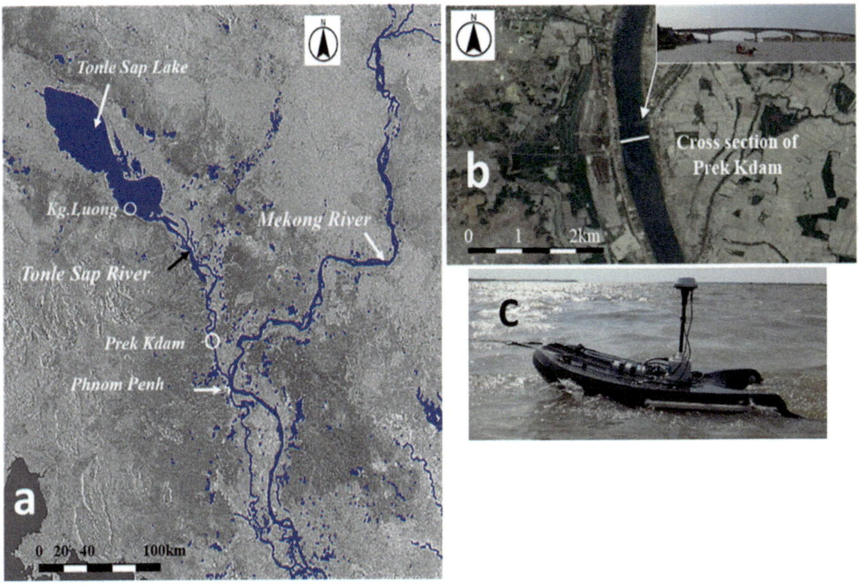

Fig. 12.1 (**a**) Location of the study site, (**b**) Prek Kdam observation section. The cross section of ADCP observation is around 200 m south of the bridge. (**c**) HydroBoard-II equipped with ADCP and RTK-GPS

coordinate measurements (SonTek 2013). The measurement accuracy of the flow velocity is 1% of the measurement value or 0.2 cm/s; the flow velocity measurement range is ±10 m/s; and the flow velocity measurable water depth is 0.2–40 m. M9 automatically switches 1.0 and 3.0 MHz and divide cell depth according to water depth (SonTek 2013).

The study sites of TSL and TSR are presented in Fig. 12.1a. Prek Kdam with around 600 m of river width and 10–16 m of water depth was selected as the current profile monitoring point in our study as presented in Fig. 12.1b. Observations were taken by using a small boat to tow a float (HydroBoard-II) equipped with ADCP and RTK GPS (Fig. 12.1c). The cross section is located around 200 m south of a bridge crossing TSR. Prek Kdam is an important water level observation station in TSR because it connects TSL with a 1D model developed by Yoneda et al. (2019) under the SATREPS project. It also works as the downstream boundary for the 2D model developed in Chap. 13 (see also Chap. 14 for details) and a validation point for the moving boat-based river velocity derived by subtracting the velocity of the boat from that of the river. ADCP equipped with single or differential GPS can calculate the flow velocity using two methods: GPS reference and bottom tracking (BT) reference. If BT observation is performed normally, BT reference discharge is applied. If not, a GPS reference is used. In the case of ADCP integrated with RTK GPS, if reception is normal, a GPS reference is used. Observations were made from March 2017 to March 2018 with an approximate frequency of once every 2 weeks. Twenty-two observations were made, with a round-trip survey of Prek Kdam carried

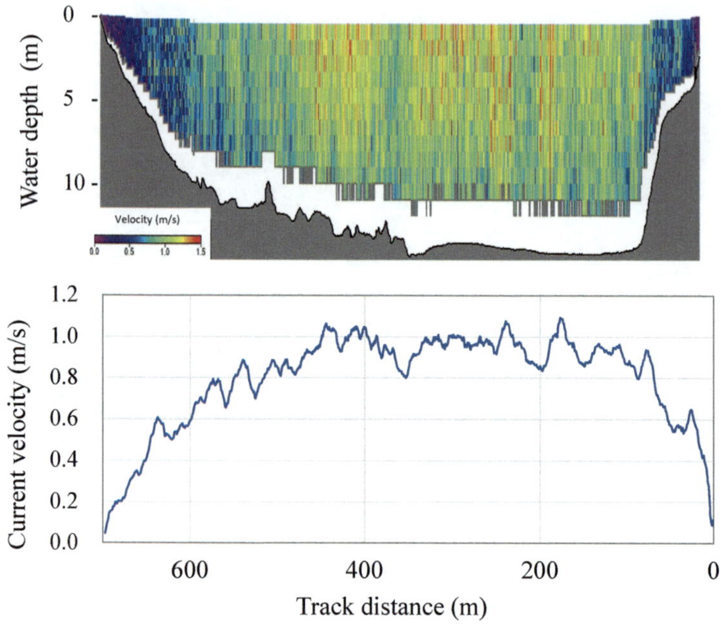

Fig. 12.2 Flow velocity distribution. (**a**) Vertical profile. (**b**) horizontal profile in Nov. 25, 2017, forward observation; the observed discharge was 6512 m^3/s

out for each. Normally, in the moving boat survey, observation is carried out for the forward trip and the backward trip; thus, the survey is called a round-trip survey for evaluating the observation reliability and accuracy. Figure 12.2 illustrates the vertical and horizontal profile of velocity distribution on November 25, when the largest discharge of 6512 m^3/s was observed.

12.3 Accuracy of Discharge Measurement

A summary of the observational results is presented in Table 12.1. The discharge presents the GPS referenced value obtained by RTK GPS, in principle. However, when GPS quality is not RTK but single, the discharge is shown the value obtained by BT reference. Accuracy of ADCP observation is generally evaluated in terms of the deviation of observed discharge over one or two round trips. Since USGS guidelines (Mueller et al. 2013) recommend using 5% or less as accurate data, the coefficient of variation (CV) of each observed discharge by the round-trip was rated as either S, A, or B, corresponding to CVs of less than 5%, 5–10%, or more than 10%, respectively. There were 17 times of S-ratings, four times of A, and only one time of B out of 22 times of observation.

Table 12.1 ADCP discharges, coefficient of variations (CVs), rating and GPS quality

Date	GPS referenced discharge			CV (%)	Rating	Beam separation		GPS quality
	Forward (m^3/s)	Backward (m^3/s)	Mean (m^3/s)			Forward (%)	Backward (%)	
2017/3/24	631	605	618	2.1	S	0.4	0.0	RTK
2017/5/13	539	439	489	10.2	B**	0.7	9.5	RTK
2017/5/27	−431	−502	−467	7.6	A**	4.8	4.5	RTK
2017/6/11	−216	−262	−239	9.6	A**	39.7	19.6	RTK
2017/6/24	−843	−725	−784	7.5	A**	29.3	10.2	RTK
2017/7/7	616	672	644	4.3	S	2.7	0.1	RTK
2017/7/22	−6173	−5546	−5860	5.4	A**	41.1	44.8	RTK
2017/8/13	−5581	−5664	−5623	0.7	S	7.1	11.6	RTK
2017/8/20	−2705	−2850	−2778	2.6	S	1.2	1.9	RTK
2017/9/2	−395	−404	−400	1.1	S	0.6	5.2	RTK
2017/9/15	2433	2427	2430	0.1	S	0.2	0.8	RTK
2017/10/1	1503	1544	1524	1.3	S	0.0	0.0	RTK
2017/10/14	5045*	*4891	4968	1.5	S	0.5	1.9	Single
2017/11/11	5239	5211	5225	0.3	S	6.3	2.5	RTK
2017/11/25	6512	6600	6556	0.7	S	0.4	0.9	RTK
2017/12/10	5086	4804	4945	2.9	S	3.3	19.3	RTK
2017/12/24	4749	4669	4709	0.8	S	0.6	0.2	RTK
2018/1/7	3681*	*3686	3684	0.1	S	0.0	0.2	Single
2018/1/20	2693	2764	2729	1.3	S	1.8	1.2	RTK
2018/2/4	1777	1736	1757	1.2	S	0.3	0.0	RTK
2018/2/18	699	689	694	0.7	S	0.4	0.3	RTK
2018/3/9	1179	1178	1179	0.0	S	0.0	0.0	RTK

*: Bottom tracking referenced discharge is used because GPS quality is single
**: The observations rated A or B

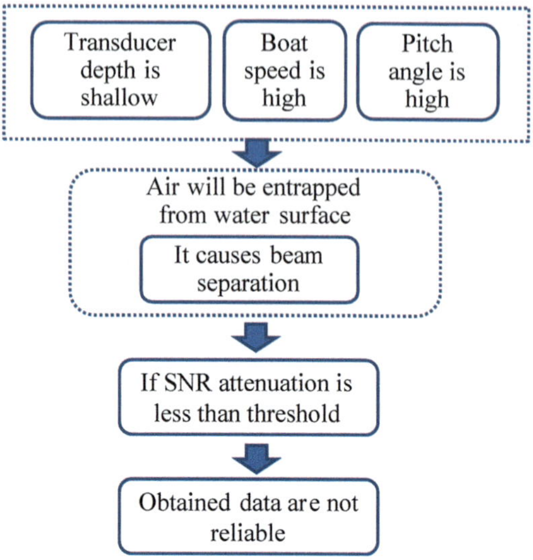

Fig. 12.3 Flow of SNR attenuation caused by beam separation

We analyzed the cause of accuracy degradation of five times of obserbations rated A and B. All the observations rated A or B have indicated a large beam separation percentage that is caused by air entrainment in either or both of forward and backward trips. The signal noise ratio (SNR) attenuations of the four ADCP beams were crucial to finding causes of accuracy degradation, which might include air entrainment, high concentrations of floating matter, and riverbed vegetation. If the boat speed is far higher than flow velocity, and/or the pitch and roll angle of HydroBoard-II are extremely high, the entrained air between water and the transducer can significantly attenuate the SNR signal (SonTek 2013).

The flow of SNR attenuation as a result of air entrainment is illustrated in Fig. 12.3 (Fujii et al. 2019). The transducer depth, boat speed, and sway (pitch and roll) of HydroBoard-II are related factors that can cause air entrainment between the transducer and water. Data can be used even in the cases of frequent beam separation warnings if the attenuation of SNR is above the allowable value; however, if the attenuation is lower than its threshold of 5 dB, the reliability of the data cannot be guaranteed. Figure 12.4 is an example of the SNR signal attenuation pattern during the June 11 forward observation when a beam separation warning occurred in 39.7% of its sample. Figure 12.4b illustrates SNR attenuation pattern when the beam is not separated, and the four beams remained unseparated and reached the river bottom together. By contrast, in Fig. 12.4a, one of the beams has separated at a shallow layer due to air entrainment. During the survey on June 11 forward trip, the mean boat speed was 1.01 m/s, and the pitch and roll were large. However, as the transducer depth was as shallow as 2 cm, many times of SNR attenuation were caused by air entrainment.

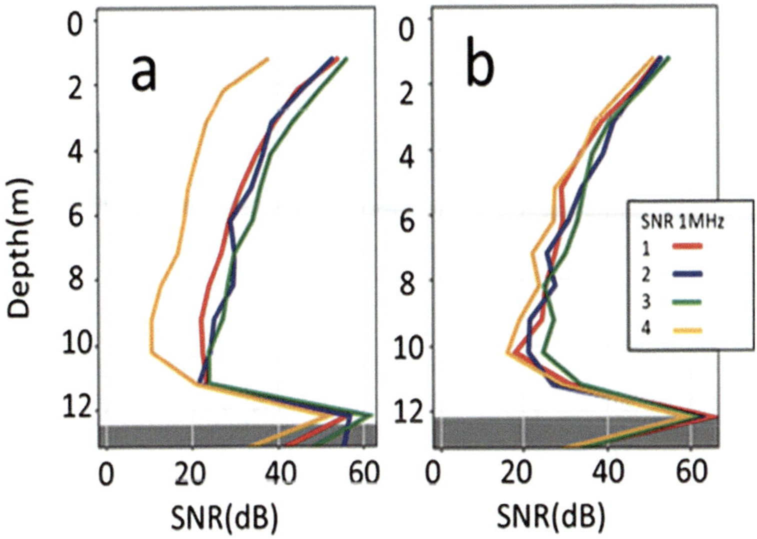

Fig. 12.4 SNR attenuation pattern (June11 forward). (**a**) one beam has separated due to air entrainment. (**b**) four beams remain together

12.4 Hydraulic Characteristics at Prek Kdam

The flow discharge (mean of the forward and backward observations) at Prek Kdam from March 2017 to March 2018 is illustrated in Fig. 12.5a. Out of the 22 observations, the flow reversal was observed from May 27 to September 2, except July 7. The discharge of -5860 m^3/s on July 22 and -5623 m^3/s on August 13 indicate large reverse flow. Since the measurement interval between July 22 and August 13 is 3 weeks, the maximum flow estimated to have occurred in early August has not been observed. A maximum normal flow of 6556 m^3/s, corresponding to a mean flow velocity of 0.88 m/s, was observed on November 25. The water level fluctuations at Prek Kdam and Kampong Luong (hereafter refer to as Kg. Luong, a representative TSL water level station) are illustrated in Fig. 12.5b. Four periods of water level reversal occurred from the latter half of May to early August 2017, reflecting the complex flow characteristics of TSR. However, once the water level of Kg. Luong exceeded that of Prek Kdam at the end of August, the flow condition remained constant until the following wet season.

The relationship between our water level and discharge observations at Prek Kdam is illustrated in Fig. 12.5c. The rating curve follows a counterclockwise loop over time that can be divided into three regimes depending on the water level and discharge characteristics: (1) a rising period from March to August; (2) high-water level (WL) period from August to October, and (3) a falling period from November to March. It is noteworthy that the discharge in period three changes significantly despite the small change in WL. Although the rating curve follows this

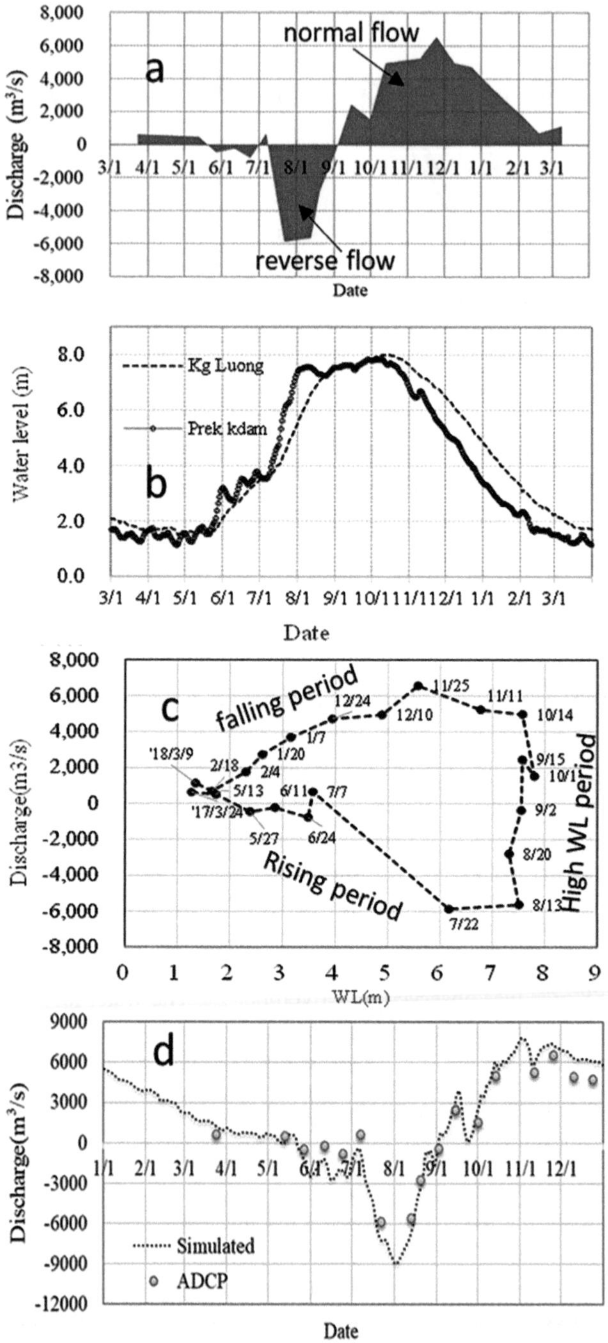

Fig. 12.5 Hydraulic characteristics of ADCP discharge at Prek Kdam. (**a**) ADCP discharge (March 2017–March 2018). (**b**) Water level at Prek Kdam and Kg. Luong (March 2017–March 2018). (**c**) Relationship between the water level and discharge. (**d**) Comparison of simulated and observed water levels in 2017

loop, it is difficult to derive an analytical relation between the WL and flow discharge (H-Q formula) in the high-water-level period, although Inomata and Fukami (2008) and Kummu et al. (2014) have both proposed such formulas (Fujii et al. 2019).

To understand ADCP accuracy, we compared ADCP results rated "S" with those produced by a 1D hydraulic model that was calibrated using the ADCP data observed in 2002 under the MRC WUP-JICA project (MRC 2004; Yoneda et al. 2019). Figure 12.5d compares simulated discharge between 1D model and 17 times of ADCP observed data rated "S." The Nash–Sutcliffe model efficiency coefficient (NSE) between the ADCP data at Prek Kdam and the 1D model results is 0.92, suggesting that the ADCP data are accurate enough for the hydraulic model calibration and validation.

Key Points

- We carried out the 1-year discharge observation from March 2017 to March 2018 at Prek Kdam in the Tonle Sap River by ADCP equipped with RTK. Out of 22 times of round-trip observations, the coefficient of variation (CV) less than 5%, which is recommended by the USGS, was 17 times; 5–10%, 4 times; and 10–15%, 1 time. Almost all observations rated A or B have shown large beam separation and abnormal SNR attenuation lower than the allowable level that was caused by air entrainment between water and transducer. To avoid accuracy degradation by air entrainment, it is crucial for one to monitor the transducer depth in the moving boat measurement.
- Discharge and velocity profile observation at Prek Kdam revealed the unique flow characteristics with respect to the relationship between the WL and discharge in the transition period from reverse to normal flow. Thus, the rating curve at Prek Kdam has indicated a counterclockwise loop over time that can be divided into three regimes depending on the water level and discharge characteristics: a) rising period from March to August; b) high-WL period from August to October; and c) falling period from November to March.
- From the comparison of ADCP discharge with that of 1D hydraulic models, ADCP data are accurate and reliable enough for calibration and validation of hydraulic models.

References

Binh DV, Kantoush S, Sumi T. Changes to long-term discharge and sediment loads in the Vietnamese Mekong Delta caused by upstream dams. Geomorphology. 2020;353:107011.

Darby SE, Hackney CR, Leyland J, Kummu M, Lauri H, Parsons DR, Best JL, Nicholas AP, Aalto R. Fluvial sediment supply to a mega-delta reduced by shifting tropical-cyclone activity. Nature. 2016;539:276–9.

Fujii H, Garsdal H, Ward P, Ishii M, Morishita K, Boivin T. Hydrological roles of the Cambodian floodplain of the Mekong River. Int J River Basin Manage. 2003;1:253–66.

Fujii H, Nakamura T, Ly S, Lun S, Heng S, Fujihara Y, Hoshikawa K, Nakata M. Flow observation of the Tonle Sap River by ADCP and cause of decrease in observation accuracy. J Rainwater Catch Syst. 2019;24(2):17–25.

Gordon, RL. Acoustic Doppler current profilers. Principles of operation: a practical primer. Second edition for broadband ADCPs. RD Instruments; 1996; p. 57.

Hackney C, Best J, Leyland J, Darby S, Parsons D, Aalto R, Nicholas A. Modulation of outer bank erosion by slump blocks: disentangling the protective and destructive role of failed material on the three-dimensional flow structure. Geophys Res Lett. 2015;42:10663–70. https://doi.org/10.1002/2015GL066481.

Hackney, C.R., Darby S., Parsons., et al. (2018) The influence of flow discharge variations on the morphodynamics of a diffluence–confluence unit on a large river. Earth Surf Process Landf 43: 349-362.

Hidayat H, Vermeulen B, Sassi MG, Torfs PJJF, Hoitink AJF. Discharge estimation in a backwater affected meandering river. Hydrol Earth Syst Sci. 2011;15(8):2717–28.

Inomata H, Fukami K. Restoration of historical hydrological data of Tonle Sap Lake and its surrounding areas. Hydrol Process. 2008;22:1337–50.

Kummu M, Tes S, Yin S, Adamson P, Józsa J, Koponen J, Richey J, Sarkkula J. Water balance analysis for the Tonle Sap Lake–floodplain system. Hydrol Process. 2014;28:1722–33.

MRC. Study on hydro-meteorological monitoring for water quality rules in Mekong River basin, final report. Vol. II, paper III, hydrological monitoring; 2004.

MRC. Overview of the hydrology of the mekong basin. Mekong River Commission, Vientiane, Lao PDR; 2005. p. 1–73.

Mueller D, Wagner C, Rehmel M, Oberg K, Rainville F. Measuring discharge with acoustic doppler current profilers from a moving boat, USGS, Chapter 22 of Section A, Surface-Water Techniques Book3, Applications of hydraulics; 2013. p. 1–95.

Nguyen NT, Nakajo S, Mukunoki T. Effects of current on sediment transport at Dinh An estuary, Mekong River, Vietnam. J JSCE B3 (Ocean Eng). 2015;71(2):I_790–5.

Oberg KA, Mueller DS. Validation of streamflow measurements made with acoustic Doppler current profilers. J Hydraul Eng. 2007;33:1422–32. https://doi.org/10.1061/(ASCE)0733-9429133:12(1421).

Parsons DR, Jackson PR, Czuba JA, Engel FL, Rhoads BL, Oberg KA, Best JL, Mueller DS, Johnson KK, Riley JD. Velocity Mapping Toolbox (VMT): a processing and visualization suite for moving vessel ADCP measurements. Earth Surf Process Landf. 2013;38:1244–60.

Sassi MG, Hoitink AJF, Vermeulen B, Hidayat. Discharge estimation from H-ADCP measurements in a tidal river subject to sidewall effects and a mobile bed. Water Resour Res. 2011;47 (W06504):1–14. https://doi.org/10.1029/2010WR009972.

SonTek. River Surveyor S5/M9 system manual firmware version 4.02:1-232; 2013.

Tanaka T, Yoshioka H, Siev S, Fujii H, Fujihara Y, Hoshikawa K, Ly S, Yoshimura C. An integrated hydrological-hydraulic model for simulating surface water dynamics in a tropical lake and its floodplain, water. MDPI. 2018;10:1213.

Yoneda I, Fujii H, Fujihara Y. Improvement of hydrological and hydraulic model applying GPCP and its characteristics in the Tonle Sap Lake. J Rainwater Catch Syst. 2019;25(1):23–31. (In Japanese with English abstract).

Chapter 13
Mathematics and Numerics of a Two-Dimensional Local Inertial Equation

Hidekazu Yoshioka and Tomohiro Tanaka

13.1 Shallow Water Models

Modeling and analysis of surface water flows, such as river floods, flows in lakes and seas, and water distributions in irrigation networks, have been of central importance in a wide variety of problems related to water resources and the environment. Such water flows are often very complex and prevent us from utilizing analytical approaches except for a few cases like the classical dam-break problems and shock waves (Szymkiewicz 2010; Toro 2001). Numerical computation of hydrodynamic equations has, therefore, been a major approach to understanding dynamics of surface water flows.

In addition, it is important to evaluate how much complexity is necessary to resolve the target problem. Models based on the three-dimensional incompressible Navier–Stokes equation are physically reasonable (Munoz and Constantinescu 2020), but its numerical computation can be time-consuming even with modern computational architecture. Typically, shallow surface (whose water depth is sufficiently smaller than the horizontal flow area) water flows can be viewed as flows under the hydrostatic assumption where the pressure is a linear function of water depth: The hydrodynamics of such flows are reasonably described using the shallow water equations (Szymkiewicz 2010; Toro 2001). Surface water flows occurring in Tonle Sap Lake (TSL) and its associated floodplains are examples to which the shallow water theory based on the hydrostatic assumption applies.

The shallow water equations are a system of nonlinear hyperbolic partial differential equations (PDEs) governing the water depth and horizontal momentum fluxes

---

H. Yoshioka (✉)
Shimane University, Matsue, Japan
e-mail: yoshih@life.shimane-u.ac.jp

T. Tanaka
Kyoto University, Kyoto, Japan

C. Yoshimura et al. (eds.), *Water and Life in Tonle Sap Lake*,
https://doi.org/10.1007/978-981-16-6632-2_13

or velocities and can simulate transient flows involving shock and rarefaction waves. These waves are not important in applications where the flood extent is of more interest, such as the TSL case. In such cases, the local inertial equation (LIE) (Bates et al. 2010) serves as a simpler and more efficient alternative. This is also a system of nonlinear hyperbolic PDEs, but it has less mathematical and physical complexities than shallow water equations. In this chapter, we first present 2D LIE for describing the shallow surface water flows. Second, we explain its discretization and review computational accuracy and stability.

13.2 2D Local Inertial Equation

13.2.1 Formulation

We consider shallow surface water flows occurring in the horizontal domain Ω with the conventional 2D Cartesian x - y coordinates. The bed elevation of the domain Ω is assumed to be time-independent and is denoted as $z = z(x, y)$.

The time is denoted as $t \geq 0$. The water depth at the time t and the horizontal point $(x, y) \in \Omega$ is the non-negative quantity and is denoted as $h = h(t, x, y)$. The momentum fluxes, which are products of the water depth and horizontal flow velocities, are denoted as $(p, q) = (p(t, x, y), q(t, x, y))$ in a similar way. The rainfall and evaporation from the water body are denoted as $r = r(t, x, y)$ and $e = e(t, x, y)$, respectively. The gravitational acceleration is denoted as g.

2D LIE contains the continuity equation of the mass balance,

$$\frac{\partial h}{\partial t} + \frac{\partial p}{\partial x} + \frac{\partial q}{\partial y} = r - e, \tag{13.1}$$

and momentum equations of the horizontal momentum balance in the x-direction,

$$\frac{\partial p}{\partial t} + gh\left[\frac{\partial (h + z)}{\partial x} + \frac{n^2 p|p|}{h^{10/3}}\right] = 0 \tag{13.2}$$

and in the y-direction,

$$\frac{\partial q}{\partial t} + gh\left[\frac{\partial (h + z)}{\partial y} + \frac{n^2 q|q|}{h^{10/3}}\right] = 0. \tag{13.3}$$

Here, the conventional Manning's formula with the roughness coefficient $n > 0$ is assumed in the friction slope terms (the last terms) of momentum Eqs. (13.2) and (13.3). Equations (13.1)–(13.3) are simultaneously solved for $t > 0$ in the domain Ω starting from an initial condition at the time $t = 0$ and the boundary condition along the boundary of Ω.

2D LIE is a system of hyperbolic PDEs. Therefore, its local behavior can be understood by studying wave structures induced by the reduced system,

$$\frac{\partial h}{\partial t}+\frac{\partial p}{\partial x}+\frac{\partial q}{\partial y}=0 \tag{13.4}$$

and

$$\frac{\partial p}{\partial t}+gh\frac{\partial h}{\partial x}=0 \quad \text{and} \quad \frac{\partial q}{\partial t}+gh\frac{\partial h}{\partial y}=0. \tag{13.5}$$

System Eqs. (13.4)–(13.5) are 2D LIE without the source and sink terms on a flat and smooth topography. We further assume that the flow is only in the x-direction and obtain the system

$$\frac{\partial h}{\partial t}+\frac{\partial p}{\partial x}=0 \quad \text{and} \quad \frac{\partial p}{\partial t}+gh\frac{\partial h}{\partial x}=0. \tag{13.6}$$

This reduced system has been studied by Martins et al. (2016). They discussed the Riemann problem of this system on a wet bed and theoretically demonstrated that its solutions contain the shock waves with a speed scaled as $O(\sqrt{gh})$. There are two important findings from this simplified analysis. First, the system Eq. (13.6) and 2D LIE Eqs. (13.1)–(13.3) cannot represent rarefaction waves. Second, the speeds of shock waves are smaller than the shallow water equations (Toro 2001). Therefore, the applicability of the LIEs seems to be more limited than that of the shallow water equations.

The brief analysis here is plain but somewhat different from problems encountered in applications where the friction slope terms are not negligible. In such cases, analytical approaches are merely successful, and we must rely on some numerical method. Refer to Tanaka et al. (2018a) for further details.

13.2.2 Numerical Discretization

We present a semi-implicit finite difference scheme to discretize 2D LIE Eqs. (13.1)–(13.3). This scheme uses a 2D structured rectangular grid (Fig. 13.1, Tanaka et al. (2018b), which harmonizes with the commonly available raster data. There exist two important points in the scheme. The first point is the spatiotemporally staggered discretization of the continuity and momentum equations, with which 2D LIE can be numerically integrated in a simple way. The second point is the semi-implicit discretization of the friction slope terms with the help of the local exact solutions to auxiliary ordinary differential equations. This technique allows us to handle the friction slope terms, terms often triggering numerical instability, in a stable and efficient manner. This kind of numerical techniques for handling the

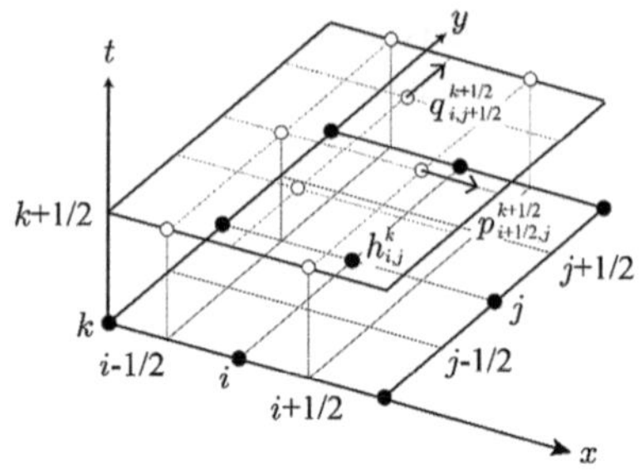

Fig. 13.1 A sketch of the staggered grid

nonlinear source terms has been utilized for numerical discretization of other problems where the nonlinearity is important but may trigger numerical instability (Li et al. 2010; Akhmouch and Benzakour Amine 2017). Besides, our numerical scheme requires the use of neither nonlinear iteration methods such as the Newton method nor linear solvers such as the Gauss–Seidel method. This means that our scheme is easily parallelized. Finally, the proposed scheme does not require added artificial viscosity terms employed in conventional shallow water solvers.

The details of the discretization of the scheme are presented following Tanaka et al. (2018b) with minor corrections in the temporal integration procedure. The quantity Q evaluated at time $t = k\Delta t$ at location $(x, y) = (i\Delta x, j\Delta y)$ is denoted as $Q^k_{i,j}$, where i, j, and k are integers or half integers numbering the locations and the time steps. Equation (13.1) is discretized as follows:

$$\frac{h^{k+1}_{i,j} - h^k_{i,j}}{\Delta t} + \frac{p^{k+1/2}_{i+1/2,j} - p^{k+1/2}_{i-1/2,j}}{\Delta x} + \frac{q^{k+1/2}_{i,j+1/2} - q^{k+1/2}_{i,j-1/2}}{\Delta y} = r^k_{i,j} - e^k_{i,j}. \tag{13.7}$$

Set the positivity-preserving nonlinearly reconstructed water depths,

$$\tilde{h}^k_{i+1/2,j} = \max\left\{h^k_{i,j} + z_{i,j}, h^k_{i+1,j} + z_{i+1,j}\right\} - \max\left\{z_{i,j}, z_{i+1,j}\right\} \tag{13.8}$$

and

$$\tilde{h}^k_{i,j+1/2} = \max\left\{h^k_{i,j} + z_{i,j}, h^k_{i,j+1} + z_{i,j+1}\right\} - \max\left\{z_{i,j}, z_{i,j+1}\right\}. \tag{13.9}$$

Then, rewrite (13.2) at the location $((i + 0.5)\Delta x, j\Delta y)$ in the semi-discrete form

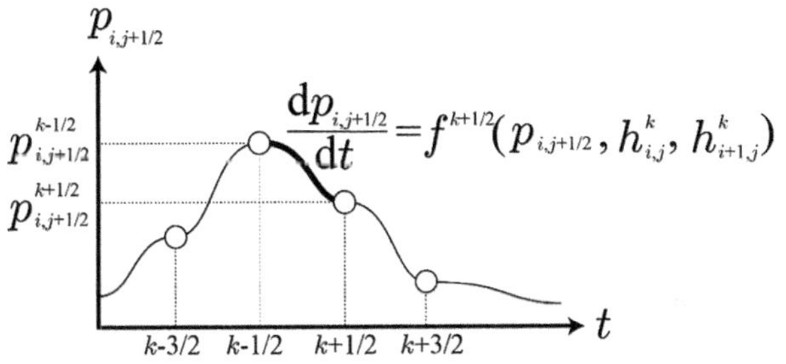

Fig. 13.2 A conceptual diagram of the temporal discretization method

$$\frac{\mathrm{d}p_{i+1/2}}{\mathrm{d}t} + g\tilde{h}^k_{i+1/2,j}\left[\frac{\left(h^k_{i+1,j} + z_{i+1,j}\right) - \left(h^k_{i,j} + z_{i,j}\right)}{\Delta x} + \frac{n^2 p_{i+1/2}\left|p_{i+1/2}\right|}{\left(\tilde{h}^k_{i+1/2,j}\right)^{10/3}}\right]$$
$$= 0. \tag{13.10}$$

This is a nonlinear ordinary differential equation that can be solved exactly in the time interval $(k - 1/2)\Delta t \le t \le (k + 1/2)\Delta t$ to obtain the updated value $p^{k+1/2}_{i+1/2,j}$ employing $p^{k-1/2}_{i+1/2,j}$ as the initial condition (Fig. 13.2). To find the exact solution to the local problem, set the following coefficients:

$$A_{i+1/2,j,k} = g\tilde{h}^k_{i+1/2,j}\frac{\left(h^k_{i+1,j} + z_{i+1,j}\right) - \left(h^k_{i,j} + z_{i,j}\right)}{\Delta x}, \tag{13.11}$$

$$B_{i+1/2,j,k} = \frac{gn^2}{\left(\tilde{h}^k_{i+1/2,j}\right)^{7/3}} > 0,\ C_{i+1/2,j,k} = \frac{A_{i+1/2,j,k}}{B_{i+1/2,j,k}}, \tag{13.12}$$

$$D_{i+1/2,j,k} = \frac{\sqrt{C_{i+1/2,j,k}} + p^{k-1/2}_{i+1/2,j}}{\sqrt{C_{i+1/2,j,k}} - p^{k-1/2}_{i+1/2,j}},\ E_{i+1/2,j,k}$$
$$= \frac{1}{\sqrt{C_{i+1/2,j,k}}}\tan^{(-1)}\left(\frac{p^{k-1/2}_{i+1/2,j}}{\sqrt{C_{i+1/2,j,k}}}\right). \tag{13.13}$$

Here, we temporally assume $C_{i+1/2,j.k} > 0$.

What we will do here is to directly solve Eq. (13.10). If $A_{i+1/2,j,k} = 0$, we use

$$p^{k+1/2}_{i+1/2,j} = \frac{p^{k-1/2}_{i+1/2,j}}{1 + B_{i+1/2,j,k}\Delta t\left|p^{k-1/2}_{i+1/2,j}\right|}, \tag{13.14}$$

where some subscripts have been omitted for the sake of simplicity. Similarly, if $A_{i+1/2,j,k} > 0$ and $p^{k-1/2}_{i+1/2,j} \le 0$, we use

$$p_{i+1/2,j}^{k+1/2} = \sqrt{C_{i+1/2,j,k}} \frac{D_{i+1/2,j,k} e^{-2\sqrt{C_{i+1/2,j,k}} B_{i+1/2,j,k} \Delta t} - 1}{1 + D_{i+1/2,j,k} e^{-2\sqrt{C_{i+1/2,j,k}} B_{i+1/2,j,k} \Delta t}}. \tag{13.15}$$

If $A_{i\ +\ 1/2,\ j,\ k} > 0$, $p_{i+1/2,j}^{k-1/2} > 0$ and $B_{i\ +\ 1/2,\ j,\ k} \Delta t \le E_{i\ +\ 1/2,\ j,\ k}$, we use

$$p_{i+1/2,j}^{k+1/2} = \frac{1}{\sqrt{C_{i+1/2,j,k}}} \tan\left(\sqrt{C_{i+1/2,j,k}} \left(E_{i+1/2,j,k} - B_{i+1/2,j,k} \Delta t\right)\right). \tag{13.16}$$

Finally, if $A_{i\ +\ 1/2,\ j,\ k} > 0$ and $p_{i+1/2,j}^{k-1/2} > 0$ and $B_{i\ +\ 1/2,\ j,\ k} \Delta t > E_{i\ +\ 1/2,\ j,\ k}$, we use

$$p_{i+1/2,j}^{k+1/2} = \sqrt{C_{i+1/2,j,k}} \frac{D_{i+1/2,j,k} e^{-2\sqrt{C_{i+1/2,j,k}} \left(B_{i+1/2,j,k} \Delta t - E_{i+1/2,j,k}\right)} - 1}{1 + D_{i+1/2,j,k} e^{-2\sqrt{C_{i+1/2,j,k}} \left(B_{i+1/2,j,k} \Delta t - E_{i+1/2,j,k}\right)}}. \tag{13.17}$$

The representations of $A_{i\ +\ 1/2,\ j,\ k} > 0$ are now completed. If $A_{i\ +\ 1/2,\ j,\ k} < 0$, we can simply apply the transformations $p_{i+1/2,j}^{k-1/2} \to -p_{i+1/2,j}^{k-1/2}$ and $C_{i\ +\ 1/2,\ j,\ k} \to -C_{i\ +\ 1/2,\ j,\ k}$. The remaining momentum equation (Eq. 13.3) is discretized the same way.

The initial and boundary conditions can be directly specified along the spatio-temporal boundary. They are often problem-dependent and are omitted here. Finally, note that we frequently encounter wet–dry interfaces along which the water depth vanishes ($h = 0$). Managing the wet–dry interfaces in a numerically stable manner is another key for successful numerical computation. This point is not discussed in this chapter but will be found in Chap. 14, where 2D LIE is applied to the integrated hydrological hydraulic simulation of TSL and its flood plains.

13.3 Accuracy and Stability of the Numerical Schemes

Computational accuracy of the present and earlier finite difference schemes is briefly discussed in this section. Tanaka et al. (2018b) examined the scheme against the 1D benchmark test of constant wave propagation (Bates et al. 2010) where an exact solution is available. The computational domain is the 1D channel (0, 5000) m. In this test, the wet–dry interface propagates along the x-direction. The wavefront is concave because of the friction effect. Bates et al. (2010) examined the scheme against a variety of spatial and temporal resolutions and found that the presented scheme can compute numerical solutions without any computational breakdowns and that they converge toward the exact solution. The computational results demonstrated that the concave profile of the exact solution is correctly reproduced by the scheme. Previous finite difference schemes, which are the fully implicit scheme (Tanaka et al. 2018c) and the conventional semi-implicit scheme (Tanaka et al. 2018b), have been examined and compared. The result indicates that the computational performances of these schemes are comparable with each other.

The scheme presented in Sect. 13.2.2 is the most physically consistent because it does not use the future values in the discretization at each time step as in the previous ones (Tanaka et al. 2018b, c). In fact, the scheme can reproduce a uniform flow limit even under an exceptionally large time step in the sense of Xia and Liang (2018). This consistency property, mentioned as the asymptotic-preserving property in computational physics, implies that a scheme equipped with this property has lower risks of generating physically unrealistic water flows than those of schemes without it.

Stability analysis against a perturbed uniform flow demonstrated was conducted to see the computational stability of the presented and previous schemes. The flow considered here is unidirectional, and the 1D counterpart of the LIE was employed. The initial condition is a uniform flow on a slope with a constant gradient. Manning's roughness coefficient of the channel is 0.03 $m^{-1/3}$ s. The initial water depth was set at 0.1 m. The perturbation with a size of 0.01% of the initial water depth and line discharge to both the lower and upper boundaries was added to the initial condition. A scheme was then judged to be stable if it recovered the original uniform flow as the time elapsed with the relative errors of discharge and water depth less than 0.1% from those of the uniform flow. Otherwise, the scheme was judged to be unstable. The computational results demonstrated that the conventional semi-implicit scheme has a narrower stability region than that designated by the classical CFL condition (Tanaka et al. 2018b, c), whereas the others did not. The fully implicit scheme had higher stability than the presented scheme. Therefore, the presented scheme is a physically consistent, sufficiently accurate, and stable scheme for numerically solving 2D LIE. The computational results hence support the applicability of the 2D LIE joint with the finite difference scheme explained in this chapter to simulation of shallow surface water flows in TSL.

As described, 2D LIE is useful for simulating shallow surface water flows. The numerical scheme explained in this chapter is used in the numerical computation in Chaps. 14, 26, and 46. Regarding applications of the 1D and 2D LIEs, you may refer to Bates et al. (2010) and articles cited there. We believe that 2D LIE will continuously serve as a simple and efficient tool for modeling and analysis of shallow surface water flows, such as those of TSL. Finally, both numerical schemes (Tanaka et al. 2018b, c) reviewed in this section perform comparably well in applied problems of TSL without significant computational difficulties.

Key Points

- The 2D local inertial equation is a simple and efficient shallow water model.
- The staggered nature of the scheme reasonably manages the benchmark test cases, demonstrating the scheme's applicability to surface water flows in TSL.
- Its computation can be carried out using a stable and accurate numerical scheme.

References

Akhmouch M, Benzakour Amine M. A time semi-exponentially fitted scheme for chemotaxis-growth models. Calcolo. 2017;54:609–41. https://doi.org/10.1007/s10092-016-0201-4.

Bates PD, Horritt MS, Fewtrell TJ. A simple inertial formulation of the shallow water equations for efficient two-dimensional flood inundation modelling. J Hydrol. 2010;387(1–2):33–45. https://doi.org/10.1016/j.jhydrol.2010.03.027.

Li Y, Lee HG, Jeong D, Kim J. An unconditionally stable hybrid numerical method for solving the Allen–Cahn equation. Comput Math Appl. 2010;60:1591–606. https://doi.org/10.1016/j.camwa.2010.06.041.

Martins R, Leandro J, Djordjević S. Analytical and numerical solutions of the Local Inertial Equations. Int J Non Linear Mech. 2016;81:222–9. https://doi.org/10.1016/j.ijnonlinmec.2016.01.015.

Munoz DH, Constantinescu G. 3-D dam break flow simulations in simplified and complex domains. Adv Water Resour. 2020;2020:103510. https://doi.org/10.1016/j.advwatres.2020.103510.

Szymkiewicz R. Numerical modeling in open channel hydraulics. Dordrecht: Springer; 2010. https://doi.org/10.1007/978-90-481-3674-2.

Tanaka T, et al. An integrated hydrological-hydraulic model for simulating surface water flows of a shallow lake surrounded by large floodplains. Water. 2018a;10(9):1213. https://doi.org/10.3390/w10091213.

Tanaka T, Yoshioka H, Siev S, Fujii H, Ly S, Yoshimura C. A consistent finite difference local inertial model for shallow water simulation. Hydrol Res Lett. 2018b;13(2):28–33. https://doi.org/10.3178/hrl.13.28.

Tanaka T, Yoshioka H, Siev S, Fujii H, Ly S, Yoshimura C. Performance comparison of the three numerical methods to discretize the local inertial equation for stable shallow water computation. In: Li L, et al., editors. AsiaSim 2018: methods and applications for modeling and simulation of complex systems. Singapore: Springer; 2018c. p. 451–65. https://doi.org/10.1007/978-981-13-2853-4_35.

Toro EF. Shock-capturing methods for free-surface shallow flows. New York: Wiley; 2001.

Xia X, Liang Q. A new efficient implicit scheme for discretising the stiff friction terms in the shallow water equations. Adv Water Resour. 2018;117:87–97. https://doi.org/10.1016/j.advwatres.2018.05.004.

Chapter 14
Numerical Simulation of Hydrodynamics Using the Two-Dimensional Local Inertial Equation

Tomohiro Tanaka and Hidekazu Yoshioka

14.1 Selection of a Two-Dimensional Hydraulic Model

Large-scale shallow lakes in Southeast Asia are exposed to severe flood/drought risks. Furthermore, environmental pollution is serious in these lakes because of poor wastewater management. To assess their impacts and design adaptation plans, it is essential to understand the hydrodynamics of a lake under various hydrological and/or social conditions, for which mathematical models of simulating lake hydrodynamics play an important role. Several models have been developed and applied to Tonle Sap Lake (TSL), most of which (except for the latest study by Yu et al. (2019)) targeted not only TSL but the whole/lower Mekong River (MR) basin (Dutta et al. 2007; Hai et al. 2008; Try et al. 2018). Dutta et al. (2007) and Hai et al. (2008) employed shallow water equations (SWEs) as the governing hydrodynamic equations that are exactly derived from the Navier–Stokes equations through vertical integration. This is often the case in hydrodynamic simulations; however, it requires careful treatment of the advection term, which sometimes interrupts its operational use. Try et al. (2018), by contrast, employed the diffusive wave equation (DWE), which neglects the advective and local inertial terms of SWEs. This equation is simpler than the SWEs and is sometimes used for river flow simulations, but it remains challenging for numerical stability as it requires a finer time step because of the nature of the parabolic equation. Besides, the modelers must make significant

Supplementary Information The online version of this chapter (https://doi.org/10.1007/978-981-16-6632-2_14) contains supplementary material, which is available to authorized users.

T. Tanaka (✉)
Kyoto University, Kyoto, Japan
e-mail: tanaka.tomohiro.7c@kyoto-u.ac.jp

H. Yoshioka
Shimane University, Matsue, Japan

C. Yoshimura et al. (eds.), *Water and Life in Tonle Sap Lake*,
https://doi.org/10.1007/978-981-16-6632-2_14

efforts in constructing the triangle mesh in the finite element method adopted by Hai et al. (2008) Therefore, the use of existing methods in operation and/or assessment under various scenarios is not always easy because of high computational costs and efforts, especially when only limited computer/human resources are available.

As introduced in Chap. 13, owing to its intermediate form between SWEs and DWE, the local inertial equation (LIE), which neglects only the advective term of SWEs, demonstrates sufficient accuracy while avoiding the complexity in implementation (Hunter et al. 2007). As the spatial scale of TSL is sufficiently larger than that of water depth, we employ the two-dimensional LIE (2D LIE) for simulating the horizontal hydrodynamics of TSL.

14.2 Implementation of 2D LIE in TSL

2D LIE is constructed in and around TSL (inside the thick brown line in Fig. 14.1) with the Cartesian orthogonal coordinate system used in Chap. 13. The bed elevation of each mesh is provided from the digital elevation model of SRTM embedded with the bathymetry data of TSL and the Tonle Sap River (TSR). The model is driven by appropriate initial and boundary conditions. The initial condition is set as a spatially uniform water level with a similar area to a typical one in the dry season. Boundary conditions are the tributary river discharges and the water stage at the downstream edge of TSR in the study area. They are provided from state-of-the-art hydrological and river routing models. Tributary river discharges (the light-green boxes in Fig. 14.1) are provided as the lateral inflows of the continuity equation from the Geomorphology-Based Hydrological Model (GBHM) (Yang et al. 2002) that simulates hydrological processes in the brown area in Fig. 14.1 (see Chap. 8 for the model setup). By contrast, river flow data from the main MR are used to prescribe the water level at the Prek Kdam station (PK) (see Fig. 14.1) downstream of TSR. In the rainy season, backwater from MR causes flooding in TSL and is thus an important boundary condition. The boundary water level is provided by MIKE 11 to simulate a one-dimensional river network (the blue lines in Fig. 14.1). See Fujii et al. (2003) for details of the model setup of MIKE 11.

The integrated model was validated against the observed water level data at the Kampong Luong (KL) and Kampong Chhang stations (Fig. 14.1). The observed and simulated water levels are illustrated in Fig. 14.2. During the observation period, the water levels at both stations are quite accurately reproduced using 2D LIE driven by the boundary conditions from GBHM and MIKE 11. The error magnitude for the whole period is -1.12 to 0.622 m, and that of the annual maximum water level is -0.223 to 0.0288 m. These errors are acceptably small because the water levels dynamically vary several meters at these stations. The simulated flood area was also validated using the satellite images of the flood area by MODIS and Landsat from 2000 to 2003 (see Tanaka et al. (2018) for more details).

Fig. 14.1 Simulation domain and cascade system using GBHM, MIKE 11, and 2D LIE. Brown area, simulation domain with GBHM; blue lines, 1-D flow network with MIKE11; and brown line, the boundary of the simulation domain with 2D LIE (Tanaka et al. 2018)

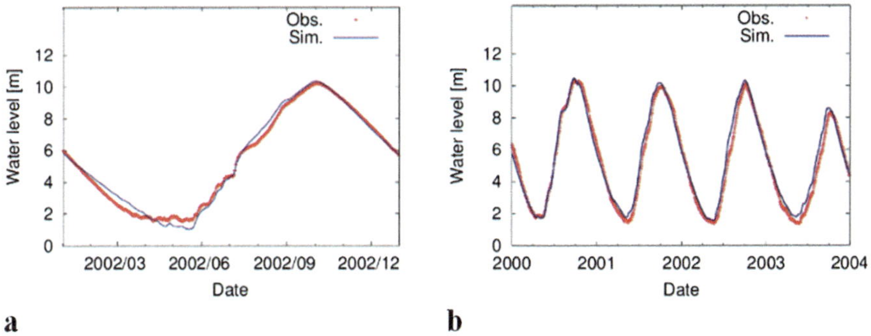

Fig. 14.2 Observed (red) and simulated (blue) water level at (**a**) the Kampong Chhnang and (**b**) Kampong Luong stations (Tanaka et al. 2018)

14.3 Prediction of the Lake Water Level

The sensitivity analysis indicated a high impact of the water level at the PK station in TSR on the water level at KL. The backwater flow from TSR propagates to TSL in the rainy season; therefore, if this applies to many parts of the lake, using the observed water level at PK station, the distributed lake water level can be estimated even without any numerical simulations. This subsection investigates the feasibility of a simple estimation of the lake water level using the observed water level at PK based on flood-delay analysis (FDA) explained below.

Given that 2D LIE validated above well-reproduced lake water level from July 1998 to December 2003, we calculated the root mean square error (RMSE) between the observed water level at PK and the simulated water level at each cell of the model, shifting PK water level data from that of 1 day to that of 30 days. Figure 14.3 (a) and (b) indicate the delay for which the RMSE is the smallest and the resulting RMSE value, respectively. The lag time with the smallest RMSE is 5–10 days inside the lake, where the RMSE is smaller than 30 cm. Notably, it is far earlier (1–5 days) in the outside of the lake; however, the RMSE value is much higher (1–5 m). More outside areas are much harder to predict only from the river observation. This is considered because the target period included the dry season during which these cells are dry. Future research will split the analysis into shorter periods with similar hydrodynamics such as the rainy and dry seasons or each month. According to the cross-correlation analysis (CCA) (Fig. 12.4 in Chap. 12) that does not use hydrodynamic simulation, a long-term flood pulse at Phnom Penh Port propagates to PK and KL after 6 and 23 days, respectively, implying that the duration of flood propagation between PK and KL is around 17 days. The difference from FDA may be due to their

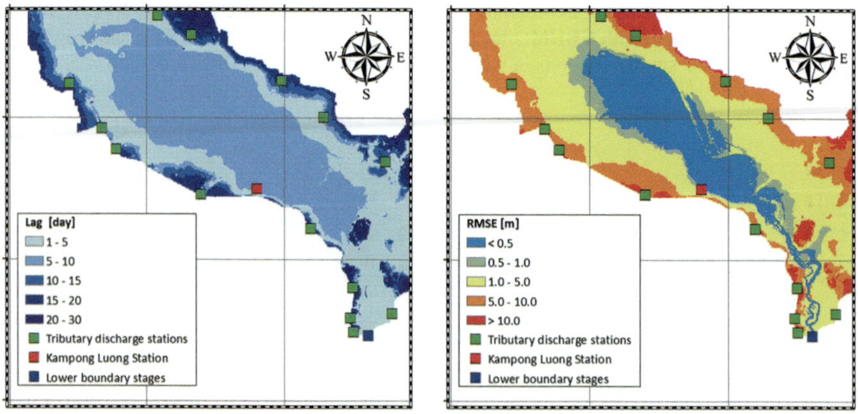

a b

Fig. 14.3 (**a**) Time lag in the water level change at Prek Kdam (observation) and each location in the lake (simulation) with (**b**) the smallest root mean square error (RMSE) among 1- to 30-day lags from July 1998 to December 2003

methodological difference; i.e., FDA used the simulated water level without any filters because its purpose is to explore the predictability of the water level over the lake. By contrast, CCA decomposed the simulated water level into long-term periodic waves and short-term fluctuation for improved understanding of both forms of wave propagation. The FDA here is, therefore, consistent with CCA. FDA in our application indicated that the water level in the main lake, which is wet throughout the year, can be reasonably predicted using the observed water level of the several precedent days at PK.

14.4 Automatic Domain Updating for Accelerating Simulation

The 2D LIE model can achieve a computationally efficient numerical simulation of large-scale horizontal hydrodynamics. Typical flood simulators calculate the momentum and continuity equations at all the meshes or only wet cells by detecting the wet cells at each time step. Solving the equations, as well as even simply checking whether the water depth is non-negative, is costly (Medeiros and Hagen 2014). Recently, the automatic domain updating (ADU) algorithm, which automatically traces the flood area boundaries (yellow cells in Fig. 14.4) at every time step and skips any access to dry cells (green cells in Fig. 14.4), has been developed by Tanaka et al. (2019) for more efficient river flooding simulation. However, this algorithm may not be applicable to lake flow simulation because direct rainfall/evapotranspiration to dry cells would affect the lake flow, which is normally negligible for simulating river overflow. This impact on flow simulation in the TSL is investigated, and the improvement of computational efficiency is evaluated in the next subsection.

The simulated water levels at KL with/without the ADU are compared in Fig. 14.5. The result with the ADU is consistent with that of the original model without the ADU (the error magnitude is 0.15% in the dry season and 0.87% in the rainy season, respectively); errors due to implementing the ADU (red) are thus not visible during most of the simulation period (except for the dry season in 2001 and 2003).

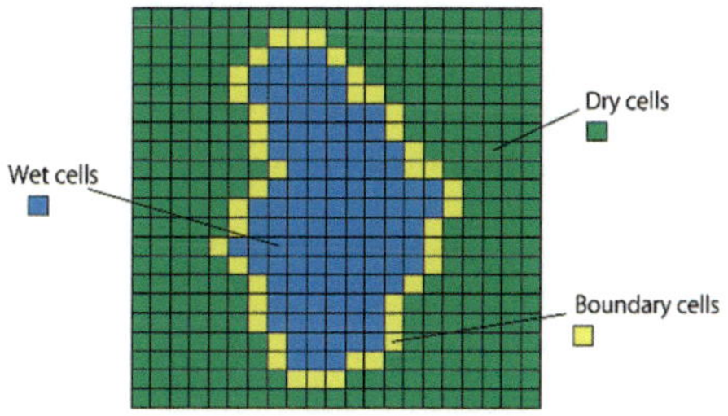

Fig. 14.4 Schematic drawing of the ADU algorithm. The momentum and mass balance equations are solved in wet (blue) and boundary (yellow) cells. When a boundary cell is flooded, it becomes a wet cell, and its surrounding dry cells become new boundaries. (Tanaka and Yoshioka 2020)

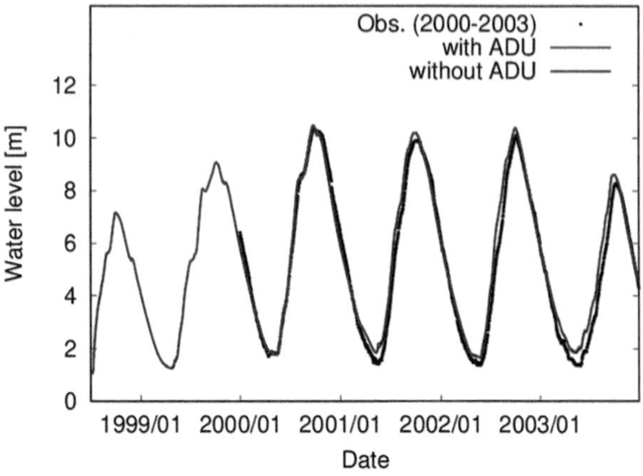

Fig. 14.5 The observed and simulated water level at the Kampong Luong station with and without ADU (Tanaka and Yoshioka 2020)

Like the other water bodies in Southeast Asia, the TSL expands in the rainy season, and it is characterized by not only rainfall but also the unique backwater from the MR, as indicated in the sensitivity analysis. At the beginning of the rainy season, rainfall on the dry cells is neglected; after the flood area expands together with the simulation domain, as illustrated by blue cells in Fig. 14.4, more rainfall is considered (around 70% of the total rainfall is considered in the later rainy season, whereas only 40% is considered in the beginning). This mechanism contributed to the smaller impact of missing rainfall on simulation results. As a result, the impact of employing the ADU on the lake flow simulation is particularly small in TSL.

The 2D LIE with/without ADU was compiled with Intel® C++ compiler 15 and carried out on a Linux machine with an Intel® Xeon® processor (CPU E5-2680, 2.8 GHz). The computational time for the 5.5-year simulation from July 1998 to December 2003 was 43.0 and 91.7 h with and without the ADU on a single core, respectively, indicating 2.1 times faster computation when ADU was used. The simulation with ADU is 1.8 times and 1.4 times faster than that without ADU on 8 and 48 cores, respectively. Therefore, ADU more significantly improves the computational efficiency of 2D LIE as the computational environment becomes less massive. Our computational results suggest that the ADU is still an effective method for flood simulation involving lake flow dynamics, especially under limited computational resources. In summary, we found that the ADU algorithm enhances the numerical modeling of lake flow dynamics by significantly reducing the computational burden in exchange for the small reduction in the accuracy in the case of the large-scale lake flow simulation in semi-arid areas. 2D LIE and ADU are implemented in C++ code (Data 14.1).

Key Points

- The two-dimensional local inertial equation (2D LIE) combined with hydrological/one-dimensional river flow models reproduced the gauged water level and flood area in Tonle Sap Lake with reasonable accuracy.

- Comparison between the simulated water level by 2D LIE and the observed downstream (Prek Kdam) water level revealed that the water level over the main lake is predictable from several days to a few weeks ahead by using the observed water level at the Prek Kdam station.
- The ADU algorithm enhanced the numerical modeling of lake flow dynamics by significantly reducing the computational burden in exchange for the small reduction in the accuracy in the case of the large-scale lake flow simulation in semi-arid areas.

References

Dutta D, Alam J, Umeda K, Hayashi M. A two-dimensional hydrodynamic model for flood inundation simulation: a case study in the lower Mekong river basin. Hydrol Process. 2007;21:1223–37.

Fujii H, Garsdal H, Ward P, Ishii M, Morishita K, Boivin T. Hydrological roles of the Cambodian floodplain of the Mekong River. Int J River Basin Manage. 2003;1:253–66.

Hai PT, Masumoto T, Shimizu K. Development of a two-dimensional finite element model for inundation processes in the Tonle Sap and its environs. Hydrol Process. 2008;22:1329–36.

Hunter NM, Bates PD, Horritt MS, Wilson MD. Simple spatially-distributed models for predicting flood inundation: a review. Geomorphology. 2007;90(3–4):208–25.

Medeiros SC, Hagen SC. Review of wetting and drying algorithms for numerical tidal flow models. Int J Numer Methods Fluids. 2014;71:473–87.

Tanaka T, Yoshioka H. An application of the automatic domain updating to the Tonle Sap Lake, Cambodia. Hydrological Res Lett. 2020;14(2):68–74.

Tanaka T, Yoshioka H, Siev S, Fujii H, Fujihara Y, Hoshikawa K, Ly S, Yoshimura C. An integrated hydrological-hydraulic model for simulating surface water flows of a shallow lake surrounded by large floodplains. Water. 2018;10:1214.

Tanaka T, Tachikawa Y, Ichikawa Y, Yorozu K. An automatic domain updating method for fast 2-dimensional flood-inundation modelling. Environ Model Softw. 2019;116:110–8.

Try S, Lee G, Yu W, Oeumg C, Jang C. Large-scale flood-inundation modeling in the Mekong River basin. J Am Society Civil Eng. 2018;23(7):05018011.

Yang D, Herath S, Musiake K. A hillslope-based hydrological model using catchment area and width functions. Hydrol Sci J. 2002;47(1):49–65.

Yu W, Kim Y, Lee D, Lee G. Hydrological assessment of basin development scenarios: impacts on the Tonle Sap Lake in Cambodia. Quat Int. 2019;503:115–27.

Chapter 15
Hydrodynamic Properties Characterized by Two-Dimensional Hydraulic Simulation

Kim Lengthong, Chhuon Kong, Tomohiro Tanaka, and Hidekazu Yoshioka

15.1 Water Volume-Inundation Area

Flood dynamic characteristics of Tonle Sap Lake (TSL) were evaluated from the daily simulation result from 1998 to 2003 using the 2D-LIE model (see Chap. 14 for details). Initially, basic variables such as spatial depth, daily water volume, and water balance were analyzed. Consequently, the inundation area of TSL was derived from the output of 2D-LIE, where the area with a flood depth over 0.5 m was defined to be inundated. The depth less than 0.5 m was considered to be a non-inundated area because it simulated an unacceptable large inundated area. Thus, 0.5 m was considered to be the threshold for the computation. Figure 15.1 illustrates the relationship between the lake volume (km^3) and the lake area (km^2) from July 1998 to December 2003 (66 months in total). The area of the lake shrinks during the dry season when the water outflows from TSL to the Mekong River (MR; see also Chap. 1).

In some months of the study period, the surface area of TSL shrank to the minimum value of approximately 2800 km^2, and the volume also dropped to 2.5 km^3. By contrast, the inundation area reaches the maximum value of 14,900 km^2 when the volume of the lake rises to 86 km^3. Comparing both seasons, the inundation area during the rainy season is around four to eight times larger than

Supplementary Information The online version of this chapter (https://doi.org/10.1007/978-981-16-6632-2_15) contains supplementary material, which is available to authorized users.

K. Lengthong (✉) · C. Kong
Institute of Technology of Cambodia, Phnom Penh, Cambodia

T. Tanaka
Kyoto University, Kyoto, Japan

H. Yoshioka
Shimane University, Matsue, Japan

C. Yoshimura et al. (eds.), *Water and Life in Tonle Sap Lake*,
https://doi.org/10.1007/978-981-16-6632-2_15

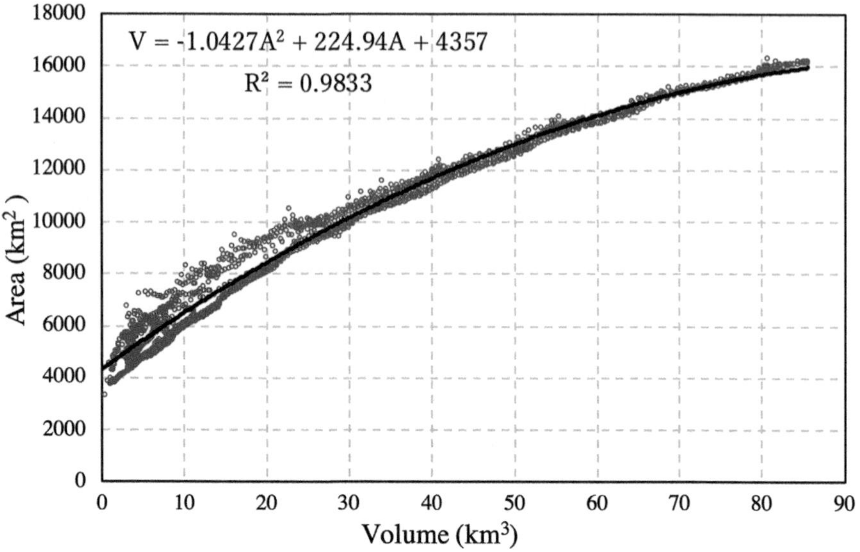

Fig. 15.1 The mean daily relationship between the lake volume *V* (km^3) and the inundation area *A* (km^2) simulated using the two-dimensional model LIE (gray dots). The black line indicates the quadratic function fitted (see the function and coefficient of determination in the figure)

that during the dry season, and it accounts for approximately 8% of the total area of Cambodia, which is 181,035 km^2.

The significant relationship between lake volume and water level at the water stage at Kampong Luong is presented in Fig. 15.2. From the figure, the water depth varies from 1.5 m to around 6 m for the first 30 km^3 of lake volume. After this point, the volume tends to have a gradual increase when the depth at Kampong Luong station slightly rises. Consequently, the area of the lake substantially expands to respond to a shallow depth, large volume.

15.2 Monthly Water Balance

Water balance is one of the main characteristics of lakes. In TSL, the major components of the water balance include water from various sources (e.g., inflow and outflow via rivers, precipitation and evaporation; see also Chaps. 1, 6, and 8). The possible methods described in the previous studies including the analysis of the water balance equation using digital bathymetric model (Kummu et al. 2014) and the application of the Google Earth engine (GEE) and WEAP model (Mab and Kositsakulchai 2020) were conducted. In addition to the annual water balance from a previous study (Kummu et al. 2014), the monthly water balance of TSL derived from the output of LIE is summarized in Table 15.1. For the period between May and September, the positive balance of TSL occurs during the rainy season,

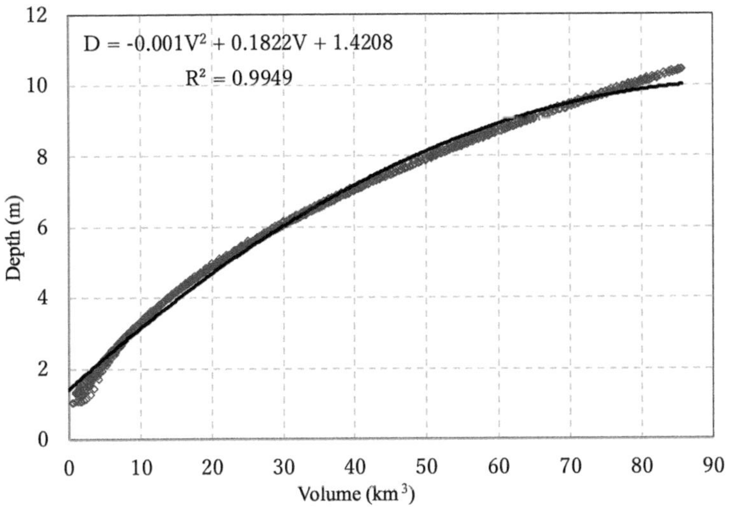

Fig. 15.2 Variation of the lake volume V (km^3) and water depth at Kampong Luong D (m) simulated using the two-dimensional model LIE (gray dots). The black line indicates the quadratic function fitted (see the function and coefficient of determination in the figure)

Table 15.1 The monthly water balance of Tonle Sap Lake output from the LIE model

	IN (km^3)				OUT (km^3)			
Month	Inflow	Prep	TSR IN	Total IN	TSR OUT	Eva	Total OUT	Balance
Jan	1.5	0.1	0.0	1.6	−10.8	1.1	11.9	−10.3
Feb	0.9	0.1	0.0	1.0	−6.0	1.6	7.6	−6.6
Mar	0.7	1.3	0.0	2.0	−3.9	2.1	5.9	−4.0
Apr	1.8	1.7	0.0	3.5	−2.2	2.6	4.8	−1.3
May	2.9	3.2	0.8	6.8	−1.7	2.3	4.0	2.9
Jun	3.7	3.2	6.6	13.5	−0.2	2.7	3.0	10.5
Jul	4.6	3.8	14.4	22.8	−0.5	2.3	2.8	20.0
Aug	5.9	4.0	13.9	23.8	−1.3	1.7	3.0	20.8
Sep	7.2	4.5	7.2	18.8	−2.4	1.4	3.7	15.1
Oct	8.4	5.0	0.0	13.4	−24.2	1.9	26.1	−12.7
Nov	5.8	1.6	0.0	7.5	−24.9	1.8	26.7	−19.2
Dec	3.0	0.2	0.0	3.2	−18.5	1.1	19.6	−16.4
Annual	46.4	28.6	42.9	117.9	−96.5	22.6	119.1	−1.2

whereas the negative balance takes place from October until April. During the rainy season, the lake and its sub-catchments receive a significant amount of precipitation along with the water inflow from the MR via the TSR into the lake. During this period, the lake functions as the water storage to reserve the inflow. Table 15.1 indicates that the inflow via the TSR contributes more or less 50% of the total inflow into the lake. The interaction flow between TSL and the MR remains the most influential during the dry season when the reverse flow occurs. A large portion of water, which accounts for 80–90%, discharges from the lake between October and April.

15.3 Distribution of Depth and Velocity

The seasonal change in the depth in TSL is drastic because of the inflow and outflow fluctuations, leading to the extended area and volume. The output of monthly depth, by contrast, can be illustrated as a spatial variation over the lake surface. Selection of the seasonal change of lake depth represented the four crucial periods. The water depth during July, October, February, and May represented the situation during the early rainy season, peak rainy season, mid-dry season, and the driest season, respectively. The spatial distribution of lake depth in the seasons is illustrated in Fig. 15.3.

During the rainy season, the lake system receives the inflow from its tributaries, floodplain, MR, and direct precipitation on the lake. Figure 15.3a indicates that the

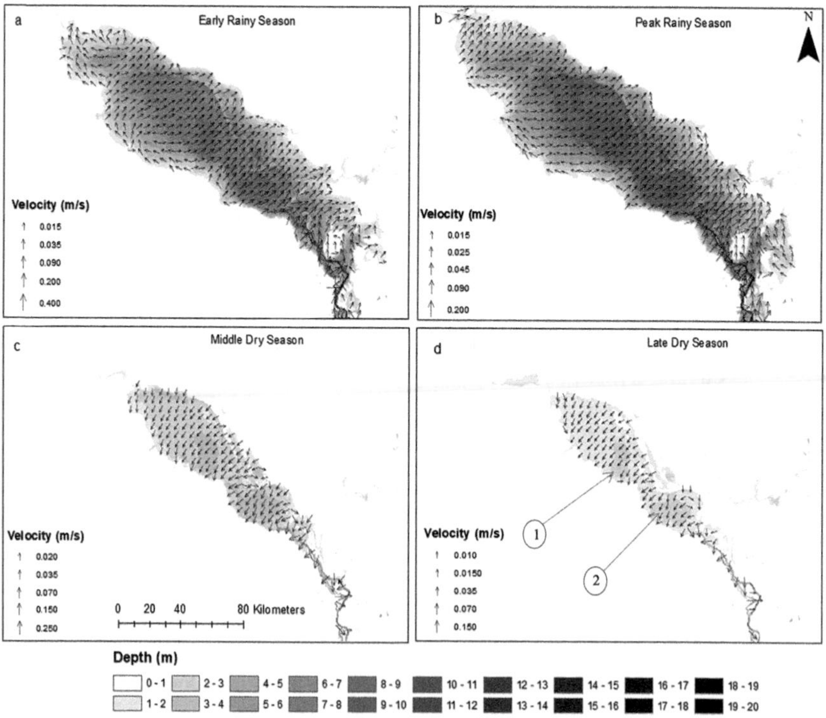

Fig. 15.3 The seasonal spatial depth and velocity of TSL from the 2D-LIE model (**a**) the depth during the early rainy season, (**b**) the depth during peak rainy season, (**c**) the depth during the dry season, and (**d**) the depth during the driest season

lake water stage ranges from 3 m around the boundary and considerably increases to approximately 8 m in the middle part of the lake.

Based on Fig. 15.3b, the average depth of the lake body is around 9 m, and the inundated floodplain enlarged in this peak period. The maximum water depth of the lake reaches 11 m before it starts to decrease. The backwater flow occurs during late November when the water outflows from the lake into the MR through the TSR. In the dry season, the lake flow discharges to the TSR, and the depth decreases to approximately 6–7 m.

Consequently, as illustrated in Fig. 15.3c, the overall water depth in February is around 2–3 m. The water depth is the lowest in May, the driest period when the lake depth around the center drops to around 1.0 m or less (Fig. 15.3d). However, the two locations (around Kaoh Kaev village in Pursat, location 1, and Chhnok Tru in Kampong Chhnang, location 2), the water remains 2–3 m in depth (Fig. 15.3d).

Since the flow direction of TSL seasonally changes and interacts with the MR, Fig. 15.3 also illustrates the direction of velocity. For visibility, the velocity was displayed at every 5000 m by 5000 m grid. Figure 15.3a indicates that the water flows from tributaries into the lake at a velocity between 0.015 m/s to 0.4 m/s with the direction toward the Northeast. By contrast, the direction of velocity indicates the various directions during the dry season, for instance, in May (Fig. 15.2d). The magnitude of velocity significantly decreases to a value between 0.01 m/s and 0.15 m/s. See Image 15.1 for a sample animation of the flow fields from 2000 to 2003.

15.4 Hydraulic Retention Time

The hydraulic retention time of a lake is one of the most useful pieces of information about the lake's hydrodynamics. The assumption of average lake retention time (T_0) is determined by the ratio of average lake volume (km^3) to the average inflow into the lake (km^3/year). As described in the previous section, TSL is considered to be a highly dynamic lake because the lake's volume during rainy season is about 20 times larger than dry season. Indeed, the seasonal retention time of the lake should be valuable for full understanding of the circumstance of the lake, during both the dry season and the rainy season. The retention time was calculated quarterly at the end of March, June, September, and December. Consequently, we obtained the distribution of retention time of the lake over the year. As presented in Table 15.1, the retention times of the first quarter (Q_1) were calculated for the duration from the beginning of January to the end of March. The approach of opting for the period was based on the inflow and reverse flow from the lake (Shivakoti et al. 2020). The discharge of the lake to the MR starts at some point in early October and continues until late May or early June when the rainy season reoccurs.

Table 15.2 The seasonal variation of the retention time of Tonle Sap Lake

Year	Retention time (days)				Average
	Q_1[a]	Q_2	Q_3	Q_4	
1999	35.77	82.85	214.98	86.50	105.12
2000	105.12	138.33	241.63	93.44	144.54
2001	84.31	87.60	221.55	95.99	122.27
2002	221.92	88.70	217.54	118.62	161.69
2003	81.76	55.11	188.70	80.30	101.47
Average	105.85	90.52	216.81	94.90	127.02

[a]Q_1 is the period between January and March; Q_2, from April to June; Q_3, from July to September; and Q_4, from October to December

The average retention time of TSL over 5 years is around 127 days (~4 months and 7 days). For the first quarter of each year (Q_1), it was considered as the period of water flowing back to the MR. In Q_2, the retention time of the lake is the shortest compared to those in other periods of the year. This happens because the volume of the lake shrinks to its minimum, while inflow from its tributaries remains. From early July to the end of September, the peak season (Q_3), the retention time is much longer before it starts to drop during the fourth quarter Q_4 (it becomes shorter). In Q_3, the longer retention time is probably due to the higher water surface level of the MR that blocks the water from flowing out. Although the retention time from Table 15.2 was derived from the result of the 2D-LIE model simulation, it revealed a strange transition between Q_4 of 2001 and Q_1 of 2002. In the actual condition, the retention time of Q_1 should be shorter than that of Q_4 because Q_1 is the period of the dry season during which the water is discharged from the lake to MR. This anomaly is due to less inflow into the lake during the first quarter of 2002. Although the retention time is the ratio of lake volume and total inflow, the inflow during this period of 2002 significantly dropped compared to the same period of other years of study. Moreover, the discharge flowing into the MR of that period remained its trend. As we can see, the retention time of TSL significantly changes based on the season and flow interaction between the lake and MR.

Furthermore, the average retention time of TSL is compared with that of other lakes in Fig. 15.4. Compared with other lakes, the retention time of TSL is considerably short. Although a lake with larger volume and depth tends to have a longer retention time than a smaller lake (Kvarnäs 2001), this assumption could not apply for the case of TSL. The position of TSL's retention time locates almost at the bottom part of the scatter graph (see Fig. 15.4), which means a short period compared to other lakes that have smaller area but have larger hydraulic retention time, whereas most of the lakes scatter at the right side of the graph. The shorter retention time of TSL is probably affected by larger inflow during the rainy season and outflow during the dry season and the resulting dramatic seasonal variation of the volume itself, featuring the uniqueness of TSL.

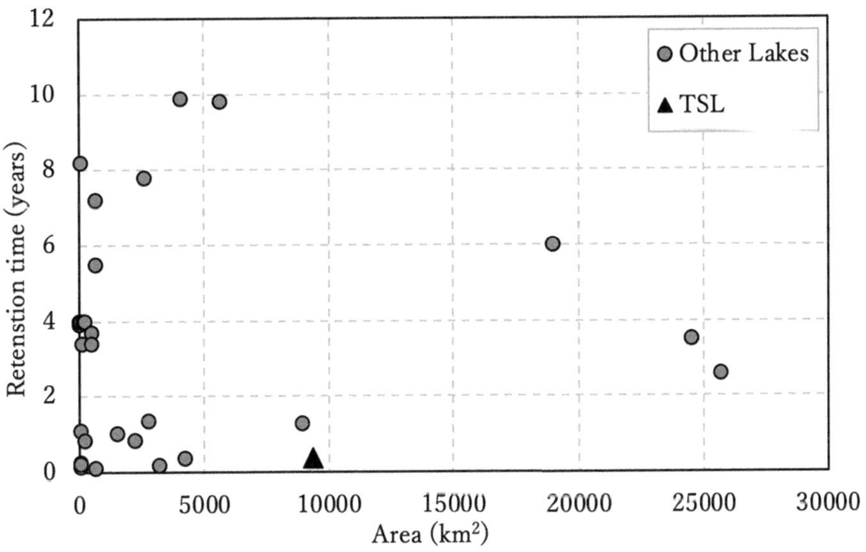

Fig. 15.4 The average retention time of TSL compared to that of some other lakes around the world

Key Points

- In TSL, the minimum inundation area and water volume appear to be approximately 2200 km^2 and 2.5 km^3, respectively, and the average depth is less than 2.0 meters. By contrast, the area and volume reach the peaks in October, and they are approximately 15,000 km^2 and 85 km^3, respectively.
- The positive water balance of TSL is between May and September, and the negative value starts from October to April.
- The average hydraulic retention time of TSL is 4 months and 7 days.
- The retention time of TSL is considerably shorter compared to that of many other shallow lakes over the world because of larger inflow during the rainy season and the outflow during the dry season and the drastic seasonal variation of the water volume.

References

Kummu M, Tes S, Yin S, Adamson P, Józsa J, Koponen J, Sarkkula J. Water balance analysis for the Tonle Sap Lake-floodplain system. Hydrol Process. 2014;28(4):1722–33.

Kvarnäs H. Morphometry and hydrology of the four large lakes of Sweden. Ambio. 2001;30(8): 467–74.

Mab P, Kositsakulchai E. Water balance analysis of tonle sap lake using weap model and satellite-derived data from google earth engine. Science and Technology Asia. 2020;25(4):45–58.

Shivakoti BR, Pham NB, Seingheng H, Yoshimura C, et al. Environmental changes in Tonle Sap Lake and its floodplain: status and policy recommendations. Japan: Institute for Global Environmental Strategies (IGES), Tokyo Institute of Technology (Toyko Tech), and Institute of Technology of Cambodia (ITC); 2020.

Chapter 16
Flow Regime of a Floating Village Using a Three-Dimensional Hydraulic Model

Takashi Nakamura, Hideto Fujii, Ly Sarann, Lun Sambo, Heng Sokchhay, Yoichi Fujihara, and Keisuke Hoshikawa

16.1 Importance of the Flow Regime in Floating Villages

Tonle Sap Lake (TSL) provides not only the largest fishery in Cambodia but also places of residence (see also Chaps. 1–4 and 31–34 for the detailed benefits of the TSL ecosystem). More than one million people live on the floodplain that is routinely flooded in the wet season (Keskinen 2006). To cope with the seasonal fluctuation of the water level in the fishery, many people gain substantial benefits from living in floating houses. The people in floating houses form communities, with approximately 90 villages in TSL (Johnstone et al. 2013, see also Chaps. 2 and 40). Figure 16.1 illustrates an example of a floating village at Chhnok Tru. This floating village is one of the largest villages in TSL and is located in the area connecting TSL and Tonle Sap River (TSR).

Because such floating villages do not have water and sanitation infrastructure, waste and excreta are directly dumped into lake water underneath the house. For protecting the resident from health damage caused by the polluted water, a hydraulic flow simulation must be implemented. The simulated flow enables us to assess many processes of the pollution: tracing transport of the pollution due to the flow,

T. Nakamura (✉)
Tokyo Institute of Technology, Tokyo, Japan
e-mail: tnakamur@depe.titech.ac.jp

H. Fujii
Yamagata University, Tsuruoka, Japan

L. Sarann · L. Sambo · H. Sokchhay
Institute of Technology of Cambodia, Phnom Penh, Cambodia

Y. Fujihara
Ishikawa Prefectural University, Nonoichi, Japan

K. Hoshikawa
Toyama Prefectural University, Imizu, Japan

C. Yoshimura et al. (eds.), *Water and Life in Tonle Sap Lake*,
https://doi.org/10.1007/978-981-16-6632-2_16

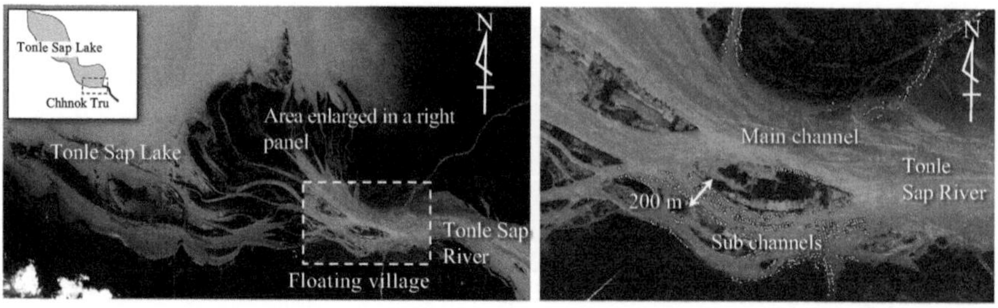

Fig. 16.1 Location of Chhnok Tru floating village. In the right panel, white dots represent the location of floating houses found in a satellite image

evaluating the period required for dilution, grasping a speed of sedimentation of the waste, predicting erosion of toxic substances using flows on the bed, etc. Because the floating village tends to be located where topography changes in a small area such as a small meandering channel and river mouth of tributaries, the flow in the village has a complex profile that cannot be expressed as a depth-averaged value in the two-dimensional (2D) model. Therefore, especially for the floating villages, a three-dimensional (3D) hydraulic model seems to realize the most powerful simulation. To demonstrate how we conduct the 3D simulation and what we can expect as the results, an actual simulation of the Chhnok Tru floating village is described in this chapter.

16.2 Three-Dimensional Hydraulic Model

Although many 3D hydraulic models have been individually developed with various assumptions and diverse numerical methods, so-called non-hydrostatic models have an advantage to solve a complex water flow in the village for the following reason. Because the proper pressure gradients can be solved independently of the water-surface slope, the non-hydraulic model is expected to achieve a more precise simulation of the flow often found in a water area adjacent to the village: a secondary (spiral) flow in a meandering channel, a flow separation in a confluence zone, a vertical circulation flow in the strong wind, etc. One of the advantages of the 3D model is that vertical transport of suspended/dissolved substances can be assessed with the calculated flow field. Because vertical transport is controlled both by the strength of turbulence and the existence of thermal stratification, the 3D model is requested to have a turbulence model (e.g., so-called k-ε/k-ω RANS models) and consider a vicissitude of the thermal stratification according to meteorological conditions.

Tokyo Institute of Technology–WAter Reservoir Model (TITech-WARM) is one of the 3D non-hydrostatic hydraulic models that satisfy the afore-mentioned requests. After TITech-WARM was proposed as a numerical hydraulic model to solve a 3D water flow in a river channel, it has been improved and applied to various types of water flow: a saltwater intrusion into a river channel (Xu et al. 2012), transport of dissolved oxygen and assessment of hypoxia (Xu et al. 2013), wind-driven current and generation of thermal stratification in a lake (Nakamura et al. 2013; Nakamura and Iwata 2016), sediment transport in a water reservoir (Takahira et al. 2013), and so on.

By solving the incompressible Navier–Stokes equations supplemented with k-ε turbulence model, TITech-WARM estimates temporal and spatial changes of water flow velocity $\mathbf{u} = (u, v, w)$, pressure p, turbulence kinetic energy k, turbulence energy dissipation rate ε, and water temperature T:

$$\frac{D\mathbf{u}}{Dt} = \frac{\partial}{\partial x}\left(K_H \frac{\partial \mathbf{u}}{\partial x}\right) + \frac{\partial}{\partial y}\left(K_H \frac{\partial \mathbf{u}}{\partial y}\right) + \frac{\partial}{\partial z}\left(K_V \frac{\partial \mathbf{u}}{\partial z}\right) - \frac{\Delta p}{\rho} + \frac{\mathbf{F}_E}{\rho}, \tag{16.1}$$

$$\frac{Dk}{Dt} = \frac{\partial}{\partial x}\left(\frac{K_H}{\sigma_k}\frac{\partial k}{\partial x}\right) + \frac{\partial}{\partial y}\left(\frac{K_H}{\sigma_k}\frac{\partial k}{\partial y}\right) + \frac{\partial}{\partial z}\left(\frac{K_V}{\sigma_k}\frac{\partial k}{\partial z}\right) + P_k - G_k - \varepsilon, \tag{16.2}$$

$$\begin{aligned}\frac{D\varepsilon}{Dt} &= \frac{\partial}{\partial x}\left(\frac{K_H}{\sigma_\varepsilon}\frac{\partial \varepsilon}{\partial x}\right) + \frac{\partial}{\partial y}\left(\frac{K_H}{\sigma_\varepsilon}\frac{\partial \varepsilon}{\partial y}\right) + \frac{\partial}{\partial z}\left(\frac{K_V}{\sigma_\varepsilon}\frac{\partial \varepsilon}{\partial z}\right)\\ &\quad + \left(C_1(P_k - (1 - C_3)G_k) - C_2\varepsilon\right)\frac{\varepsilon}{k},\end{aligned} \tag{16.3}$$

$$\frac{DT}{Dt} = \frac{\partial}{\partial x}\left(\frac{K_H}{\sigma_T}\frac{\partial T}{\partial x}\right) + \frac{\partial}{\partial y}\left(\frac{K_H}{\sigma_T}\frac{\partial T}{\partial y}\right) + \frac{\partial}{\partial z}\left(\frac{K_H}{\sigma_T}\frac{\partial T}{\partial z}\right) + \phi_{Heat}, \tag{16.4}$$

$$\frac{\partial u}{\partial x} + \frac{\partial v}{\partial y} + \frac{\partial w}{\partial z} = 0, \tag{16.5}$$

$$\begin{aligned}P_k &= K_V\left(\left(\frac{\partial u}{\partial z}\right)^2 + \left(\frac{\partial v}{\partial z}\right)^2\right), G_k = -\frac{g}{\rho}\frac{K_V}{\sigma_t}\frac{\partial \rho}{\partial z}, K_H = 0.01D^{4/3}, K_V\\ &= K_{mol} + C_\mu \frac{k^2}{\varepsilon},\end{aligned} \tag{16.6}$$

where x and y are horizontal coordinates; z is a vertical coordinate; and u, v, and w are flow velocity components in the x, y, and z directions, respectively. K_H and K_V are kinematic viscosity in the horizontal and vertical directions, respectively. K_H is determined according to Richardson's 4/3 law with a characteristic length D (horizontal grid size), whereas K_V is estimated as a combination of molecular viscosity K_{mol} and turbulence viscosity. The k-ε turbulence model is used, and the turbulence viscosity is evaluated from turbulence kinetic energy k and dissipation rate ε. σ_k, σ_ε, σ_t, σ_T, C_1, C_2, C_3, and C_μ are parameters in the k-ε turbulence model, and standard values can be employed for them (Nakamura 2020). ρ is the density of water and is estimated from water temperature according to the equation of state of water. In the solution of water temperature T, to consider a development/disappearance of the thermal stratification, influence by the atmosphere is taken into account by adding a source term φ_{Heat} into Eq. (16.4). φ_{Heat} is a heat flux between the atmosphere and water body and is modeled as a combination of "incoming shortwave radiation φ_s," "outcoming longwave radiation φ_l," and "latent and sensible heat transfer (φ_c+φ_e)" with meteorological observation data,

$$\varphi_{Heat} = \beta(1 - \alpha)\, \varphi_s - \varphi_l - (\varphi_c + \varphi_e), \tag{16.7}$$

where α and β are the reflection ratio (albedo) and the absorption ratio for the shortwave solar radiation at the water surface, respectively. Observative data are inputted for φ_s, whereas φ_l and (φ_c + φ_e) are estimated from observational data of air temperature with some empirical equations such as the Swinbank and Rowher's equation. $\mathbf{F}_E$ represents extra forces affecting the water. In addition to a bottom frictional force and gravity, wind's frictional force is modeled in $\mathbf{F}_E$. Together with a set of the above equations, the following conservative equation is solved to trace a temporal/spatial change of the water level $h(t, x, y)$.

$$\frac{\partial h}{\partial t} + \frac{\partial}{\partial x}\left(\int_b^h u dz\right) + \frac{\partial}{\partial y}\left(\int_b^h v dz\right) = 0, \tag{16.8}$$

where $b(x, y)$ is a river bed level. Equations (16.1)–(16.8) are solved in a similar algorithm with a Simplified MAC scheme (Nakamura 2020).

TITech-WARM employs a unique computational grid system, named Soroban (the Japanese word for abacus). The gridpoint can be moved vertically, and the vertical positions of the grid points can be rearranged like a bead moving up/down in an abacus. During the computation, the grid system automatically gathers the grid points into space where the vertical change of water temperature or other substances is large. Thus, TITech-WARM tends to provide a more accurate simulation in comparison with hydraulic models with a conventional grid system.

Generally, 3D hydraulic simulation is a time-consuming task. In some cases, it takes a day to simulate a change of flow in a day. To accelerate the calculation and shorten the time of simulation, TITech-WARM employs parallel computing with a domain decomposition technique. First, the original water area is split into some smaller areas called "sub-domains." Then, by calculating the sub-domains using some central processing units (CPUs) in parallel, the time of simulation is ideally shortened in an inverse proportion to the number of CPUs. In many cases, by using 16 CPUs, the time of simulation could be shortened to one-tenth of a non-parallel case.

16.3 Preparation of Computational Conditions

On a wide floodplain of TSL, many small water channels form a complex network with anabranches. From the satellite images, it could be easily found that almost all the floating villages are located either in the channel or a transition area between the channel and TSL. In general, in those locations, flow speed in the village is mainly controlled by the magnitude of the water current flowing from the outside through the channel. Furthermore, in the village, spatial fluctuations of the channel's topography generate a 3D complex flow locally changing the flow's speed and direction. Thus, to apply a hydraulic model to the floating villages, we have to make a certain effort to prepare carefully both the bathymetry and discharge conditions for input to the model.

16.3.1 Bathymetry

A high-resolution bathymetric map should be made to restore the small water channels in the simulation. In the case of Chhnok Tru, the map was prepared as a Digital Elevation Model (DEM) having a mesh size of 10 m, because a main part of the village locates in a sub-channel with a 200 m width (see Fig. 16.1). Figure 16.2 illustrates the process of making DEM. First, a satellite image analysis was conducted with Sentinel 2 Multispectral Instrument (MSI) data. Based on the Normalized Digital Water Index (NDWI), the location of a water surface in TSL and the water channel networks were detected:

$$\mathrm{NDWI} = \frac{\rho_{\mathrm{Green}} - \rho_{\mathrm{NIR}}}{\rho_{\mathrm{Green}} + \rho_{\mathrm{NIR}}}, \tag{16.9}$$

where ρ_{Green} and ρ_{NIR} are reflectance for the green and near-infrared (NIR) bands, respectively. Mekong River Commission had made a comprehensive bathymetric

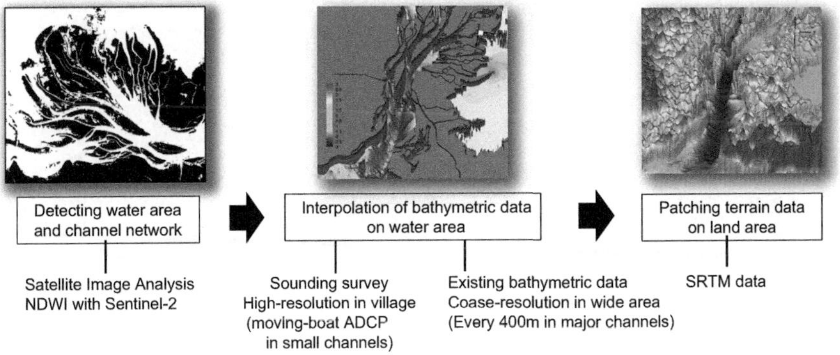

Fig. 16.2 Preparation of Digital Elevation Model (DEM) in Chhnok Tru floating village

map for TSR (Mekong River Commission (MRC) and Ministry of Public Work and Transport of Cambodia 1999). However, its spatial resolution (every 400 m in the longitudinal direction of the channel) is not enough to represent the fine topographical change in the village, and the depth data are provided only for the main channel, not for the small channels where the floating houses are. Therefore, to complement the lack of data, a sounding survey was conducted for the small channels in March 2018. In the survey, the Acoustic Doppler Current Profiler (ADCP) installed on a boat was used to measure the depth (Chap. 12). By driving the boat in a zigzag movement, the spatially dense data were observed over the whole area of the village. On the meshes categorized as water area with the NDWI, the elevation of the bed surface is interpolated using both the bathymetric map and the sounding survey. After the interpolation, the elevation of the land surface was patched from "Shuttle Radar Topography Mission 3—arc second global digital elevation model (SRTM3)" (Jarvis et al. 2008), and the final product of the DEM was generated for the area.

16.3.2 Discharge at the Boundary

Small channels form a dendriform network upstream of the Chhnok Tru, and water flows into the village from a plurality of channels. As a boundary condition of the 3D simulation, we need to know the discharge flowing from the individual channel. To estimate the discharge, we applied a 2D hydraulic model into a wide area (20 km × 9 km) containing both the Chhnok Tru floating village and the upstream dendriform channel network (Fig. 16.3). In the 2D simulation, a fine mesh size of 10 m was used to represent the small water channels. To overcome a huge computational effort due to the fine mesh, we applied a newly developed 2D model that can solve the shallow water equation and accelerate the simulation to 100 times faster than a conventional model by using General Purpose computing on Graphics

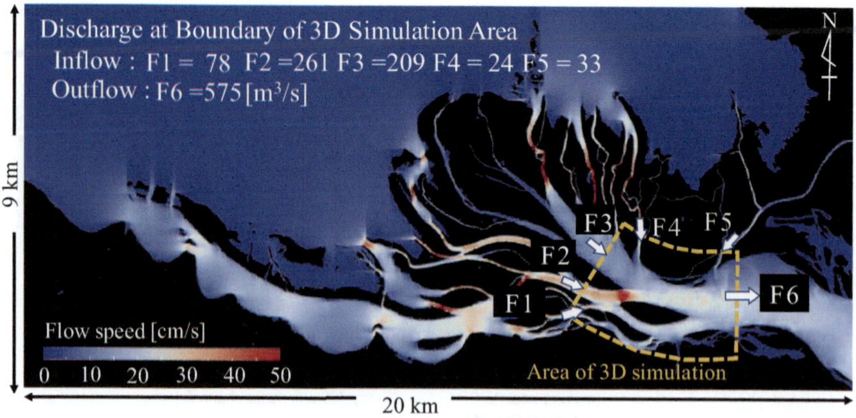

Fig. 16.3 Calculated result of 2D hydraulic simulation

Processing Units technique (Nakamura et al. 2019). The simulation was conducted for 1 week in March 2018. The observed water level at Kompong Luong in TSL was used as a boundary condition of the upstream end. After confirming that the calculated result agrees well with the discharge observed in some channels with ADCP, the discharge flowing from each channel into the village was estimated by integrating the flow velocity at the boundary of the area of 3D simulation (shown using white arrows in Fig. 16.3).

16.4 3D Flow Simulation of Chhnok Tru Area

With the prepared bathymetric and discharge data, a 3D flow simulation was conducted with TITech-WARM. A computational domain was set to a 3.5 km section of TSR, which includes both the main channel and sub-channels. Horizontal mesh size was set to 20 m, and 20 grid points were used in the vertical direction independently of water depth. The computation was conducted for 1 week, which is the same period calculated by 2D hydraulic simulation. Until the end of the simulation, 25 h were spent on parallel computation using four PCs (Intel core i7-3930K CPU). Discharge and water level at the boundary were fixed to the values calculated by 2D simulation. As meteorological input data (wind speed/direction, air temperature, humidity, cloudiness, and solar radiation intensity), monthly averaged hourly data were used. Although locations of all houses (1362 houses) in the village were read from a satellite image on Google Earth (captured on 2017/1/1), influences of the houses on a water flow were not modeled in the simulation.

Figure 16.4 illustrates a calculated result of longitudinal flow velocity at the water surface. Black dots show locations of a floating house, and black figures and arrows represent discharge and flow direction estimated from the 3D simulation, respectively. As illustrated in the figure, spatial change of depth and fork/joint of channels vary flow speed drastically from place to place. Rapid flow is found, and its speed reaches 45 cm/s in the outside of the village, whereas the flow moderates in the sub-channel where most of the floating houses are located, and its speed is 30 cm/s at most. In the downstream area in the village, there is a dead water region due to a dead-end canal. Because of easy access to the land through the canal, many floating houses gather in the dead water region.

Figure 16.5 illustrates the flow velocity profile at three different cross sections. The location of each cross section is depicted by a red dashed line in Fig. 16.4. In each panel, a color contour represents a normal (longitudinal) component of flow, whereas black vectors show the flow component parallel to a cross section. It is found that the flow has a spatial profile that can be calculated from a 3D solution: attenuation of flow speed near the bottom due to friction and generation of circular flow due to the strong wind blowing (clearly shown in the C–C′ cross section). It must be pointed out that those 3D flow profiles (in other words, vertical change of flow) cannot be calculated using a 2D hydraulic model; thus, the 3D flow profiles are crucial to assess environmental problems such as wastewater pollution. Pathogen

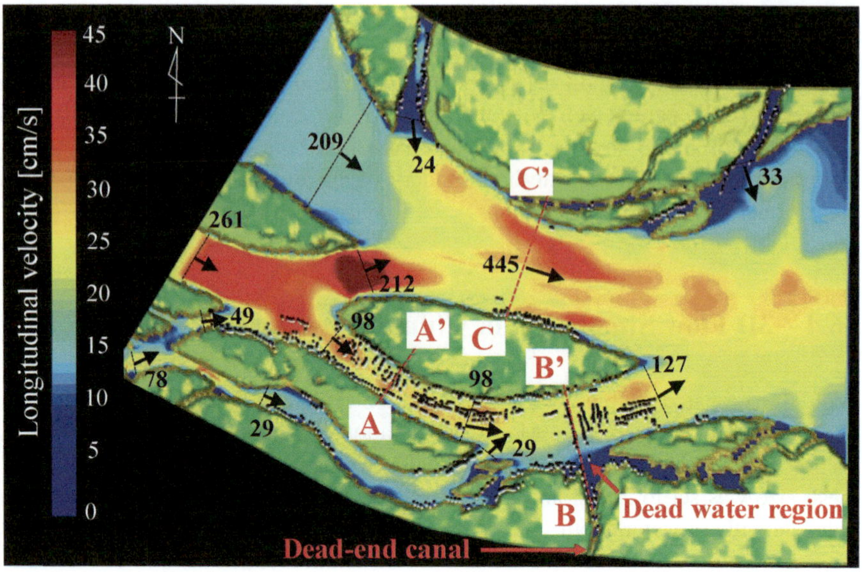

Fig. 16.4 3D simulation result of longitudinal flow velocity at the water surface in March 2018. Black numbers represent volumetric flow rates [m^3/s] at each cross section

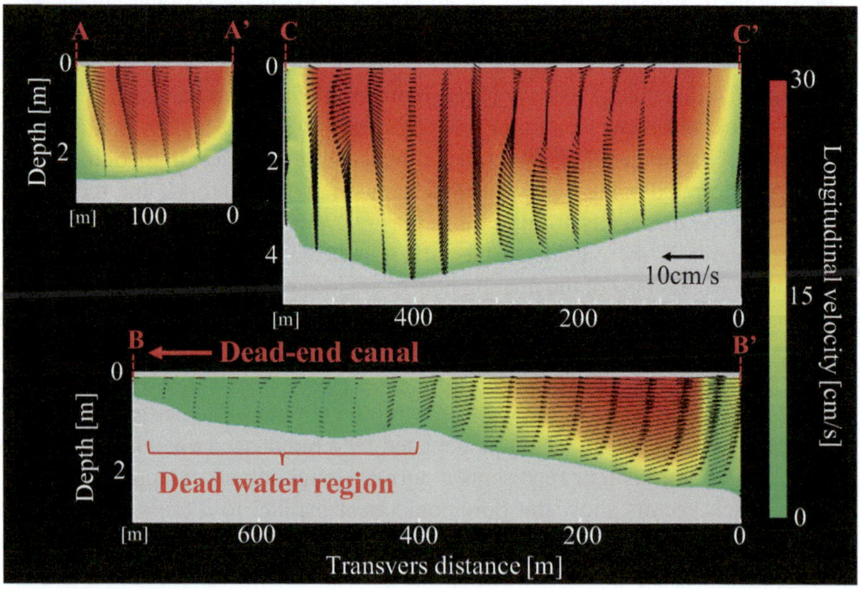

Fig. 16.5 Flow velocity in the three cross sections in March 2018

contamination in the lake water should be a cause to make diarrhea the major disease in the floating village. To assess the risk of infection, the transport of pathogens should not be considered with the depth-averaged flow but with the flow at the water surface, because people can be infected with pathogens by taking/drinking the lake water from the water surface. In Chap. 41, we introduce an application of 3D flow simulation to the risk assessment of pathogen infection. Not only the pathogen but also the 3D flow profiles should also be useful to assess soil sedimentation or heavy metal accumulation on the bed surface because the erosion/resuspension rate does not depend on the speed of depth-averaged flow but on the flow speed on the bed surface.

Key Points

- Tokyo Institute of Technology–WAter Reservoir Model (TITech-WARM) was applied to a floating village, Chhnok Tru. TITech-WARM is a 3D non-hydrostatic hydraulic model with a turbulence model (e.g., so-called k-ε/k-ω RANS models) that considers thermal stratification.
- 3D flow simulation of Chhnok Tru clearly showed drastically slower flow velocity in a dead water region, where many floating houses gather due to easy access to the land through the dead-end canal.
- Simulating a three-dimensional profile of the flow must be helpful to assess environmental problems to which the vertical profile of flow velocity is important, e.g., the risk assessment of pathogen infection, soil sedimentation, and/or heavy metal accumulation on the bed surface.

References

Jarvis A, Reuter HI, et al. Hole-filled SRTM for the globe version 4. available from the CGIAR-CSI SRTM 90 m database. 2008. http://srtm.csi.cgiar.org. Accessed Jun 2017.

Johnstone G, Puskur R, et al. Tonle Sap scoping report. CGIAR Research Program on Aquatic Agricultural Systems, Penang, Malaysia 2013. Project Report: AAS-2013-28; 2013.

Keskinen M. The lake with floating villages: socioeconomic analysis of the Tonle Sap Lake. Water Resour Dev. 2006;22:463–80.

Mekong River Commission (MRC) and Ministry of Public Work and Transport of Cambodia. Hydrographic atlas Mekong River in Cambodia, Mekong River Commission and Ministry of Public Work and Transport of Cambodia, Tech. Rep. 1999.

Nakamura T. User Manual of TITech-WARM. 2020. http://nakalab.depe.titech.ac.jp/Lab/?page_id=671. Accessed 22 May 2020.

Nakamura T, Iwata K. Three-dimensional numerical flow analysis of saline intrusion in the northeastern shallow area of Lake Ogawara. J Jpn Society Civil Eng Ser. 2016;B1 72: I_697–702.

Nakamura T, Toda M, et al. Development of a new 3D numerical water reservoir flow model "TITech-WARM" and application to the Kamafusa lake. J Jpn Society Civil Eng Ser. 2013;B1 69:I_775–80.

Nakamura T, Murakami S, et al. Efficacy of a GPGPU-acceleration to inundation flow simulation in Tonle Sap Lake in Cambodia. Eng J. 2019;23:151–69.

Takahira K, Wang M et al. Numerical study on Seasonal Stratified Flow and Sediment Transport in Joumine Reservoir Tunisia, PROCEEDINGS OF THE 35TH IAHR WORLD CONGRESS, VOLS III AND IV, Chengdu, China, Paper Num. W-1322; 2013.

Xu X, Nakamura T, et al. Modeling of saline water movement in tone river estuary using three-dimensional CIP-SOROBAN model. Thai Environ Eng J. 2012:111–5.

Xu X, Ishikawa T, et al. Three-dimensional modeling of hydrodynamics and dissolved oxygen transport in Tone River Estuary. J JSCE. 2013;1:194–213.

Part IV
Sediment Dynamics

Chapter 17
Sediment and Suspended Solids: Spatiotemporal Dynamics

Sokly Siev, Rina Heu, Heejun Yang, Ty Sok, Uk Sovannara, Rajendra Khanal, Chantha Oeurng, Hul Seingheng, and Chihiro Yoshimura

17.1 Sediment Balance and Geomorphology

Sediment is important to sustain the geomorphology of Tonle Sap Lake (TSL) and its floodplain by providing essential nutrients to sustain the productive ecosystem (Kummu et al. 2008). The TSL and its floodplains connecting to the Mekong mainstream through the Tonle Sap River (TSR; see also Chap. 1 for details) are dependent mainly on the Mekong sediment regimes because the sediment transport from the Mekong River Basin is the main source of sediment for TSL (Kummu et al. 2008), while the surrounding tributaries of TSL also contribute substantial sediment (see also Chap. 8 for the hydrology of the inflow river basins). The net sediment accumulation in TSL in the last ca. 5500 years is only 0.5–0.7 m (Kummu et al. 2008; Penny et al. 2005), which was approximately eight to ten times smaller than that in Bosten Lake (5.9 m) in China in the last ca. 5400 years (Wünnemann et al. 2006) and approximately five times smaller than that in Galapagos Lake (2.95 m) in

S. Siev (✉) · H. Seingheng
Institute of Technology of Cambodia, Phnom Penh, Cambodia

Ministry of Industry, Science, Technology and Innovation, Phnom Penh, Cambodia
e-mail: siev.sokly@misti.gov.kh

R. Heu · T. Sok · C. Oeurng
Institute of Technology of Cambodia, Phnom Penh, Cambodia

H. Yang
Kyoto University, Kyoto, Japan

U. Sovannara · C. Yoshimura
Tokyo Institute of Technology, Tokyo, Japan

R. Khanal
Tokyo Institute of Technology, Tokyo, Japan

Policy Research Institute, Kathmandu, Nepal

C. Yoshimura et al. (eds.), *Water and Life in Tonle Sap Lake*,
https://doi.org/10.1007/978-981-16-6632-2_17

the west of South of America in the last ca. 6200 years (Conroy et al. 2008). The rate of sediment accumulation may vary owing to the lake water level, climate condition, and other variable depositional conditions.

Presently, the sedimentation rate in TSL is still very slow at around 0.1–0.16 mm/year in the dry season (Tsukawaki 1997). In 1997–2003, the lake received 5.1 MT/year of sediment from the Mekong River and 2.0 MT/year from the tributaries of the lake and released only 1.4 MT/year of sediment back to the Mekong River on average (Kummu et al. 2008). This means that around 80% of the sediment derived from the Mekong River and the lake's tributaries remains stored in the lake and its floodplain. The annual variation of sediment flux into the lake is significant, for example, 3.5 MT in 1998 and over 9 MT in 2000 (Kummu et al. 2008). This variation is controlled by flood pulse (e.g., seasonal water level and water volume) influenced by Mekong River. The sediment flux to the lake becomes largest in August, when approximately 2.5 MT of suspended solids are transported from the Mekong River to the lake via TSR and is trapped mainly by riparian and floodplain environments (Kummu et al. 2008). The high suspended solid concentration during the dry season, when the lake is very shallow, is caused by active resuspension of the sediment (see Chap. 19).

Isotopes and elemental concentrations in lake sediment cores indicated the timing when the lake first began to receive water and sediment input via Mekong River, which were the period between 4450 and 3910 cal year BP (Day et al. 2011). Moreover, Mekong River captured the lake causing erosion of 10 km^3 of sediment in the lake-wide during the period of drying and declining sea level (postdating ~6200 cal years BP) (Darby et al. 2020). The capturing by Mekong and wave reworking erosion might have lowered or shifted the lake's bottom elevation (by at least 1.2 m) (Darby et al. 2020). This process also implies the vulnerability of the Chaktomuk connection (the confluence between Mekong River and TSR) to changes in hydro-climatic conditions, underscoring the importance of maintaining natural river flow dynamics at the Chaktomuk connection to support the TSL's unique and highly valuable ecosystem (Darby et al. 2020). The long-term landscape evolution was thus punctuated by a rapid river capture and erosion of the lake area, which established the flourishing ecosystem (Darby et al. 2020).

17.2 Sediment

Investigation of sediment dynamics is crucial for the environmental management in TSL and could provide insights to understanding of the flood-pulse-influenced shallow lakes elsewhere in the world (e.g., Poyang Lake and Dongting Lake in China and Salada Lake in Argentina). For this reason, we conducted an extensive survey to collect samples at 39 inundated points in TSL and its floodplain during September 2016, December 2016, and March 2017. The bottom sediment was collected from the surface of the lake's bottom (e.g., 1 cm) using Satake's core sampler (with the inner diameter of 5.5 cm and 50 cm length). The collected samples were analyzed to determine the water content, organic content, particle size distribution, and density. Sediment grain sizes were determined using the laser diffraction method with a Laser Diffraction Particle Size Analyzer SALD3000 (refer to Siev et al. 2018 for the details). The following paragraphs explain and describe the finding of sediment's characteristics and dynamics in terms of seasonality (e.g., every 3 months), locations (e.g., TSL and floodplain), and sediment inputs influenced by the flood pulse of Mekong River.

The results indicated that bottom sediments in TSL have different characteristics from those in the floodplain. The water content in TSL bottom sediments (range: 37.9–55.6%) was greater than that in the floodplain (11.1–54%) over the sampling period. The average wet density of the sediment in TSL (1.3 g/cm^3) was lower than that in the floodplain (1.65 g/cm^3). The average organic content (as determined by loss of ignition) of TSL and floodplain sediments were 6.5% and 9.5%, respectively (Siev et al. 2018), which are comparable to 8% in Chungpyung Lake's sediment in Korea and 6% in Jamsil submerged dam's sediment of the Han River in Korea (Kim et al. 2003). In TSL and its floodplain, the majority of sediment grain is fine silt and clay (>60% in mass). The average proportions of sand, silt, and clay were: 10.7%, 67.4%, and 21.8% in TSL for September; 5.6%, 69.2%, and 25% in TSL for December; 13.8%, 64.9%, and 21.2% in TSL for March; 1.7%, 64.5%, and 33.5% in the floodplain for September and December; and 4.7%, 78.7%, and 16.4% in the suspended solids for June, respectively. Median grain sizes (D50) of the sediment and suspended solids between TSL and the floodplain were not substantially different, ranging from 5 to 6 μm and 7 to 8 μm in TSL and the floodplain, respectively.

The sediment texture (i.e., clay, silt, and sand contents) implies the hydrodynamic and sediment dynamic conditions in TSL and its floodplain. Fine silt and clay are often suspended in the water column under turbulent conditions, and, consequently, sediment under harsh hydrodynamic conditions tends to be relatively coarser in the bottom sediment (e.g., sand). For this reason, the ranges (e.g., 80–100%, 50–80%, 20–50%, and 0–20%) of content clay in the deposited sediment are used and classified based on the level of hydrodynamic conditions, namely, calm/low (I), moderate (II), high (III), and very high (IV), respectively. By contrast, ranges of sand content (e.g., A for 90–100%, B for 50–90%, C for 10–50%, and D for 0–10%) are used and classified for rough textural deposited sediment in different hydrodynamic conditions (Fig. 17.1, Pejrup 1988).

The texture of the bottom/deposited and suspended sediment is reflective of the patterns and hydrodynamic conditions in TSL. Higher proportions of fine-grained sediment were observed in the floodplains that have a relatively calm hydrodynamic condition compared to that in TSL owing to inundated vegetation. According to the Pejrup diagram, the characteristic of the sediment in TSL was subject to the high (III) and very high (IV) hydrodynamic conditions, whereas that in the floodplain mostly was plotted into groups D–III, where sand contents were less than 10%, indicating calmer hydrodynamic conditions. Similarly, the sand contents in TSL in September and December are slightly elevated compared with those in the floodplain, except at several points (one point in September and four points in December, located near the shore of TSL) assigned to group C-III. In March, most of the textural composition of the sediment of TSL was classified as groups C-III and C-IV, where the sand

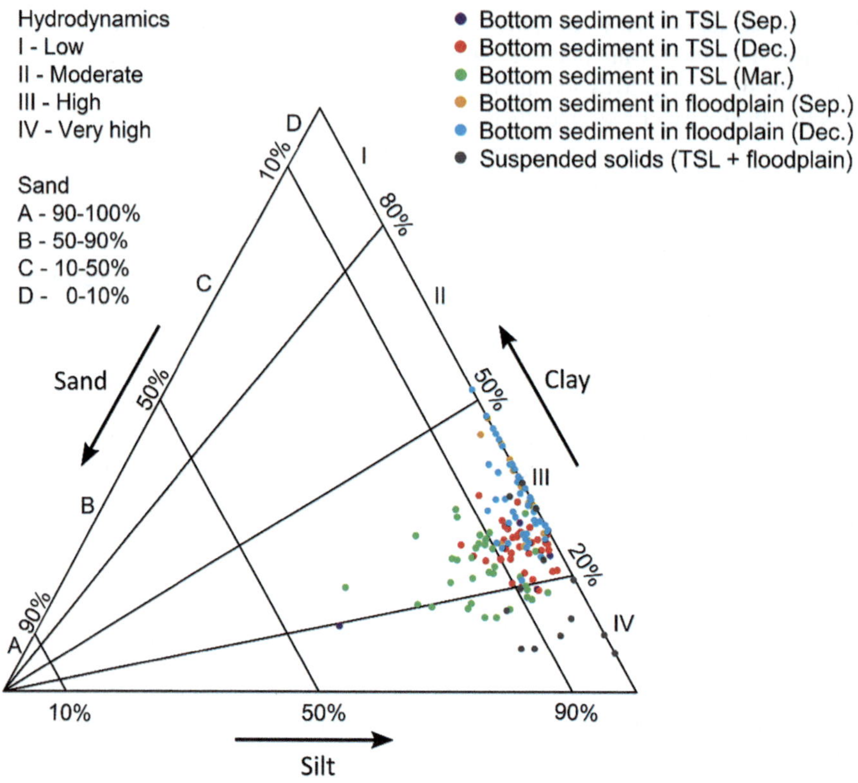

Fig. 17.1 Pejrup diagram illustrating textural composition (sand, silt, and clay) of the bottom sediment and suspended solids in TSL. The triangle was divided into 16 groups. Letters (A, B, C, or D) indicate sand content, and numbers (I, II, III, or IV) indicate hydrodynamic conditions. This figure is adapted from Siev et al. (2018)

contents were higher than 10%. The differences between December and March included sand content and hydrodynamic conditions, where the water depth and water level fluctuated greatly between the dry and wet season (Campbell et al. 2009; Kummu et al. 2014). Thus, degrees of the hydrodynamic level in TSL were categorized as high based on the textural composition, and the floodplain area showed a calmer hydrodynamic condition, facilitating the sedimentation process. Overall, this pattern implies that the resuspension process is intense in TSL during the dry season when the shallow water is mixed well by waves generated by a large fetch (see resuspension process in detail in Chap. 19).

Sediment dynamics in TSL are influenced by seasonal variation in the water level and sediment transport from Mekong River via TSR and the surrounding tributaries of TSL, and sedimentation and resuspension occur simultaneously. June is typically the beginning of the flooding season, when the water depth in the lake progressively increases because of water inputs from the TSR and lake's tributaries. During the flooding in September, sediment load input into the lake ranges from approximately

5.1 MT (Kummu et al. 2008) to 6.3 MT (Lu et al. 2014), resulting in a higher net sedimentation rate in the floodplain. December is the receding flood season, where the flood water is slowly released back to Mekong River through TSR. The release of sediment load outflow was relatively low compared (approximately 20%) to that of the inflow (Kummu et al. 2008). The sediment is trapped primarily by the vegetation at the lake's shoreline and in the floodplain (Kummu et al. 2008; Siev et al. 2018). Thus, the sedimentation process is more dominant in December, as indicated by the decline of the total suspended solid (TSS) concentration and the lack of input of sediment load from the TSR during this time (see Sect. 17.3). By contrast, the resuspension process is more dominant in March, reflected by the highest TSS concentration observed among our sampling periods. This is because the water depth during that time was the lowest among the sampling periods, and the previously deposited and fine-grained sediments are possibly and easily re-suspended by winds and human activities (see Chap. 19).

To date, studies on sediment transport/load in TSL are commonly based on statistical analysis (e.g., Kummu et al. 2008; Lu et al. 2014) rather than the process-based approach. Thus, the sediment balance of TSL still remains unclear. Therefore, further studies must be undertaken to fully understand the sediment balance of TSL, the sedimentation–resuspension process, and its relation to the Mekong River Basin by the process-based and/or integrated approach. Change in sediment load due to forest cover change and hydropower dams should be quantified and taken into account, as it could cause an imbalance in the sediment, resulting in environmental, ecological, and geomorphology changes in TSL and its floodplain in short- and long-term periods.

17.3 Suspended Solids

To understand the spatiotemporal dynamics of suspended solids/sediment, we explain how they vary across seasons, water depths, and water environments and how they are relevant to sedimentation and resuspension processes in TSL based on field survey and statistical analysis, respectively. The abovementioned survey also collected 250 water samples from different water environments/locations (tributaries, lake, and floodplains), water layers (surface and sub-surface), and seasons (high- and low-water periods). All the samples were subjected to analyze and identify TSS concentration and organic content (Siev et al. 2018).

TSS concentration varied among seasons and water environments ranging from approximately 1–752 mg/L (Figs. 17.2 and 17.3). The average TSS in September in Chhnok Tru and its tributaries was higher than those in Kampong Luong and TSL. In December, averages were relatively low across tributaries, lake, and floodplains. In March, the averages of TSS were in following order: TSL > floodplains > tributaries. In June, the averages of TSS showed no clear pattern and no similar value among water environments (tributaries, lake, and floodplains). That being said, some floodplain areas (e.g., Kampong Luong and Chhnok Tru) showed a lower

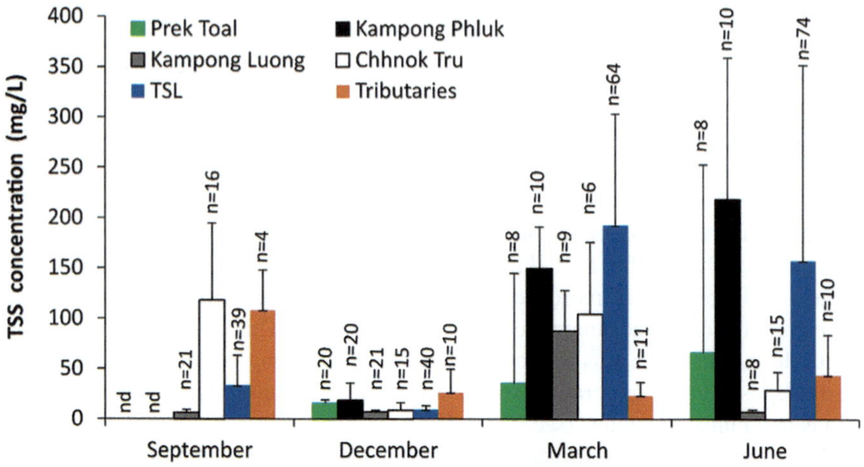

Fig. 17.2 Seasonal and spatial variation in TSS concentration in TSL, 4 floodplain areas (Prek Toal, Kampong Luong, Kampong Phluk, and Chhnok Tru), and 4–11 surrounding tributaries of TSL (the value is the average of the 4–11 tributaries). Some sampling locations in those floodplains were under water in both wet and dry seasons, and some were only in the wet season. Error bars, standard deviation; nd, not determined. This figure is adapted from Siev et al. (2018)

average of TSS compared to that in TSL and tributaries, whereas other areas (e.g., Kampong Phluk and Prek Toal) in the floodplain showed a higher one (Fig. 17.2).

High TSS concentration in September was observed at the southern part of the lake, which is the junction to TSR indicating that substantial sediment was entered from the TSR and gradually spread into the lake (Figs. 17.2 and 17.3). The TSS showed a low concentration in December, ranging from 3 to 11 mg/L across the surface area of the lake with a slight difference between the bottom and the surface (Fig. 17.3). In addition, TSL exhibited a higher TSS concentration across the lake in March with a low concentration only near the shore of the Prek Toal area, whereas higher concentrations were observed in the Kampong Phluk and Kampong Luong areas (Fig. 17.3). The TSS concentrations decreased in June (Fig. 17.3) in the southern part and shoreline of TSL compared to those in March, whereas the concentrations remained high in the Kampong Phluk area. The reason for that might be due to human activities. That being said, Kampong Phluk area is considerably one of the most visited places by tourists among other floodplain areas because of its attractive floating villages and flooded forest (Fig. 17.3). However, a study should be conducted to confirm whether a tourist activity will have a significant impact on sediment dynamics (e.g., resuspension and sedimentation).

During the high-water periods (September and December), results from the Pearson pair-wise correlation test indicated that TSS did not significantly correlate with the resuspension rate ($r = 0.11$, $p = 0.5$) but significantly positively correlated with the net sedimentation rate ($r = 0.41$, $p = 0.02$) and is significantly positively correlated with a gross sedimentation rate ($r = 0.33$, $p = 0.07$). This indicated that the sedimentation process was more dominant than the resuspension process and that

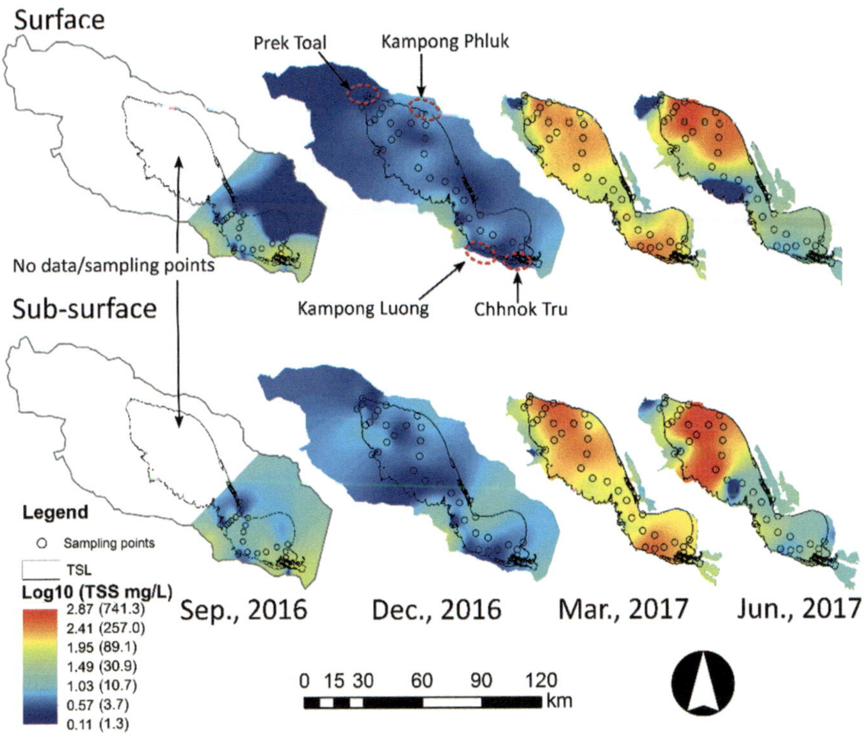

Fig. 17.3 Spatial and temporal distributions of TSS concentration during the sampling period (2016–2017). The first row shows the TSS concentration at the water surface, whereas the second row shows the concentration approximately 3 m below the water surface. Note that values and colors are shown on a logarithmic scale, whereas the exact values are indicated in parentheses. This figure is adapted from Siev et al. (2018)

deposited sediments were from the inflow of sediments rather than the lake bottom sediment during the high-water period. By contrast, the TSS concentration had significant positive correlations with the gross sedimentation rate ($r = 0.56$, $p < 0.001$), resuspension rate ($r = 0.45$, $p = 0.002$), and net sedimentation rate ($r = 0.56$, $p < 0.001$) during the low-water period (March and June).

This indicated that the resuspension process was more dominant, and the total settling flux was contributed from the lake bottom sediment (approximately 90%; see in Chap. 19) in the low-water period. Overall, sediment dynamics in TSL has two distinct characteristics, sink (sedimentation) during the high-water period and source (resuspension) during low-water period. Although the understanding and knowledge of sediment dynamics in TSL is gradually increased and receiving more attention, further study should focus also on sediment-related processes and their relation to the productivity in TSL (e.g., water quality, nutrient dynamics, biochemical process, and ecological process), which would improve our current knowledge and understanding toward the sustainable environmental management and conservation in TSL.

Key Points

- The majority sediment in TSL and its floodplain is fine silt and clay (>60% in mass), and the median sediment grain size (D50) is mostly 5–8 μm.
- Suspended solids/sediment show a large spatiotemporal variation in concentration ranging from 1 to 752 mg/L, caused by seasonal variation in the water level and sediment load from Mekong River and the lake's tributaries.
- In the wet season, TSS concentration is characterized by water level, sediment load inputs, and different water environments owing to its hydrodynamic conditions (e.g., tributaries, lake, and floodplain areas).
- During the high-water period, TSL and the floodplain play a role as a sink of sediment, where mostly fine-grained sediment is deposited.
- In the subsequent low-water period, the deposited sediments play a role as a source of suspended solids in the water column.

Acknowledgments Authors would like to thank relevant people who directly and indirectly supported this study and field survey.

References

Campbell IC, Say S, Beardall J. Tonle Sap Lake, the heart of the lower Mekong. In: The Mekong. Elsevier; 2009. p. 251–72. https://doi.org/10.1016/B978-0-12-374026-7.00010-3.

Conroy JL, Overpeck JT, Cole JE, Shanahan TM, Steinitz-Kannan M. Holocene changes in eastern tropical Pacific climate inferred from a Galápagos lake sediment record. Quat Sci Rev. 2008;27: 1166–80. https://doi.org/10.1016/j.quascirev.2008.02.015.

Darby SE, Langdon PG, Best JL, Leyland J, Hackney CR, Marti M, Morgan PR, Ben S, Aalto R, Parsons DR, Nicholas AP, Leng MJ. Drainage and erosion of Cambodia's great lake in the middle-late Holocene: the combined role of climatic drying, base-level fall and river capture. Quat Sci Rev. 2020;236:106265. https://doi.org/10.1016/j.quascirev.2020.106265.

Day MB, Hodell DA, Brenner M, Curtis JH, Kamenov GD, Guilderson TP, Peterson LC, Kenney WF, Kolata AL. Middle to late Holocene initiation of the annual flood pulse in Tonle Sap Lake. Cambodia J Paleolimnol. 2011;45:85–99. https://doi.org/10.1007/s10933-010-9482-9.

Kim LH, Choi E, Stenstrom MK. Sediment characteristics, phosphorus types and phosphorus release rates between river and lake sediments. Chemosphere. 2003;50:53–61. https://doi.org/10.1016/S0045-6535(02)00310-7.

Kummu M, Penny D, Sarkkula J, Koponen J. Sediment: curse or blessing for Tonle Sap Lake? Ambio. 2008;37:158–63. https://doi.org/10.1579/0044-7447(2008)37.

Kummu M, Tes S, Yin S, Adamson P, Józsa J, Koponen J, Richey J, Sarkkula J. Water balance analysis for the Tonle Sap Lake-floodplain system. Hydrol Process. 2014;28:1722–33. https://doi.org/10.1002/hyp.9718.

Lu X, Kummu M, Oeurng C. Reappraisal of sediment dynamics in the Lower Mekong River. Cambodia Earth Surf Process Landforms. 2014;39:1855–65. https://doi.org/10.1002/esp.3573.

Pejrup M. The triangular diagram used for classification of estuarine sediments: a new approach. In: Tide-influenced sedimentary environments and facies. Dordrecht, Netherlands: Springer; 1988. p. 289–300. https://doi.org/10.1007/978-94-015-7762-5_21.

Penny D, Cook G, Sok SI. Long-term rates of sediment accumulation in the Tonle Sap, Cambodia: a threat to ecosystem health? J Paleolimnol. 2005;33:95–103.

Siev S, Yang H, Sok T, Uk S, Song L, Kodikara D, Oeurng C, Hul S, Yoshimura C. Sediment dynamics in a large shallow lake characterized by seasonal flood pulse in Southeast Asia. Sci Total Environ. 2018;631–632:597–607. https://doi.org/10.1016/j.scitotenv.2018.03.066

Tsukawaki S. Lithological features of cored sediments from the northern part of The Tonle Sap Lake, Cambodia. In: The International Conference on Stratigraphy and Tectonic Evolution of Southeast Asia and the South Pacific. Bangkok, Thailand; 1997. pp. 232–239.

Wünnemann B, Mischke S, Chen F. A Holocene sedimentary record from Bosten Lake, China. Palaeogeogr Palaeoclimatol Palaeoecol. 2006;234:223–38. https://doi.org/10.1016/j.palaeo.2005.10.016.

Chapter 18
Total Suspended Solid Dynamics Revealed by Long-Term Satellite Image Analysis

Keisuke Hoshikawa, Yoichi Fujihara, Sokly Siev, Seiya Arai, Takashi Nakamura, Hideto Fujii, Ty Sok, and Chihiro Yoshimura

18.1 Suspended Solids in a Shallow Lake and Its Relation to Flood Pulse

A few studies have investigated the TSS concentration in Tonle Sap Lake (TSL) through sampling surveys. Kummu et al. (2008) conducted a comprehensive survey of TSL and its surrounding floodplains in almost every month from July 2001 to June 2002. The survey revealed a detailed seasonal pattern in TSS concentration and its difference in the lake and floodplain areas. Chap. 17 indicates the spatial distribution of TSS concentrations, sedimentation/resuspension rates, and their seasonal changes based on water and bottom sediment samples collected at 39 points along TSL every 3 months from September 2016 to July 2017.

In our investigation, satellite-borne optical remote sensing, which can observe more than 100 km in width at once, demonstrated a high potential for long-term and

K. Hoshikawa (✉)
Toyama Prefectural University, Imizu, Japan
e-mail: hoshi@pu-toyama.ac.jp

Y. Fujihara
Ishikawa Prefectural University, Nonoichi, Japan

S. Siev
Institute of Technology of Cambodia, Phnom Penh, Cambodia

Ministry of Industry, Science, Technology and Innovation, Phnom Penh, Cambodia

S. Arai · T. Nakamura · C. Yoshimura
Tokyo Institute of Technology, Tokyo, Japan

H. Fujii
Yamagata University, Tsuruoka, Japan

T. Sok
Institute of Technology of Cambodia, Phnom Penh, Cambodia

C. Yoshimura et al. (eds.), *Water and Life in Tonle Sap Lake*,
https://doi.org/10.1007/978-981-16-6632-2_18

quasi-continuous monitoring of TSS concentrations. Long-term monitoring of suspended sediment concentrations (SSCs) with satellite images was also carried out in the Lower Mekong River Basin. The application of remote sensing to TSL requires a high temporal resolution and consideration of the representativeness of the estimated spatial distribution of TSS concentrations. A complex range of factors determines TSS concentration in large and shallow lakes, whereas the SSC in the Mekong River is strongly determined by the flow discharge (Lu et al. 2014). Shear stress on bottom sediments by wind-induced waves and water currents are the main causes of resuspension in shallow water bodies (Lund-Hansen et al. 1999; Chung et al. 2009). Interactions among the lake water, floodplain, and river water (caused by flood pulses) lead to a supply of sediments, dilution of TSS concentration or resuspension. The spatial distribution of TSS concentrations in large and shallow lakes should be more complicated and changing frequently and rapidly.

This chapter aims to introduce an application of satellite images to clarify the short- and long-term dynamics of TSS concentration and possible factors governing TSS concentration in surface waters, such as water exchange with the surrounding floodplain, current of floodwater from the Mekong River to the lake, and disturbance of water surface by wind. Daily reflectance data across 5531 days from January 1, 2003, to December 31, 2017, provided by the Moderate Resolution Imaging Spectroradiometer (MODIS) Aqua satellite were used to indicate detailed temporal changes in TSS concentrations and their representative spatial distribution during four periods within a year.

18.2 Estimation of TSS Concentration Using Satellite Images

TSL indicates a wide range of fluctuations in TSS concentrations from nearly 0 mg/L to more than 700 mg/L (see Chap. 17). For water with a wide range of SSC fluctuations, linear regressions between indices deriving from the reflectance of multiple bands and the natural SSC logarithm indicate a high correlation. In other words, SSC is given by a natural exponential function of reflectance indices (Long and Pavelsky 2013).

Multispectral images acquired by Landsat sensors are widely used in various fields, including estimation of water quality. Sensors aboard Landsat-5 launched in 1984 and its successors have been observing land surface every 16 days with basically the same resolution and observation ranges of wavelength enabling long-term observation on land surface. Fleifle (2013) and Dang et al. (2018) investigated suspended sediment load or SSC in the Mekong River and its delta using Landsat-5 TM images and Landsat-7 ETM+ images.

The application of remote sensing to TSL requires a very high frequency of data acquisition. The spatial distribution of TSS in TSL changes largely day by day. Although it is necessary to obtain a representative distribution from satellite images on as many days as possible, a large part of the surface of a lake is covered by clouds most of the days, especially in the rainy season. High frequency of data acquisition increases the number of available images with a small portion of cloud cover. In addition, the observation period is also important to trace the long-term changes in TSS concentration. The surrounding environment of TSL has changed rapidly since 2000 as reflected by an increase of GDP of Cambodia from 3.768 billion USD in 2000 to 27.089 billion USD in 2019 (World Bank n.d.). Such rapid economic development may impose a burden on water quality. Furthermore, the number of constructed hydropower dams in the Mekong River Basin has increased from 7 to 66 in the period from 2000 to 2017 (Lin and Qi 2019). Therefore, it is necessary to cover the period at least since early 2000s in our investigation.

MODIS Terra and Aqua, launched in 1999 and 2002, respectively, observe almost the whole surface of the earth daily at 250, 500, or 1000 m spatial resolutions. By contrast, the Landsat series with a return period of 16 days have the longest observation period since 1972. A combination of Multispectral Imagers (MSI) on Sentinel-2A and -2B, which were launched in 2015 and 2017, respectively, provides visible and near-infrared images with a 10 or 20 m resolution every 5 days. Therefore, MODIS provides information most suitable among other satellite images for this study, which aims to cover both representative spatial distribution and long-term changes in TSS.

In this study, MYD09GA Version 6 data with daily surface spectral reflectance of MODIS Aqua from bands 1 to 7, as provided by NASA, were used. TSS concentrations were estimated by a natural exponential function using the MODIS Aqua MYD09GA images. Although the MODIS Terra satellite began operation in 2000 (2 years before MODIS Aqua), the MODIS Terra data were not used because the rate of Terra decay was greater than that of Aqua (NASA n.d.).

MYD09GA images contain reflectance (band 7; wavelength: 2105–2155 nm) that has already been corrected by consideration of atmospheric conditions, such as gases, aerosols, and Rayleigh scattering (Vermote and Wolfe 2015). In addition, the pixels with SWIR reflectance over 0.015 were excluded.

The selected 25 sampling points of TSS surveyed in December 2016, March 2017, and June 2017 (from Chap. 17) were combined with reflectance resulting in 14 available points on 8 days after exclusion of combinations for which SWIR reflectance of MYD09GA had a value larger than 0.015. Equation (18.1) was derived by log approximation with the 25 combinations of TSS measurement values and MYD09GA red-band (band 1) reflectance. R-squared is 0.92.

$$TSS_{\mathrm{MODIS1}} = 2.12\exp\left(26.3\ \mathrm{ref}_1\right) \tag{18.1}$$

In Eq. (18.1), ref_1 is the reflectance value of MODIS bands 1, and TSS_{MODIS1} is the TSS concentration values estimated from the reflectance values of MODIS bands 1. The Root Mean Square error for all 25 measurement values is 14.1 mg/L, whereas it is 6.0 mg/L for values less than 250 mg/L.

Another exponential regression between MODIS band 1 and TSS concentration in the TSL by Wang et al. (2020) resulted in different coefficients, that is, 10.32 and 21.7, instead of 2.12 and 26.3, respectively. The regression by Wang et al. (2020) provided larger values of TSS than those provided by Eq. (18.1), especially in the range with small reflectance. This was caused by differences in relation between measured TSS and reflectance. Further, TSS measurement and analysis are required to validate this difference.

18.3 Spatial Distribution of TSS Concentration

18.3.1 Outflow Period

TSL is connected to the Mekong River through the Tonle Sap River. Inflow from the Mekong River to the lake occurs during the period from June to September, whereas outflow from the lake flows to the Mekong River occurs from October to May of the following year (see Chaps. 1, 11, and 15).

Figure 18.1 illustrates daily TSS_{MODIS1} median in April–March, July, and August for each pixel throughout the research period (2003–2017). Compared with Fig. 17.3 in Chap. 17, which was derived through interpolation of measurement values, clearer spatial patterns are observed in Fig. 18.1. Figure 18.1 indicates spatial distributions that appear most frequently, whereas Fig. 17.3 illustrates spatial distribution at each moment. A smaller area of high concentration of TSS more than 500 mg/L in Fig. 18.1 compared with that in Fig. 17.3 is due to the effect of temporal averaging of reflectance.

The highest TSS concentrations during the outflow period (April–May) appear near the center of the northern lake, and the TSS concentration decreased toward the lakeshore (Fig. 18.1a). Macrophytes, which are observed along the shore of the permanent lake, were one of the factors corresponding to a lower TSS concentration.

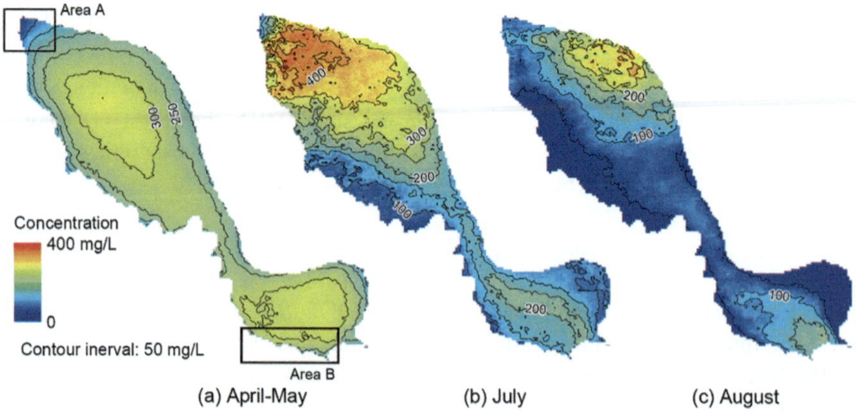

Fig. 18.1 Spatial distribution of the estimated TSS concentration in TSL's outflow (April to March) and inflow (July and August) periods. (Originally published as Fig. 9 of Hoshikawa et al. (2019). The three of seven periods in the original figure are extracted. The original figure is shaded in black and white)

Macrophytes with dense stems lower the bottom sediment turbulence due to flow velocity reduction (Horppila et al. 2013).

Another factor for lowering TSS concentration near the shore during the outflow period should be water flow from flooded forests and floodplain areas. The lowest TSS concentration appeared at the northwestern end of the lake (Area A). In this area, there is the mouth of Sangkae River, which runs through a flooded forest more than 50 km. By contrast, TSS concentration often exceeded 250 mg/L around the southern end of the lake (Area B), where the width of flooded forest is less than 2 km, smallest around the lake.

The mean TSS concentration in the flooded forest area was only one-third of that in the lake (Sarkkula et al. 2004). When the TSL water level began to decrease after the inflow period, water that spilled over onto the floodplain area began to recede into the lake. Flood waters entering the subsurface near the landward end of the floodplain flowed back to the permanent lake together with contributions from upland groundwater along the permanent lake shore after a decrease in water level (Burnett et al. 2017). Dilution by these water flows should have decreased TSS concentration in the shore areas of the permanent lake.

A higher density of vegetation lowered the TSS concentration on the floodplain. Suspended sediments were primarily trapped by vegetation at the lake edge and on the floodplain (Kummu et al. 2008). Sedimentation rates in the forested floodplain around TSL were higher than those in the lake. In addition, resuspension rates in the forested flood plain were lower than those in the lake. The flooded forest showed the highest reduction in sediment resuspension (26%), followed by shrubland (13%) and grassland (4%) in comparison with the open water areas (see Chap. 19). This fact indicates that floodplain deforestation may increase TSS concentrations in the permanent lake area.

18.3.2 *Inflow Period*

During the inflow period from July to September, the patterns of spatial distribution of TSS concentration were different from those of the outflow distribution. In July, a clear protrusion of areas with high TSS concentrations was observed from the Tonle Sap River mouth toward the northern lake (Fig. 18.1b). The highest TSS concentrations were observed near the northwestern end of the lake. The protrusion of areas with high TSS concentrations from the Tonle Sap River mouth was observed even in August, although by that time, the TSS concentrations had decreased. The highest TSS concentrations in the northern section were observed along the northeastern shore (Fig. 18.1c). Inundated forests over 20 km in width cover the southwestern shore. In comparison, the inundated forests on the northeastern side are only around 5 km in width due to land use change from forest to agricultural fields. Resuspension may have occurred in shallowly flooded agricultural areas or bare land intensely.

The spatial distribution of TSS concentrations in August indicated that the effect of suspended sediments supplied by inflow from the Mekong River on the TSS concentration of the TSL surface water was limited. The suspended sediment concentration in the Tonle Sap River at the Phnom Penh Port during the inflow period had a high positive correlation with the amount of inflow. Thus, TSS concentrations were the largest in August (Lu et al. 2014). However, TSS concentrations near the Tonle Sap River mouth were lower in August than those in July. This indicates that TSS concentration on the surface of TSL does not depend on the amount of TSS supplement from the Tonle Sap River.

18.4 Water Depth and Resuspension

The depth of the lake is an important factor affecting wind and inflow current-induced resuspension. The ratio of the contribution of wind and inflow to TSS concentration varied by season and year, depending on the amount of inflow.

Figure 18.2 illustrates the relations between the water level at Kampong Luong and estimated TSS concentration in non-drought years (a non-drought year is defined when a maximum water level of the lake exceeds 8 m in a particular year, whereas less than 8 m is defined as a drought year. There are 11 years in which the maximum water level exceeded 8 m between 2000 and 2017). The highest and lowest TSS concentrations were observed at the beginnings of inflow and outflow periods, respectively. According to results of field measurement indicated in Fig. 19.1 (see Chap. 19), resuspension rates in these periods (June and December) were also the highest and lowest. This suggests that TSS concentration is highest and lowest in not only surface area that can be observed by remote sensing but also all the layers from the bottom to the surface of the lake's water column.

There are significant negative correlations between the water levels and TSS concentrations during the outflow periods. Wind is the major cause of sediment resuspension during this period. The frequency distributions of wind speed and average wind speed were almost constant throughout the research period. This caused an increase of TSS concentration along almost the same path every year. Short-term fluctuation of TSS may be caused by daily fluctuation of wind speed and human activities. The slightly higher path in 2017 is related to lower water levels in the previous inflow period, as discussed in the next section.

The range of inter-annual fluctuation in the TSS concentration from July to September was larger than that in other seasons. This indicates that TSS concentration during this period was largely determined by the amount of inflow. The high concentration of TSS was mainly caused by resuspension by water current, whereas wind resuspension also should have occurred, especially in areas close to coast. The annual inflow amount from the Mekong River to TSL fluctuated from 24.5 to 54.0 km^3 during the period from 1997 to 2004 (Kummu et al. 2014), whereas the average wind speed did not show significant seasonal and annual variations from 2006 to 2017. Therefore, the contribution of wind-induced resuspension to the TSS concentration is dependent only on the water level or depth, which resulted in a lower TSS concentration limit throughout the year. The seasonal and interannual

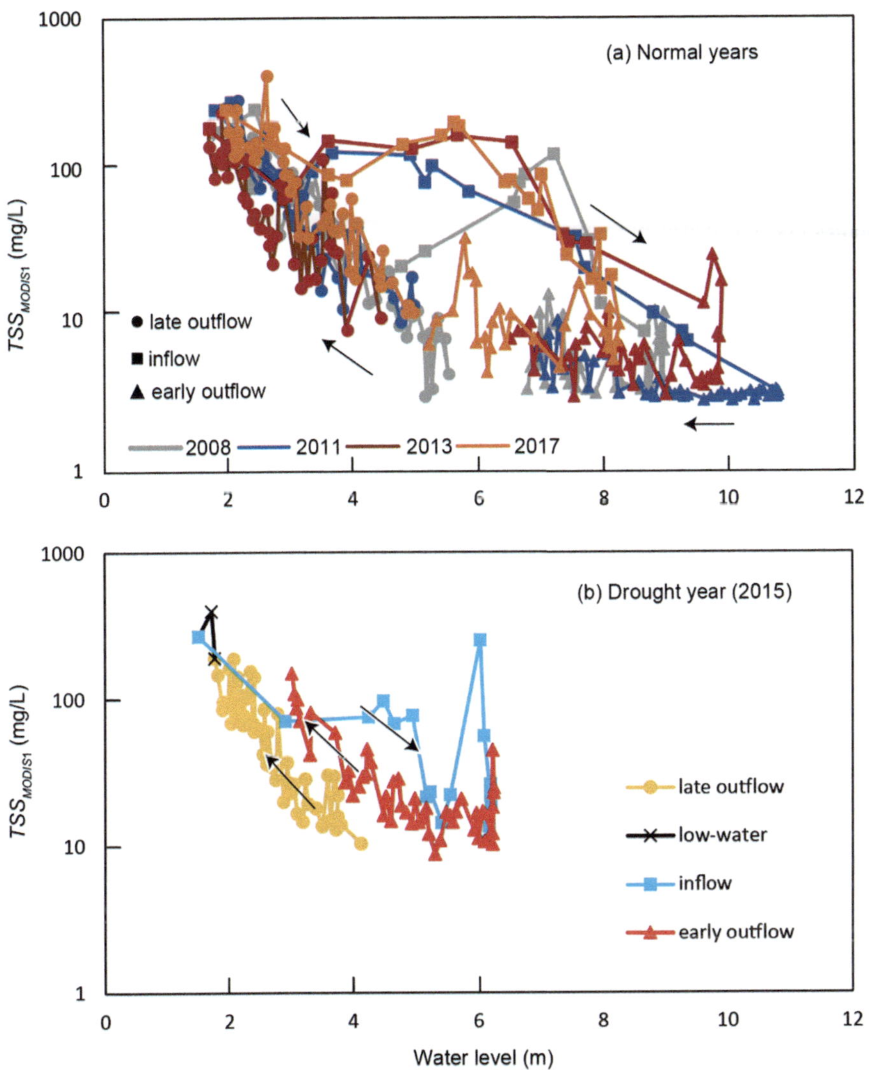

Fig. 18.2 The relation between the water level and the median of TSS concentration in normal years and drought years. The panel (**a**) contains a few plots for the low-water period, which are behind other plots. (Originally published as Fig. 7 of Hoshikawa et al. (2019). Symbols and colors to indicate periods are modified from the original)

fluctuations in the inflow current caused the annual fluctuation in the path of decrease of TSS concentration. The anthropogenic bottom sediment disturbance may also have caused fluctuations (see Chap. 19 regarding sediment resuspension and its affecting factors). Regardless, the effect of resuspension on the TSS concentrations due to the inflow current at the water surface is weakened as the water level exceeds approximately 6 m.

18.5 Long-Term Trends and a Recent Anomaly

No clear long-term TSS concentration increase trends related to growing anthropogenic pressures around the lake were observed (Fig. 18.3). The lake volume increased from 1.3 km^3 in the dry season to as much as 75 km^3 in the rainy season, depending on the flood intensity (Chap. 15). Because of such a large flood pulse, nutrients that cause blooming of phytoplankton flowing into the lake from areas surrounding the lake may have been washed out every year.

A decline in the flood pulse, which decreases the amount of inflow, may change the relationship between TSS concentration and water depth during the outflow periods. After an extraordinary low peak of water level in 2015, TSS concentration at each water level was higher than those of previous years and resulted in the highest TSS concentration in the period from 2003 to 2017, during April 2016 (Fig. 18.3). One of the reasons for the apparently higher TSS concentration as estimated from MODIS may be a bloom of phytoplankton. The concentrations of chlorophyll-a and total plankton biomass are increased in TSL during the late outflow period (Sarkkula et al. 2004). High concentration of chlorophyll-a (Chl-a; see Chap. 31) in water significantly increases reflectance of wavelength around 700 nm (Gin et al. 2002). Reflectance observed by band 1, 620–670 nm may, therefore, be partially affected. Further research on the relation between the Chl-a concentration and water depth of TSL is required to test this hypothesis (see also Chap. 31 for spatiotemporal variation of Chl-a in TSL).

A TSS concentration decline at the water surface by sediment trapping was not detected in this study, despite a rapid increase in dams since the 1990s. Studies on the long-term changes in sediment load in the Mekong River basin indicate that dams constructed in the basin trap a large portion of sediments originating upstream (Kummu and Varis 2007; Walling 2008). A model simulation indicated that cumulative sediment trapping by dams will be 96% after completion of 38 dams (Kondolf et al. 2014). Runoff regulation by dams, which decreases runoff during flood seasons (Kummu and Varis 2007), may increase local TSL sediment resuspension by decreasing water levels during flood seasons.

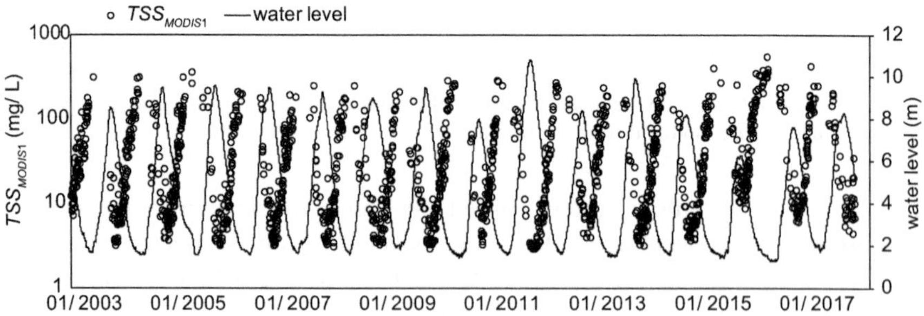

Fig. 18.3 Time series of the estimated TSS (median) and water level at Kampong Luong. (Originally published as Fig. 5 of Hoshikawa et al. (2019))

Key Points

- The median of TSS concentration ranged from 4 to 420 mg/L based on the estimation using MODIS band 1 (visible red), which showed R-squared of 0.92 and a root mean square error of 14.1 mg/L.
- The potential for resuspension was strongly governed by the water depth.
- Seasonal changes in spatial distribution of TSS concentration were clearly revealed using MODIS daily reflectance data.
- The low TSS concentration appeared in areas adjacent to flooded forests when the lake water was flowing out to the Mekong River.
- High TSS concentration appeared in dry seasons after rainy seasons when maximum water levels were lower than normal years.

References

Burnett WC, Wattayakorn G, Supcharoen R, Sioudom K, Kumd V, Chanyotha S, Kritsananuwat R. Groundwater discharge and phosphorus dynamics in a flood-pulse system: Tonle Sap Lake, Cambodia. J Hydrol. 2017;549:79–91.

Chung EG, Bombardelli FA, Schlado SG. Sediment resuspension in a shallow lake. Water Resour Res. 2009;45:W05422. https://doi.org/10.1029/2007WR006585.

Dang TD, Cochrane TA, Arias ME. Quantifying suspended sediment dynamics in mega deltas using remote sensing data: a case study of the Mekong floodplains. Int J Appl Earth Obs Geoinf. 2018;68:105–15.

Fleifle AE. Suspended sediment load monitoring along the Mekong River from satellite images. Earth Sci Clim Change. 2013;4(6):160. https://doi.org/10.4172/2157-7617.1000160.

Gin KYH, Koh ST, Lin II. Study of the effects of suspended marine clay on the reflectance spectra of phytoplankton. Int J Remote Sens. 2002;23(11):2163–78.

Horppila J, Kaitaranta J, Joensuu L, Nurminen L. Influence of emergent macrophyte (Phragmites australis) density on water turbulence and erosion of organic-rich sediment. J Hydrodyn. 2013;25(2):288–93.

Hoshikawa K, Fujihara Y, Siev S, et al. Characterization of total suspended solid dynamics in a large shallow lake using long-term daily satellite images. Hydrol Process. 2019;33(21): 2745–58. https://doi.org/10.1002/hyp.13525.

Kondolf GM, Rubin ZK, Minear JT. Dams on the Mekong: cumulative sediment starvation. Water Resour Res. 2014;50(6):5158–69.

Kummu M, Varis O. Sediment-related impacts due to upstream reservoir trapping, the lower Mekong River. Geomorphology. 2007;85:275–93.

Kummu M, Penny D, Sarkkula J, Koponen J. Sediment: curse or blessing for Tonle Sap Lake? AMBIO J Hum Environ. 2008;37(3):158–63.

Kummu M, Tes S, Yin S, Adamson P, Józsa J, Koponen J, Richey J, Sarkkula J. Water balance analysis for the Tonle Sap Lake–floodplain system. Hydrol Process. 2014;28:1722–33.

Lin Z, Qi J. A new remote sensing approach to enrich hydropower dams' information and assess their impact distances: a case study in the Mekong River basin. Remote Sens. 2019;11(24): 3016. https://doi.org/10.3390/rs11243016.

Long CM, Pavelsky TM. Remote sensing of suspended sediment concentration and hydrologic connectivity in a complex wetland environment. Remote Sens Environ. 2013;129:197–209.

Lu XX, Kummu M, Oeurng C. Reappraisal of sediment dynamics in the Lower Mekong River, Cambodia. Earth Surf Process Landforms. 2014;39:1855–65.

Lund-Hansen L, Petersson M, Nurjaya W. Vertical sediment fluxes and wave-induced sediment resuspension in a shallow-water coastal lagoon. Estuar Coast. 1999;22:39–46.

NASA. Is there a difference in Terra vs Aqua AOD? n.d. https://darktarget.gsfc.nasa.gov/content/there-difference-terra-vs-aqua-aod. Accessed 28 Sept 2020.

Sarkkula J, Baran E, Chheng P, Keskinen M, Koponen J, Kummu M. Tonle Sap Pulsing System and fisheries productivity. Contribution to the XXIXe International Congress of Limnology (SIL 2004); Lahti, Finland, 2004, 8–14 August 2004

Vermote E, Wolfe R. MOD09GA MODIS/Terra Surface Reflectance Daily L2G Global 1kmand 500m SIN Grid V006 [Data set]. NASA EOSDIS LP DAAC; 2015. https://doi.org/10.5067/MODIS/MOD09GA.006.

Walling DE. The changing sediment load of the Mekong River. AMBIO J Hum Environ. 2008;37(3):150–7.

Wang Y, Feng L, Liu J, Hou X, Chen D. Changes of inundation area and water turbidity of Tonle Sap Lake: responses to climate changes or upstream dam construction? Environ Res Lett. 2020;15(9):90940a1.

World Bank. GDP (current US$)—Cambodia. n.d. https://data.worldbank.org/indicator/NY.GDP.MKTP.CD?end=2019&locations=KH&start=1993&view=chart. Accessed 14 Apr 2021.

Chapter 19
Sediment Resuspension and Its Relation to Flood Pulse

Uk Sovannara, Sokly Siev, Sato Michitaka, Rajendra Khanal, Ty Sok, Sive Thea, Sophal Try, Chantha Oeurng, and Chihiro Yoshimura

19.1 Sediment Resuspension Rate

The limnology of shallow lakes is strongly influenced by the dynamics of sediment resuspension, possibly resulting in high turbidity, promotion of internal loading of nutrients, and phytoplankton dominance. The upper layer of the sediments in shallow lakes is subjected to the exchange of free oxygen and substances contained in sediment (i.e., nutrients) with the water column through physical, chemical, and biological processes (Blom et al. 1992), thereby being related to the overall structure of the lake ecosystem.

Influenced by the monotonal flood pulse, sediment processes in TSL exhibit strong seasonality (see Chaps. 17 and 18). Sedimentation was dominant during the high-water period (e.g., Sep.–Dec.), whereas the dominant process shifts to resuspension during the low-water period (e.g., Mar.–Jun.). The resuspension rate (RR) in TSL is much higher than the gross sedimentation rate (SR) during the low-water period (Siev et al. 2018). Although the inflow of sediments to TSL during

U. Sovannara (✉) · S. Michitaka · C. Yoshimura
Tokyo Institute of Technology, Tokyo, Japan

S. Siev
Institute of Technology of Cambodia, Phnom Penh, Cambodia

Ministry of Industry, Science, Technology and Innovation, Phnom Penh, Cambodia

R. Khanal
Tokyo Institute of Technology, Tokyo, Japan

Policy Research Institute, Kathmandu, Nepal

T. Sok · S. Thea · C. Oeurng
Institute of Technology of Cambodia, Phnom Penh, Cambodia

S. Try
Kyoto University, Kyoto, Japan

C. Yoshimura et al. (eds.), *Water and Life in Tonle Sap Lake*,
https://doi.org/10.1007/978-981-16-6632-2_19

the high-water period contributed to sedimentation flux in TSL, most of the sedimentation flux in TSL was derived from the lake bed (approximately 90%) through resuspension in the low-water period. High resuspension rates in the lake during the low-water period indicate that little of the material that settles during the high-water period is permanently deposited. Therefore, sedimentation and resuspension in TSL are almost in balance (Siev et al. 2018), explaining the reason for the very low annual sedimentation rate (0.05–2.55 mm/year) in the TSL (Kummu et al. 2008; Penny et al. 2005; Tsukawaki 1997).

In our survey, RR was determined in September ($n = 25$) and December ($n = 8$) in 2016 and in March ($n = 34$) and June ($n = 6$) in 2017 in four floodplain areas around TSL (i.e., Kampong Phluk, Prek Toal, Kampong Luong, and Chhnok Tru) using sediment traps following Gasith (1975). Kampong Phluk and Prek Toal are located in the northern part of TSL, whereas Kampong Luong and Chhnok Tru were located in the southern part. The observed RR indicated a significant seasonal variation (three-way ANOVA, $p < 0.001$, $F = 9.77$) and spatial variation (three-way ANOVA, $p < 0.01$, $F = 4.74$). Higher RR was observed during the low-water period (March–June). The average water depths (±standard error) at the 39 sampling locations across the lake between 2016 and 2017 were 0.9 ± 0.6 m in September, 2.7 ± 0.6 m in December, 3.9 ± 1.4 m in March, and 4.7 ± 0.8 m in June, corresponding to average resuspension rates of 115 ± 66 g/m^2/day, 93 ± 128 g/m^2/day, 260 ± 246 g/m^2/day, and 348 ± 227 g/m^2/day, respectively. The highest resuspension rate was observed at Kampong Phluk (412 ± 278 g/m^2/day), an area located in the northern part of the lake, followed by Kampong Luong (245 ± 192 g/m^2/day), Chhnok Tru (191 ± 198 g/m^2/day), and Prek Toal (78 ± 119 g/m^2/day) (Fig. 19.1).

The water depth seasonally changed, resulting in seasonal variation of the dynamic ratio and water volume of the lake. The concentrations of total suspended solids (TSS) in TSL were positively correlated with RR ($r = 0.45$, $p < 0.01$) during the low-water period, whereas they showed no significant correlation ($r = 0.11$, $p > 0.05$) during the high-water period (Siev et al. 2018). This indicates that sediment resuspension in the TSL strongly contributed to TSS concentration in the water column during the low-water period. As resuspension-induced TSS concentrations in the TSL increase greatly during the low-water period, light availability is more likely to be a limiting factor for the pelagic primary production. In addition, during the low-water period, sediment resuspension can also affect nutrient cycling in TSL. For instance, phosphorus concentration in the lake water significantly depends on the interactions between the sediment and water because the phosphorus pool in the sediment is generally over 100 times higher than that in the lake water (Søndergaard et al. 2003). High TSS concentration would also be a constraint for the treatment for water supply, requiring high doses of flocculants and coagulants (e.g., aluminum and ferric salts; Siev et al. 2018).

The resuspension rate in TSL was found to be higher than those in some small shallow lakes such as Wingra Lake, Hiidenvesi Lake, Lammijarv Lake, and Peipsi Lake but lower than that in Taihu Lake (Fig. 19.2, Siev et al. 2018). A variety of factors influence sediment resuspension (e.g., lake shape, water depth, wind-wave

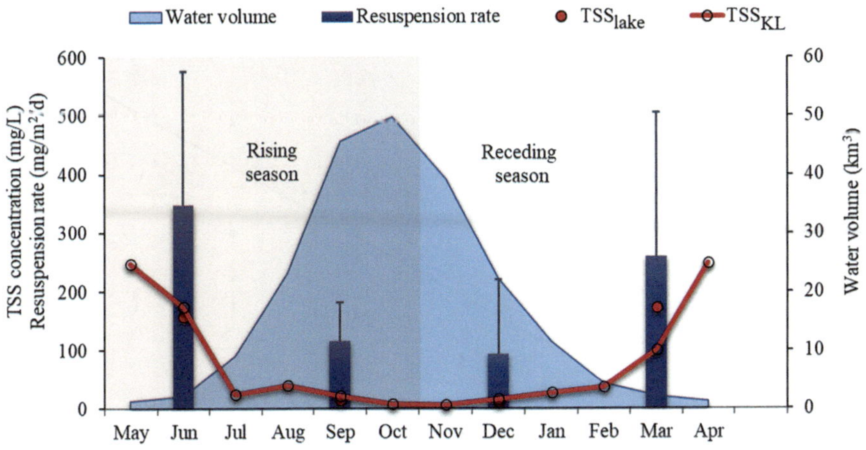

Fig. 19.1 Seasonal changes in resuspension rates measured. The average concentration of TSS measured at Kampong Luong (TSS_{KL}, 1995–2010)[2], monthly average water volume in the lake (1995–2010)[2,3], whole lake average concentration of TSS (TSS_{lake}) calculated as the average concentrations of TSS from all the sampled points in September and December 2016 and in March and June 2017 (2016–2017)[1], and the sediment resuspension rate measured with 73 traps installed at different locations in TSL (2016–2017)[1]. [1]Siev et al. (2018); [2]MRC database; [3]Kummu et al. (2014). The error bars represent the standard deviations

action, vegetation; Zhu et al. 2015; Siev et al. 2018), as elaborated in the following sections.

19.2 Sediment Resuspension in Relation to Flood Pulse

The seasonal flood pulse distinguishes hydrodynamic conditions in TSL from those of typical shallow lakes, driving seasonal changes in lake bathymetry and thus its surface area and depth distribution. Such changes significantly affect sediment processes, particularly resuspension (Siev et al. 2018). The dynamic ratio has been used in analyzing the relationship between potential wave disturbance and sediment resuspension, with a higher ratio generally relating to higher RR (Bachmann et al. 2000; Zhu et al. 2015; Siev et al. 2018). For example, Fig. 19.2 indicates a positive correlation between the dynamic ratio and the resuspension rate in lakes in the world and TSL. As TSL is a shallow and flat-bottomed lake, the seasonal variation in the water depth between the low- and high-water periods directly corresponds to the lake volume and surface area. Consequently, the mean monthly dynamic ratio of the lake also changes seasonally from 14 in the high-water period to 128 in the low-water period, corresponding to the increment in TSS concentration (Fig. 19.3).

The bathymetry of TSL is a key factor for the higher sediment resuspension than that of the other shallow lakes mentioned above because sediment resuspension is

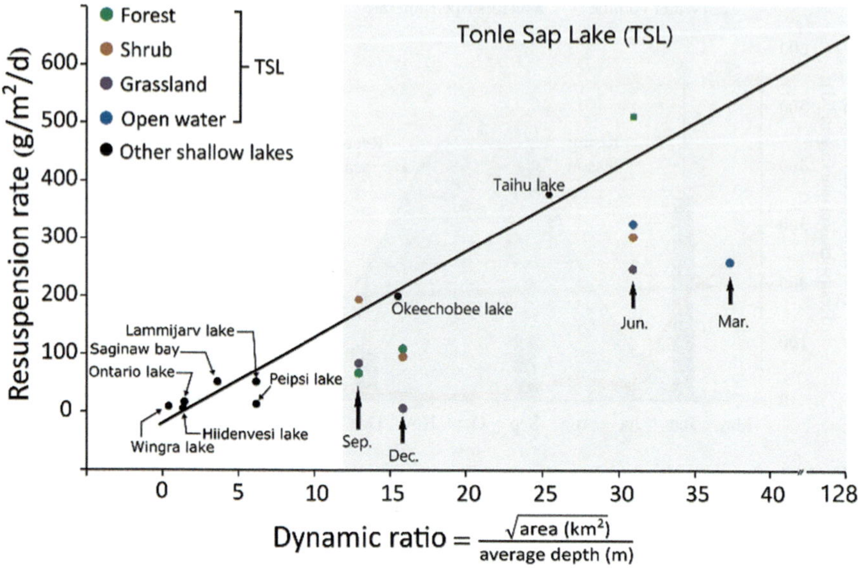

Fig. 19.2 Relation between the dynamic ratio and the resuspension rate in lakes in the world (modified from Zhu et al. 2015, Siev et al. 2018). The plots of TSL include results from the four sampling periods (September and December 2016 and March and June 2017). The gray area represents the range of the dynamic ratio of TSL calculated using data from MRC's database and Siev et al. (2018). Markermeer Lake (not plotted in the above figure) has a dynamic ratio < 10 and a resuspension rate of ~1000

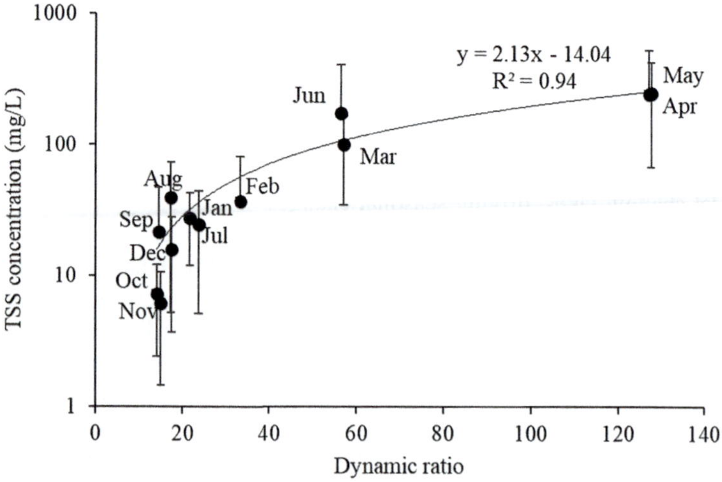

Fig. 19.3 Monthly average TSS concentration at Kampong Luong against the dynamic ratio of TSL. The surface area of the TSL was calculated using the method in Kummu et al. (2014) based on the database of the Mekong River Commission. The error bars represent standard errors

likely to occur in lakes with low water depths and large open areas (Evans 1994; Zhu et al. 2015; Siev et al. 2018). In general, wave generation, which affects the intensity of resuspension, depends on the energy transfer from the wind to the water surface, which is a function of lake exposure to the water surface, e.g., fetch (the unobstructed distance over which the wind can blow), wind characteristics (e.g., speed, direction, and duration), and water depth (Evans 1994; Gloor et al. 1994; Fagherazzi and Wiberg 2009). However, RR in TSL during the low-water period (March and June) is lower than in Taihu Lake (dynamic ratio, 25; RR, ~400 g/m^2/day) and Markermeer Lake (dynamic ratio, <10; RR, ~1000 g/m^2/day) although those lakes have smaller dynamic ratios, indicating that other factors affect resuspension. For instance, while the dynamic ratio is smaller, the estimated annual sediment RR in Markermeer Lake is higher than that in Taihu Lake, which was due to stronger near-bottom currents in Markermeer Lake (Zhu et al. 2015). In TSL, which is influenced by the flood pulse, seasonal variations in water depth appear to be a moderator determining the wave-imposed hydrodynamic shear stress on the lake bed.

19.3 Factors Affecting Sediment Resuspension

In addition to water level fluctuations and the dynamic ratio, there are some other factors affecting resuspension in TSL, namely, wind, vegetation, bioturbation, and human activities, and their relative importance varies with water level as discussed below.

19.3.1 Wind

Wind-induced water disturbance is considered to be one of the most important factors affecting resuspension in shallow lakes. The simultaneous action of wind-induced currents and surface waves leads to an increase in the bottom shear stress, which may exceed the critical shear stress and cause resuspension (Churchill et al. 2004; Hawley et al. 2004). According to Sheng and Lick (1979), wind-induced waves account for up to 70% of sediment resuspension in shallow lakes. Wind also affects phosphorus dynamics in numerous typical shallow lakes that have no substantial seasonal variation in water depth (e.g., Taihu Lake, China). In a shallow lake characterized by the flood pulse such as TSL, wind-induced turbulence accounts for resuspension during the low-water period, whereas it is less important during the high-water period. The shallowness and large area of TSL during the low-water period, as indicated by the dynamic ratio (Fig. 19.3), tends to provide sufficient wind fetch to generate turbulence in the bottom layer, thereby promoting sediment resuspension and increasing turbidity.

Wind data on the lake were not available. Wind speeds and wind directions in the wet and dry seasons at Siem Reap Airport (2006–2018), approximately 15 km away from TSL, showed the wind condition to be predominately moderate (wind speed: 2–6 m/s, accounting for >72% of the time), with occasional strong winds (wind speed: >6 m/s, <5% of the time). Wind speed, however, did not differ between those two periods, although the wind direction reversed its path governed by two monsoon periods characterized by two distinct seasons (i.e., south-west monsoon from mid-May to early October and north-east from November to March; Uk et al. 2018; see also Chap. 6 regarding climate and rainfall in Cambodia).

TSS concentration in TSL has been found to be an exponential function of wind speed and water depth, indicating a stronger effect of wind on TSS concentration during the low-water season (Sato et al. 2021). For example, higher TSS concentrations tend to be observed under higher wind speed and a shallower condition. However, such empirical relationships between the wind speed and TSS concentration differ spatiotemporally over the lake. In general, during the low-water period in the whole TSL, TSS was related to wind speed (W) and water depth (h) by the empirical equation $TSS = f(W^m, h^{-n})$ ($m > 0$, $n > 0$). Wind speeds as low as 3 m/s were found to cause resuspension in Lake Filso (mean depth: 1.03 m), whereby resuspension increased with higher wind speed. In Lake Taihu (mean depth: 1.9 m), long-term moderate wind (2–6 m/s) keeps sediment suspended in the water column (Chao et al. 2017). Because wind speed and water depth in TSL vary from 2 to 6 m/s and from 1 to 10 m, respectively, TSS is found to spatiotemporally vary in TSL, with generally higher TSS concentration during the low-water period than during the high-water period.

19.3.2 Vegetation

Vegetation in lakes mitigates the effects of wind-induced waves and thus reduces RR in vegetated areas of TSL compared to that in areas of open water (Siev et al. 2018 and citations therein). Suspended solids in TSL are trapped by vegetation and settle out on the floodplain, up to 80% of the annual sediment influx (Kummu et al. 2008). Aquatic vegetation can potentially increase friction and resistance against wind and flow depending on its structure, density, height, and stem diameter.

The ecosystem of TSL is composed of at least eight subsystems: permanent waterbody, tributary, seasonally inundated forest, shrubland, grassland, receding and floating rice fields, seasonally flooded crop field, and marsh swamp (Lamberts 2001, Koponen et al. 2010). In addition to the natural vegetation, several other crops are cultivated in the floodplain, for example, rice, vegetables, lotus, and corn (Koponen et al. 2010). The TSL's floodplain consists of grassland (49.1%), shrub (30.2%), inundated forest (1.5%), water and soil (19.1%), and urban area (0.1%) (Baran 2005).

The average sediment rate and its proportion to the average gross sediment rate depend on vegetation types in TSL (Table 19.1). Compared to forest-dominated

Table 19.1 Average sediment resuspension rate (RR) in TSL and its proportion to average gross sedimentation rates (RR/SR) in different vegetation types compared with the open water. (Adopted from Arias et al. (2013) and Siev et al. (2018))

Vegetation				
Type	Height (m)	Average RR (mg/m^2/day)	Average RR/SR ratio (%)	Average reduction of RR/SR (%)
Grassland	1.7–3.4	117 ± 146	81	4
Shrub	4.4–6.6	178 ± 205	73	13
Inundated forest	8.2–11.2	215 ± 365	59	26
Open water	–	307 ± 370	85	0

Note: Values following "±" represent standard deviation

areas, open water has higher RR or SR, thereby confirming the effect of vegetation on resuspension in TSL. For example, the inundated forest reduces the resuspension of sediment by up to 26%, which is the highest reduction of sediment resuspension among those of other vegetation types.

19.3.3 Bioturbation

Sediment resuspension in TSL might also be caused by aquatic organisms to some extent. For example, benthivorous fish and burrowing benthic animals, which feed and move on the surface layer of sediment, can cause resuspension and increase TSS concentration (Scheffer et al. 2003). In TSL, the low-water period is the time of massive fish migrations from floodplain habitats into the permanent water body and from the lake to Tonle Sap River. The extensive and diverse migration between TSL and Mekong River has been reported for more than 200 fish species (Lamberts 2001). Additionally, experimental results from Scheffer et al. (2003) suggested that the presence of benthivorous fish may help prevent consolidation of lake sediment that, in the absence of fish, would be sufficiently stabilized during the period with little wave action. Consolidation of sediment during such quiet periods may apparently allow the sediment to become firm enough to resist the shear stress caused by waves during windy periods. Such potential effect of fish on sediment resuspension has not been investigated in TSL, but it should not be ignored and needs further investigation.

19.3.4 Human Activity

Human activities could potentially affect the resuspension of sediment although, to date, no clear data are available to show how it affects RR in TSL. The relevant activities include fishing, boat for transportation, tourism, and selling goods

(Fig. 19.4). Population growth in the TSL basin has resulted in an increase in food demand and exerts pressure on the lake's natural resources (Uk et al. 2018; see also Chap. 22). According to Hortle (2007), TSL is surrounded by one of the most densely populated regions in Cambodia. Approximately five million people live around the lake (Ministry of Planning 2013), with livelihoods relying on resources and services provided by the TSL ecosystem. In addition, this ecosystem suffers from indiscriminate fishing at the largest scale in the world (McCann et al. 2015). Thus, these human activities are possibly causes of sediment resuspension where and when the water depth is shallow enough for people to create turbulence on the sediment surface.

Key Points

- Sediment resuspension in TSL is a seasonal process, which occurs dominantly during the low-water period.

Fig. 19.4 Running boats in TSL causing sediment resuspension. (Photos taken by authors)

- Sediment in TSL and its floodplain serves as a major source of suspended solids during the low-water period, contributing to the increment of TSS concentration in the water column.
- During the low-water period when the lake is shallow, as well as in large open areas, the wind-induced wave would create sufficient turbulence in the bottom layer to stir the lake bed, resulting in resuspension.
- Flood pulse substantially influences sediment dynamics in TSL, especially resuspension, as a distinct seasonal process, which makes TSL a unique ecosystem.
- Water level and floodplain vegetation significantly affect sediment resuspension.
- To date, no clear evidence is available to show how wind, bioturbation, and human activities affect sediment resuspension in TSL; therefore, further investigation is needed.

Acknowledgments The authors would like to thank the Mekong River Commission (MRC) for providing data for this work.

References

Arias ME, Cochrane TA, Norton D, Killeen TJ, Khon P. The flood pulse as the underlying driver of vegetation in the largest wetland and fishery of the Mekong Basin. Ambio. 2013;42(7):864–76.

Bachmann RW, Hoyer MV, Canfield DE. The potential for wave disturbance in shallow Florida Lakes. Lake Reserv Manag. 2000;16:281–91.

Baran E. Cambodian inland fisheries: facts, figures and context. Phnom Penh, Cambodia: WorldFish Center; 2005.

Blom G, Van Duin EHS, Aalderink RH, Lijklema L, Toet C. Modelling sediment transport in shallow lakes—interactions between sediment transport and sediment composition; 1992. p. 153–66.

Chao J-Y, Zhang Y, Kong M, Zhuang W, Wang L-M, Shao K-Q, Gao G, Shao Q, Gao G. Long-term moderate wind-induced sediment resuspension meeting phosphorus demand of phytoplankton in the large shallow eutrophic Lake Taihu. PLoS One. 2017;12:1–15.

Churchill JH, Williams AJ, Ralph EA. Bottom stress generation and sediment transport over the shelf and slope off of Lake Superior's Keweenaw peninsula. J Geophys Res Oceans. 2004;109 (C10):C10S04.

Evans RD. Empirical evidence of the importance of sediment resuspension in lakes. Hydrobiologia. 1994;284:5–12.

Fagherazzi F, Wiberg PL. Importance of wind conditions, fetch, and water levels on wave-generated shear stresses in shallow intertidal basins. J Geophys Res. 2009;114:F03022.

Gasith A. Tripton sedimentation in eutrophic lakes—simple correction for the re-suspended matter. Verh Int Ver Limnol. 1975;109:445–54.

Gloor M, Wtiest A, Mtinnich M. Benthic boundary mixing and resuspension induced by internal seiches. Hydrobiologia. 1994;284:59–68.

Hawley N, Lesht BM, Schwab JD. A comparison of observed and modeled surface waves in southern Lake Michigan and the implications for models of sediment resuspension. J Geophys Res Oceans. 2004;109(C10)

Hortle K. Consumption and the yield of fish and other aquatic animals from the Lower Mekong Basin. MRC Technical Paper. No. 16; 2007. p. 87.

Koponen J, Lamberts D, Sarkkula J, Inkala A, Junk W, Halls A, Kshatriya M. Primary and fish production report. Vientiane, Laos PDR; 2010.

Kummu M, Penny D, Sarkkula J, Koponen J. Sediment: curse or blessing for Tonle Sap Lake? Ambio. 2008;37(3):158–63.

Kummu M, Tes S, Yin S, Adamson P, Józsa J, Koponen R, J., Sarkkula, J. Water balance analysis for the Tonle Sap Lake-floodplain system. Hydrol Process. 2014;28:1722–33.

Lamberts D. Tonle Sap fisheries: a case study on floodplain gillnet fisheries. Bangkok, Thailand: RAP Publication; 2001.

McCann KS, Gellner G, McMeans BC, Deenik T, Holtgrieve G, Rooney N, Hannah L, Cooperman M, Nam S. Food webs and the sustainability of indiscriminate fisheries. Can J Fish Aquat Sci. 2015;665:656–65.

Ministry of Planning. Cambodia Inter-censual Population Survey (2013). Spatial distribution and growth of population. 2013. Cambodia, Phnom Penh.

Penny D, Cook G, Sok SI. Long-term rates of sediment accumulation in the Tonle Sap, Cambodia: a threat to ecosystem health? J Paleolimnol. 2005;33:95–103.

Sato M, Khanal R, Uk S, Siev S, Sok T, Yoshimura C. Impact of wind on the spatio-temporal variation in concentration of suspended solids in Tonle Sap Lake, Cambodia. Earth. 2021;2: 424–39.

Scheffer M, Portielje R, Zambrano L. Fish facilitate wave resuspension of sediment. Limnol Oceanogr. 2003;48(5):1920–6.

Sheng YP, Lick W. The transport and resuspension of sediments in a Shallow Lake. J Geophys Res. 1979;84(C4):1809–26.

Siev S, Yang H, Sok T, Uk S, Song L, Kodikara D, Oeurng C, Hul S, Yoshimura C. Sediment dynamics in a large shallow lake characterized by seasonal flood pulse in Southeast Asia. Sci Total Environ. 2018;631–632:597–607.

Søndergaard M, Jensen JP, Jeppesen E. Role of sediment and internal loading of phosphorus in shallow lakes. Hydrobiologia. 2003;506–509:135–45.

Tsukawaki S. Lithological features of cored sediments from the northern part of The Tonle Sap Lake, Cambodia. In: The International Conference on Stratigraphy and Tectonic Evolution of Southeast Asia and the South Pacific. Bangkok, Thailand. 1997. p. 232–239.

Uk S, Yoshimura C, Siev S, Try S, Yang H, Oeurng C, Li S, Hul S. Tonle Sap Lake: current status and important research directions for environmental management. Lakes Reserv Res Manag. 2018;23:177–89.

Zhu M, Zhu G, Nurminen L, Wu T, Deng J, Zhang Y, Qin B, Ventelä A-M. The influence of macrophytes on sediment resuspension and the effect of associated nutrients in a Shallow and Large Lake (Lake Taihu, China). PLoS One. 2015;10:e0127915.

Chapter 20
Sediment Loads from Tonle Sap Lake Tributaries

Sok Ty, Ich Ilan, Ky Sereyvatanak, Oeurng Chantha, Song Layheang, and Chihiro Yoshimura

20.1 Sediment Load

The input of sediment to the Tonle Sap Lake (TSL) is one of the key factors, together with the flood pulse, that contribute to the high productivity of the lake (see also Chap. 31 regarding primary production). Local people unofficially claim that this lake is gradually filling up with sediment due to the increase of sediment load from the catchment (Sarkkula et al. 2003). Understanding the sources, load, and balance of sediment in the TSL is critically important to addressing this concern. The Mekong River is responsible for the majority (~72%) of the sediment delivered to TSL, whereas the tributaries of the TSL account for around 28% (Kummu et al. 2005, 2008). The average suspended sediment flux into the TSL from the Mekong and lake's tributaries is 5.1 Mt/year and 2.0 Mt/year, respectively (Kummu et al. 2005; see also Chap. 17). Far from assisting resource management, the debate surrounding sedimentation in TSL and sediment input from each major tributary has proved inconclusive or, at worst, misleading. Ongoing changes in the sediment regime due to basin developments, especially those associated with hydropower reservoirs and land cover changes, will affect the sedimentation and erosion process at the catchment scale, affecting the TSL basin (Kummu et al. 2005). Understanding the sediment load and yield from a river catchment is beneficial for management purposes because it is often needed to estimate the quantity of sediment delivered to the downstream reservoir/lake and estimate the lake pollution loading.

S. Ty · I. Ilan · K. Sereyvatanak · O. Chantha (✉) · S. Layheang
Institute of Technology of Cambodia, Phnom Penh, Cambodia
e-mail: chantha@itc.edu.kh

C. Yoshimura
Tokyo Institute of Technology, Tokyo, Japan

C. Yoshimura et al. (eds.), *Water and Life in Tonle Sap Lake*,
https://doi.org/10.1007/978-981-16-6632-2_20

20.2 Application of the Sediment Model

This section describes the hydrology and sediment transport from 1985 to 2015 in Sen River and Chinit River catchments, which are major tributaries among 11 tributaries of the TSL Basin (Fig. 20.1). The largest sub-basin of TSL is the Sen River catchment, which covers approximately 16,000 km^2 (~20% of total TSL basin) with a length of 520 km, whereas the Chinit catchment covers approximately 8300 km^2 (~10% of total TSL basin) with a length of 265 km from the north and middle of Cambodia to the end at TSL. The topography of the catchments can be categorized as mountainous (at the maximum elevation of 750 m for Sen and 680 m for Chinit) but flat in the downstream part. The basins receive precipitation from upper and middle reaches via tributary streams and meander through an incised valley and lower reach; they enter a floodplain and TSL. The majority of the sub-basins consist of lowlands with elevations of less than 100 m above the mean sea level and gentle slopes. Tropical monsoons dominate the climate in both catchments, and the rainfall slightly increases with elevation (see Chap. 6 for climate and rainfall in TSL).

The Soil and Water Assessment Tool (SWAT) model can be used to predict flow, sediment load, and nutrient amount, and water quality in large complex catchments worldwide with different soils, land use, and management conditions over long periods (Neitsch et al. 2011). In the SWAT model, sediment transport and yield are estimated using the Modified Universal Soil Loss Equation (Singh et al. 2012). In the Mekong River basin, including the TSL basin, SWAT has been previously applied to estimate the runoff from climatic parameters and use the estimated runoff as the inflow to the basin simulation model (e.g., Oeurng et al. 2019). However, besides the capability of estimating runoff from climatic data, SWAT can also be widely applied in the world to understand sub-basin erosion; sediment transport, including channel degradation, deposition; and water quality. Therefore, the SWAT model was applied to simulate the long-term streamflow and sediment load to better understand the sediment load that contributes to TSL.

The spatial dataset used in SWAT simulation consisted of the Digital Elevation Model with a resolution of 30 × 30 m, soil type, and land use/land cover in 2002 obtained from the Mekong River Commission (MRC). For both catchments, climate data were available from 1980 to 2015. For the Sen River catchment, the observed daily rainfall was available between 1980 and 2015, and the observed daily streamflow was available for the periods between 2001 and 2013 at Stung Sen City (Fig. 20.1). Sediment data were available from 2004 to 2008. For the Chinit River catchment, the observed daily streamflow from 1997 to 2015 at Kampong Thmar and sediment data were available from 2005 to 2008 (Fig. 20.1). Ground-observed rainfall and streamflow were obtained from the Department of Hydrology and River Works of the Ministry of Water Resources and Meteorology (MOWRAM). Sediment data were also obtained from MOWRAM under the MRC framework: Water Quality Monitoring Network.

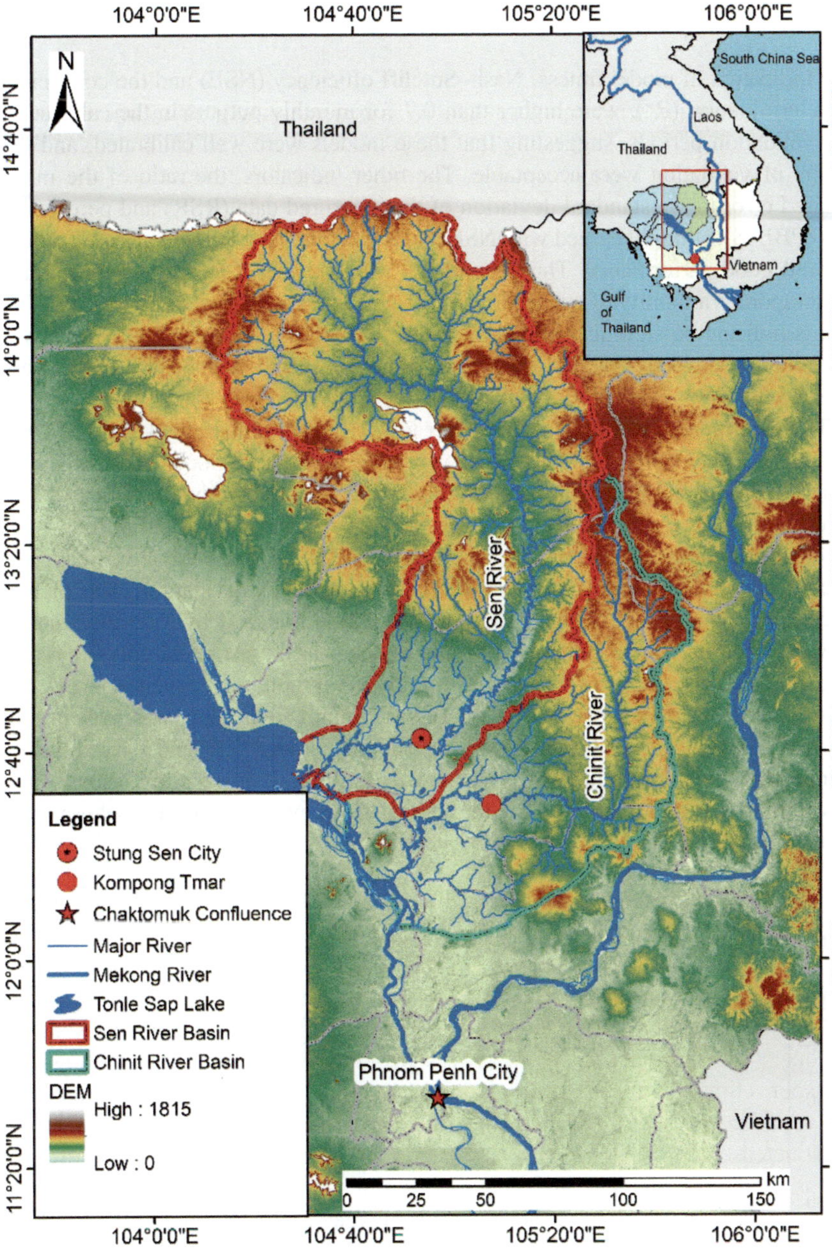

Fig. 20.1 Sen and Chinit River catchments, the major tributaries of TSL

20.3 Modeled Streamflow

The indicators of model fitness, Nash–Sutcliff efficiency (NSE) and the coefficient of determination (R^2), were higher than 0.7 for monthly outputs in the calibration and validation periods, suggesting that these models were well calibrated, and the results of validation were acceptable. The other indicators, the ratio of the mean squared error to the standard deviation of the measured data (RSR) and percentage bias (PBIAS), strongly agreed with NSE and R^2 and indicated that the SWAT model showed good performance. Thus, the model was considered to properly reproduce the temporal variability of streamflow in the Sen and Chinit River catchments. The results indicate no significant trend in water discharge of the Sen and Chinit River catchments between 1985 and 2015 and between 1990 and 2015, respectively (Fig. 20.2).

The maximum streamflow of Sen River was 910 m^3/s in the rainy season of 1999 and 2000, which were considered to be wet years and caused floods downstream of both Sen and Chinit just before entering the TSL. The streamflow decreased as low as approximately 7 m^3/s in every dry season. The specific water yield of the Chinit River catchment was 430 mm/year, whereas the specific water yield of the Sen River catchment was 388 mm/year (90% of the Chinit River discharge). The annual discharge in the Sen River catchment was highly dominated by the rainy season flow, 67% of the annual flow, whereas the dry season flow covered only 33%. For the Chinit River catchment, the low streamflow commonly happened in every dry season and was approximately 6 m^3/s. The maximum streamflow was 596 m^3/s in the rainy season of 2011, which was considered a wet year and caused heavy inundation in the lowland area. The annual discharge during rainy season flow was highly dominated by 74% of the annual flow, whereas the dry season flow covered only 26% in the Chinit River catchment.

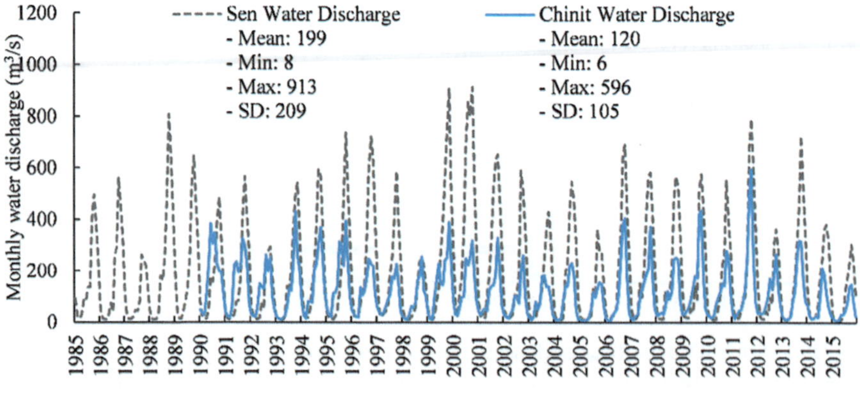

Fig. 20.2 Temporal variability of monthly streamflow contributing to TSL and its floodplain in the eastern part from the Sen River catchment from 1985 to 2015 and the Chinit River catchment from 1990 to 2015 (model outputs)

20.4 Estimated Sediment Load

All indicators (NSE and R^2) suggested that the SWAT model showed satisfactory and reasonable performance in its simulation of monthly sediment loads. For Sen River, R^2 was 0.78, and NSE was 0.70 for the model calibration (2004–2006), followed by 0.83 of R^2 and 0.70 of NSE for model validation (2007–2008). All these values of the indicators are shown in a good class. For Chinit River, R^2 was 0.82, and NSE was 0.66 for the model calibration (2005–2006), whereas R^2 was 0.77, and NSE was 0.71 for model validation (2007–2008). All of these values of the indicators are shown in a good class. The monthly and seasonal sediment load and water discharge were distinguished (Fig. 20.3). The annual average water discharge and sediment loads were estimated to be 6200 ± 2000 million cubic meters (MCM) and 0.38 ± 0.13 million tons per year (Mt/year), respectively, between 1985 and 2015 for the Sen catchment (Fig. 20.4a). In the wettest over the study period, year 2000, the water discharge was approximately 13,300 MCM, and the highest sediment load of 0.88 Mt. to the TSL was confirmed. 3000 MCM of water discharge transported a minimum annual sediment load of 0.2 Mt. in 1998. The average annual discharge and sediment load were estimated to be 3813 ± 1000 MCM and 0.24 ± 0.1 Mt/year, respectively, from 1990 to 2015 for the Chinit River catchment (Fig. 20.4b). 1600 MCM of water discharge transported the minimum sediment load of around 0.1 Mt in 2015. This value was the lowest peak at the outlet of Chinit basin in this period. Moreover, in the wettest year of the study period, 2011, the discharge was 5800 MCM, and it carried the highest sediment load of 0.46 Mt to TSL, which was the highest peak at the outlet of the Chinit River basin in that period.

The average sediment yields in the Sen and Chinit River catchments were found to be 24 t/km²/year and 29 t/km²/year, respectively. This result aligned with those of Sarkkula et al. (2003), Kummu et al. (2005), and Kummu et al. (2008), who

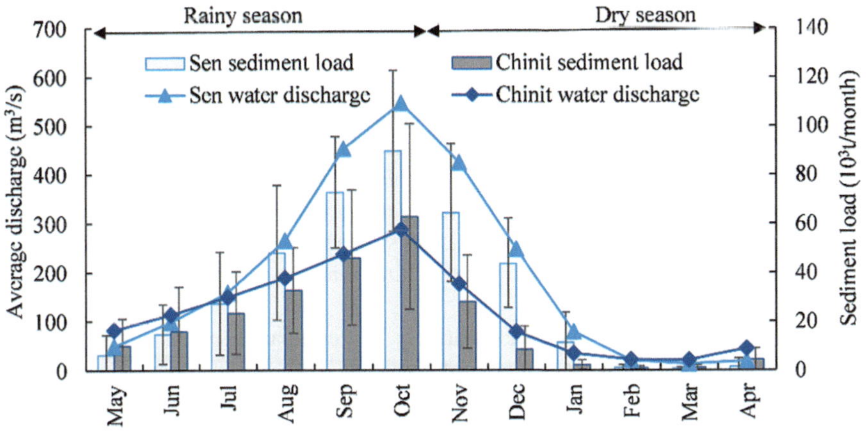

Fig. 20.3 Average monthly sediment load and water discharge contributing to Tonle Sap Lake from the Sen River catchment for 1985–2015 and the Chinit River catchment for 1990–2015

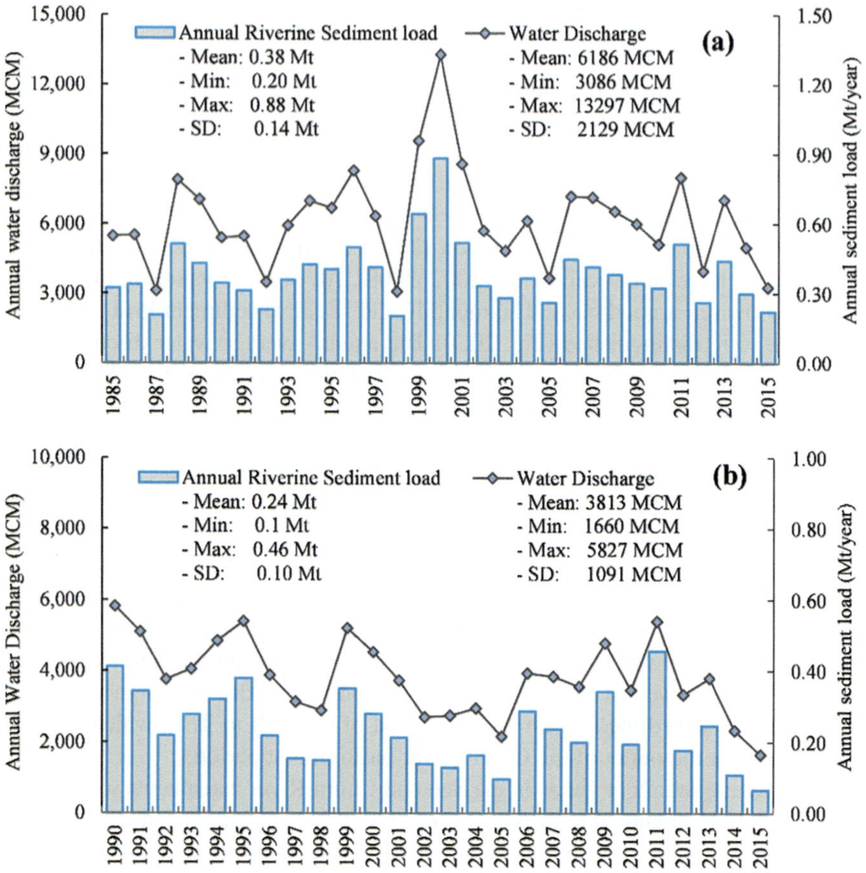

Fig. 20.4 Annual sediment load and water discharge contributing to Tonle Sap from (**a**) the Sen River catchment for 1985–2015 and (**b**) the Chinit River catchment for 1990–2015

estimated that 2.0 Mt (equal to 25.3 t/km^2/year of sediment yield) of sediment is delivered each year from the 11 tributaries that discharge into the TSL. The sediment yield in both catchments was strongly correlated with the water discharge. However, other factors affect the erosion rate, for example, the soil type (Dystric Leptosol) and land use/land cover (intensive agricultural land and deciduous). The results also indicate that the sub-basins covered by a forest generally have low sediment yields, whereas the sub-basins covered by agricultural land or grass have high sediment yields.

The average sediment yield was relatively low compared to that in the Mekong Basin (200 t/km^2/year by Milliman and Syvitski (1992)) and other tropical river basins such as the Irrawaddy (634 t/km^2/year), Purus of Amazon Basin (81 t/km^2/year), Salween in Myanmar (307 t/km^2/year) (Latrubesse et al. 2005), and Red River (600 t/km^2/year) (Dang et al. 2010) and the mean specific sediment yield of Asian and South-Eastern Asian rivers (380–600 t/km^2/year) (Milliman and Meade 1983).

The sediment derived directly from the steep slope of the two catchments may be stored temporarily at the slope base, rather than moving directly through the stream system. It is more likely that sediment particles might be stored as colluvium before moving into the stream. For the lower part of both catchments, sediment particle loads may be stored as alluvium in the river floodplain, riverbed, or bank of the stream. The upper part of the two basins is covered by various forest types. However, further study would enable an in-depth understanding of the derived sediment loads and yields and allow the identification and prioritization of sediment sources of these catchments within the TSL basin. Lastly, continuous monitoring of sediment load in all the major tributaries should be given more attention because the sediment load data in those tributaries remain scarce and/or non-existent. Quantifying the sediment load from all the tributaries is a piece of puzzle to fully understanding sediment balance in the lake.

Key Points

- Understanding the sediment load from river catchments of TSL is important for accurate estimation of sediment and carry-on pollutants transported into the lake.
- The annual average sediment loads were estimated at 0.38 ± 0.13 Mt/year for Sen River between 1985 and 2015, whereas Chinit River was estimated to deliver an average annual sediment load of 0.24 ± 0.1 Mt/year from 1990 to 2015 to TSL.
- From the exported sediment load, the average sediment yields in the Sen and Chinit catchments were estimated to be 24 t/km^2/year and 29 t/km^2/year, respectively.
- This study provides better understanding of the sediment load from the two major tributaries and provides the baseline information required to devise the sustainable catchment management plan of TSL.

References

Dang TH, Coynel A, Orange D, et al. Long-term monitoring (1960–2008) of the river sediment transport in the Red River Watershed (Vietnam): temporal variability and dam-reservoir impact. Sci Total Environ. 2010;408(20):4654–64.

Kummu M, Koponen J, Sarkkula J. Assessing impacts of the Mekong development in the Tonle Sap Lake. Paper presented at the proceedings of the international symposium on role of water sciences in Transboundary River basin management. 2005.

Kummu M, Penny D, Sarkkula J, et al. Sediment: curse or blessing for Tonle Sap Lake? AMBIO J Hum Environ. 2008;37(3):158–63.

Latrubesse EM, Stevaux JC, Sinha R. Tropical rivers. Geomorphology. 2005;70(3–4):187–206.

Milliman JD, Meade RH. Worldwide delivery of river sediment to the oceans. J Geol. 1983;91(1): 1–21.

Milliman JD, Syvitski JP. Geomorphic/tectonic control of sediment discharge to the ocean: the importance of small mountainous rivers. J Geol. 1992;100(5):525–44.

Neitsch SL, Arnold JG, Kiniry JR et al. Soil and water assessment tool theoretical doc-umentation version 2009. 2011.

Oeurng C, Cochrane TA, Chung S, et al. Assessing climate change impacts on river flows in the Tonle Sap Lake Basin, Cambodia. Water. 2019;11(3):618.

Sarkkula J, Kiirikki M, Koponen J et al. Ecosystem processes of the Tonle Sap Lake. Paper presented at the Ecotone II-1 workshop, Phnom Penh/Siem Reap, Cambodia; 2003.

Singh A, Imtiyaz M, Isaac R, et al. Comparison of soil and water assessment tool (SWAT) and multilayer perceptron (MLP) artificial neural network for predicting sediment yield in the Nagwa agricultural watershed in Jharkhand, India. Agric Water Manag. 2012;104:113–20.

Chapter 21
Physico-Chemical Properties of Suspended Solids and Sediment

Winarto Kurniawan, Chompey Den, Uk Sovannara, Sokly Siev, Phat Chanvorleak, Ty Boreborey, Eden M. Andrews, Kuok Fidero, and Hirofumi Hinode

21.1 Importance of Suspended Solids and Sediment

Suspended solids and sediment play essential roles in the aquatic system because they provide substrates for aquatic plants and microbes. They also play a significant role in the retention and transportation of essential elements for aquatic life. This role is achieved by their ability to adsorb ions and compounds, including nutrients (e.g., phosphate and nitrate) and contaminants (e.g., heavy metals and pesticides) on their surface, and they retain or transport those elements during their settlement or movement. Because of these properties, it is possible to understand the condition and status of an aquatic system by analyzing suspended solids and sediment. Therefore, this chapter will discuss the physico-chemical properties of suspended solids and sediment in Tonle Sap Lake (TSL) with insight into the overall condition of the lake.

W. Kurniawan (✉) · C. Den · U. Sovannara · E. M. Andrews · H. Hinode
Tokyo Institute of Technology, Tokyo, Japan
e-mail: kurniawan.w.ab@m.titech.ac.jp

S. Siev · K. Fidero
Institute of Technology of Cambodia, Phnom Penh, Cambodia

Ministry of Industry, Science, Technology and Innovation, Phnom Penh, Cambodia

P. Chanvorleak · T. Boreborey
Institute of Technology of Cambodia, Phnom Penh, Cambodia

C. Yoshimura et al. (eds.), *Water and Life in Tonle Sap Lake*,
https://doi.org/10.1007/978-981-16-6632-2_21

21.2 Grain Size and the Crystalline Property

Knowledge of the type of suspended solids and sediment from a particular lake is vital to understand their influence on lake habitants. To investigate the properties of TSL's suspended solids and sediment, samples of 27 sites were collected and analyzed (Fig. 21.1). Sediment samples were dried at 105 °C for 24 h prior to analysis. As mentioned in Chap. 17, the grain size of TSL sediment was fine, with the main constituent being silt; this was consistent with findings of the previous study (Sarkkula et al. 2010). The average proportion of sand, silt, and clay varied according to season due to suspended solids' settling and resuspension (Chap. 17). Clay and silt from the sediment are considered to make up a large portion of suspended solids, as such fine solids tend to be suspended in lake water. Clay is known to show substantial adsorption capability; thus, the clay constituent in the suspended solids can play a significant role in the transportation and retention of various principal elements.

Powder X-ray Diffractometry (XRD) indicated a slight variance in the crystallinity of sediment among the sampling sites. The major crystalline structure of quartz was detected in all samples, together with some clay minerals (smectite and illite) and various other minor structures. The presence of quartz in all of the samples is considered to be the result of erosion from surface soil of alluvial deposits surrounding catchments of TSL and from the erosion of metamorphic and granitic rock bodies distributed in the Mekong River basin (Kummu et al. 2008; Tsukawaki et al. 1994).

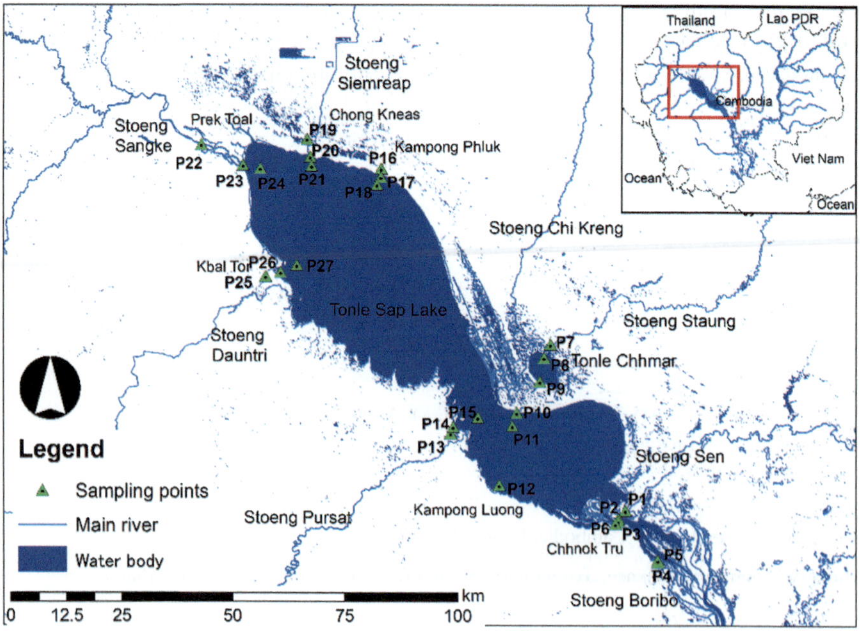

Fig. 21.1 Location of sampling sites

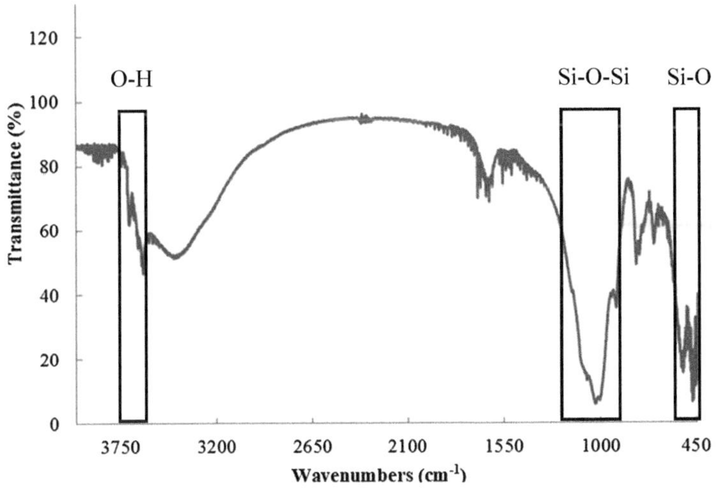

Fig. 21.2 Example of FTIR spectrum of TSL sediment (sample from P2 at the border between the Lake and Tonle Sap River; see Fig. 21.1)

By contrast, illite also originates from the Mekong River, which flows to TSL through Tonle Sap River (TSR) during the rainy season (Okawara and Tsukawaki 2002). This hypothesis was confirmed by the presence of illite in the sediment of TSR. However, illite has not been detected in the sediment and soil samples from other rivers and the alluvial plains around the lake.

Fourier Transform Infrared Spectrophotometry (FTIR) further confirmed the presence of siliceous material in the sediment of TSL. As illustrated in Fig. 21.2, the presence of absorbance peak in the wavenumber range from 800 to 1200 cm^{-1}, which is attributable to Si–O–Si bonds, was observed. Particularly, the peak around a wavenumber of 900 cm^{-1}, attributable to kaolinite (illite), reconfirmed the presence of clay minerals in the sediment as shown by grain size analysis and XRD analysis. The analysis of heat-treated suspended solid samples showed similar spectra, which further confirmed the presence of clay minerals in the suspended solid samples.

These findings prove the presence of clay minerals in the suspended solids and sediment in TSL. Further studies on the influence of those clay materials in the transportation and retention of compounds such as phosphorus and heavy metals are necessary to investigate their relation to the water quality of TSL.

21.3 Thermal Property

Thermogravimetric-Differential Thermal Analysis (TG-DTA) was used to study the thermal behavior of the samples. TG was used to study the weight losses that had occurred during the heating of the sample, whereas DTA was used to study the heat flow from and to the sample during heating.

The TG-DTA curve of sediment sampled from Site P2 (Fig. 21.3) looked similar to that of sediment from other sites. Three significant weight losses were observed in the TG curve. These losses are observed from around room temperature to 200 and 450 °C. The first weight loss was associated with an endothermic peak in the DTA curve and can thus be attributed to the release of gas and moisture trapped in the pores of particles or in the inter-particulate spaces. Based on the high percentage of weight loss (around 5%), most weight loss could be attributed to moisture release.

The second weight loss was associated with an exothermic peak in the DTA curve, suggesting oxidation. Siev et al. (2018) reported that TSL sediments contained 4.1% of organic material, closely resembling the weight loss shown in the figure (3.8%); thus, this result justifies the attribution of this weight loss to the oxidation of organic content in the samples. Given the broad exothermic peak in the DTA curve, it can be concluded that the organic contents in the samples consist of various chemical compounds.

The third weight loss was associated with another endothermic peak in the DTA curve. This peak indicates carbonates' presence, including calcium carbonate in bivalve shells, which can be identified by visual observation. Calcium carbonate decomposes at elevated temperatures (600–900 °C) into calcium oxide and carbon dioxide, which likely contributes to the third weight loss. However, since the weight

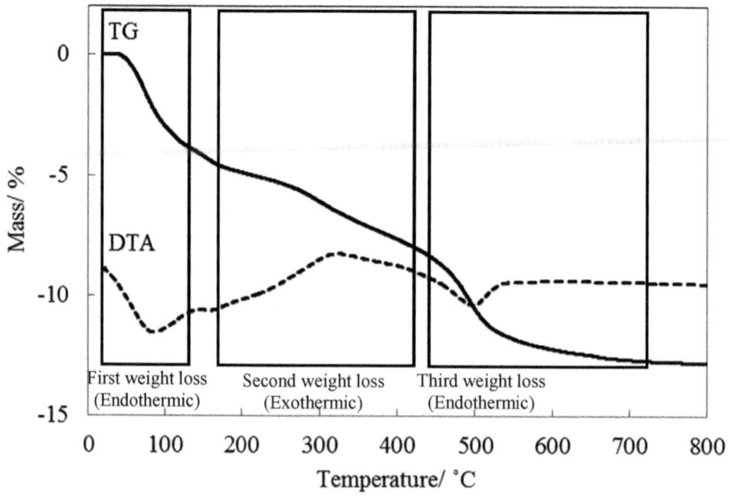

Fig. 21.3 A typical TG-DTA curve of TSL sediment (sample from P2 at the border between the Lake and Tonle Sap River; see Fig. 21.1)

loss started from 450 °C, other carbonates such as sodium carbonate might also be present.

Furthermore, analysis of suspended solids indicated loss on ignition ranging from 2.5% to 87.0%, with an average value of 31.0%. This result indicated that most suspended solid constituents were unburnable materials, as described in the previous section. The burnable constituent of suspended solids mainly originates from organisms such as vegetation, phytoplankton, and aquatic fauna.

21.4 Elemental Constituents

X-ray fluorescence spectrometry (XRF) was used to investigate the elemental composition of suspended solids and sediment. For elemental analysis of heavy metals with concentrations lower than the detection limit of XRF, Atomic Absorbance Spectroscopy (AAS) was used. According to method 3050B of the US Environmental Protection Agency (EPA), the measurement samples for AAS analysis were prepared by digestion of solid samples.

The elemental analysis indicated that the main constituents of sediment samples were silicon (Si), aluminum (Al), and iron (Fe). The average concentrations of those major constituents are Si, 34.3%; Al, 8.3%; and Fe, 3.9%. It is natural to observe Si, Al, and Fe in the sediment given their most abundant quantity in the earth's crust besides oxygen (Lutgens and Tarbuck 2000). Results from other lakes (Badovci Lake in Kosovo and Skaldar Lake in the Albania–Montenegro border) indicated that Al and Fe are the major components in the sediment, even though the concentrations of Al and Fe in those lakes were lower than those in TSL (1.7% and 1.9% for Badovci Lake and 1.4% and 1.7% for Skaldar Lake, respectively) (Shehu et al. 2020). The high concentration of those elements, especially Fe, in TSL could be due to the difference in geological properties of those lakes.

Phosphorus (P) was also detected in some of the samples, with the highest concentration of 14.0%. The sedimentary P might be in organic form in sediments. However, as the organic contents measured in sediments were relatively low, there should be other origins of P in the sediment. As mentioned in Chap. 24, in March, when the P concentration of lake water was the highest, particulate P contributed significantly to the total P concentration. Therefore, it was considered that P existed as a phosphate ion, which is adsorbed on the surface of sediment from lake water. Adsorption of the significant amount of phosphate ions can only occur when the concentration of phosphate ions is high in the lake water. The high concentration of P in TSL sediment implies the possibility of an increased concentration of phosphate ions in some parts of the lake, as discussed later in this chapter.

Besides Fe, manganese (Mn) was detected in the sediment, although Mn only existed in a relatively low content (<0.6%). Compared to the Mn concentration in lake water at the same sampling site, Mn in sediment exists mainly as a naturally occurring mineral, not because of adsorption from lake water. This is because the Mn concentration in lake water samples taken from this site was low, and the presence of

Mn at low concentration as a naturally occurring mineral is a commonly found phenomenon (Lopez et al. 2006), indicating its low health risks to human populations and the ecosystem.

In addition to P, the spatial distributions of major elements (Si, Al, Fe, and Ca) were confirmed. The spatial distribution of Si indicated that the regions near estuaries of tributaries indicated high Si content. This is due to the flow of tributaries, which transports suspended solids to the estuaries. Moreover, high concentrations of Si were found in the southern part of TSL due to the transport of suspended solids and sediment with the water exchange between TSL and TSR.

The concentration of Fe was found to be higher in the northern part of the lake. Moreover, the comparison to Fe concentrations in sediment in various lakes such as Bafa Lake in Turkey, Badovci Lake in Kosovo, and Skaldar Lake in the Albania–Montenegro border and average value from lakes in Poland (Bafa Lake = 2.9%, Badovci Lake = 1.9%, Skaldar Lake = 1.7%, average of lakes in Poland = 1.5%) indicated that Fe concentrations in TSL are higher (Algül and Beyhan 2020; Bojakowska 2016; Shehu et al. 2020). This comparison could suggest that besides the natural sources, Fe in TSL could also come from human activity, such as corrosion and abrasion of human infrastructure. However, the geo-accumulation index analysis for Fe in TSL indicated that the value was negative (between −1 and −7) for all sites. The present value is still lower than the world surface rock average; therefore, the pollution of Fe from anthropogenic sources is unlikely.

The Ca content is well distributed throughout TSL. This Ca content was considered to originate from bivalve shells as visually observed in the sediment samples and from Ca containing minerals such as zeolitic minerals, commonly found in the natural environment.

The spatial distribution of P indicated a higher concentration of phosphorus in the northern part of the lake (Fig. 21.4). The hot spot for P concentration is within the estuaries of tributaries and major floating villages such as Chong Kneas, Kampong Phluk, Prek Toal, and Kbal Tor (Fig. 21.4). The high P concentration in sediment from this area is likely to originate from high concentrations of phosphate ions in the lake water there. Comparison of the spatial distribution of P in sediment to P in water, as illustrated in Chap. 24 (Fig. 24.2), indicated a similar pattern, thus supporting the hypothesis that P in the sediment is derived from the adsorption of P in the lake water. Notably, the high phosphorus concentration in the northern part of the lake can be attributed to human activities. Aside from the P from household sewage, P from fertilizer used in the agricultural sector is generally the major contributor. According to Senevirathne et al. (2010), between 1990 and 2009, agricultural land in the northern part of the lake increased significantly (34%), potentially contributing to the increase in P concentration in the lake water in this region.

The elemental analysis using AAS confirmed that only lead (Pb), besides Fe and Mn, existed at concentrations higher than that stipulated in the EPA standard in the sediment samples. Zinc (Zn), copper (Cu), chromium (Cr), and cadmium (Cd) were also detected but at a lower concentration than the EPA standard. Furthermore, the geo-accumulation indices for Zn, Cu, Cr, and Cd are in the range of −3 to −7, −3 to

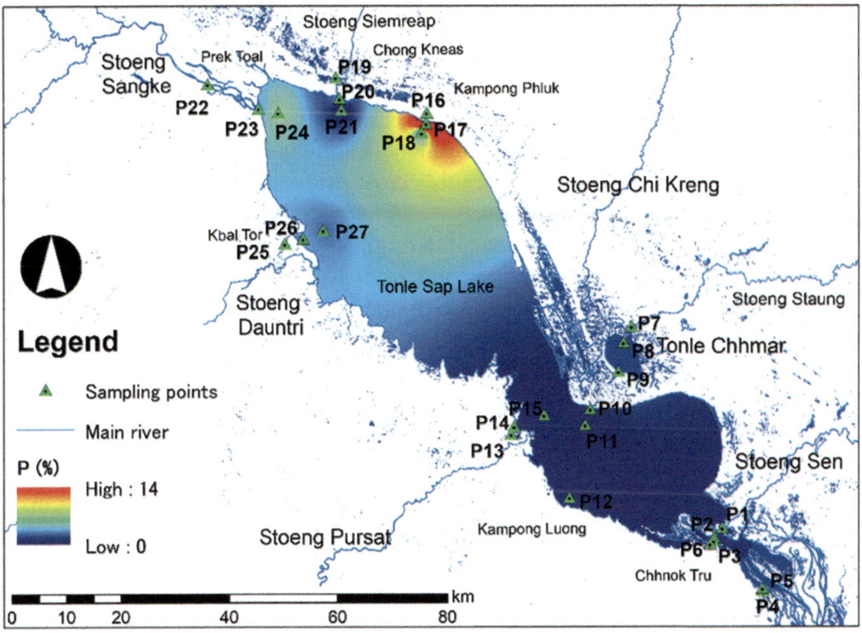

Fig. 21.4 Spatial distribution of sedimentary phosphorus in TSL

−9, −6 to −10, and −3 to −7, respectively, confirming that these elements' concentrations are lower than the world surface rock average, thus suggesting that pollution from anthropogenic sources is unlikely. The spatial distributions of Pb in sediment and lake water did not resemble each other; thus, Pb in sediment and Pb in lake water did not influence each other, and Pb in sediment is likely to originate from naturally occurring insoluble minerals, thus posing low threats to both humans and the environment. Still, the geo-accumulation index analysis indicated that the index for Pb was in the range of 0 to 1, indicating low to moderate pollution; thus, ongoing monitoring of lead concentration in lake water and sediment is necessary.

However, the additional analysis on Pb concentration of the lake water indicated an average concentration of 0.10 mg/l, which was higher than that in EPA standards for lake water. This fact could pose a hazard to human health and the lake ecosystem. This high concentration of Pb might be attributed to anthropogenic activities such as runoff from non-point sources and tributaries of TSL in the form of pesticides, waste from batteries, additives in gasoline and pigment, and exhaust from automobiles (Jaishankar et al. 2014). Another possibility is that Pb contamination could also originate from the Mekong River basin. The concentration of Pb in the rainy season (average value: 0.15 mg/l) was higher than that in the dry season (average value: 0.10 mg/l). During the rainy season, the water flows from the Mekong River into TSL, suggesting that the difference in Pb concentration between the rainy and dry seasons is due to the flow from the Mekong River. Nonetheless, more effort for continuous monitoring of Pb concentration and elucidation of the cause is necessary.

Key Points

- Suspended solids and sediment from TSL consist mainly of clay minerals in the form of smectite and illite, which might promote the retention and transport of important minerals.
- Major elements of TSL sediment are silicon, aluminum, and iron, which are also the significant constituents of the earth's crust, with an average concentration of 34.3%, 8.3%, and 3.9%, respectively.
- Ca was also observed and considered to originate from bivalve shells as visually observed in the sediment samples and from Ca containing minerals such as zeolitic minerals, which are commonly found in the natural environment.
- Geo-accumulation index analysis indicated that despite TSL has high iron concentrations compared to the values found in other lakes, the pollution of Fe from anthropogenic sources is unlikely.
- High phosphorus content (up to 14.0%) was detected in sediment from the northern part of the lake, which is attributed to anthropogenic activities.
- High lead concentration, with an average value of 36.7 mg/kg in sediment, was found. A separate finding of high lead concentration in lake water indicated a potential risk to human and environmental health, and continuous monitoring of lead concentration in lake water and sediment and additional research are necessary.

References

Algül F, Beyhan M. Concentrations and sources of heavy metals in shallow sediments in Lake Bafa, Turkey. Sci Rep. 2020;10(1):1–12.

Bojakowska I. Phosphorus in lake sediments of Poland—results of monitoring research. Limnol Rev. 2016;16(1):15–25.

Jaishankar M, Tseten T, Anblagan N, Mathew BB, Beeregowda KN. Toxicity, mechanism and health effects of some heavy metals. Interdiscip Toxicol. 2014;7(2):60–72.

Kummu M, Penny D, Sarkkula J, Koponen J. Sediment: curse or blessing for Tonle Sap Lake? Ambio. 2008;37:158–63.

Lopez P, Navarro E, Marce R, Ordoñez J, Caputo L, Armengol J. Elemental ratios in sediments as indicators of ecological processes in Spanish reservoirs. Limnetica. 2006;25(1–2):499–512.

Lutgens FK, Tarbuck EJ. Essentials of geology. 7th ed. Upper Saddle River: Prentice Hall; 2000.

Okawara M, Tsukawaki S. Composition and provenance of clay minerals in the northern part of Lake Tonle Sap, Cambodia. J Geogr. 2002;111(3):341–59.

Sarkkula J, Koponen J, Lauri H, Virtanen M. Origin, fate and impacts of the Mekong sediment. Vietiane, Mekong River Commission; 2010.

Senevirathne, N., Mony, K., Samarakoon, L., Hazarika, M.K. (2010) Land use/land cover change detection of tonle sap watershed, Asian Conference of Remote Sensing, Cambodia

Shehu I, Stafilov T, Faiku F. Assessment of heavy metal concentrations with fractionation method in sediments and waters of the Badovci Lake (Kosovo). J Environ Public Health. 2020; https://doi.org/10.1155/2020/3098594.

Siev S, Yang H, Sok T, Uk S, Song L, Kodkara D, Oeurng C, Hul S, Yoshimura C. Sediment dynamics in a large shallow lake characterized by seasonal flood pulse in Southeast Asia. Sci Total Environ. 2018;631-632:597–607.

Tsukawaki S, Okawara M, Lao KL, Tada M. Preliminary study of sedimentation in Lake Tonle Sap, Cambodia. J Geogr. 1994;103:623–36.

Part V
Physicochemical Water Quality

Chapter 22
Basin-Wide Distribution of Water Quality

Mong Marith, Uk Sovannara, Sok Ty, Kaing Vinhteang, Oeurng Chantha, Rajendra Khanal, and Chihiro Yoshimura

22.1 Seasonality and Spatial Distribution of Water Quality

Water quality is an important factor that determines the state of an aquatic environment, having significant effects not only on the environmental health of the ecosystem but also on the livelihood of people relying on its resources and services. In the case of Tonle Sap Lake (TSL), the ecosystem has been providing numerous benefits, and there exist remarkable interactions among various domains (see Chap. 1). In this context, this chapter assesses the long-term trend and status as well as seasonal and spatial variation of water quality in the TSL ecosystem including its tributaries between 1995 and 2010, and then relevant potential factors affecting water quality are discussed.

A total of 16 physicochemical parameters at 6 available monitoring stations located in the TSL basin were selected during the investigated period from the Mekong River Commission (MRC)'s database. Five of the six stations are located in the tributaries, while one station represents the lake body at Kampong Luong (KL). Two of the tributary stations, Bak Prea (BP) and Phnom Krom (PKr), are at the furthermost upstream area of the lake in Siem Reap and Battambang, respectively.

M. Marith · S. Ty · O. Chantha · C. Yoshimura
Institute of Technology of Cambodia, Phnom Penh, Cambodia

U. Sovannara (✉)
Tokyo Institute of Technology, Tokyo, Japan

K. Vinhteang
Tokyo Institute of Technology, Tokyo, Japan

Institute of Technology of Cambodia, Phnom Penh, Cambodia

R. Khanal
Tokyo Institute of Technology, Tokyo, Japan

Policy Research Institute, Kathmandu, Nepal

C. Yoshimura et al. (eds.), *Water and Life in Tonle Sap Lake*,
https://doi.org/10.1007/978-981-16-6632-2_22

Three of them are located downstream of the lake, namely, Phnom Penh Port (PPP) located at the confluence between Tonle Sap River (TSR) and Mekong River (MR), Kampong Chhnang (KC), and Prek Kdam (PK) located in TSR.

Influenced by the regional monsoonal climate and the monotonal flood pulse, TSL exhibits distinguishing characteristics between seasons not only in hydrology but also in biogeochemistry. The average monthly concentrations for 1995–2010 show that the water quality in TSL varied over the year (Table 22.1). To understand the seasonality, cluster analysis (CA) was applied to the normalized data of the 16 selected water quality parameters at each sampling station. The result confirmed the temporal variability of sampling sites over the lake, and months were classified into different clusters at all stations (Table 22.2). In the result from CA, Cluster I included November, December, January, February, and March, corresponding to the beginning of the dry season in Cambodia and representing the transition period when the lake discharges water into its only outlet TSR to MR. The water depth in the lake started to gradually recede, reaching a low level in the dry season.

Cluster II included April, May, and June, corresponding to the late dry and early wet seasons. In the dry season, when the lake's water depth is shallow and stagnant while the temperature is high, water pollution by human becomes worse, especially in densely populated areas (Lamberts 2001; MRC 2010). Rainwater and fishery activities also contributed to the increase in nutrient concentrations in the dry season (Takahashi et al. 2002). The early monsoon rainwater in April to May flows through towns and fields carrying agrochemical, wastewater, and other wastes into the lake. It also raises the concentration of the nutrient by stirring the anaerobic bottom sludge and eluting the ammonium ions into the lake water. Additionally, when the fishery activities peak in January, thousands of people gather to make a traditional Cambodian cheese called "prohok," leaving fish gut and blood into the water. These wastes settle at the bottom of the lake. These fish wastes start to decompose as the temperature continues to rise toward the dry season, and this consumes oxygen and increase the level of ammonium ions.

Cluster III included the remaining months (July, August, September, and October), corresponding to the wet season. This is the period when the flow reversal in TSR occurs, introducing the backward water from MR as the new water source for TSL in addition to precipitation. The incoming water from MR has different water quality from the lake and its tributaries (e.g., Ca^{2+}, Mg^{2+}), possibly a factor partly distinguishing the water quality in the lake during the wet season from in the lake during the dry season (e.g., biological production and sedimentation). From the perspective of water quality assessment, the results from CA suggested that sampling during only three sampling periods in a year may be sufficient to assess the temporal variability of water quality in TSL.

The seasonal variability of water quality at the tributary monitoring stations indicated the influence of other factors that dynamically change relative to the season (e.g., rainfall, agriculture, flood pulse). The results from CA were also different among stations, implying that the water quality in the tributaries shows the spatial distribution in terms of seasonal patterns (Table 22.2). For example, although BP and PKr are both located in the upstream region of the lake, the cluster analysis grouped the sampling month differently. Likewise, for the stations along TSR, which

Table 22.1 The monthly average concentration of the water quality parameters monitored at Kampong Luong in TSL between 1995 and 2010. Data were obtained from MRC's database

Months	Temp (°C)	pH (–)	TSS (mg/L)	EC (mS/m)	Ca^{2+} (meq/L)	Mg^{2+} (meq/L)	Na^{+} (meq/L)	K^{+} (meq/L)	Alk (meq/L)	Cl^{-} (meq/L)	SO_4^{2-} (meq/L)	NO_3–N (mg/L)	NH_4^{+}–N (mg/L)	TP (mg/L)	DO (mgL)	COD_{Mn} (mg/L)
Jan	28	6.9	27	8.7	0.39	0.23	0.24	0.04	0.66	0.09	0.10	0.15	0.08	0.09	6.3	4.8
Feb	30	7.3	37	8.1	0.38	0.17	0.24	0.05	0.61	0.09	0.11	0.06	0.08	0.05	6.7	5.7
Mar	31	7.0	101	9.0	0.33	0.20	0.26	0.05	0.56	0.10	0.12	0.39	0.14	0.11	7.0	5.7
Apr	32	6.9	249	8.5	0.27	0.19	0.35	0.06	0.50	0.12	0.17	0.60	0.11	0.17	5.3	5.8
May	31	7.0	248	8.7	0.30	0.16	0.35	0.07	0.46	0.16	0.22	0.55	0.05	0.26	6.0	5.1
Jun	31	7.0	174	11.0	0.44	0.24	0.34	0.05	0.63	0.14	0.27	0.26	0.06	0.15	5.6	4.2
Jul	31	7.1	25	11.2	0.53	0.25	0.26	0.04	0.73	0.10	0.21	0.18	0.10	0.06	6.2	3.6
Aug	30	7.1	40	10.5	0.57	0.25	0.23	0.04	0.84	0.08	0.14	0.16	0.09	0.08	5.3	2.6
Sep	30	6.9	22	9.6	0.53	0.25	0.17	0.03	0.80	0.06	0.11	0.08	0.09	0.03	5.7	3.2
Oct	31	7.0	7	8.6	0.44	0.22	0.17	0.04	0.68	0.06	0.12	0.05	0.07	0.04	5.7	3.1
Nov	30	6.9	6	7.4	0.37	0.20	0.16	0.03	0.57	0.05	0.11	0.07	0.06	0.03	6.6	3.6
Dec	28	6.8	16	7.8	0.34	0.22	0.19	0.04	0.58	0.08	0.13	0.07	0.08	0.09	6.0	4.4
Average	**30**	**7.0**	**77**	**9.1**	**0.41**	**0.22**	**0.24**	**0.04**	**0.64**	**0.09**	**0.15**	**0.21**	**0.08**	**0.10**	**6.0**	**4.3**

Table 22.2 Seasonality in water quality in TSL based on cluster analysis (CA). The signs (○, •, Δ, and ◇) represent identical groups of the sampling period in terms of water quality. Note that CA was applied to each site separately; thus, an identical sign does not mean similar water quality among the stations

Month		Monitoring station					
		Bak Prea (BP)	Phnom Krom (PKr)	Kampong Luong (KL)	Kampong Chhnang (KC)	Prek Kdam (PK)	Phnom Penh Port (PPP)
Dry season	Nov	○	Δ	○	◇	•	•
	Dec	○	Δ	○	◇	○	•
	Jan	Δ	Δ	○	○	○	•
	Feb	•	Δ	○	○	○	•
	Mar	•	Δ	○	○	○	•
	Apr	•	•	•	•	Δ	•
Wet season	May	Δ	•	•	•	Δ	Δ
	Jun	Δ	•	•	Δ	◇	Δ
	Jul	Δ	•	•	Δ	◇	○
	Aug	Δ	○	Δ	Δ	◇	○
	Sep	○	○	Δ	◇	•	○
	Oct	○	○	Δ	◇	•	•

experienced flow reversal between TSL and MR annually, a year can be classified into three or four groups, and the water quality exhibited different characteristics based on the seasons and locations.

22.2 Long-Term Trend and Status of Water Quality

Long-term trend analysis revealed that the water quality in the TSL basin has degraded over the period 1995–2010 (Table 22.3), which suggested an increasing pressure on the lake ecosystem. For example, significantly increasing trends were detected in most physicochemical parameters measured in TSL at KL. For the upstream part of the basin, significantly increasing trends in EC and NH_4^+-N were observed at PKr and SO_4^{2-} at BP, respectively. For the downstream stations, including KC, PK, and PPP, significant trends in TP, SO_4^{2-}, Cl^-, and Na^+ were found at all three sites. EC and TP also showed significant increasing trends at KC and PPP. Significant increasing trends in TSS were also observed at PK and KC.

Table 22.4 presents the average concentrations of major ions in TSL between 1995 and 2010 and other freshwater lakes in the world. In general, the concentrations (meq/L) of Ca^{2+} (0.41), Mg^{2+} (0.22), Na^+ (0.24), K^+ (0.04), Cl^- (0.09), and SO_4^{2-} (0.150) are within the range of the world lakes (Table 22.4). The water quality in TSL is presumed to result from a mixture of water in the basin and backward water from MR (Oyagi et al. 2017), possibly driving clear seasonal patterns of water quality in TSL.

Table 22.3 Long-term trends of water quality parameters in the TSL basin based on the Seasonal Kendall test from 1995 to 2010. Note that temperature was excluded from the table because there was no significant trend detected at all monitoring stations. The arrows represent significant increasing or decreasing trends at p-value <0.05. The value below each arrow represents the calculated Sen's slope ($\times 10^{-4}$ per year). Data of water quality record were obtained from MRC's database

Monitoring station	pH	TSS	EC	Ca^{2+}	Mg^{2+}	Na^{+}	K^{+}	Alk	Cl^{-}	SO_4^{2-}	NO_3^{-}–N	NH_4^{+}–N	TP	DO	COD_{Mn}
Upstream															
BP						↗			↗	↗					
						0.7			0.6	0.5					
PKr			↗		↗			↗		↗		↗			
			16		0.7			0.9		0.6		0.35			
Lake station															
KL		↗	↗	↗		↗	↗	↗	↗	↗	↘		↗		↘
		36	5	0.3		0.1	0.02	0.2	0.1	0.2	0.1		0.2		2.9
Downstream															
KC		↗	↗			↗	↗		↗	↗	↘		↗		
		46	3			0.1	0.01		0.1	0.2	0.2		0.14		
PK	↘	↗				↗			↗	↗			↗	↗	↘
	0.6	42				0.1			0.1	0.2			0.13	1.4	1.6
PPP	↘		↗			↗			↗	↗		↗	↗		
	0.6		3			0.1			0.1	0.2		0.04	0.16		

Table 22.4 Major ions in various freshwater lakes in the world

Lake	Location	Mean concentration of ions (meq/L)					
		Ca^{2+}	Mg^{2+}	Na^{+}	K^{+}	Cl^{-}	SO_4^{2-}
Tonle Sap Lake	Cambodia	0.41	0.22	0.24	0.04	0.09	0.150
Lake Santa Maria del Oro[a]	Central Mexico	1.05	5.66	8.20	0.48	6.86	0.030
Taihu Lake[b]	China	1.597	0.59	2.81	0.13	2.61	1.597
Lake Bosomtwee[c]	Ghana	0.15	1.15	11.18	1.13	2.67	0.119
Loktak Lake[d]	India	0.48	0.63	0.34	0.23	0.65	0.002
Sella Lake[e]	India	0.06	0.08	0.00	0.00	0.03	0.003
Pandoh Lake[f]	India	0.78	0.27	0.19	0.17	0.07	0.057
Lake Chilwa[g]	Malawi	0.95	0.65	14.29	0.32	9.67	0.379
Inle Lake[h]	Myanmar	1.83	1.36	0.44	0.43	0.076	0.033
Rara Lake[i]	Nepal	0.57	0.66	0.02	0.03	0.08	0.004
Lake Onego[j]	Russia	0.22	0.15	0.04	0.05	0.05	0.067

[a] Caballero et al. (2013)
[b] Tao et al. (2013)
[c] Amankwaa et al. (2020)
[d] Mayanglambam and Neelam (2020)
[e] Deka et al. (2015)
[f] Ramanathan (2007)
[g] Missi and Atekwana (2020)
[h] Akaishi et al. (2006)
[i] Gurung et al. (2018)
[j] Efremova et al. (2019)
Data for Tonle Sap Lake were obtained from MRC's database for the period 1995–2010

The water quality status in TSL between 1995 and 2010 was also compared against the recommended guideline recommended by MRC for the protection of aquatic life, human health, eutrophication, and agriculture. Overall, the pH value varied from 6.0 to 8.4, having an average value of 7.0. The variation of pH value was within the recommended range for the protection of human health and the protection of aquatic life (pH between 6 and 9; Table 22.1). However, the values of EC, NH_4^+-N, and COD_{Mn} were not compliant with the recommended values for the protection of human health (Table 22.1; Fig. 22.1b, c). Likewise, DO sometimes fell as low as 1.7 mg/L, far below the guideline value for the protection of human health (DO $\geq$6 mg/L) and the protection of aquatic life (DO > 5 mg/L; Table 22.1; Fig. 22.1a). The maximum TP concentration recorded was 0.99 mg/L, which was far above the guideline value of eutrophication (TP < 0.13 mg/L; Table 22.1, Fig. 22.1d), indicating occasional/seasonal problem of eutrophication (see Chap. 24 for details regarding phosphorus dynamics). The values of EC varied between 3.5 and 21.2 mS/m, with an average value of 9.1 mS/m (Table 22.1). These values fall outside the recommended ranged values recommended by MRC for the protection of human health (70 mS/m < EC < 150 mS/m). However, the water quality as indicated by EC values (Table 22.1) showed no problem for agricultural use both for general irrigation (<70 mS/m) and paddy rice (<200 mS/m). TSL water pollution has threatened the drinking water supply for local people,

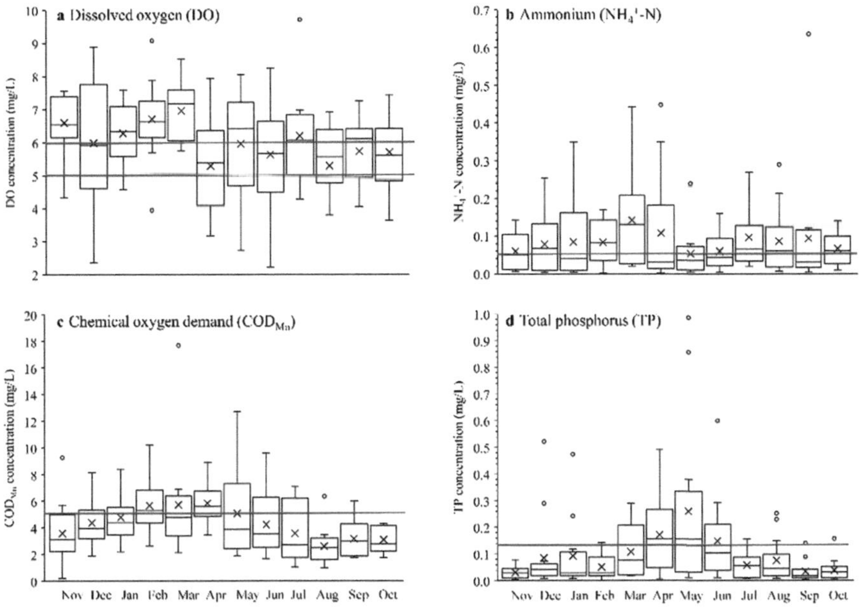

Fig. 22.1 Variation of the water quality parameters that are not compliant with the water quality guideline for (**a**, **b**, and **c**) the protection of human health, (**b**) protection of aquatic life, and (**d**) the protection from eutrophication proposed by MRC. Note: EC was not plotted because the values were not compliant with the recommended range for the protection of human health for all months. Red lines represent the recommended threshold

while eutrophication has resulted in an explosive development of harmful invasive plant species such as water hyacinth (Kuenzer 2013).

22.3 Major Factors Affecting Water Quality

The significant increasing trends in major ions (e.g., SO_4^{2-}, Cl^-, Na^+), nutrients (e.g., TP, NH_4^+-N), and EC and Alk in the TSL ecosystem (Table 22.3) indicate increasing anthropogenic pressure on water quality in TSL.

Over the past decades, many changes have been occurring in the ecosystem of TSL as well as its adjacent ecosystem (e.g., MR basin). The total population in Cambodia has greatly increased over time, doubling within just a few decades from roughly seven million after the collapse of the Khmer Rouge regime in 1979 to over 15 million in 2015. This trend is predicted to reach 22.5 million by 2050 (Fig. 22.2; Uk et al. 2018). In fact, TSL is surrounded by the most densely populated region of Cambodia (Hortle 2007), with approximately five million people inhabiting on and around the lake (see Chapter 1). The average annual population growth in the TSL basin between 1998 and 2008 was estimated at 1.4% (Keskinen et al. 2013). People

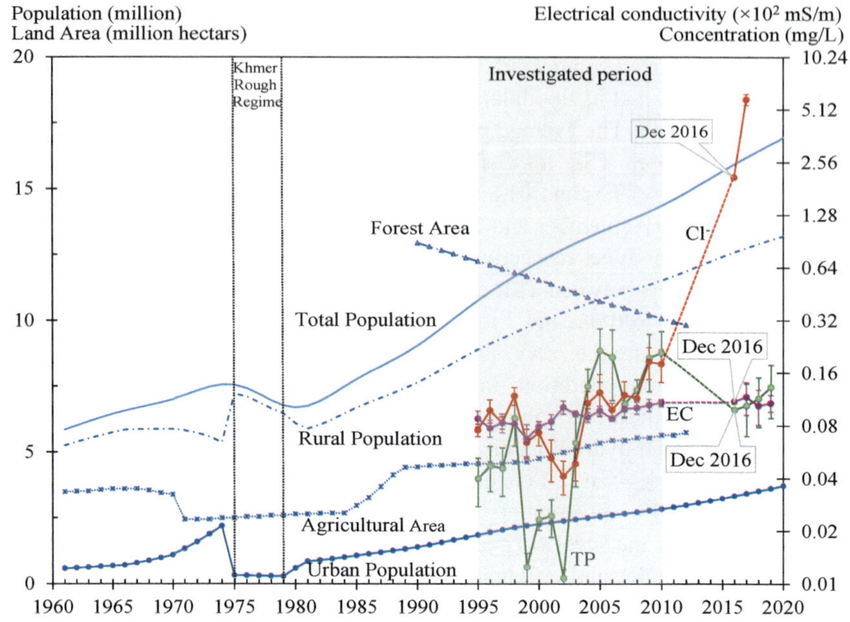

Fig. 22.2 Population and land use trends in Cambodia, including the TSL region, adapted from UK et al. (2018), and some key parameters' trends in TSL in 1995–2010. Error bars represent standard errors, and gray shade represents the investigated period

also dwell directly on the lake water in floating villages in and around the inundated forest (see Chap. 1), discharging wastewater directly into the lake without treating it.

The demographic change in the TSL basin has intensified human demands, consequently leading to significant land use and land cover changes. Significant expansion of the agricultural lands and decline of forested land areas have been observed in TSL. The upstream area of the TSL basin is occupied mainly by agricultural land area and some rapid urbanization in which the population growth rate is fast, while the downstream area is dominated by forest, but the forest cover in the area has been reduced drastically (Senevirathne et al. 2010). Around the lake, the flooded forest has also declined drastically (see Chap. 47). Such significant change negatively affected the water quality and the health of the lake and its tributaries.

Agriculture contributes to environmental problems through the application of fertilizers and pesticides, deforestation, and irrigation. Previous studies revealed that EC, TSS, and TP had a strong positive correlation with agricultural land but a strong negative relation with forest cover (Sobonn et al. 2007; Tong and Chen 2002). In the lower MR basin, which includes the TSL ecosystem, it has been reported that 75% of TP is largely attributable to intensive agricultural activities, mainly from the increasing trend of fertilizer uses (Liljeström et al. 2012). Pollution from other agricultural chemicals (i.e., pesticides) is also potentially substantial in the TSL basins (Uk et al. 2018; see also Chaps. 36 and 37).

Nonetheless, insufficient data make it challenging for quantitative assessment to better understand the ecosystem processes. To date, for instance, water quality data covering all the subbasin of TSL have not been reported. Thus, it is essential that more systematic regular monitoring efforts are made to collect water quality parameters (e.g., nutrients, DO, COD) to gain deeper insight into the water quality status in the basin and better basin management plan and environmental policy in the face of accelerated development (e.g., urbanization) and increasing anthropogenic pressures (e.g., population growth) on the TSL's ecosystem.

Key Points

- Water quality of the TSL basin exhibits the distinct seasonality in characteristics as cluster analysis identified three or four distinct groups of months. Nutrients, suspended solids, chloride, and sodium in TSL show higher concentration in the late dry and early wet seasons than during the late wet and early dry season.
- Water quality in the TSL basin has degraded in 1995–2010 as indicated by significantly increasing trends in water quality parameters, including major ions, electrical conductivity, and nutrients in TSL and the tributaries.
- Water quality degradation in TSL was potentially attributable to anthropogenic stresses, namely, land use/cover change, fishery activities, urbanization, and population growth.
- It is important to conduct systematic monitoring and water quality assessment and make the best use of them for understanding environmental change and science-based lake basin management.

Acknowledgments The authors thank the Mekong River Commission (MRC) for the provision of the observed data for this work.

References

Akaishi F, Satake M, Otaki M, Tominaga N. Surface water quality and information about the environment surrounding Inle Lake in Myanmar. Limnology. 2006;7(1):57–62.

Amankwaa G, Yin X, Zhang L, Huang W, Cao Y, Ni X. Hydrochemistry and multivariate statistical analysis of the quality of water from Lake Bosomtwe for agricultural and human consumption. J Water Supply Res Technol Aqua. 2020;69(7):704–19.

Caballero M, Rodriguez A, Vilaclara G, Ortega B, Roy P, Lozano-García S. Hydrochemistry, ostracods and diatoms in a deep, tropical, crater lake in Western Mexico. J Limnol. 2013;72(3)

Deka JP, Tayeng G, Singh S, Hoque RR, Prakash A, Kumar M. Source and seasonal variation in the major ion chemistry of two eastern Himalayan high altitude lakes, India. Arab J Geosci. 2015;8: 10597–610.

Efremova TA, Sabylina AV, Lozovik PA, Slaveykova VI, Zobkova MV, Pasche N. Seasonal and spatial variation in hydrochemical parameters of Lake Onego (Russia): insights from 2016 field monitoring. Inland Waters. 2019;9(2):227–38.

Gurung S, Gurung A, Sharma CM, Jüttner I, Tripathee L, Bajracharya RM, Raut N, Pradhananga P, Sitaula BK, Zhang Y, Kang S, Guo J. Hydrochemistry of Lake Rara: a high mountain lake in western Nepal. Lakes Reserv. 2018;23(2):1–11.

Hortle, K. (2007). Consumption and the yield of fish and other aquatic animals from the lower Mekong Basin. MRC technical paper. No. 16, p. 87.

Keskinen, M., Kummu, M., Salmivaara, A., Paradis, S., Lauri, H., de Moel, H., Ward, P., & Sokhem, P. (2013). Tonle Sap now and in the future? Final report of the Exploring Tonle Sap Future study, Aalto University and 100Gen Ltd. With Hatfield Consultants Partnership, VU University Amsterdam, EIA Ltd. and Institute of Technology of Cambodia, in partnership with Tonle Sap Authority and Supreme National Economic Council. Water & Development Publication WD-11, Aalto University, Espoo, Finland. pp. 84.

Kuenzer C. Threatening Tonle Sap: challenges for Southeast Asia's largest freshwater lake. Pacific Geography. 2013;40:29–31.

Lamberts D. Tonle Sap fisheries: a case study on floodplain gillnet fisheries. Bangkok: RAP Publication; 2001.

Liljeström I, Kummu M, Varis O. Nutrient balance assessment in the Mekong Basin: nitrogen and phosphorus dynamics in a catchment scale. Int J Water Resour Dev. 2012;28:373–91.

Mayanglambam B, Neelam SS. Physiochemistry and water quality of Loktak Lake water, Manipur, India. Int J Environ Anal Chem. 2020:1–24.

Missi C, Atekwana EA. Physical, chemical and isotopic characteristics of groundwater and surface water in the Lake Chilwa Basin, Malawi. J Afr Earth Sci. 2020;162:103737.

MRC. (2010). Assessment of basin-wide development scenarios: impacts on the Tonle Sap ecosystem. Basin Development Plan Program, phase 2. 10. Vientiane, Lao PDR.

Oyagi H, Endoh S, Ishikawa T, Okura Y, Tsukawaki S. Seasonal changes in water quality as affected by water level fluctuations in Lake Tonle Sap, Cambodia. Geogr Rev Jpn Ser B. 2017;90(2):53–65.

Ramanathan AL. Seasonal variation in the major ion chemistry of Pandoh Lake, Mandi district, Himachal Pradesh, India. Appl Geochem. 2007;22(8):1736–47.

Senevirathne, N., Mony, K., Samarakoon, L., & Hazarika, M. K. (2010). Land use/land cover change detection of Tonle Sap Watershed, Cambodia. 31st Asian Conference on Remote Sensing 2010, ACRS 2010, 852–857.

Sobonn, R., Paringit, E., & Nadaoka, K. (2007). Monitoring and assessing the effects of land use/land cover on water quality of Tonle Sap Lake, Cambodia, by spatial data analysis and remote sensing approach. Proceedings of the 28th Asian Conference on Remote Sensing (ACRS 2007).

Takahashi Y, Doi R, Enomoto H. The environment. In: The Kingdom of Cambodia: from reconstruction to sustainable development. Tokyo: JICA; 2002. p. 280–315.

Tao Y, Yuan Z, Fengchang W, Wei M. Six-decade change in water chemistry of large freshwater Lake Taihu, China. Environ Sci Technol. 2013;47(16):9093–101.

Tong STY, Chen W. Modeling the relationship between land use and surface water quality. J Environ Manag. 2002;66(4):377–93.

Uk S, Yoshimura C, Siev S, Try S, Yang H, Oeurng C, Shangshang L, Hul S. Tonle Sap Lake: current status and important research directions for environmental management. Lakes Reserv. 2018;23(3):177–89.

Chapter 23
Basic Physicochemical Water Quality: Spatiotemporal Distribution

Uk Sovannara, Khoeurn Kimleang, Taing Chanreaksmey, Sokly Siev, Rajendra Khanal, Sok Ty, Sive Thea, Oeurng Chantha, and Chihiro Yoshimura

23.1 Importance of Physicochemical Water Quality

The management of the ecological integrity of an aquatic ecosystem requires monitoring and assessment of water quality (Strobl and Robillard 2008; Patil et al. 2012; see also Chap. 22). The basin-wide distribution of and the potential factors affecting the water quality in the TSL basin have been discussed elsewhere using historical data from the Mekong River Commission between 1995 and 2010 (Chap. 22). In this chapter, we present the spatiotemporal distribution of basic physicochemical water quality in the lake based on our recent extensive field surveys (2016–2019; Fig. 23.1), covering the low-water period (March and June) and the high-water period (September and December). The survey monitored a total of ten physicochemical parameters important for the maintenance of the health of the lake ecosystem: water temperature, pH, electrical conductivity (EC), total dissolved solids (TDS), dissolved organic carbon (DOC), particulate organic matter (POM), total suspended solids (TSS), oxidation–reduction potential (ORP), dissolved oxygen (DO) and its saturation (%), and turbidity.

U. Sovannara (✉) · C. Yoshimura
Tokyo Institute of Technology, Tokyo, Japan

K. Kimleang · T. Chanreaksmey · S. Ty · S. Thea · O. Chantha
Institute of Technology of Cambodia, Phnom Penh, Cambodia

S. Siev
Institute of Technology of Cambodia, Phnom Penh, Cambodia

Ministry of Industry, Science, Technology and Innovation, Phnom Penh, Cambodia

R. Khanal
Tokyo Institute of Technology, Tokyo, Japan

Policy Research Institute, Kathmandu, Nepal

C. Yoshimura et al. (eds.), *Water and Life in Tonle Sap Lake*,
https://doi.org/10.1007/978-981-16-6632-2_23

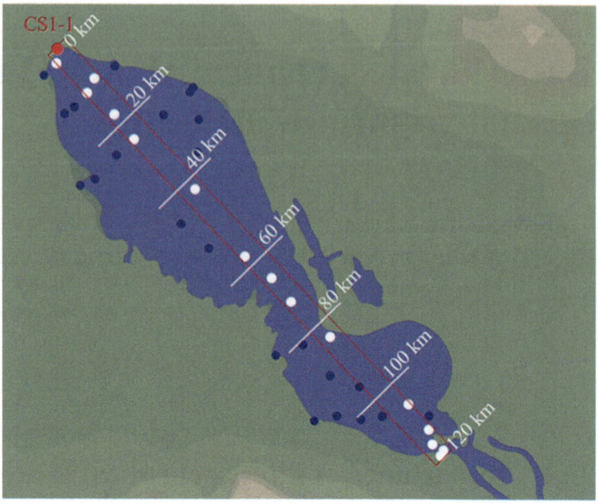

Fig. 23.1 Sampled sites in the water quality survey in TSL. All dots represent the sampling sites and were plotted as depth-averaged for describing the longitudinal distribution in the following figures. White and red dots denote the samplings used for vertical profile plotting (color charts) using Ocean Data View software (Schlitzer 2021) in the following sections. The red dot is the starting point (CS1–1) of the longitudinal transect (width of 5 km)

The water temperature, pH, and ORP play an important role in controlling the physicochemical and biological processes in lakes (e.g., Brezonik and Arnold 2011; Shao and Chu 2013). For instance, the water temperature affects the availability of DO, which is an important water quality parameter to elucidate the lake ecosystem as it closely relates to photosynthesis and heterotrophs, while pH and ORP affect the chemical speciation, solubility, and bioavailability of heavy metals and nutrients. DO availability affects the metabolism of biota, the mineralization pathways of organic matter, and various other abiotic redox-dependent physicochemical and mineralogical processes (Froelich et al. 1978; Torres et al. 2013; Diaz 2016). EC is an estimator of total dissolved salt in water and serves as a useful tool to indicate the overall anthropogenic effect (Bhateria and Jain 2016). TDS represents the portion of organic and inorganic substances contained in the water (Rice et al. 2012). TSS and turbidity affect light transmission through the water as well as the water temperature (e.g., heat energy absorbed by suspended matters; Ellis 1936; Reid 1961; Ryder and Pesendorfer 1989). Dissolved and particulate organic matters are also important components of freshwater chemistry, playing essential roles in various limnological processes (e.g., light penetration, photosynthesis, and photochemistry, Lindell and Rai 1994; Morris et al. 1995; Williamson et al. 1999; Kjelland et al. 2015) and strongly regulating the carbon and energy cycle of freshwater ecosystems (Battin et al. 2009).

23.2 Water Temperature

The water temperature in TSL varied within the ranges of 25.6–36.9 °C during the investigated period (Table 23.1), with the depth-averaged (±standard deviation) value of 29.2 ± 0.4 °C during the high-water period and 30.4 ± 0.7 °C during the

Table 23.1 Whole-lake depth-averaged values of basic water quality of Tonle Sap Lake

		Water depth [m]	Temperature [°C]	EC [μS/cm]	TDS [mg/L]	Turbidity [NTU]	pH	ORP [mV]	DOsat [%]	DO [mg/L]
2016										
December	Min	3.6	25.6	26	16	2.6	6.5	19	19	1.5
	Mean	4.8	27.0	101	60	16.0	7.5	234	81	6.4
	Max	8.8	29.4	144	86	53.4	8.3	472	110	8.7
	SD	0.9	0.9	26	15	9.2	0.4	109	24	1.9
2017										
March	Min	0.5	27.1	63	38	–	6.5	125	18.8	1.4
	Mean	1.3	30.1	135	80	–	8.1	202	96.2	7.2
	Max	3.7	34.9	283	168	–	9.1	280	147.8	10.7
	SD	0.5	1.3	25	15	–	0.5	28	23.3	1.7
June	Min	1.6	28.9	40	24	–	6.7	19	29.4	2.2
	Mean	2.8	31.0	145	86	–	7.7	188	87.0	6.4
	Max	5.5	35.7	211	126	–	8.9	250	152.7	10.9
	SD	0.7	1.5	36	21	–	0.6	41	22.4	1.6
September	Min	3.0	27.7	33	20	3	5.9	-157	1	0.1
	Mean	6.7	29.9	126	75	115	7.4	134	95	7.2
	Max	9.9	35.3	345	205	4000	8.4	292	138	10.2
	SD	0.8	0.9	42	25	421	0.4	76	26	1.9
December	Min	2.4	26.1	46	28	0	6.1	-207	1	0.1
	Mean	5.8	28.6	97	59	39	7.1	100	69	5.3
	Max	8.5	31.8	393	240	4000	7.9	358	106	8.5
	SD	1.0	0.5	44	27	223	0.4	84	34	2.7
2018										
March	Min	0.5	27.4	33	20	–	5.1	-212	3	0.2
	Mean	1.8	29.4	115	69	–	7.2	37	70	5.3
	Max	4.3	31.8	340	206	–	8.8	208	123	9.1
	SD	0.7	0.8	45	27	–	0.5	91	34	2.6

(continued)

Table 23.1 (continued)

		Water depth [m]	Temperature [°C]	EC [µS/cm]	TDS [mg/L]	Turbidity [NTU]	pH	ORP [mV]	DOsat [%]	DO [mg/L]
June	Min	0.4	28.1	11	7	20	5.6	-122	2	0.2
	Mean	1.1	29.9	205	123	435	7.5	173	65	4.9
	Max	4.5	34.9	610	366	4000	8.3	317	120	8.4
	SD	0.7	0.9	113	68	533	0.3	94	30	2.3
September	Min	2.5	29.0	71	42	1	6.4	-98	4	0.3
	Mean	8.7	29.8	104	62	10	7.5	298	87	6.6
	Max	11.5	31.9	113	71	99	8.3	364	114	8.6
	SD	0.9	0.5	3	2	8	0.4	51	19	1.4
December	Min	2.8	27.8	59	36	2	5.8	202	8	0.6
	Mean	3.8	29.4	116	70	12	7.4	257	94	7.2
	Max	6.4	31.3	173	104	91	8.9	325	132	10.0
	SD	0.6	0.6	22	13	13	0.6	20	31	2.4
2019										
March	Min	0.5	27.9	71	41	12	5.3	124	30	2.2
	Mean	1.8	30.6	131	77	242	7.7	213	109	8.1
	Max	4.3	36.9	195	113	4000	9.5	283	221	16.1
	SD	0.7	2.1	27	15	506	0.8	28	39	2.6
June	Min	0.1	27.8	62	35	36	6.5	80	34	2.6
	Mean	1.0	31.2	168	98	893	7.6	311	100	7.4
	Max	3.6	35.2	396	230	2556	8.1	368	122	8.9
	SD	0.8	1.6	47	28	630	0.3	30	17	1.2

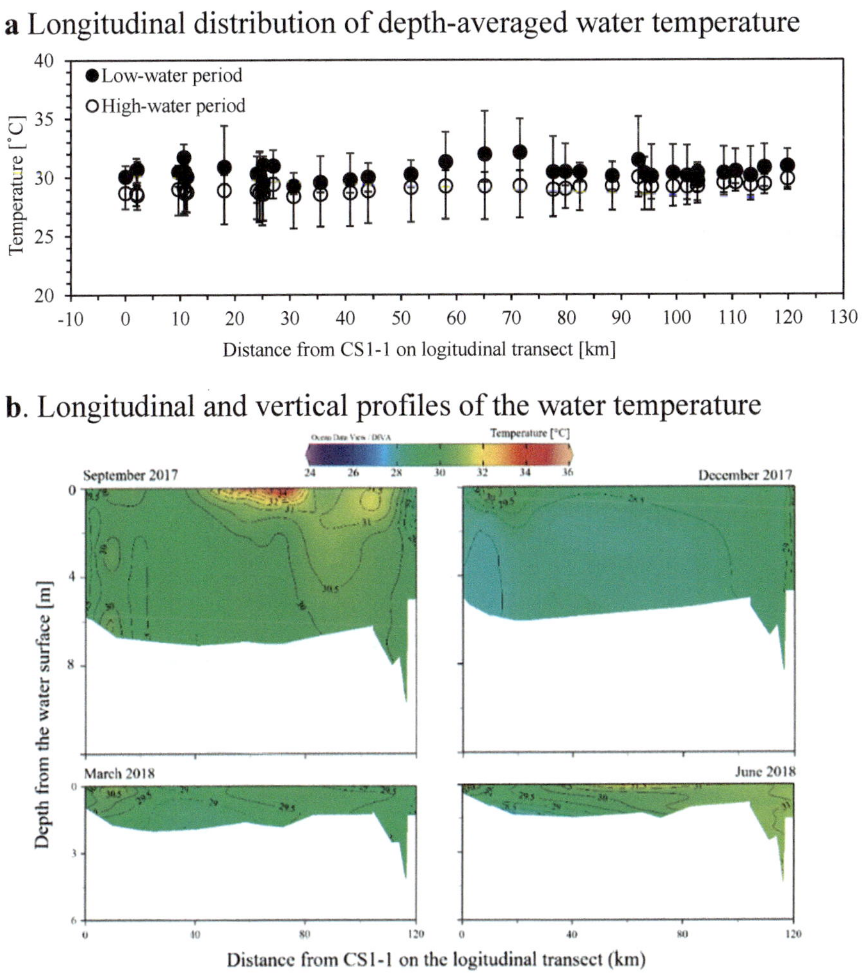

Fig. 23.2 (**a**) Longitudinal distribution of depth-averaged water temperature in TSL during the investigated period (2016–2019), with error bars representing the minimum and maximum depth-averaged temperature. (**b**) Longitudinal and vertical profile of water temperature in the high-water period (September and December 2017) and the low-water period (March and June 2018)

low-water period, respectively (Fig. 23.2a presents depth averages). This range was wider than a reported range of 28–33 °C (Sarkkula et al. 2003) and 24.1–32.3 °C (Burnett et al. 2017). Various factors affect the water temperature in lakes. These factors include the climatic and geomorphic drivers, such as air temperature, solar radiation, humidity, wind, weather, turbidity/suspended solids, stream bank vegetation, water inflow (river, stormwater, and groundwater), lake surface area, and water depth (Ellis 1936; Reid 1961; Ryder and Pesendorfer 1989; Edinger et al. 1968).

In the TSL region, the amount of solar radiation is not much different between the monsoon wet (June to November) and the monsoon dry (December to May) seasons

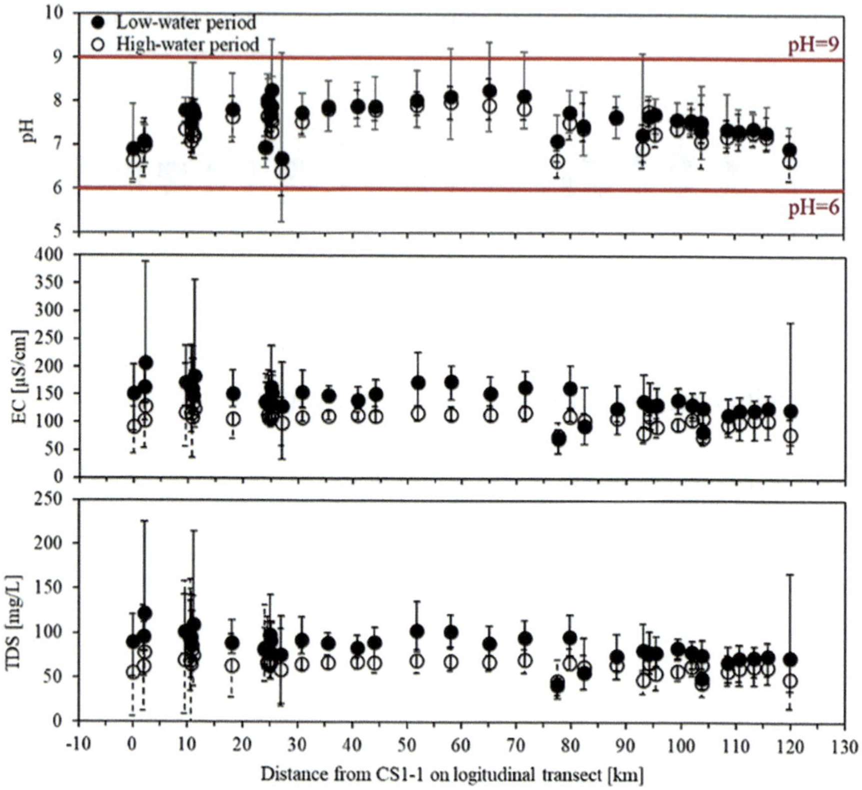

Fig. 23.3 Spatiotemporal variability of the depth average of pH, EC, and TDS. Error bars represent the minimum and maximum depth averages

(Chap. 6) as the typical rainfall in the rainy season is a downpour lasting only for 1 or 2 h a day. The air temperature in Cambodia peaked during the end of the monsoon dry season (e.g., March to April, the low-water period), and the air temperature was the lowest around December to January (minimum 20 °C in January, the high-water period; Uk et al. 2018). Interestingly, the results also showed a region of low water temperature (approx. 29 °C) from 30 to 40 km along the longitudinal transect. This spatial pattern was also confirmed by the satellite-based observation (Chap. 7) and is possibly caused by groundwater input (recirculated and upland groundwater; Burnett et al. 2017). The vertical profiles confirmed that there was generally no thermal stratification throughout a year (Fig. 23.2b) as shown also by the previous study (Oyagi et al. 2017), although the result indicated the heating process possibly by solar radiation and air temperature. Typically, the vertical temperature change was generally of 1 °C or less (Holtgrieve et al. 2013), and well-mixing condition resulted from wind-induced turbulence given the shallow nature of the lake precluded thermal stratification during the low-water period.

23.3 Ions and Dissolved Solids

Related to dissolved inorganics, the ranges of pH, EC, and TDS were 5.1–9.5, 11–610 μS/cm, and 7–366 mg/L, respectively, in TSL (Fig. 23.3). Figures 23.7, 23.8, 23.9, 23.10, 23.11, 23.12, 23.13, 23.14, 23.15, 23.16 and 23.17 (Appendix) illustrates the vertical profiles of these parameters. pH showed a weak seasonal difference, while its spatial distribution appeared clear being relatively low on both sides of the lake and high in the middle. The possible processes for this distribution are active photosynthesis for high pH and anoxic condition, caused by organic pollution and low DO (Fig. 23.4), for low pH.

Unlike pH, EC and TDS showed a distinct seasonal difference. Typically, whole-lake depth-averaged levels of EC and TDS were higher during the low-water period (March and June) relative to those during the high-water period (September and December; Table 23.1). Elevated EC and TDS at the upstream area (between 0 and 30 km on the longitudinal transect) and downstream area (~120 km on the longitudinal transect; Chhnok Tru floating village) indicated pollution at both ends of the lake (Fig. 23.3). Irvine et al. (2011) also observed a peak in EC at the upstream area (Siem Reap) in April 2008. The upstream area of TSL is dominated by densely populated area (e.g., the tourism hotspots Siem Reap town and ecotourism, floating communities, and folding houses on the foreshore; see also Oyagi et al. 2017). Elevated EC and TDS during the low-water period (March and June) also reflect evaporative concentration (high water temperature; see Sect. 23.3) and resuspension of the bottom sediments. The input from groundwater discharge, precipitation, and the Mekong freshwater pulse (reversed flow) also contributed to variation in EC and TDS. The groundwater discharge was likely the cause of the decreases in EC and TDS levels during the high-water period although the groundwater has typically higher levels of EC and TDS than the lake water (Burnett et al. 2013; see also Chap. 26 regarding groundwater chemistry). This is because most of the groundwater discharge into the lake, which was highest during the high-water season (peaking in November), is probably the recirculated lake water (typically enriched in radon and low EC; see Burnett et al. 2017 for details). In November 2003, Okumura et al. (2005) recorded an EC profile at the transverse survey stations in the northern part of TSL and reported that EC was virtually uniform from the surface to the lake bed, which is consistent with our results (see Figs. 23.7, 23.8, 23.9, 23.10, 23.11, 23.12, 23.13, 23.14, 23.15, 23.16 and 23.17 in Digital Appendix).

23.4 Dissolved Oxygen, Organics, and Redox Potential

The DO conditions in TSL were highly dynamic, varying from anoxic (0.1 mg/L) to highly saturated conditions (16.1 mg/L; 221% of saturation) throughout the investigated period (Figs. 23.4 and 23.5, Table 23.1). Given the shallow and highly turbid

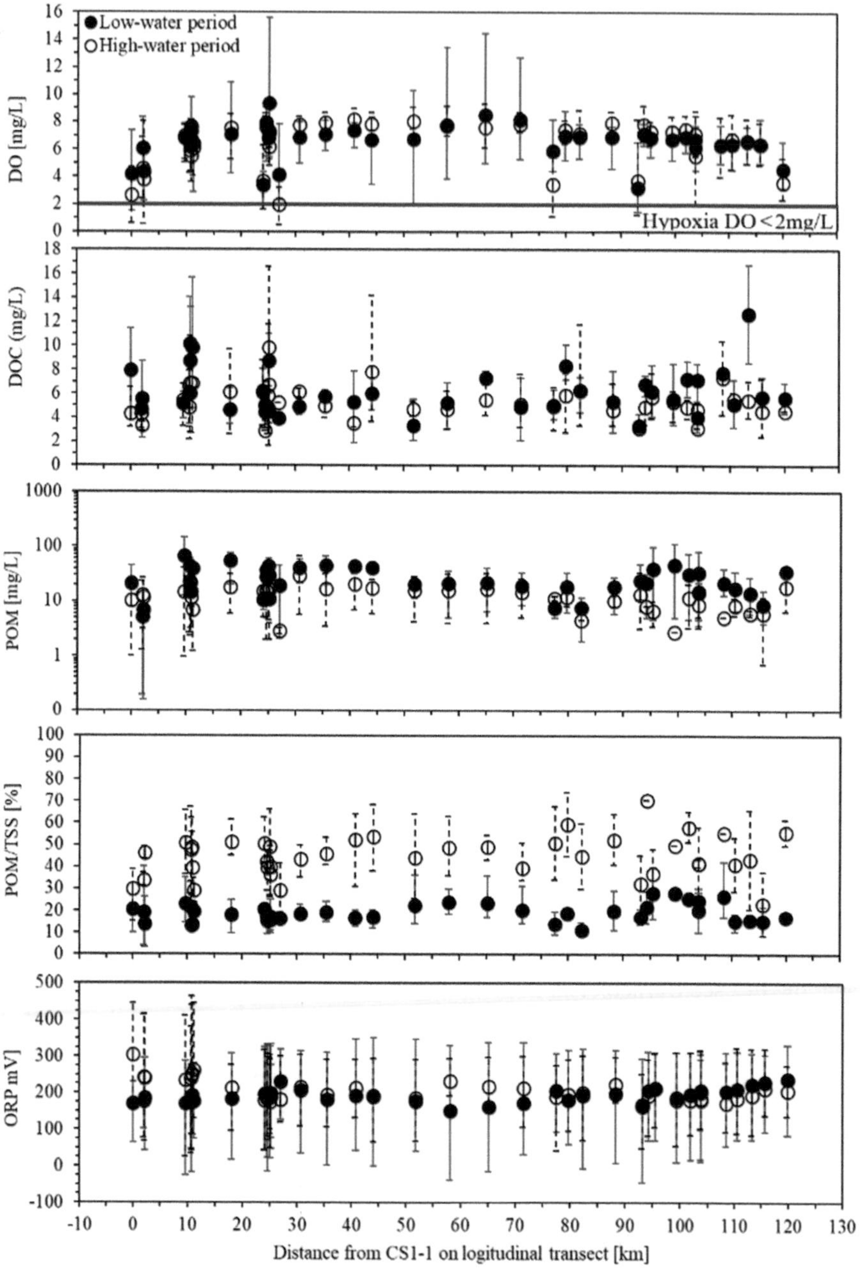

Fig. 23.4 Longitudinal distribution of depth-averaged DO, DOC, POM, and ORP in TSL during the investigated period (2016–2019; 2016–2018 for DOC and POM), with error bars representing the minimum and maximum depth averages of each parameter

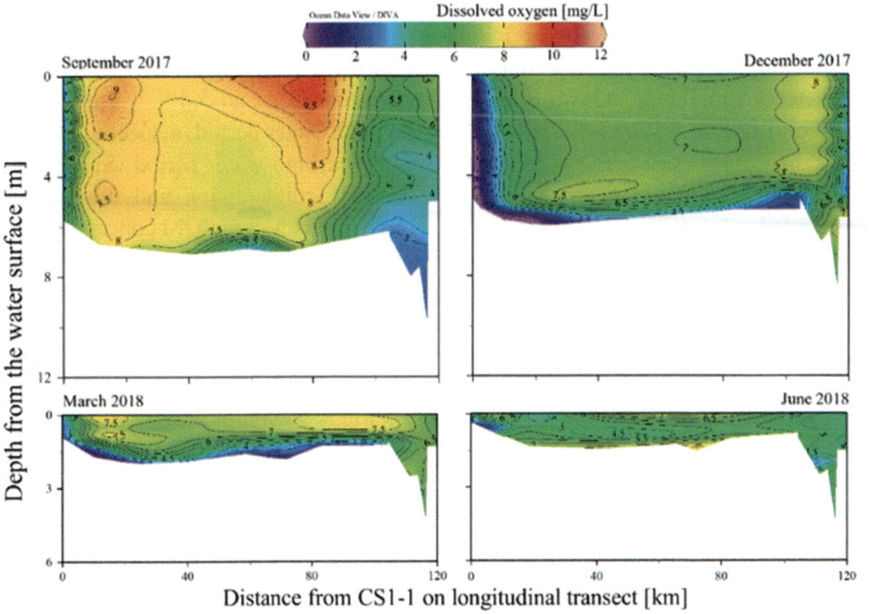

Fig. 23.5 Longitudinal and vertical profile of water DO during the high-water period (September and December 2017) and the low-water period (March and June 2018)

nature of TSL, oversaturation of DO in the water column during the low-water period was most likely attributable to wind-induced mixing and resuspension and active phytoplankton growth near the surface (Sarkkula et al. 2003, Siev et al. 2018; see also Chap. 19 for details of resuspension and Chap. 31 for primary production). Our results are in good accordance with a result reproduced by a 3D hydrodynamic model, which showed that the lake was characterized by well-oxygenated water in the lake proper and from top to bottom even during the high-water period (Sarkkula and Koponen 2003).

Although the lake was generally well-oxygenated, localized anoxic or hypoxic conditions were observed, especially in the floating villages and near the densely.

populated areas (e.g., Siem Reap), most likely attributable to organic pollution from the local human activity (e.g., direct waste disposal into the water and river inflow). Our results are consistent with previous studies. The results from a 3D modeling (Sarkkula and Koponen 2003) and continuous monitoring (Irvine et al. 2011; Holtgrieve et al. 2013) approaches showed that surface DO levels in the Siem Reap area, corresponding to 0–5 km along the longitudinal transect in our study (Fig. 23.1), could be below 4 mg/L. A continuous record of DO concentrations at four sites in TSL, two near the floating villages and two others in the flooded forest, also showed that anoxia and/or hypoxia was frequently observed irrespective of the water depths, seasons, and locations (Holtgrieve et al. 2013).

The depth-averaged concentration of DOC and POM ranged in 1.6–16.8 and 0.2–145.7 mg/L, respectively, in TSL (Fig. 23.4). The whole-lake depth-averaged (±standard deviation) concentration was 6.1 ± 2.0 mg/L for DOC and 26.7 ± 14.6 mg/L for POM during the low-water period and 5.4 ± 1.3 mg/L for DOC and 12.8 ± 5.5 mg/L for POM during the high-water period, respectively. The substantial decreases in POM concentrations in the high-water period (September and December) could be explained by effective sedimentation and dilution of suspended solids (Chap. 19). It is also noteworthy that although POM concentrations were lower, the percent contributions of POM to TSS were relatively higher during the high-water than the low-water periods (Fig. 23.4), indicating the autochthonous production of organic materials by phytoplankton during the high-water period, while resuspension of bed materials contributes to the increase of POM concentration during the low-water period.

The redox condition in TSL was highly variable, ranging from −212 mV to 472 mV over the investigated period. The whole-lake depth-averaged (±standard deviation) value of ORP was 226 ± 17 mV during the rainy season and 158 ± 28 mV during the dry season. Note that the relation of DO to biochemical oxygen demand (BOD) and chemical oxygen demand (COD) has not been investigated yet in this lake.

23.5 Turbidity

Turbidity and TSS in TSL showed high variability, fluctuating by two or more orders of magnitude throughout the investigated period (Fig. 23.6; see also Figs. 23.7, 23.8, 23.9, 23.10, 23.11, 23.12, 23.13, 23.14, 23.15, 23.16 and 23.17 and Table 23.1 in Digital Appendix). Such extreme variability was inherently linked to the flood pulse dynamics. In general, the turbidity and TSS were higher during the low-water period than the high-water period. The whole-lake depth-averaged (±standard deviation) value of turbidity and TSS was 582 ± 347 NTU and 149 ± 85 mg/L during the low-water period and 39 ± 28 NTU and 33 ± 10 mg/L, respectively.

Throughout our field survey, we observed the deep brown color of the water that results from a huge amount of muddy suspended matters (mostly clay) during March and June when the lake was relatively shallow (mean depth of 1.1–1.8 m, Table 23.1). During these periods, sediment resuspension was clearly among the factors attributable to high turbidity and TSS concentration (see also Chap. 19 for details of resuspension). The presence of algae bloom was also a contributing factor to high turbidity during the low-water period (Chaps. 28 and 31). The water transparency in TSL improved during September and December when the lake was relatively deep (mean depth of 4.8–8.7 m, Table 23.1), reducing the turbidity as a result of sedimentation taking place when the lake was relatively deeper

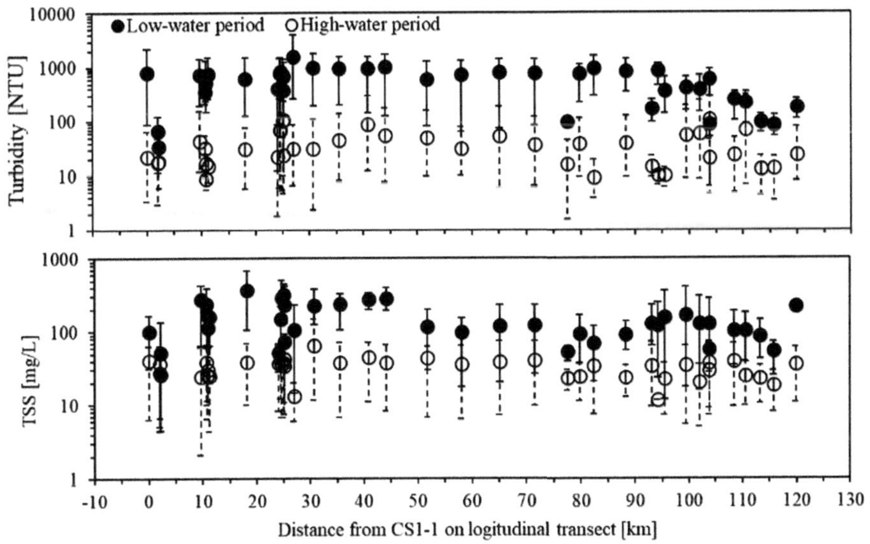

Fig. 23.6 Longitudinal distribution of depth-averaged turbidity in TSL in 2016–2019 with error bars representing the minimum and maximum depth averages

(Chap. 19). Our results were consistent with the previously reported Secchi disk readings that seasonally varied from as low as 0.02 m during the low-water period (May 2004 and May 2005) to as high as 2.7 m during the high-water period (November 2004 and December 2005; Ohtaka et al. 2010).

Key Points

- According to our survey (2016–2019), the water temperature in TSL varied from 25.6 to 36.9 °C with no occurrence of thermal stratification.
- pH, EC, and TDS varied within the range of 5.1–9.5, 11–610 μS/cm, and 7–366 mg/L, respectively.
- DO, DOC, POM, and ORP varied within the range of 0.1–16.1 mg/L, 1.6–16.8 mg/L, 0.2–145.7 mg/L, and −212–472 mV, respectively. Localized anoxia and hypoxia existed, especially in the floating villages and the densely populated northern upstream.
- Turbidity and TSS exhibited high seasonal variability, fluctuating by two or more orders of magnitude (range, 0–4000 NTU and 2–684 mg/L) with typical high values during the low-water period.

Appendix

Vertical profiles of the basic water quality in Tonle Sap Lake (TSL).

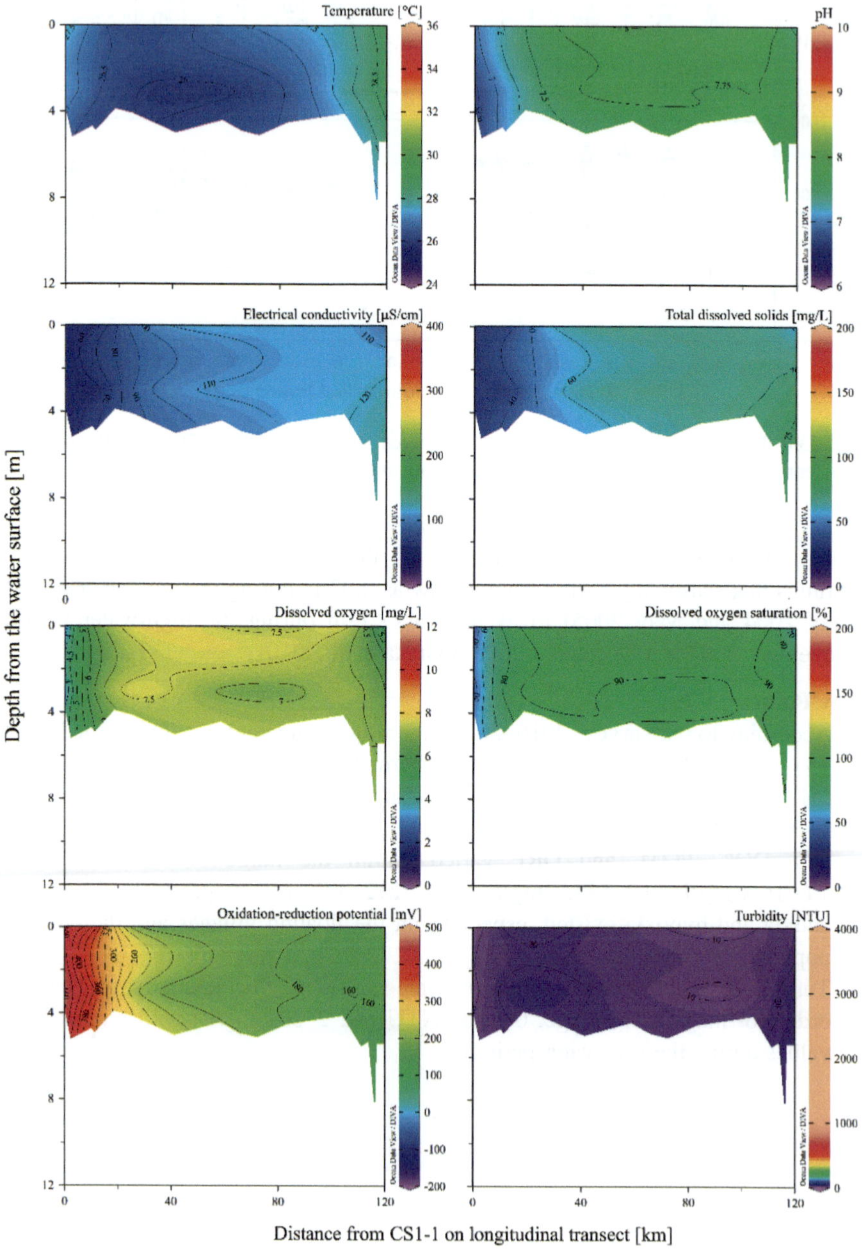

Fig. 23.7 Vertical profiles of the basic water quality in Tonle Sap Lake (TSL) in December 2016

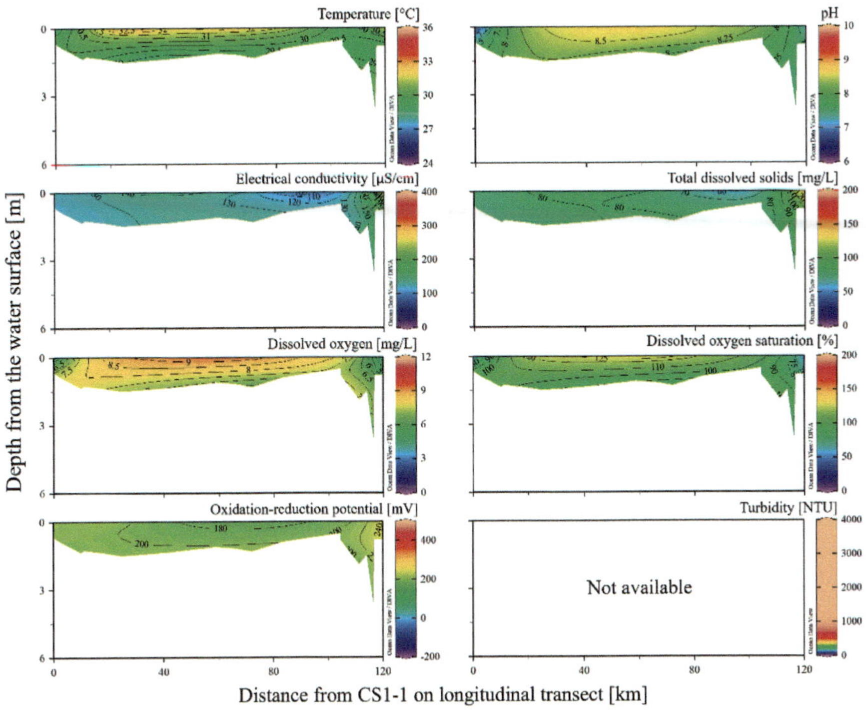

Fig. 23.8 Vertical profiles of the basic water quality in TSL in March 2017

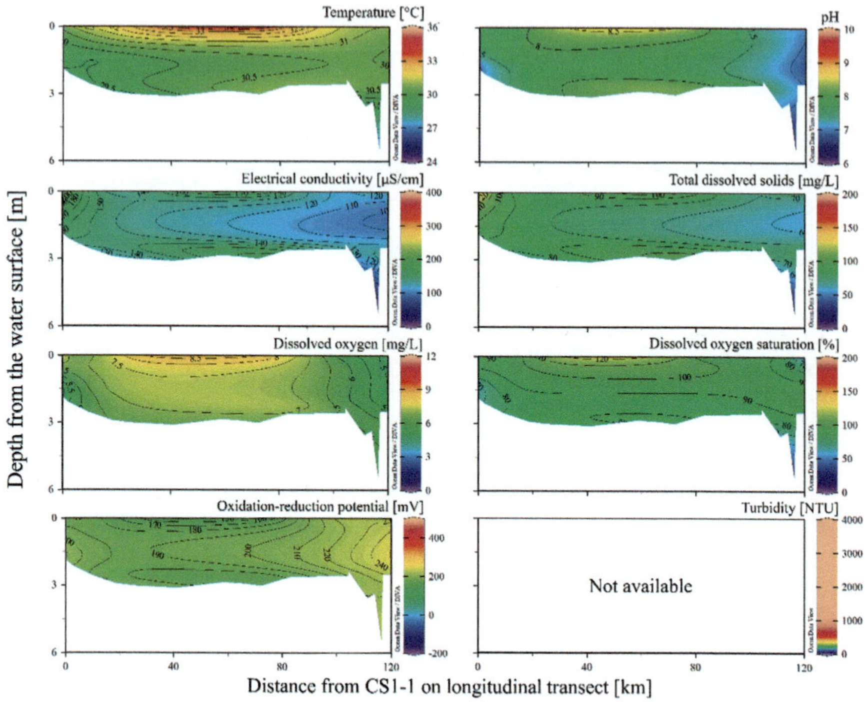

Fig. 23.9 Vertical profiles of the basic water quality in TSL in June 2017

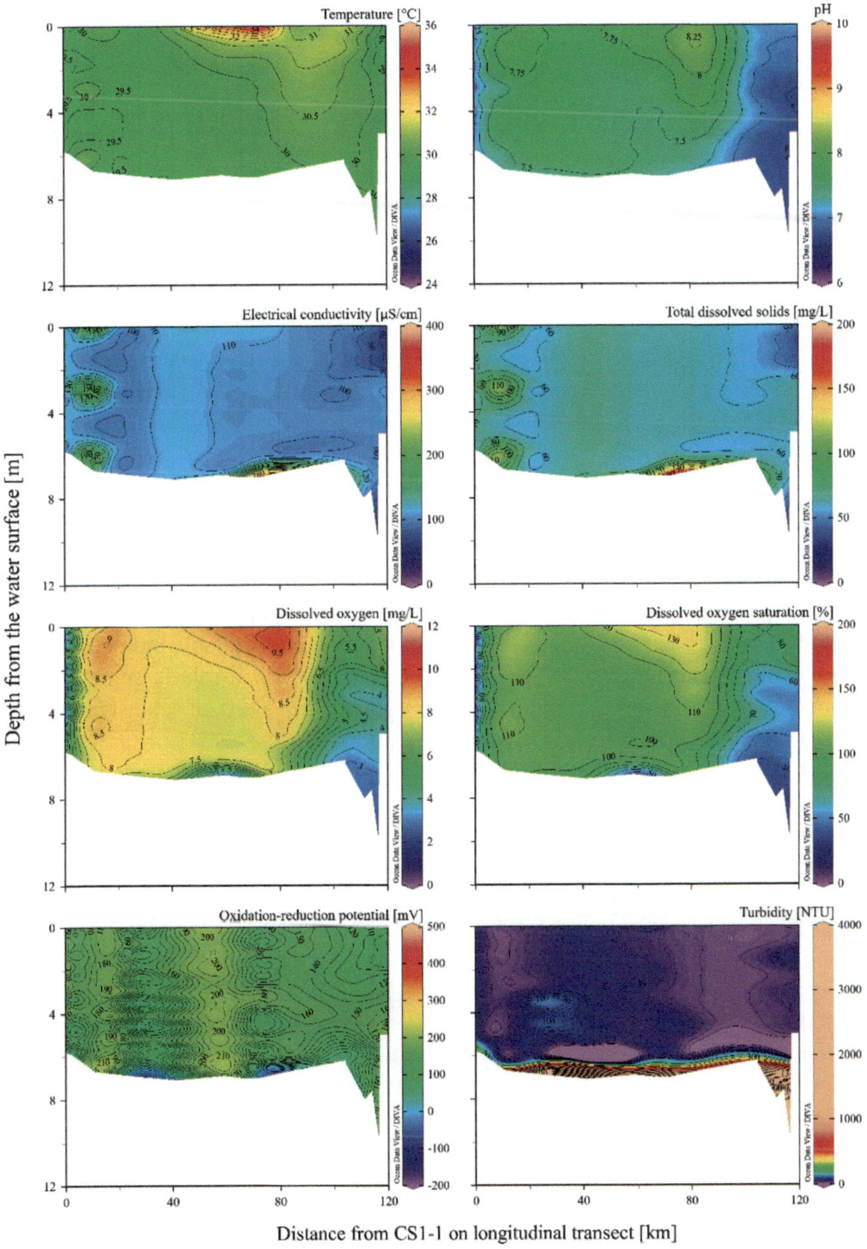

Fig. 23.10 Vertical profiles of the basic water quality in TSL in September 2017

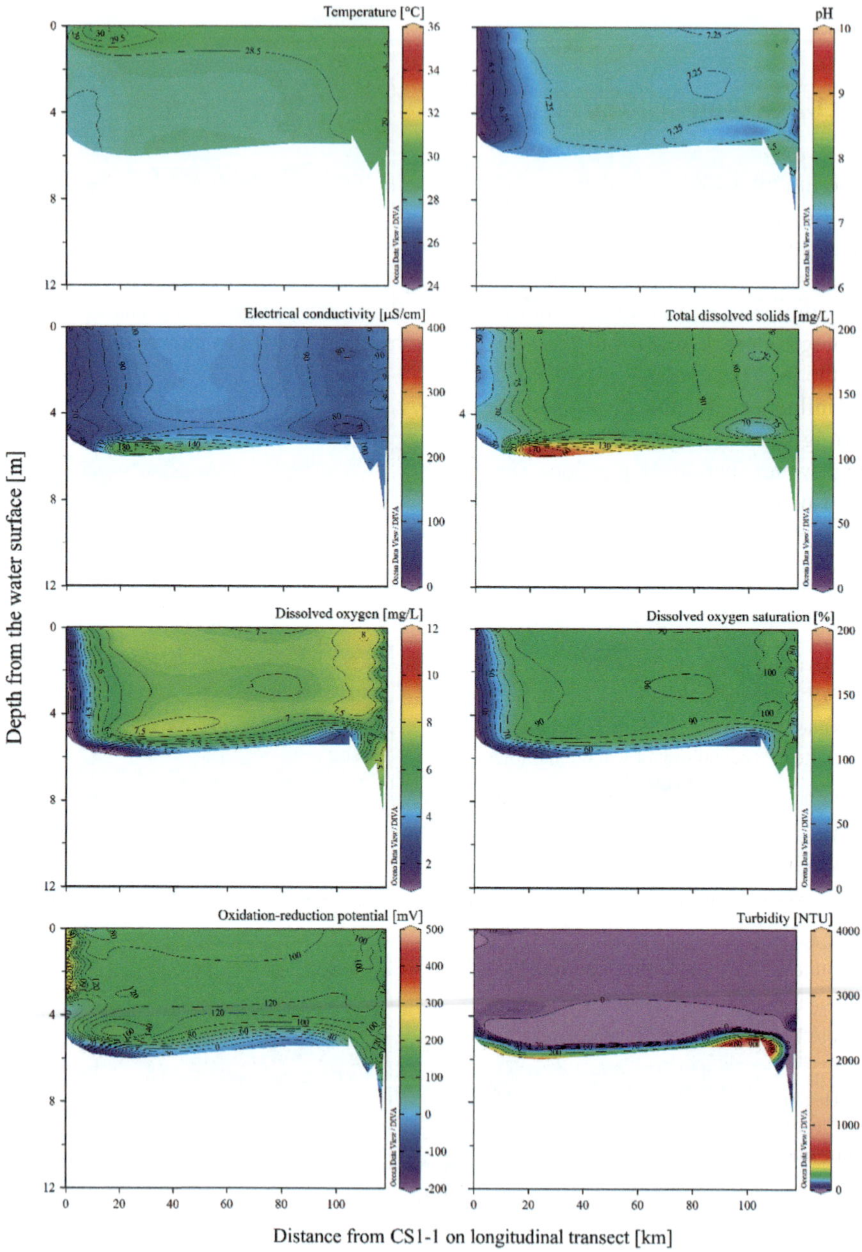

Fig. 23.11 Vertical profiles of the basic water quality in TSL in December 2017

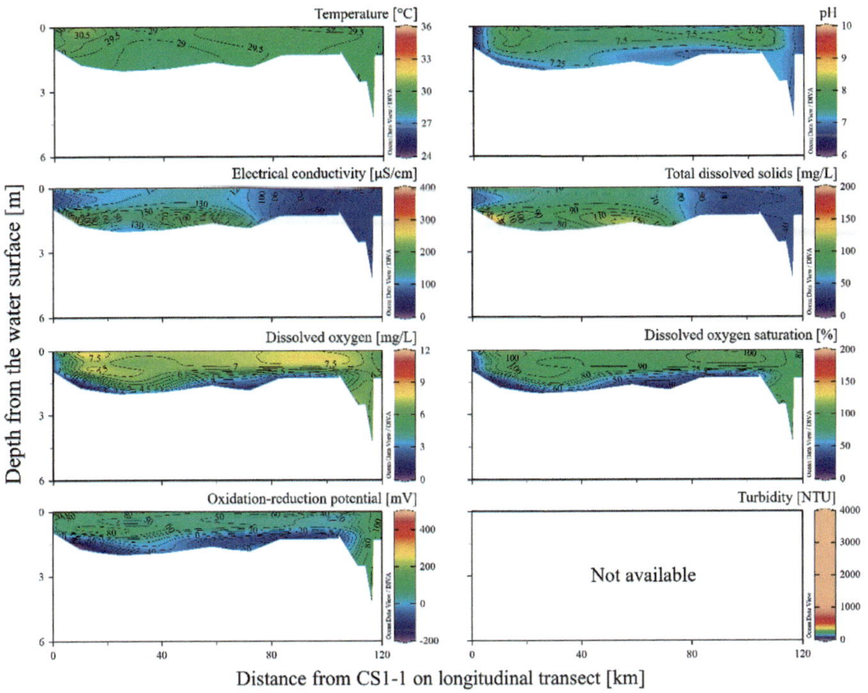

Fig. 23.12 Vertical profiles of the basic water quality in TSL in March 2018

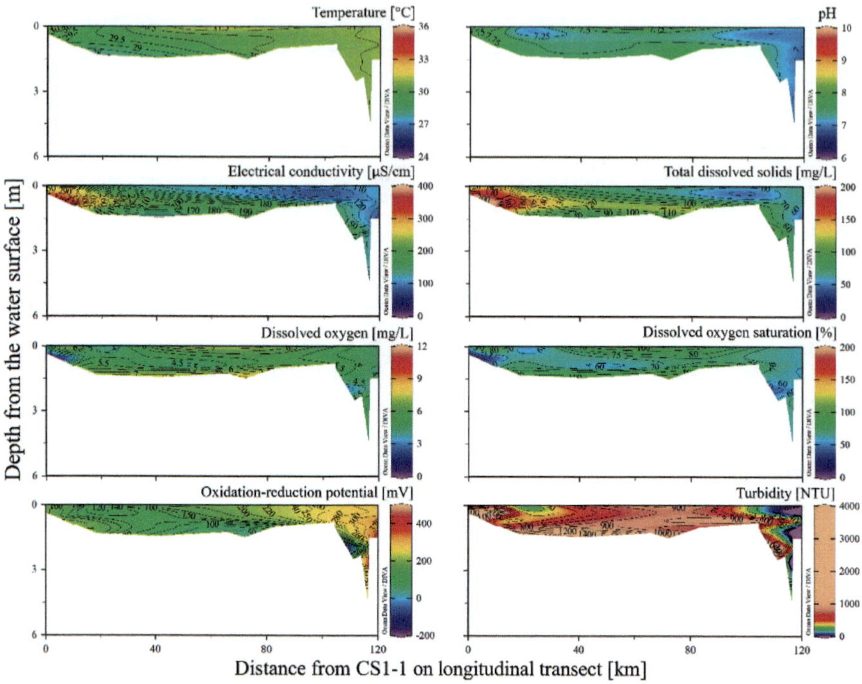

Fig. 23.13 Vertical profiles of the basic water quality in TSL in June 2018

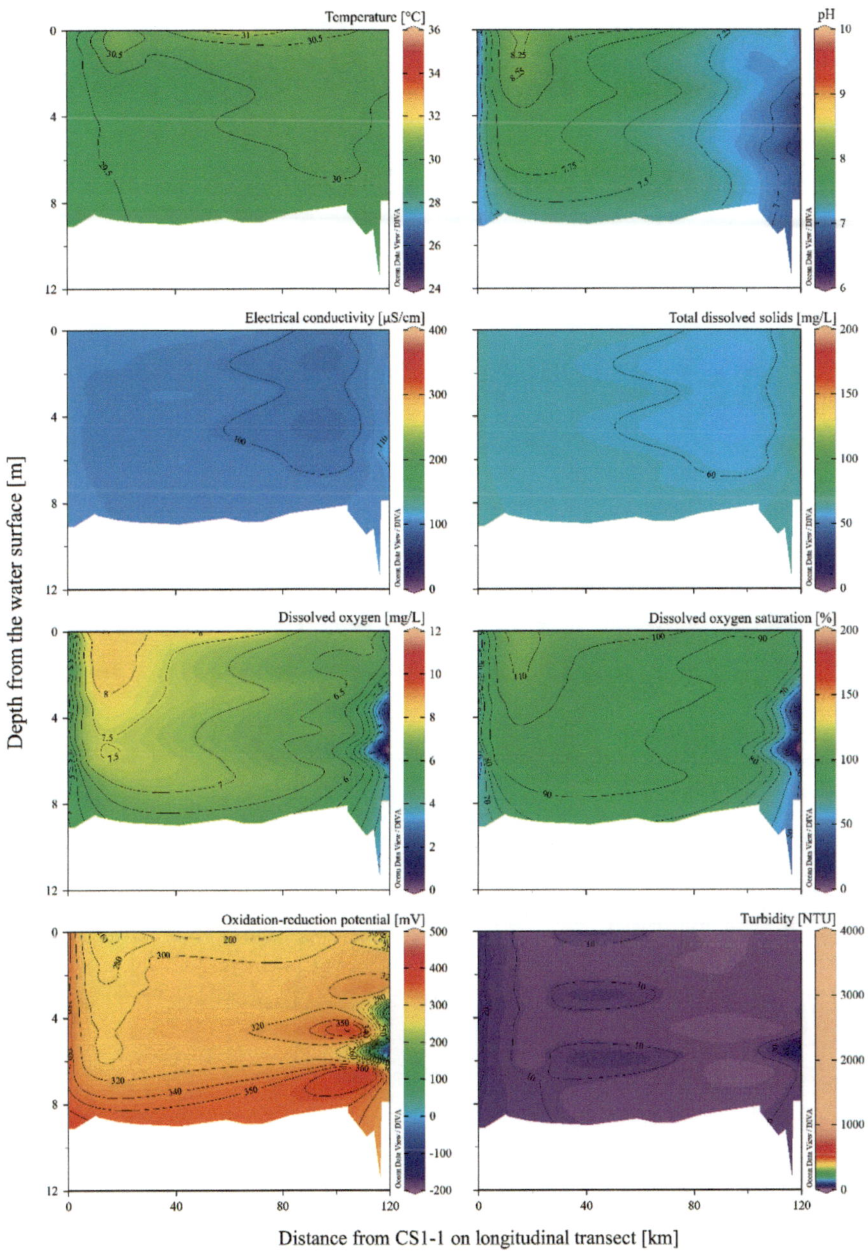

Fig. 23.14 Vertical profiles of the basic water quality in TSL in September 2018

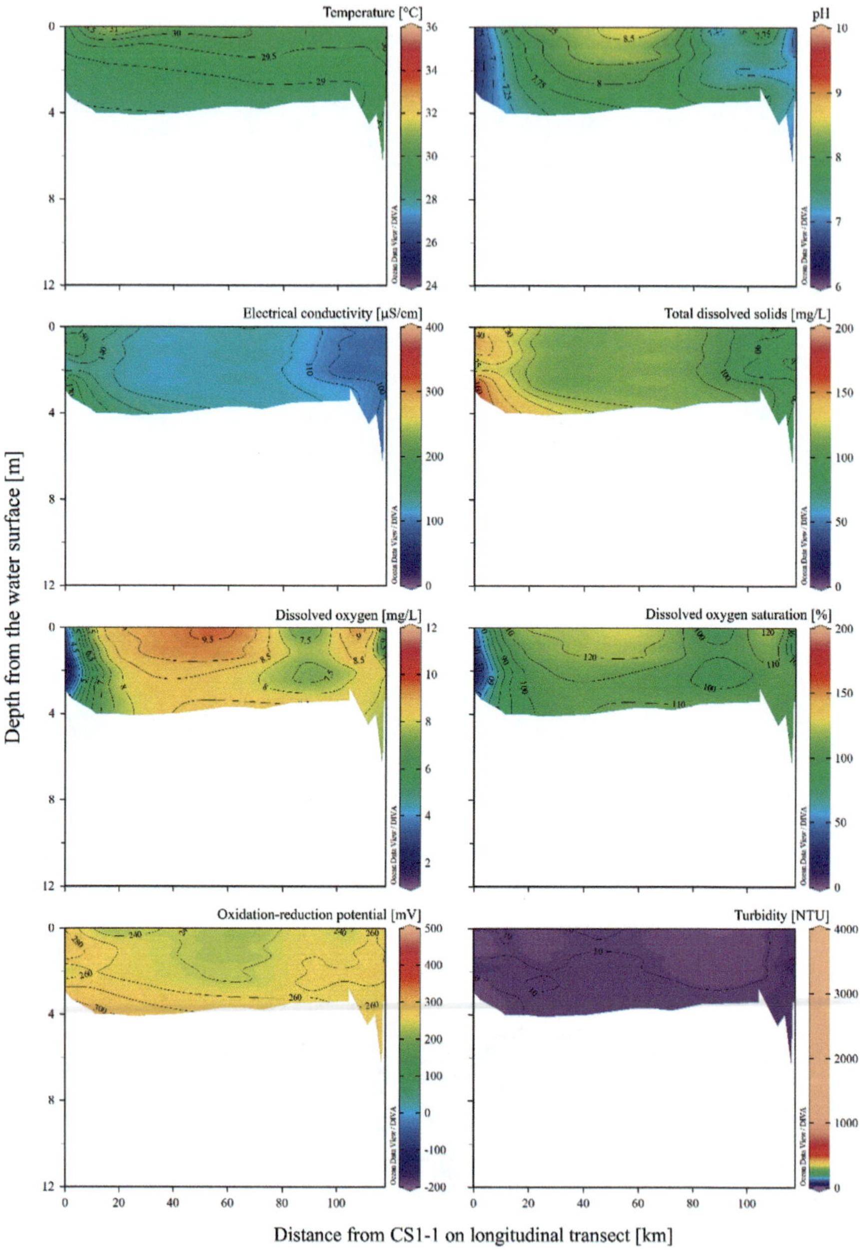

Fig. 23.15 Vertical profiles of the basic water quality in TSL in December 2018

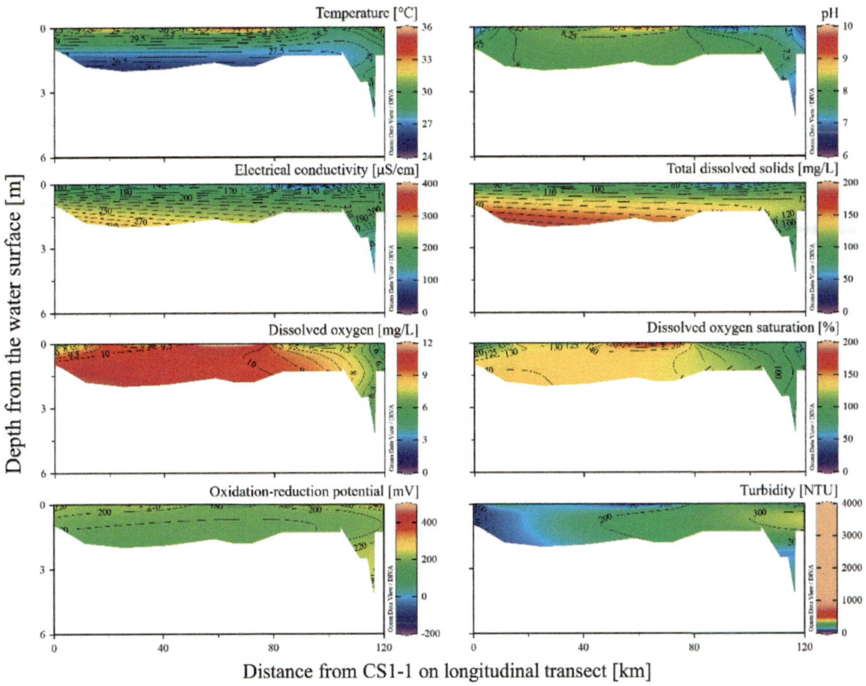

Fig. 23.16 Vertical profiles of the basic water quality in TSL in March 2019

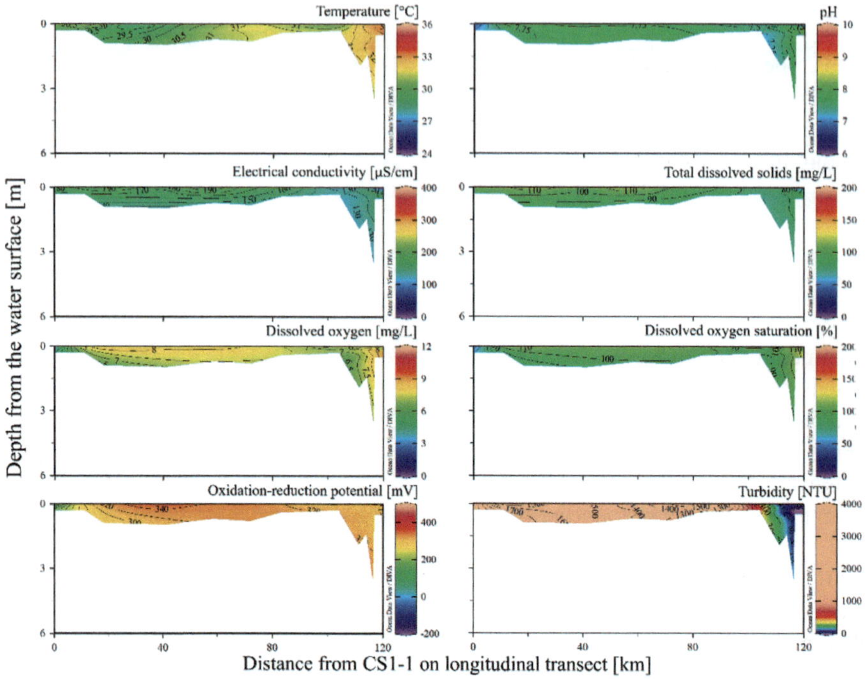

Fig. 23.17 Vertical profiles of the basic water quality in TSL in June 2019

References

Bhateria R, Jain D. Water quality assessment of lake water: a review. Sustain Water Resour Manag. 2016;2:161–73.

Battin TJ, Luyssaert S, Kaplan LA, Aufdenkampe AK, Richter A, Tranvik LJ. The boundless carbon cycle. Nat Geosci. 2009;2(9):598–600.

Brezonik PL, Arnold WA. Water chemistry: an introduction to the chemistry of natural and engineered aquatic systems. Oxford: Oxford University Press, UK; 2011. 782 p. xxiv ISBN: 9780199730728

Burnett WC, Peterson RN, Chanyotha S, Wattayakorn G, Ryan B. Using high-resolution in situ radon measurements to determine groundwater discharge at a remote location: Tonle Sap Lake, Cambodia. J Radioanal Nucl Chem. 2013;296:97–103.

Burnett CW, Wattayakorn G, Supcharoen R, Sioudom K, Kum V, Chanyotha S, Kritsananuwat R. Groundwater discharge and phosphorus dynamics in a flood-pulse system: Tonle Sap Lake, Cambodia. J Hydrol. 2017;549:79–91.

Diaz RJ. Anoxia, hypoxia, and dead zones. In: Kennish MJ, editor. Encyclopedia of estuaries. Encyclopedia of Earth Sciences Series. Dordrecht: Springer; 2016.

Edinger JE, Duttweiler DW, Geyer JC. The response of water temperatures to meteorological conditions. Water Resour Res. 1968;4(5):1137–43.

Ellis M. Erosion silt as a factor in aquatic environments. Ecology. 1936;17(1):29–42.

Froelich PN, Klinkhammer GP, Bender ML, et al. Early oxidation of organic matter in pelagic sediments of the eastern equatorial Atlantic: suboxic diagenesis. Geochim Cosmochim Acta. 1978;43:1075–90.

Holtgrieve GW, Arias ME, Irvine KN, Lamberts D, Ward EJ, Kummu M, Koponen J, Sarkkula J, Richey JE. Patterns of ecosystem metabolism in the Tonle Sap Lake, Cambodia with links to capture fisheries. PLoS One. 2013;8(8):e71395.
Irvine KN, Richey J.E, Holtgrieve GW, Sarkkula J., Sampson M (2011). Spatial and temporal variability of turbidity, dissolved oxygen, conductivity, temperature, and fluorescence in the lower Mekong River—Tonle Sap system identified using continuous monitoring. Int J River Basin Manag, 9:2, 151–168.
Kjelland ME, Woodley CM, Swannack TM, Smith DL. A review of the potential effects of suspended sediment on fishes: potential dredging-related physiological, behavioral, and transgenerational implications. Environ Syst Decis. 2015;35:334–50.
Lindell MJ, Rai H. Photochemical oxygen consumption in humic waters. Ergeb Limnol. 1994;43: 145–55.
Morris DP, Zagarese H, Williamson CE, Balseiro EG, Hargreaves BR, Modenutti B, Moeller R, Queimalinos C. The attenuation of solar UV radiation in lakes and the role of dissolved organic carbon. Limnol Oceanogr. 1995;40:1381–91.
Ohtaka A, Watanabe R, Im S, Chhay R, Tsukawaki S. Spatial and seasonal changes of net plankton and zoobenthos in Lake Tonle Sap, Cambodia. Limnology. 2010;11:85–94.
Okumura Y, Endoh S, Darith E, Oyagi H. Meteorological characteristics of Siem Reap City, Cambodia. In: Tsukawaki S, Araki Y, Oyagi H, editors. Proceedings of the First International Symposium on Evaluation of Mechanisms sustaining the biodiversity in Lake Tonle Sap. Phnom Penh: Ministry of Industry, Mines and Energy; 2005. p. 79–81.
Oyagi H, Endoh S, Ishikawa T, Okumura Y, Tsukawaki S. Seasonal changes in water quality as affected bywater level fluctuations in Lake Tonle Sap, Cambodia. Geogr Rev Jpn Ser. 2017; B2017(90):53–65.
Patil PN, Sawant DV, Deshmukh RN. Physico-chemical parameters for testing of water: a review. Int J Environ Sci. 2012;3:1194–207.
Reid GK. Ecology of inland waters and estuaries. New York: Reinhold Publishing Corp; 1961. p. 375.
Rice EW, Baird RB, Eaton AD, Clesceri LS, editors. Standard methods for the examination of water and wastewater. 22nd. American Public Health Assoc, American Water Works Assoc, Water Environment Federation; Washington, DC: 2012. (Method 2540).
Ryder RA, Pesendorfer J (1989)Large rivers are more than flowing lakes: a comparative review. In: Dodge DP, editor. Proceedings of the International Large River Symposium. Canadian Special Publication of Fisheries and Aquatic Science, vol 106, p. 65–85
Sarkkula J, Kiirikki M, Koponen J, Kummu M (2003) Ecosystem processes of the Tonle Sap Lake. In 1st workshop of ecotone phase II. Phnom Penh, Cambodia, p. 1–14.
Sarkkula J, Koponen J. Modelling Tonle Sap for environmental impact assessment and management support. Helsinki: Finnish Environment Institute and EIA Ltd., WUP-FIN Project Report for the Mekong River Commission; 2003.
Schlitzer R (2021) Ocean data view. https://odv.awi.de.
Shao H, Chu L. Water-quality engineering in natural systems: fate and transport processes in the water environment. Clean Soil Air Water. 2013;41(8):829–30.
Siev S, Yang H, Sok T, Uk S, Song L, Kodikara D, Oeurng C, Hul S, Yoshimura C. Sediment dynamics in a large shallow lake characterized by seasonal flood pulse in Southeast Asia. Sci Total Environ. 2018;631–632:597–607.
Strobl RO, Robillard PD. Network design for water quality monitoring of surface freshwaters: a review. J Environ Manag. 2008;87(4):639–48.
Torres E, Ayora C, Canovas CR, et al. Metal cycling during sediment early diagenesis in a water reservoir affected by acid mine drainage. Sci Total Environ. 2013;461–462:416–29.
Uk S, Yoshimura C, Siev S, Try S, Yang H, Ouerng C, Li S, Hul S. Tonle Sap Lake: current status and important research directions for environmental management. Lakes Reserv. 2018:177–89.
Williamson CE, Morris DP, Michael LP, Olson OG. Dissolved organic carbon and nutrients as regulators of lake ecosystems: Resurrection of a more integrated paradigm. Limnol Oceanogr. 1999:3.

Chapter 24
Nutrient Availability and Phosphorus Dynamics

Uk Sovannara, Dilini Kodikara, Kana Hashimoto, Theng Vouchlay, Marith Mong, Sokly Siev, Ty Sok, Sophal Try, Vinteang Kaing, Rajendra Khanal, Heejun Yang, Thea Seav, Chantha Oeurng, and Chihiro Yoshimura

24.1 Nutrient Availability

Flood pulse is the main driving force of Tonle Sap Lake (TSL) ecosystem's productivity (Junk et al. 1989, Lamberts 2001, Sarkkula et al. 2003). This chapter addresses the nutrient availability and phosphorus dynamics in the lake in relation to the seasonal flood pulse. In shallow water environment such as TSL ecosystem, hydrodynamics affects nutrient concentrations in the water column as well as the trophic state of the ecosystem.

The concentrations of phosphorus (P) and nitrogen (N) in TSL exhibit a strong seasonality. The data obtained from Mekong River Commission (MRC)'s database between 1995 and 2010 at Kampong Luong (KL) showed that the concentration of

U. Sovannara (✉) · D. Kodikara · K. Hashimoto · V. Kaing · C. Yoshimura
Tokyo Institute of Technology, Tokyo, Japan
e-mail: uk.s.ab@m.titech.ac.jp

T. Vouchlay
Tokyo Institute of Technology, Tokyo, Japan

Institute of Technology of Cambodia, Phnom Penh, Cambodia

M. Mong · T. Sok · T. Seav · C. Oeurng
Institute of Technology of Cambodia, Phnom Penh, Cambodia

S. Siev
Institute of Technology of Cambodia, Phnom Penh, Cambodia

Ministry of Industry, Science, Technology and Innovation, Phnom Penh, Cambodia

S. Try · H. Yang
Kyoto University, Kyoto, Japan

R. Khanal
Tokyo Institute of Technology, Tokyo, Japan

Policy Research Institute, Kathmandu, Nepal

C. Yoshimura et al. (eds.), *Water and Life in Tonle Sap Lake*,
https://doi.org/10.1007/978-981-16-6632-2_24

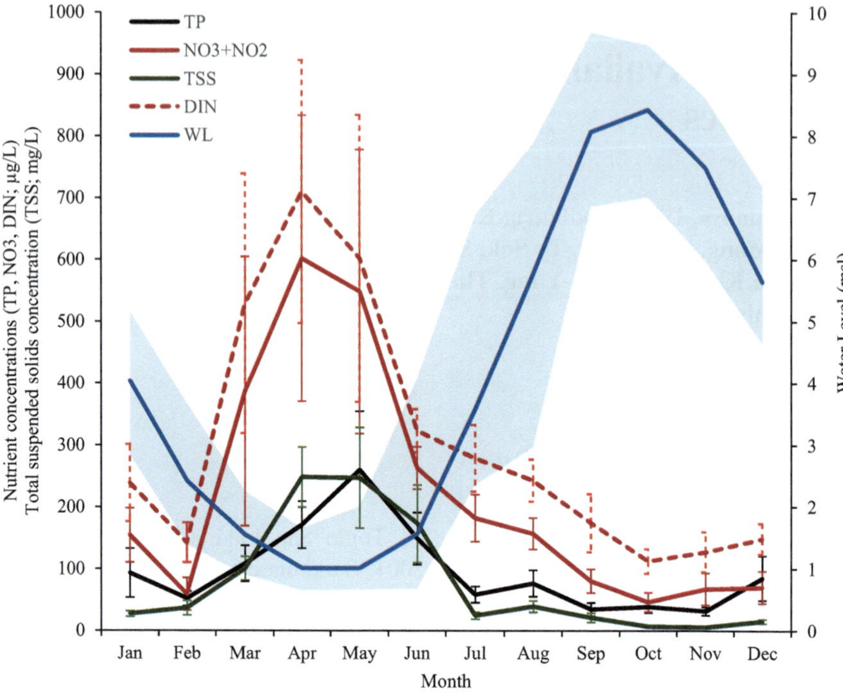

Fig. 24.1 Average monthly concentrations of TP, DIN, $NO_3 + NO_2$, TSS, and WL measured at KL in TSL during 1995–2010. The blue band represents the range of minimum and maximum monthly water level. Error bars represent standard errors. Data from the MRC's database

total P (TP, including particulate and dissolved fractions), dissolved inorganic N (DIN), nitrate + nitrite (reported here as $NO_3 + NO_2$), and ammonium (NH_4) peaked during the low-water period (e.g., April to May) when the water level (WL) in the lake is at its lowest stage and gradually decreased as the WL increased and vice versa (Fig. 24.1). The nutrient concentrations ranged greatly from 1 to 990 µg/L (average 95 µg/L) for TP, from 13 to 3144 µg/L (297 µg/L) for DIN, from 1 to 2100 µg/L (210 µg/L) for $NO_3 + NO_2$, and from 4 to 350 µg/L (84 µg/L) for NH_4.

Our field samplings during 2016–2017 also showed a spatiotemporal distribution of TP concentration (Fig. 24.2c). Average TP concentration in the surface water was higher in March (75 ± 35 µg/L) and June (64 ± 32 µg/L) than in September (30 ± 8 µg/L) and December (28 ± 11 µg/L). In June, the concentration gradient in TP was observed between the northern upstream and the southern downstream, with higher concentration at the upstream (Fig. 24.2c). This was likely caused by dilution by the reverse flow from MR to TSL. The abundance of different P forms also varied seasonally, with the total dissolved P (TDP) contributing a majority (>70%) of TP concentration during September and December, while in March and June particulate P was the dominant (>65% of TP). Unlike TP, the concentration of TDP did not change significantly, with an average of 24 ± 14 µg/L in March, 21 ± 12 µg/L in June, 23 ± 6 µg/L in September, and 19 ± 8 µg/L in December.

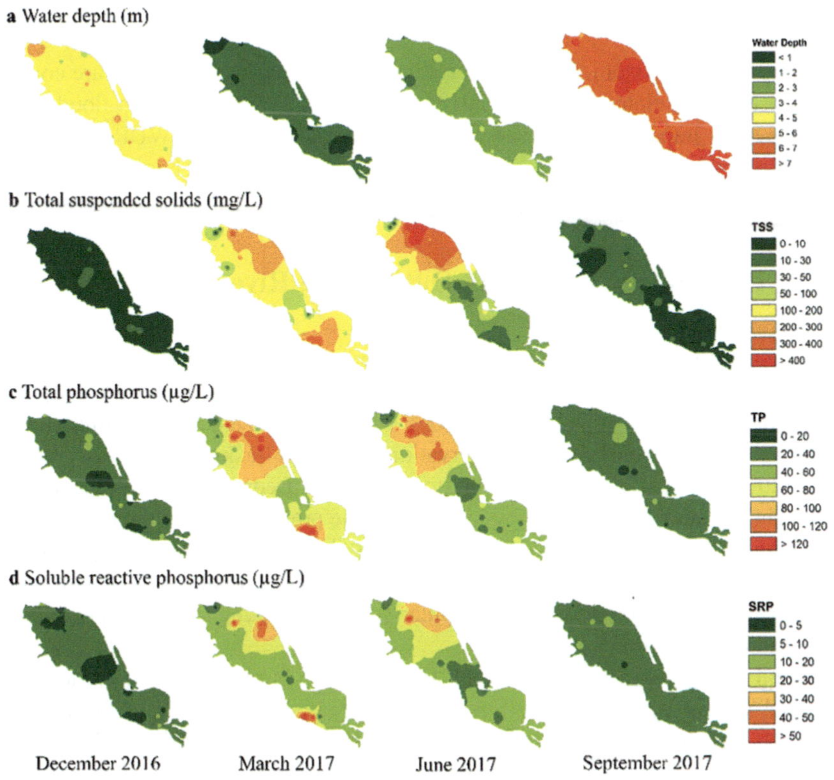

Fig. 24.2 Spatiotemporal variation of (**a**) water depth, the concentration of (**b**) TSS, (**c**) TP, and (**d**) SRP in the surface water of TSL measured during our field survey in 2016–2017

However, within TDP pool, there was a seasonal shift in the dominance of P form. The soluble reactive P (SRP), which is readily bioavailable for phytoplankton uptake, contributed >65% of TDP in March and June, while the proportion was <35% in September and December.

Although both N and P are considered the most important nutrients for aquatic life and mainly involved in the eutrophication process (Shannon and Brezonik 1972), many previous studies have indicated that P is a key factor that limits biological productivity in most freshwater lakes (Correll 1998; Filippelli 2009; Fytianos and Kotzakioti 2005; Wetzel 2001). The Redfield ratio (16 of N/P molar ratio) has been widely used as the stoichiometric reference for nutrient limitation of planktonic production in aquatic ecosystems. In case of TSL, N/P (DIN/DIP) molar ratio, calculated using data from MRC's database, WUP-FIN project, Burnett et al.'s (2017), and our recent field survey, varied from 0 to 4730 with an average value of 153. Most of the ratio (approx. 88% of the data) was above the Redfield ratio, indicating mostly N excess and thus P limited as reported also by Sakkula et al. (2003) and Burnette et al. (2017). Light limitation might occur during the low-water period due to increasing resuspension-induced TSS concentration (Figs. 24.1 & 24.2b).

In the TSL ecosystem, floodwater, sediments, and nutrient dynamics are considered closely interrelated (Uk et al. 2018), with sediments brought with floodwaters considered as important source of nutrients, especially P (MRC/WUP-FIN 2003). The following sections will discuss P balance and its relation to flood pulse.

24.2 Phosphorus Influx and Outflux Via Rivers

Influx and outflux of TP via rivers were estimated using discharge (Chap. 12) and TP concentration. The corresponding TP concentration in tributaries and TSR was obtained from MRC's database and WUP-FIN project. Due to data limitation in some tributaries, wherever the observation data was lacking, monthly average observation data of TP in the adjacent tributary was used. Atmospheric deposition of TP via rainfall was estimated by multiplying the precipitation on the lake surface, assuming its TP concentration of 26 μg/L (He et al. 2011). TP contribution from the floating villages into the lake was estimated from a unit load of 1.1 g-TP/day/capita (Takeuchi et al. 2005) and the total population of the floating villages (one million; Analy and Park 1996).

Results show that the external TP load in TSL is characterized by a strong seasonal variability (Fig. 24.3). Such seasonality was attributable to the regional monsoon climate and the seasonal flood pulse. In general, high TP influx was observed between May and September (the wet season), while TP outflux from TSL increased between October and January (the late wet season and early dry season). On an annual basis, the average TP influx to TSL system was estimated at 6428 ton/year, with a range of 4627–7632 ton/year, corresponding to the areal TP loading of 2.4 ton/year/km^2 (range value of 1.7–2.8 ton/year/km^2) for the permanent lake area of 2700 km^2. The tributary inflow accounted for a majority of TP influx to TSL, which is approximately 48–62% of its total influx. TP influx from MR via TSR due to the reversed flow was estimated at 1770 ton/year, accounting for 14–35% of the total annual TP influx. Atmospheric deposition and floating community contributed around 775 ton/year and 402 ton/year, respectively. The average TP outflux from TSL to MR via TSR was 3123 ton/year, varying between 882 and 6171 ton/year. The average monthly balance of TP flux was generally negative during the period of outflow in the late wet season and the early dry season (October to February). On average, TSL retained approximately 3305 ton/year in 1999–2003, with the interannual variation was between 1087 and 4721 ton/year.

However, TP concentration at KL showed a reversed trend from TP influx (Fig. 24.3). Despite the high TP influx between May and September, TP concentration was observed to decrease rapidly from 260 ± 94 μg/L in May to 149 ± 42 μg/L in June, reaching the level as low as 35 ± 42 μg/L in September. On the other hand, TP concentration increased greatly although the TP influx from all tributaries was greatly reduced (e.g., March to April). A possible explanation is that the lake volume

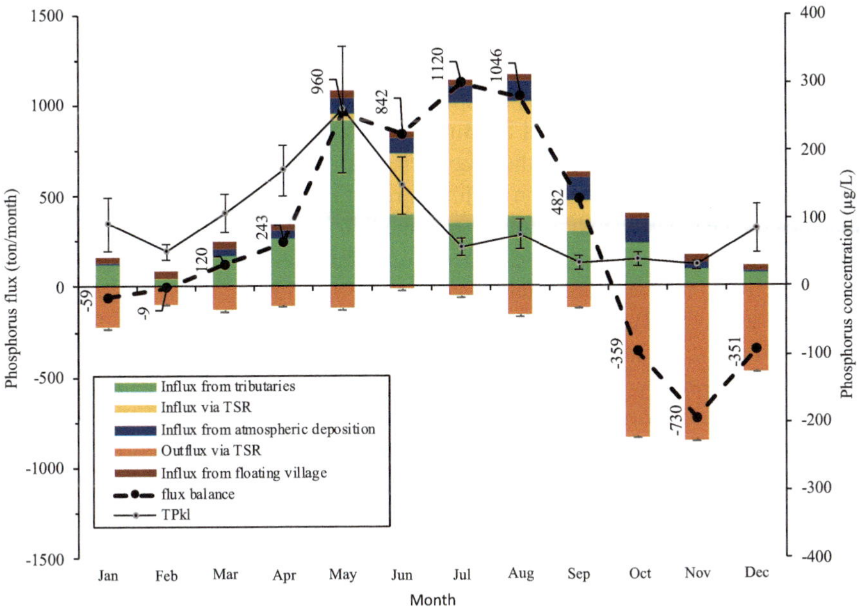

Fig. 24.3 Monthly average TP concentration at KL (1995–2010, from MRC's database) and TP influx from all tributaries including TSR, TP influx via TSR, and outflux of TSL via TSR (1999–2003 period). Error bars represent standard errors. Note: TPkl refers to the concentration of TP at KL

was lower compared to that of the wet season, and thus even a small flux from the possible sources (e.g., floating village and sediment) could raise TP concentration in the water column. During the low-water period, the shallow morphometry of the lake and wind-exposed long fetches together with fine-grained sediment could result in sediment resuspension, thereby raising TSS and P concentrations (both TP and SRP) in the water column (Figs. 24.1 and 24.2). Sedimentary TP content in TSL ranged greatly from 497 to 1481 mg/kg-dry mass with inorganic P making up a major fraction of TP (approx. 66%). The sharp increase in TP concentration during May (when the rainy season starts, the water depth in the lake remains shallow) could possibly result from both the internal and external loadings.

Nonetheless, the above estimates of P balance were possibly underestimated, both for the influx (e.g., internal loading) and outflux (e.g., phytoplankton uptake). Groundwater inflow, including recirculated lake water, might contribute nutrients such as P to TSL (Burnette et al. 2017). Contribution from sewage discharged directly into the lake was not included due to the limitation of available data. Fish production and fishery were hypothesized to play a significant role in the TSL's nutrient balance (Uk et al. 2018). A significant portion of the nutrients retained in TSL is considered to channel through the food web (e.g., fish biomass) while more than 2000 tons of migratory fish is harvested per day during the migration peak (Chapter XY), thereby contributing also to P outflux from TSL ecosystem.

24.3 Flood Pulse and Phosphorus Concentration

Nutrient delivery in flood pulse system in general, or TSL to be specific, has not been clearly determined as also noted by Burnette et al. (2017). During the wet season, when the water level/depth in the lake increases from inflowing floodwater, TSS concentration in the water column declined greatly (Figs. 24.1 and 24.2b), indicating that the change in lake depth resulting from the flood pulse was effective at removing TSS from the water column. An estimation of as much as 80% of the total incoming sediment load entering TSL annually is retained and deposited in the system (Kummu et al. 2008), mainly due to sedimentation (Siev et al. 2018). Sediment-associated nutrients deposited to the lakebed are mostly unavailable for the phytoplankton community (Campbell et al. 2009), reflecting the accumulation of nonbioavailable nutrients in the system and perhaps contributing to P storage in the lake bed. This fact is possibly one of the reasons for a large decrease in TP concentration during the high-water period (e.g., July to September) despite high TP influx (Fig. 24.3). This is the case for Lake Taihu, China, in which TP concentrations decreased significantly with the sinking particles when resuspended particles settled (Chao et al. 2017). Uptake of the readily dissolved bioavailable P by phytoplankton and other biological activities might also account for significant removal of SRP from the water column during the high-water period when the lake turbidity decreases following the sedimentation of suspended solids, allowing favorable condition for light penetration and thus primary production.

By the late wet season, TSL starts to discharge the water to MR through its sole outlet TSR, gradually decreasing WL until reaching the lowest level in the dry season. Both particulate and dissolved P (i.e., TP, SRP) and TSS in TSL peaked during the low-water period, indicating the important effects of resuspension on the ecosystem by affecting the internal nutrient loadings. Sediment resuspension is vital to the aquatic ecosystem of shallow lakes because it could affect the nutrient release, water transparency, and even the primary production (Qin, 2004).

The equilibrium P concentration (EPC_0) can be used to predict the extent to which the internal load will be released after external P load reductions and a useful tool for water managers to determine target SRP concentrations (Belmont et al. 2009). If SRP concentrations in the water column exceed the EPC_0 value, P is predicted to be retained by the sediments, while the sediments may serve as a P source if SRP concentration is lower than EPC_0. Our experimental results from a 24-h agitated incubation of sediment in TSL (nine locations) showed that EPC_0 in TSL ranged from 6 to 33 μg/l (mean value of 19 ± 3 μg/L). This range was consistent with the result of some previous studies conducted on sediments in several lakes, such as the range of 3–60 μg/L reported by Belmont et al. (2009), Shoja et al. (2017), and Zhang et al. (2011). Comparing to the bulk concentration of the corresponding SRP in the surface water, we concluded that sediment possibly releases P into the water column once sediment resuspension occurs in the following low-water period (e.g., March to June), thus serving as a P source. Therefore, resuspension-induced P release is the process behind the increment of SRP and TP in the water column during the low-water period.

Nonetheless, in a large and complex ecosystem such as TSL, many interactions among various components exist, while the gaps of knowledge remain wide. For instance, P, as well as other nutrients (e.g., N and Si), seems to have likely been influenced by the flood pulse and the regional monsoonal climate, affecting the ecosystem processes (i.e., sediment processes; water quality, microbial community). However, little is known how those processes affect nutrient dynamics (P dynamics) in TSL, let alone to quantitatively assess their effects. In addition, the internal loading of P, especially via sediment resuspension, during the low-water period seems to have significant effects on P dynamics; thus, the effects of sediment resuspension on P migration and transformation should also be taken into consideration to understand the P dynamics in the system.

Moreover, although previously overlooked, sedimentary organic P has been recently found to contribute to P bioavailability via its transformation to the reactive fractions (e.g., Shinohara et al. 2017). In our study, sedimentary organic P accounts for almost 40% of TP in sediment. When dissolved bioavailable P (i.e., SRP) is taken up and assimilated by primary producers in water, the transformation of internal sedimentary organic to inorganic P could become a crucial supplementary way for providing P for phytoplankton growth (Feng et al. 2018). Hence, knowledge of the migration and transformation of P is still incomplete, suggesting further investigation, especially in such ecosystem as TSL, which is strongly characterized by the flood pulse (deep–shallow and wet–dry conditions).

24.4 Long-Term Trend of Phosphorus Concentration

The characteristics of the TSL basin have been drastically changing (Fig. 25.3, Uk et al. 2018, and see the details also Chap. 22) and possibly also contributing to the significant upward trend of TP concentration (Fig. 24.4). This is caused by the increasing anthropogenic pressures on the TSL's ecosystem (i.e., soaring population, urbanization, hydropower dam construction) together with climate change. Between 1997 and 2010, TP concentration at measured KL has fluctuated, with increasing trend, and the trophic state of the lake with respect to TP concentration was mostly either mesotrophic or eutrophic, except for 2001–2002.

To date, there seems direct evidence to explain the sharp increase in 2003. Nevertheless, after the political instability until around 1990, economic conditions in Cambodia have considerably recovered (World Bank 2017). In January 2000, the governmental policy of Open Sky Policy has resulted in a boom in tourism in Cambodia (WEPA 2020). The numerous archaeological sites (i.e., in Siem Reap and Kampong Thom) attracted domestic and international tourists, promoting water consumption, archaic waste elimination, and rapid social mutation. Due to population growth, the population involved in agriculture and fishery rose by 130,000 and

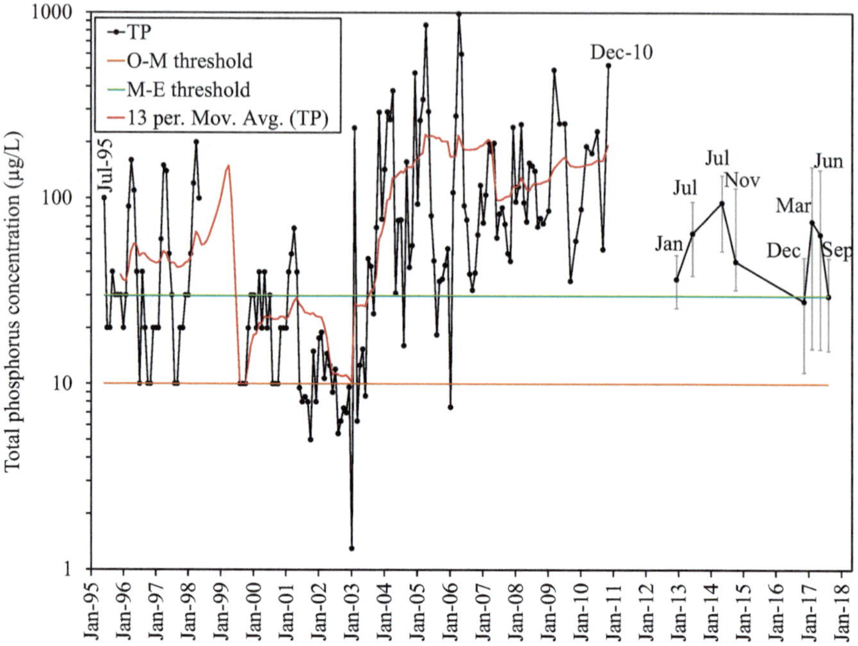

Fig. 24.4 Long-term trend of the concentration TP in TSL measured at KL (1995–2010)[1] and the average TP concentration collected from various sampling points (2013–2014)[2] and the average TP concentration collected from 36 sampling points across the lake during our recent sampling (2016–2017). Key: O, oleographic; M, mesotrophic; E, eutrophic. The red line is the 13-month moving average. Data: [1]MRC's database, [2]Burnette et al. (2017). Error bars indicate ranges

10,700 people, respectively, between 1998 and 2008, which suggests the increasing pressure on agricultural land and lake's natural resources (Keskinen et al. 2015).

Consequently, the widespread use of fertilizers in the TSL basin has been causing eutrophication in the lake, particularly in the dry season. The infrastructure for domestic waste management in urban areas has been limited, and most domestic wastewater goes directly into waterways without any treatment. Those social shifts certainly rose nutrient loading to the lake since around 2000 as indicated by TP concentration (Fig. 24.4). For instance, the massive fish kill was observed in early 2002, which was partly caused by eutrophication (WEPA 2020).

Key Points

- The concentration of nutrients (P and N) in TSL varied seasonally and spatially with peaked concentrations observed during the dry season (the low-water period).
- The concentration gradient of TP from the upstream to downstream parts of the lake during the transition period (i.e., June) is likely caused by dilution by the reverse flow from MR to TSL.

- In 1995–2014, N/P molar ratio ranged from 0.4 to 4730 (average 129), and 88% of the data were higher than the Redfield ratio, indicating P limitation most of the time in TSL.
- While the external input is an important source of P during the wet season, internal loading mainly from sediment resuspension provides an additional source of P in the water column during the low-water period.
- Given the fact that sediment–water interaction could potentially affect P dynamics, especially during the low-water period, it is crucial to understand the migration and transformation of sedimentary P species.

Acknowledgment The authors thank MRC and WUP-FIN for the provision of the observed data for this work.

References

Analy, P., & Park, N. (1996). Cambodia's great Lake: how to sustain its ecological and economic diversity. Fisheries (Bethesda), (June), 5–8.

Belmont MA, White JR, Reddy KR. Phosphorus sorption and potential phosphorus storage in sediments of Lake Istokpoga and the upper chain of lakes, Florida, USA. J Environ Qual. 2009;38:987–96.

Burnette WC, Wattayakorn G, Supcharoen R, Sioudom K, Kum V, Chanyotha S, Kritsananuwat R. Groundwater discharge and phosphorus dynamics in a flood-pulse system: Tonle Sap Lake, Cambodia. J Hydrol. 2017;549:79–91.

Campbell IC, Say S, Beardall J. Tonle Sap Lake, the heart of the lower Mekong. In: Campbell IC, editor. The Mekong: biophysical environment of an international river basin. Melbourne: Elsevier; 2009.

Chao JY, Zhang Y, Kong M, Zhuang W, Wang L-M, Shao K-Q, Gao G, Shao Q, Gao G. Long-term moderate wind-induced sediment resuspension meeting phosphorus demand of phytoplankton in the large shallow eutrophic Lake Taihu. PLoS One. 2017;12:1–15.

Correll DL. The role of phosphorus in the eutrophication of receiving waters: a review. J Environ Qual. 1998;27:261.

Feng W, Wu F, He Z, Song F, Zhu Y, Giesy JP, Wang Y, Qin N, Zhang C, Chen H, Sun F. Simulated bioavailability of phosphorus from aquatic macrophytes and phytoplankton by aqueous suspension and incubation with alkaline phosphatase. Sci Total Environ. 2018;616 (2018):1431–9.

Filippelli GM. Phosphorus cycle. Encycl Earth Sci Ser. 2009:327–30.

Fytianos K, Kotzakioti A. Sequential fractionation of phosphorus in lake sediments of Northern Greece. Environ Monit Assess. 2005;100:191–200.

He J, Balasubramanian R, Burger DF, Hicks K, Kuylenstierna JCI, Palani S. Dry and wet atmospheric deposition of nitrogen and phosphorus in Singapore. Atmos Environ. 2011;

Junk WJ, Baley PB, Asparks RE. The flood pulse concept in river-floodplain systems. Can J Fish Aquat Sci. 1989;106:110–27.

Keskinen M, Kummu M, Paradis S, Salmivaara A, Sokhem P. Using Scenarios for Information Integration and Science-Policy Facilitation: Case from the Tonle Sap Lake, Cambodia. In: Hattakas A, Vehmas J, editors. Sustainable futures in a changing climate. Proceeding of the Conference "Sustainable futures in a changing climate" 11–12 June 2014. Helsinki: Writers & Finland Futures Research Centre, University of Turku; 2015. isbn:978-952-249-303-3.

Kummu M, Penny D, Sarkkula J, Koponen J. Sediment: curse or blessing for Tonle Sap Lake? Ambio. 2008;37(3):158–63.

Lamberts D. Tonle Sap fisheries: a case study on floodplain gillnet fisheries. Bangkok: RAP Publication; 2001.

MRCS/WUP-FIN. Modelling Tonle Sap for environmental impact assessment and management support: water utilization program—modelling of the flow regime and water quality of the Tonle Sap. Helsinki: Finnish Environmental Institute; 2003. 110 pp.

Sarkkula, J., Kiirikki, M., Koponen, J., & Kummu, M. (2003). Ecosystem processes of the Tonle Sap Lake. In: 1st workshop of ecotone phase II. Phnom Penh, Cambodia. p. 1–14

Shannon EE, Brezonik PL. Relationships between lake trophic state and nitrogen and phosphorus loading rates. Environ Sci Technol. 1972;6:719–25.

Shinohara R, Hiroki M, Kohzu A, Imai A, Inoue T, Furusato E, Komatsu K, Satou T, Tomioka N, Shimotori K, Miura S. Role of organic phosphorus in a shallow eutrophic lake. Water Resour Res. 2017;53:7175–89.

Siev S, Yang H, Sok T, Uk S, Song L, Kodikara D, Oeurng C, Hul S, Yoshimura C. Sediment dynamics in a large shallow lake characterized by seasonal flood pulse in Southeast Asia. Sci Total Environ. 2018;631–632:597–607.

Shoja H, Rahimi G, Fallah M, Ebrahimi E. Investigation of phosphorus fractions and isotherm equation on the lake sediments in Ekbatan Dam (Iran). Environ Earth Sci. 2017;76(6):235.

Takeuchi T, Takahashi Y, Sina C. Sewage water quality of Phnom Penh city. J Water Environ Technol. 2005;3(1)

Uk S, Yoshimura C, Siev S, Try S, Yang H, Oeurng C, Li S, Hul S. Tonle Sap Lake: current status and important research directions for environmental management. Lakes Reserv Res Manag. 2018;23(3):1–13.

WEPA (2020). State of water environmental issues: Cambodia. Access on 23 Feb 2021 from http://www.wepa-db.net/policies/state/cambodia/index.htm

Wetzel RG. Limnology: lake and river ecosystems. 3rd ed. San Diego: Academic Press; 2001.

World Bank. Cambodia—sustaining strong growth for the benefit of all. Washington, DC: World Bank Group; 2017.

Zhang Z, Wang Z, Wang Y, Chen X, Wang H, Xu X, et al. Properties of phosphorus retention in sediments under different hydrological regimes: a laboratory-scale simulation study. J Hydrol. 2011;404(3–4):109–16.

Chapter 25
Phosphorus Dynamics: Modeling and Simulation

Theng Vouchlay, Kana Hashimoto, Uk Sovannara, Ly Sophanna, Tomohiro Tanaka, Hidekazu Yoshioka, and Chihiro Yoshimura

25.1 Modeling Phosphorus Dynamics

Phosphorus (P) is a key limiting nutrient in most freshwater lakes (Schindler et al. 2008). Excessive enrichment of P could cause various negative ecological impacts (see Chap. 24). Elucidating P dynamics in the system is of crucial importance for lake environmental management. However, monitoring of P, especially in a large lake system, is costly and needs tremendous management efforts. Water quality models describing and projecting nutrient concentration, transport, and permissible nutrient loading in lakes can resolve this issue (Bhagowati and Ahamad 2019). Water quality models can be classified into three types, namely, regression models, box-type or mass balance models, and process-based multidimensional hydrodynamic models. Both regression-based and box-type models calculate nutrient

Supplementary Information The online version of this chapter (https://doi.org/10.1007/978-981-16-6632-2_25) contains supplementary material, which is available to authorized users.

T. Vouchlay (✉)
Tokyo Institute of Technology, Tokyo, Japan

Institute of Technology of Cambodia, Phnom Penh, Cambodia
e-mail: theng.v.aa@m.titech.ac.jp

K. Hashimoto · U. Sovannara · C. Yoshimura
Tokyo Institute of Technology, Tokyo, Japan

L. Sophanna
Ministry of Environment, Phnom Penh, Cambodia

Tokyo Institute of Technology, Tokyo, Japan

T. Tanaka
Kyoto University, Kyoto, Japan

H. Yoshioka
Shimane University, Shimane, Japan

C. Yoshimura et al. (eds.), *Water and Life in Tonle Sap Lake*,
https://doi.org/10.1007/978-981-16-6632-2_25

concentration in the lake water from the lake-specific parameters on comparative studies and using steady-state assumption (e.g., Magumba et al. 2013; Håkanson et al. 2003; Wilson and Walker 1989). They do not describe spatiotemporal distribution and causal relationships. In contrast, process-based multidimensional hydrodynamic models (e.g., Zhang et al. (2008) and Huang et al. (2016)) allow us to implement these analyses. They calculate spatiotemporal changes of nutrients based on dynamic mass balance and relevant physical, chemical, and biological processes. Therefore, such process-based model clearly shows advantage over the two former types of model.

However, process-based models contain some uncertain parameters, and their application could be time-consuming due to a large dataset and heavy calculation load. Thus, the identification of major processes that influence total P (TP) concentration in lakes is a key step for their efficient applications. This chapter presents the development and application of a process-based P dynamics model to estimate TP dynamics in Tonle Sap Lake (TSL) by incorporating major P processes in the lake. The following sections illustrate the model framework and simulation results.

25.2 Phosphorus Dynamics Model for Tonle Sap Lake

Many process-based P models have been developed for freshwater lakes. However, those models have not been applied to a tropical lake having flood pulse system (i.e., TSL), which has a seasonal reversal flow and the effect of villages situated on and around the lakes. Thus, a process-based P dynamics model was developed to estimate TP dynamics in TSL by considering major P processes in the lake.

The model is governed by horizontal advection–dispersion and processes of sedimentation, internal loading, and the input of TP from tributaries, atmosphere, and villages in two dimensions by assuming complete vertical mixing as TSL is defined as a shallow lake during both the dry and the rainy seasons according to lake mixing (epilimnion) depth equation and classification of Qin et al. (2020). The dynamic mass balance of TP can be expressed as

$$\frac{\partial (hC)}{\partial t} = \frac{\partial}{\partial x}\left(D_x h \frac{\partial C}{\partial x} - v_x hC\right) + \frac{\partial}{\partial y}\left(D_y h \frac{\partial C}{\partial y} - v_y hC\right) - v_{sed}C + k_i + k_a \tag{25.1}$$

where C is the TP concentration; t is the time; v_x and v_y are the water velocity in x- and y-directions, respectively; h is the water depth; v_{sed} is the settling velocity; k_a is atmospheric loading rate; k_i is the internal loading rate; $D_x = 0.6hv_x{*}$ and $D_y = 0.6hv_y{*}$ are dispersion coefficients in x- and y-directions, respectively (Huang et al. 2016). The friction velocities, $v_x{*}$ and $v_y{*}$, are calculated by Manning equation. The internal loading rate is the total loading rate from resuspension and diffusional flux of sediment pore water P to the overlying water column. Table 25.1 summarizes the parameter ranges.

Table 25.1 Major parameters of the P dynamics model

Parameter	Symbol	Unit	Range	Reference
Settling velocity	v_{sed}	m/d	0.03–0.40	Kum and Burnett (2013); Imboden (1974)
Internal loading rate	k_i	mg/m^2/d	0.02–45.56	Ruley and Rusch (2004); Søndergaard et al. (1992)
Atmospheric loading rate (wet and dry) for tropical lakes	k_a	mg/m^2/d	0.2–0.8	Verma and Pandey (2016)

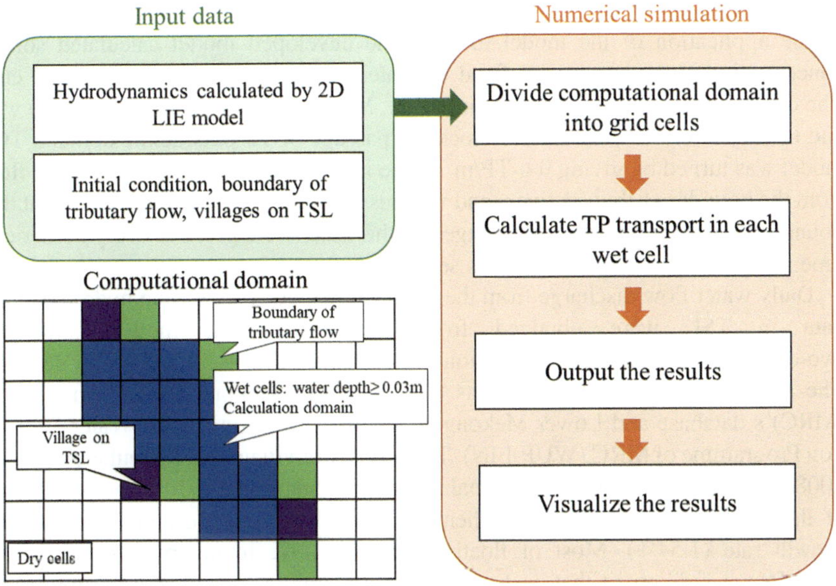

Fig. 25.1 Framework of the P dynamics model and an image of computational domain (2D LIE: two-dimensional local inertial equations)

As shown in the conceptual framework (Fig. 25.1), the major components of the model are numerical discretization, computational domain, and input data. Finite volume method is employed, and the mass conservation is locally preserved using the total variation diminishing (TVD) scheme in spatial discretization and first-order Euler forward method in temporal discretization to fasten the numerical computation and eliminate spurious numerical oscillations (Prabhakaran and Doss 2015; Smaoui et al. 2008). The monotonic upstream-centered scheme for conservation law approach is applied to TVD scheme because it is one of the most successful high-resolution schemes for hyperbolic conservation laws, which has second-order accuracy in space (Bai et al. 2018).

The evaluation of the numerical accuracy of advection–dispersion was performed against two-dimensional benchmark tests of the rotating Gaussian pulse (Pudykiewicz and Staniforth 1984) where an exact solution is available. The preliminary test with the case of TSL also confirmed the mass conservation, and an assumed bell shape distribution of the concentration was not fluctuated after a 1-month transport. The mass balance was conserved during the study period from September 1999 to December 2003.

25.3 Model Application

In the application of the model to TSL, the developed model calculated solute concentration at the wet area defined by water depth equal to or larger than 3 cm. The close-boundary condition (i.e., null flux, Miyaoka et al. 2017) between the wet and the dry interface was used to block the passage of TP outside the domain. The model was turned by giving 0 g-TP/m^3 as the initial condition and observed TP flux from the boundary tributary rivers and villages on the water for 1 year. TP flux at the boundaries of 20 tributaries and villages on the water were given at every numerical time step. The spatial resolution was set to be 500 m.

Daily water flow discharge from the boundary tributary rivers and hydrodynamic data in TSL were obtained from the hydrological models and the two-dimensional local inertial equations, 2D LIE (see Chaps. 14 and 15 for details). The observed TP concentration was obtained from Mekong River Commission (MRC)'s database and Lower Mekong Modelling Project under the Water Utilization Programme of MRC (WUP-FIN). The number of villages and population data in 2008 were obtained from the National Institute of Statistics (2010). The population of the villages in each year was then estimated based on the annual population growth rate (1.54%). Most of floating villages move to nearby flooded forest. Therefore, we assumed that each village stays within 1 km^2. The TP influx from the villages was assumed to be 0.9 g/d/capita, which is in the middle range among reported values (e.g., Mesdaghinia et al. 2015; Okada and Sudo 1986). All TP mass from the villages was assumed to accumulate on land whenever dry and be immediately released to the lake water whenever inundated.

The model was manually calibrated for the period from September 1999 to December 2001 at Kampong Luong (KGL1) and validated for the period from January 2002 to December 2003 at two locations of Kampong Luong (KGL1 and KGL2), Peam Bang (PBA), and three locations of Phnom Krom (PNK1, PNK2, and PNK3) (Fig. 25.2). The root mean square error of TP in the calibration was 15μg/L (12μg/L in the validation), while the observed TP concentration ranged from 1 to 70μg/L. The mean absolute error was 77% and 120% in the calibration and validation, respectively. The calibrated settling velocity (v_{sed}) was 0.4 m/d, and the summation of the internal and atmospheric loading rates ($k_a + k_i$) was 8.0 mg/m^2/d.

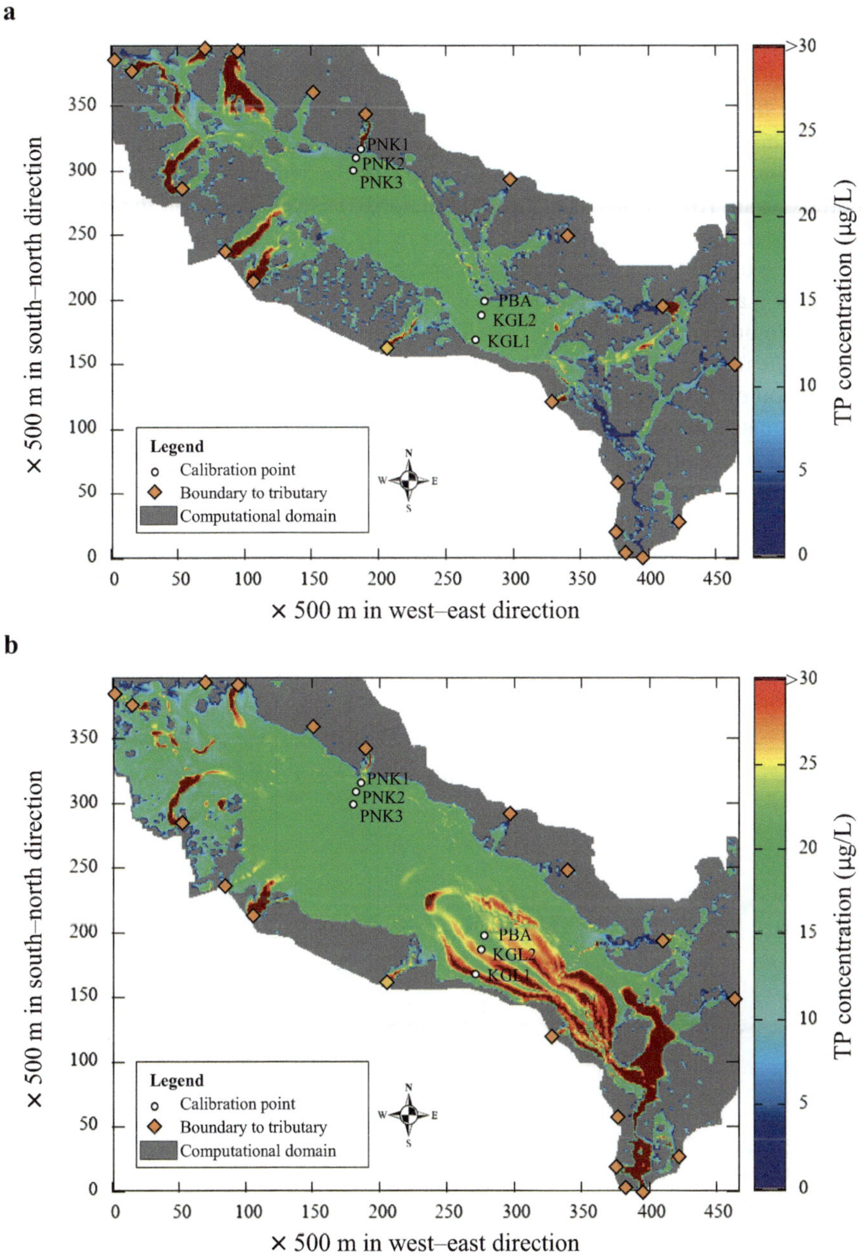

Fig. 25.2 Distribution of simulated TP concentration in TSL during—(**a**) the low-water period on May 30, and (**b**) the reversal flow period from Mekong River via Tonle Sap River to TSL on July 20, 2001

25.4 Phosphorus Dynamics: Estimation by the Model

The simulation results showed that TP influx to TSL peaked in July when the reversal flow from the Mekong River (MR) via Tonle Sap River (TSR) to TSL was the highest (Fig. 25.3). The water surface area of the lake reached the maximum in October, resulting in the highest numbers of the villages on the water and TP outflux from TSL to MR via TSR. The number of the villages on the water then started to decrease gradually in the following dry season (November–April).

The input of TP from the tributary rivers was estimated to be 137 ton/month in the dry season and 692 ton/month in the rainy season (May–October). On average, only 123 ton/month of TP flowed out from TSL during the dry season and 121 ton/month during the rainy season. Such TP outflux was one order of magnitude lower than the estimation of TP outflux in Chap. 24. This was because the TP outflux in Chap. 24 was estimated by using TP concentration of TSR, while the model simulation was based on concentration in the lake, which was estimated to be lower than that of TSR. This estimation was caused by the high settling rate and the long hydraulic retention time (i.e., 4 months and 11 days) (Figs. 25.2 and 25.3).

During the study periods, at least 171 villages stayed on the water. The average numbers of villages on the water were 324 and 513 in the dry and rainy seasons, respectively. The estimated TP influx from the villages into the lake was

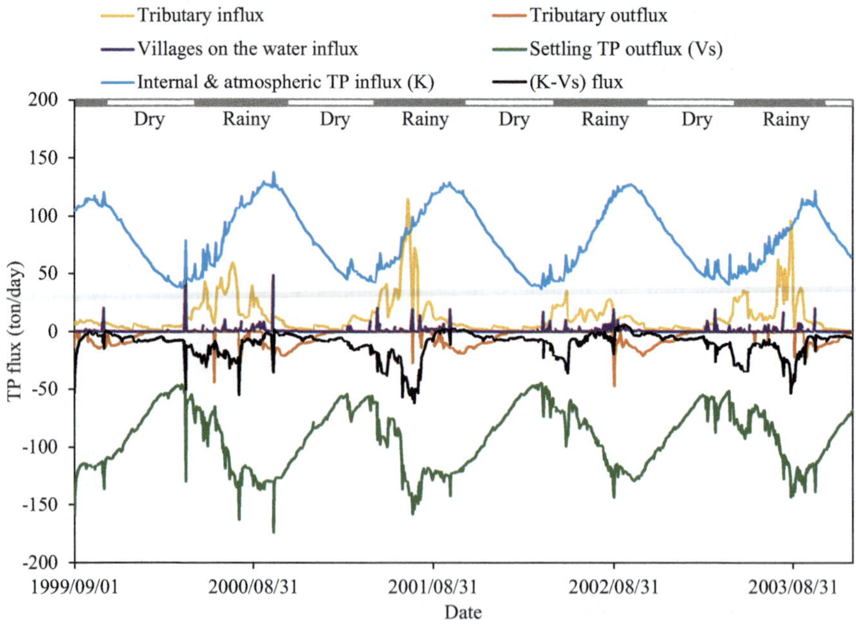

Fig. 25.3 Mass balance of TP in TSL from 1999 to 2003 (Positive value of TP flux indicates influx to TSL and negative value of TP flux indicates outflow from TSL)

approximately 18 ton/month in the dry season and 46 ton/month in the rainy season, which is equivalent to 13% and 6.7% of the total TP influx from the tributaries, respectively. Most of the time, the TP settling was higher than the internal and atmospheric loading. The total of net settling and atmospheric flux (K-Vs) was 211 ton/month and 446 ton/month in the dry and rainy seasons, respectively (Fig. 25.3).

On average, the TP mass stored in the lake water was 538 and 876 ton in the dry and rainy seasons, respectively. The water was observed to be mostly mesotrophic (10μg/L $<$ TP $\leq$ 30μg/L), and its monthly surface coverage ranged between 85% and 96%. Seasonally, the mesotrophic area was 94% in the dry season and 91% in the rainy season. Eutrophication (TP $>$ 30μg/L) expanded from 1.9% in November to 8.2% in May and contracted during the rainy season. This shift was possibly caused by the low net settling flux in the low-water period (the dry season; Fig. 25.3) and the high concentration of external input from tributaries in such period. Additionally, approximately 51% of the wet cells were dried out during the dry season, and it contained TP mass of about 77 tons or 13% of total TP mass. The dry TP was then assumed to be released back to the water due to an annual flood pulse in TSL.

During the dry season, the occurrence of eutrophication was confirmed near the outlets of the tributaries (refer Supplementary Video 25.1 for its movie). The eutrophic area was 8.5% of the inundated area in May 2001 when TSL had the highest TP influx and concentration of local tributaries in this year (Fig. 25.2a). From May to September, TSR could flow quickly into the middle of the lake, and its average TP concentration was higher than 30μg/L during the study period except in 2002. Thus, it could cause the eutrophication along that way (e.g., Fig. 25.2b). The average eutrophic area was 12% of the inundated area in July 2001 when the TP influx from TSR was the highest in this year. In the late rainy season, the lake was diluted by rainwater and showed low TP concentration in most of the tributaries.

Given such outputs, the developed P dynamics model is useful for describing the mass balance of TP processes and the spatiotemporal changes of P in TSL. As for further work, biological processes of phytoplankton uptake and release and P speciation (e.g., dissolved and particulate P) could be explicitly included to elaborate the detailed processes that determine P dynamics in the lake.

Key Points

- A two-dimensional P dynamics model was established to reproduce TP concentration in TSL by including processes of sedimentation, internal loading, and the input of TP from tributaries, atmosphere, and villages on the water.
- The model describes the spatiotemporal changes of TP and its mass balance in TSL.
- On average, more than 79% of the monthly lake area was found in mesotrophic state, and some locations reached eutrophic state during 1999–2003.
- The villages on the water were estimated to contribute approximately 13% (the dry season) and 6.7% (the rainy season) of the total TP influx from the tributaries on average during 1999–2003, which should be considered for water quality management.

- As a further work, biological processes of phytoplankton uptake and release and P speciation could be explicitly included to elaborate the detailed processes that determine P dynamics in the lake.

Acknowledgment The authors thank MRC and WUP-FIN for the provision of the observed data for this modeling work.

References

Bai F, Yang Z, Zhou W. Study of total variation diminishing (TVD) slope limiters in dam-break flow simulation. Water Sci Eng. 2018;11(1):68–74.

Bhagowati B, Ahamad KU. A review on lake eutrophication dynamics and recent developments in lake modeling. Ecohydrol Hydrobiol. 2019;19(1):155–66.

Håkanson L, Ostapenia AP, Boulion VV. A mass-balance model for phosphorus in lakes accounting for biouptake and retention in biota. Freshw Biol. 2003;48(5):928–50.

Huang L, Fang H, He G, Jiang H, Wang C. Effects of internal loading on phosphorus distribution in the Taihu Lake driven by wind waves and lake currents. Environ Pollut. 2016;219:760–73.

Imboden DM. Phosphorus model of lake eutrophication. Limnol Oceanogr. 1974;19(2):297–304.

Kum V, Burnett WC. Modeling phosphorus dynamics of Tonle Sap Lake. Int J Environ Resour. 2013;2(1):14–23. Www.Ij-er.Org

Magumba D, Maruyama A, Takagaki M, Kato A, Kikuchi M. Relationships between chlorophyll-a, phosphorus and nitrogen as fundamentals for controlling phytoplankton biomass in lakes. Environ Control Biol. 2013;51(4):179–85.

Mesdaghinia A, Nasseri S, Mahvi AH, Tashauoei HR, Hadi M. The estimation of per capita loadings of domestic wastewater in Tehran. J Environ Health Sci Eng. 2015;13(1):1–9.

Miyaoka TY, Meyer JFCA, Souza JMR. A general boundary condition with linear flux for advection-diffusion models. TEMA (São Carlos). 2017;18(2):253–72.

Okada, M., & Sudo, R. (1986). Per capita loadings of domestic wastewater. Research Report from the National Institute for Environmental Studies, Japan, No. 95.

National Institute of Statistics. 2008 Census map layer and databases. Phnom Penh: National Institute of Statistics; 2010.

Pudykiewicz J, Staniforth A. Some properties and comparative performance of the semi-Lagrangian method of Robert in the solution of the advection-diffusion equation. Atmosphere-Ocean. 1984;22(3):283–308.

Prabhakaran S, Doss LJT. Total variation diminishing finite volume schemes for one dimensional advection-diffusion equation and the relationship between flux limiter and mesh parameters. Int J Pure Appl Math. 2015;101(2):233–50.

Qin B, Zhou J, Elser JJ, Gardner WS, Deng J, Brookes JD. Water depth underpins the relative roles and fates of nitrogen and phosphorus in lakes. Environ Sci Technol. 2020;54(6):3191–8.

Ruley JE, Rusch KA. Development of a simplified phosphorus management model for a shallow, subtropical, urban hypereutrophic lake. Ecol Eng. 2004;22(2):77–98.

Schindler DW, Hecky RE, Findlay DL, Stainton MP, Parker BR, Paterson MJ, Beaty KG, Lyng M, Kasian SEM. Eutrophication of lakes cannot be controlled by reducing nitrogen input: results of a 37-year whole-ecosystem experiment. Proc Natl Acad Sci U S A. 2008;105(32):11254–8.

Smaoui H, Zouhri L, Ouahsine A. Flux-limiting techniques for simulation of pollutant transport in porous media: application to groundwater management. Math Comput Model. 2008;47(1–2): 47–59.

Søndergaard M, Kristensen P, Jeppesen E. Phosphorus release from resuspended sediment in the shallow and wind-exposed Lake Arresø, Denmark. Hydrobiologia. 1992;228(1):91–9.

Verma A, Pandey J. Atmospheric deposition coupled runoff driven shifts in nutrients and trophic status of two fresh water tropical lakes of India. Int J Appl Eng Res. 2016;11(14):8293–7.

Wilson CB, Walker WW. Development of lake assessment methods based upon the aquatic ecoregion concept. Lake Reserv Manag. 1989;5(2):11–22.

Zhang H, Culver DA, Boegman L. A two-dimensional ecological model of Lake Erie: application to estimate dreissenid impacts on large lake plankton populations. Ecol Model. 2008;214 (2–4):219–41.

Chapter 26
Chemistry of Groundwater in the Floodplain

Khy Eam Eang, Ratana Kheang, Bunhuot Ruos, Kong Chhuon, Ratino Sith, Ratha Doung, Sokly Siev, Chihiro Yoshimura, and Rajendra Khanal

26.1 Basic Property of Groundwater

The surface water quality of the Tonle Sap Lake (TSL) and its variation have attracted researchers with an increasing number of studies, while the groundwater quality in the floodplain area around TSL is rarely reported. In fact, whether the quality of the groundwater being used for households conforms to the standard of drinking water is sometimes not known. This chapter reports the quality of groundwater based on our sampling campaigns conducted twice in August (rainy season) and December (dry season) in 2019. During each sampling campaign, 23 well waters were sampled in Kampong Thom ($n = 10$), Siem Reap ($n = 7$), and Battambang ($n = 6$) (Fig. 26.1). Water samples of dug wells and bore wells were randomly gathered from households, hospitals, and schools at the edge of floodplain areas around TSL, helping us better understand and observe the hydrogeochemical variation in this region.

The well water originated from shallow groundwater sources with the well depths of 3.9–58.7 m (average well depth of 12.9 m), while the water tables ranged from 0.4

K. E. Eang (✉) · R. Kheang · B. Ruos · K. Chhuon · R. Sith · R. Doung
Institute of Technology of Cambodia, Phnom Penh, Cambodia
e-mail: khyeam@itc.edu.kh

S. Siev
Institute of Technology of Cambodia, Phnom Penh, Cambodia

Ministry of Industry, Science, Technology and Innovation, Phnom Penh, Cambodia

C. Yoshimura
Tokyo Institute of Technology, Tokyo, Japan

R. Khanal
Tokyo Institute of Technology, Tokyo, Japan

Policy Research Institute, Kathmandu, Nepal

C. Yoshimura et al. (eds.), *Water and Life in Tonle Sap Lake*,
https://doi.org/10.1007/978-981-16-6632-2_26

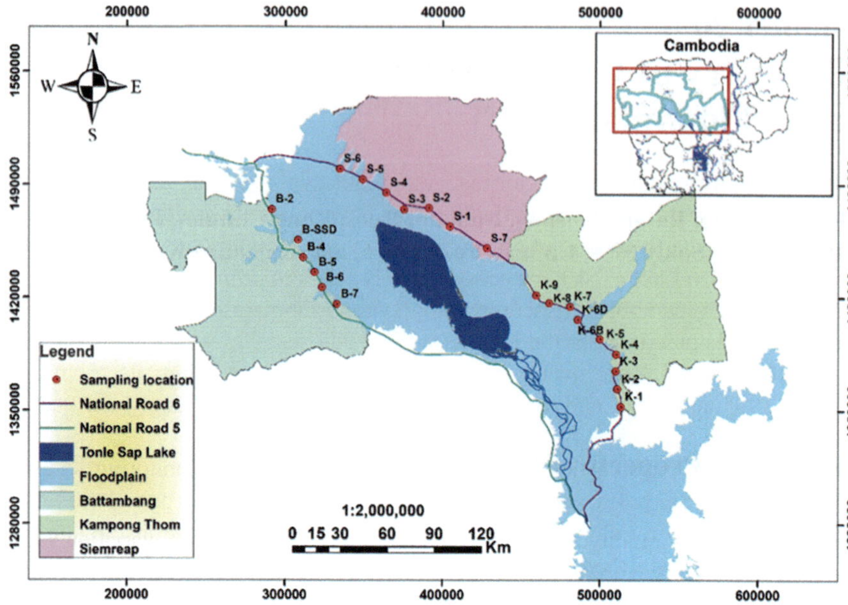

Fig. 26.1 Study area and sampling points around the Tonle Sap Lake

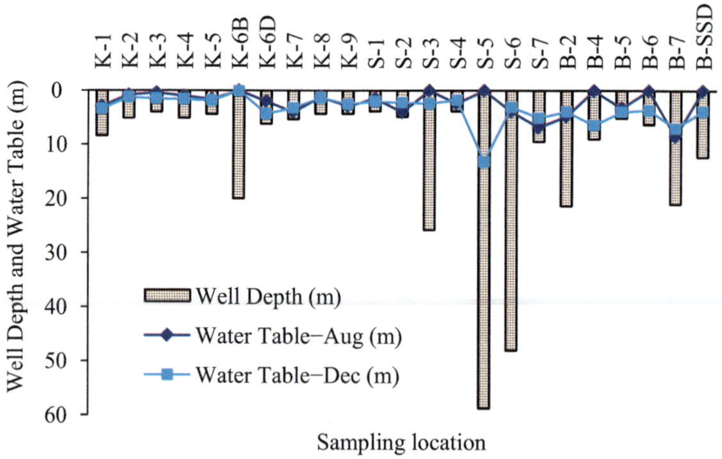

Fig. 26.2 Well depth and water table in August and December 2019 (note that no data recorded for K-6B, S-3, S-5, B-4, B-6, and B-SSD in August and no data for K-6B in December)

to 8.4 m (average water table of 3.1 m) in August and from 1.2 to 13.1 m (average water tables of 3.6 m) in December (Fig. 26.2). The water tables of some wells in rainy season were observed a bit lower than the dry season especially in sampling points at a distant edge of the floodplain, e.g., S-2, S-7, and B-7. This may be caused

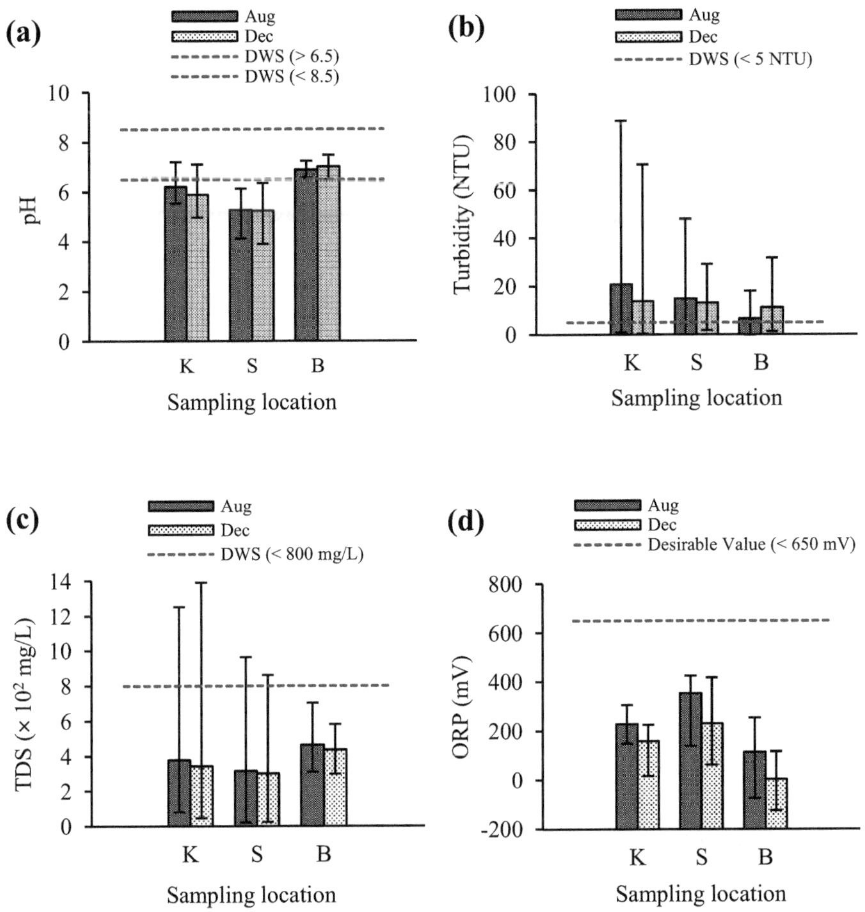

Fig. 26.3 Range and average of (**a**) pH, (**b**) Turbidity, (**c**) TDS, and (**d**) ORP in Kompong Thom (K), Siem Reap (S), and Battambang (B) (DWS: Drinking Water Standard by MIME 2004)

by our mismeasurement or the lowering due to intensive pumping, demanding some clarification in the future work.

As a result, the temperature of the groundwater ranged from 28 to 32 °C. Figure 26.3 shows the pH, total dissolved solids (TDS), turbidity, and oxidation–reduction potential (ORP) of each sample in the three provinces. In Kampong Thom province, most of the groundwater was in acidic condition, pH < 7 (Fig. 26.3a), and this acidity increases the dissolution of metals, for instance, containing excessive amounts of iron and manganese in many places (Homoncik et al. 2010; Shahin et al. 2019). TDS of the groundwater in Kampong Thom slightly varied between August and December (Fig. 26.3b). Nearly all samples were within the desirable range of drinking water standard (DWS) of Cambodia, i.e., TDS < 800 mg/L for both seasons except at K-3 (TDS $= 1251$ mg/L in August and 1390 mg/L in December), which could result from the effect of some salt minerals dissolved from soils or rocks in the well, inducing slightly salty water. The turbidity in Kampong Thom ranged from 0.8 to 89 FNU (Formazin Nephelometric Unit), with the average of 21 FNU in August and from 0.3 to 71 FNU with an average of 14 FNU in December (Fig. 26.3c). The high turbidity is caused by the presence of clay, silt, very tiny inorganic and organic matter, algae, colored dissolved organic compounds, and other microscopic organisms in the water. All groundwater samples were in oxidizing condition with the positive ORP value ranged from 18 to 306 mV, and the majority of ORP in August was greater than December, except K-6D and K-9 (Fig. 26.3d).

In Siem Reap province, pH ranged from 3.9 to 6.4, TDS from 21 to 963 mg/L, ORP from 63 to 425 mV, and turbidity from 0 to 48 FNU. The pH range was much lower than 7, which may be due to this shallower groundwater source that infiltrated through the unsaturated zone containing high carbon dioxide (CO_2). The water would be low in buffering calcium minerals but high in CO_2, inducing this low pH (Appelo and Postma 2005; Eang et al. 2018a; Eang et al. 2018b). Otherwise, the lowest pH of 3.9 in sample S-1 could also result from dissolution of the existing sulfur mineral like pyrite and result to more acidic water. In Battambang province, pH ranged from 6.5 to 7.5, TDS from 299 to 703 mg/L, turbidity from 0 to 32 FNU, and ORP from −124 to 254 mV. In this province, the pH and TDS values were generally higher than the other two provinces, which may be attributed to some carbonate minerals or salts dissolved from rocks or soils into the well water (Appelo and Postma 2005; Fitts 2002) since most of well water sources in this province were at greater depth than those of the two provinces.

26.2 Major Ions and Trace Elements

Table 26.1 summarizes the concentration of major ions (Ca^{2+}, Mg^{2+}, K^+, Na^+, Cl^-, SO_4^{2-}, and HCO_3^-). Apart of these major ions, the HCO_3^- concentrations were calculated from alkalinity, whereas the other ions were analyzed using ion chromatograph. The remarkably high concentrations of major ions during both periods in Kampong Thom were found in sample K-3, which also responded to the highest TDS, followed by samples S-7 and B-5. The concentration of each ion was noticeably varied by seasonal changes. Heavy metals such as Cd, Cu, Cr, Pb, Mn, and Fe were also measured by atomic absorption spectrophotometer (AAS). However, only the concentrations of Fe and Mn were illustrated in Fig. 26.4 as the other metals such Cr, Cd, Pb, and Cu were under the detection limit of the device (<0.001 mg/L). The concentrations of Mn and Fe for all samples in the three provinces were relatively high and ranged from 0.0 to 2.7 and from 0.0 to 24 mg/L, respectively. This could also conform that manganese often naturally occurred together with iron (Homoncik et al. 2010). Importantly, a considerable difference of Fe concentrations between August (0.0–24 mg/L) and December (0.0–0.2 mg/L) was noticed; therefore, further analysis and monitoring of Fe concentrations based on seasonal changes should be continued to elucidate relevant geochemical processes.

Table 26.1 Major ion concentrations of all the groundwater samples

Province	Season		Ca^{2+} (mg/L)	Mg^{2+} (mg/L)	K^{+} (mg/L)	Na^{+} (mg/L)	Cl^{-} (mg/L)	SO_4^{2-} (mg/L)	HCO_3^{-} (mg/L)
Kampong Thom	Aug. ($n = 10$)	Max	61	33	37	BD	68	277	344
		Min	BD	BD	BD	BD	BD	BD	2
		Mean	20	7	10	BD	11	86	82
	Dec. ($n = 10$)	Max	35	28	167	18	13	195	615
		Min	3	BD	BD	BD	BD	BD	7
		Mean	14	8	28	3	2	34	141
Siem Reap	Aug. ($n = 7$)	Max	34	12	38	12	56	216	146
		Min	BD	BD	BD	BD	BD	BD	BD
		Mean	9	3	11	2	20	35	30
	Dec. ($n = 7$)	Max	38	16	14	3	60	15	232
		Min	BD	BD	BD	BD	BD	BD	BD
		Mean	14	6	3	BD	14	2	51
Battambang	Aug. ($n = 6$)	Max	52	29	19	BD	59	317	392
		Min	9	12	BD	BD	BD	0.0	151
		Mean	30	16	4	BD	23	71	227
	Dec. ($n = 6$)	Max	40	35	13	BD	36	267	510
		Min	BD	BD	BD	BD	BD	BD	264
		Mean	16	14	5	BD	19	63	382
DWS (2004)		MPL	200	150	–	200	250	250	–
WHO (2017)		DL	–	–	–	50	250	500	–

aNote: *MPL* maximum permissible level, *DL* desirable level, *BD* below detection level

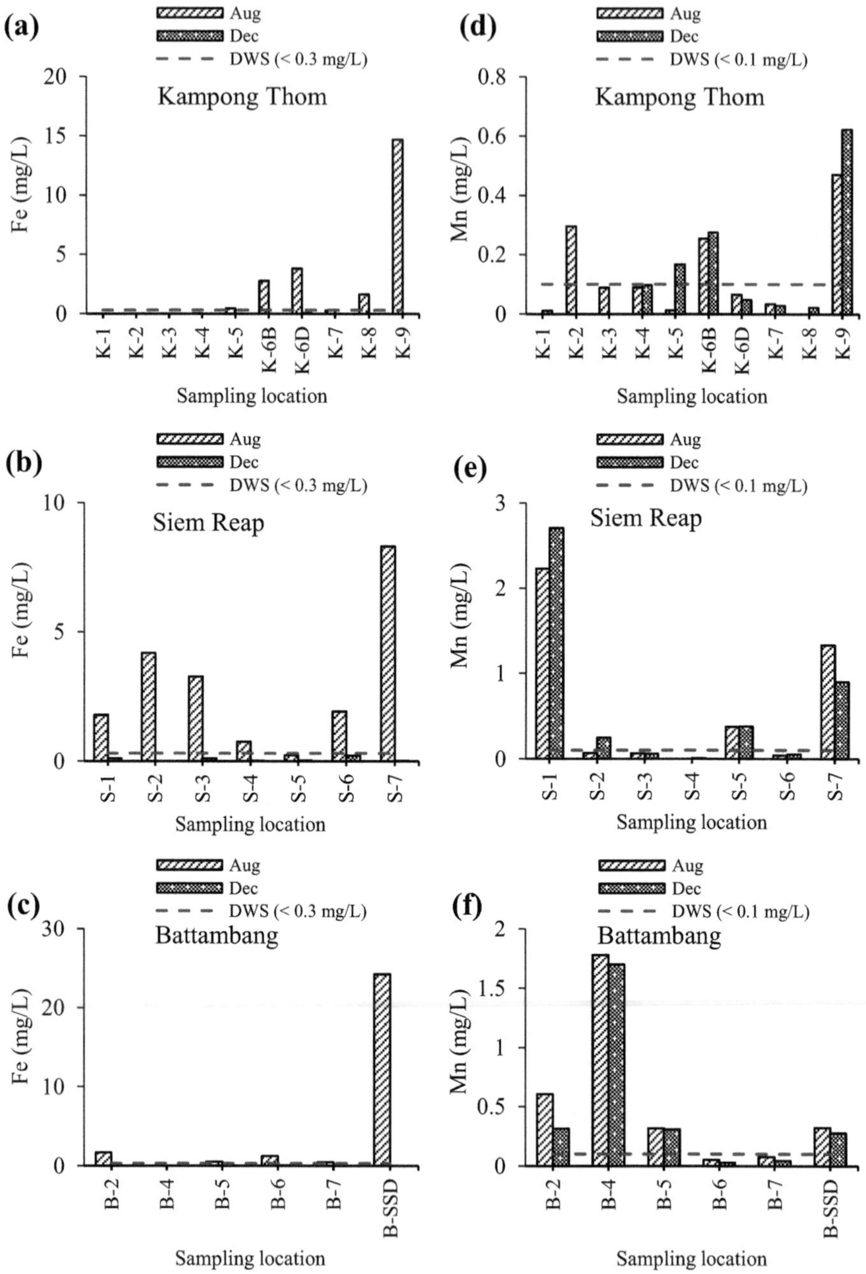

Fig. 26.4 Results of selective trace elements: (**a**)–(**c**) Fe and (**d**)–(**f**) Mn concentrations of all well water samples in the three provinces

26.3 Water/Rock Interaction

The geochemical evolution in the floodplain including the water–soil/rock interaction has not been elucidated well. The higher concentration ranges of major ions or some heavy metals could result from either water–soil/rock interaction in the well or interaction of well water with surface water in TSL that is partly contaminated. To explain and interpret more reliably, Piper/trilinear diagram and PHREEQC geochemical modeling software were used to determine the major ions representative for the hydrochemical facies or water type. It reveals that the majority of the water samples in August (rainy season) fell into blocks 5 and 1, suggesting that well water was dominated by calcium chloride (CaCl) type and calcium bicarbonate ($CaHCO_3$) type, followed by sodium chloride (NaCl), mixed calcium magnesium bicarbonate ($CaMgHCO_3$), and mixed calcium sodium bicarbonate ($CaNaHCO_3$) types (Fig. 26.5). This water type was similar to the case study of groundwater in Cu Lao Dung island surrounded by Hau River at the northern part of Vietnam (An et al. 2014) and the case study of flood groundwater and flood streamflow along the Hei River in China (Kou et al. 2019). Being slightly different from the rainy season, the majority of well waters in December (dry season) fell into $CaHCO_3$ type, whereas its minority was distributed into CaCl, mixed $CaMgHCO_3$, mixed $CaNaHCO_3$, and $NaHCO_3$ types. In short, the geochemistry is apparently influenced by carbonate minerals from natural sources, which are characterized by high levels of TDS and EC in the water.

Since significant concentrations of Ca^{2+}, HCO_3^-, Fe, and Mn were observed, the saturation index (SI) of minerals was calculated using PHREEQC (PHREEQC Interactive 3.2) to understand the contamination sources and the mineral solubility from the water–soil/rock interaction. The SI of a mineral is defined using the following formula, SI = log(IAP/K), where IAP is the ion activity product, K is the solubility product, and SI value explains the state of a mineral in the water whether it is in supersaturated (SI > 0), equilibrium/saturated (SI $= 0$), or undersaturated (SI < 0) conditions. Fig. 26.6 shows that nearly all well water in all provinces during both seasons showed SI of $CaCO_3 < 0$ (i.e., undersaturated condition with the calcite and calcite could dissolve more into the water). In addition, SI of $Mn(OH)_2 < 0$ means that manganese hydroxide was also undersaturated and able to dissolve more in water. Since the pH was not fairly high, perhaps it could not reach its precipitation point. The exception was found for $Fe(OH)_3$, whereas its SI > 0 (supersaturation) expected that iron hydroxide would precipitate at lower pH than the other minerals. Remarkably, all shallow well waters in August were mostly observed to have SI of $Fe(OH)_3 > 0$ compared to those in December. The precipitations of $Fe(OH)_3$ in August might be due to the excessive iron that was washed out from the top and subsoil by rainwater or flooding water from TSL. Nevertheless, geological formation of the soil/rock profile in the well and its mineralogical properties should be further studied for a precise clarification.

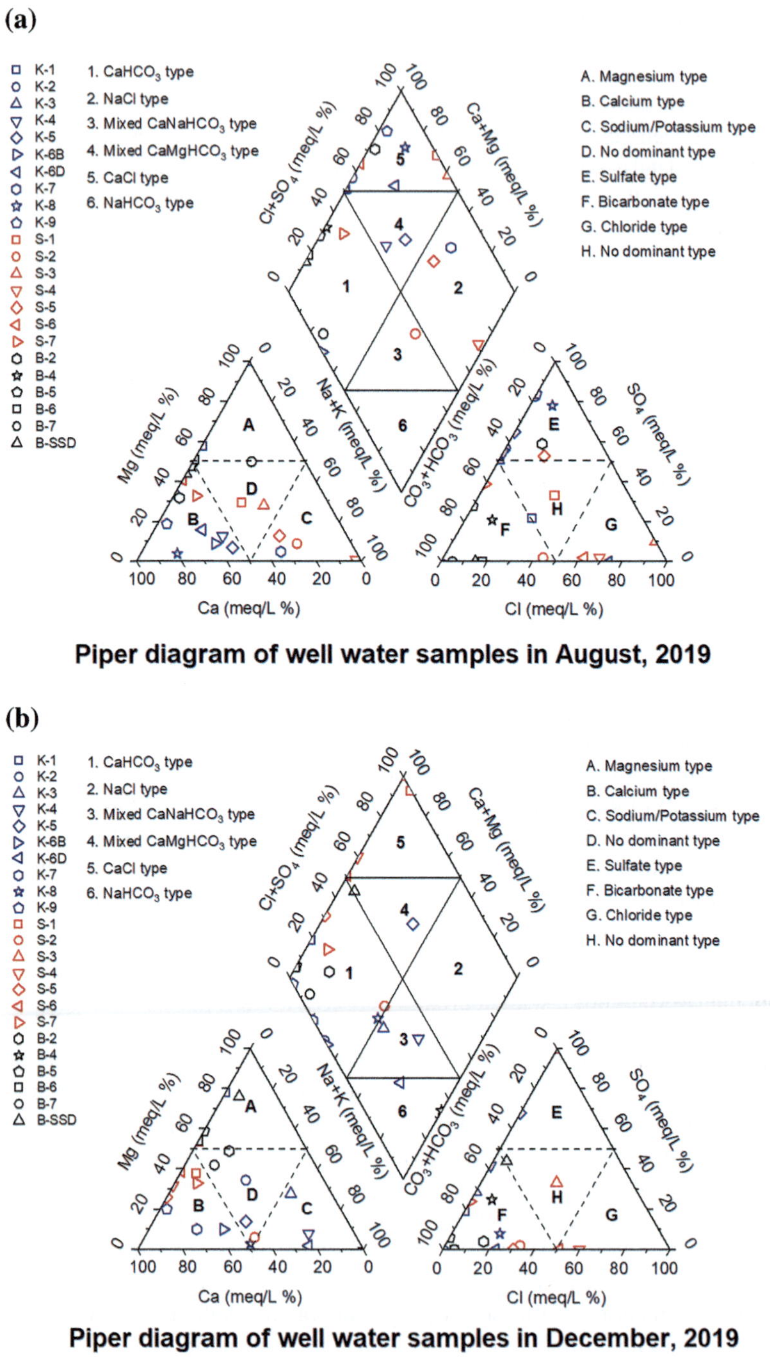

Fig. 26.5 Piper diagram of well water samples in (**a**) August and (**b**) December 2019

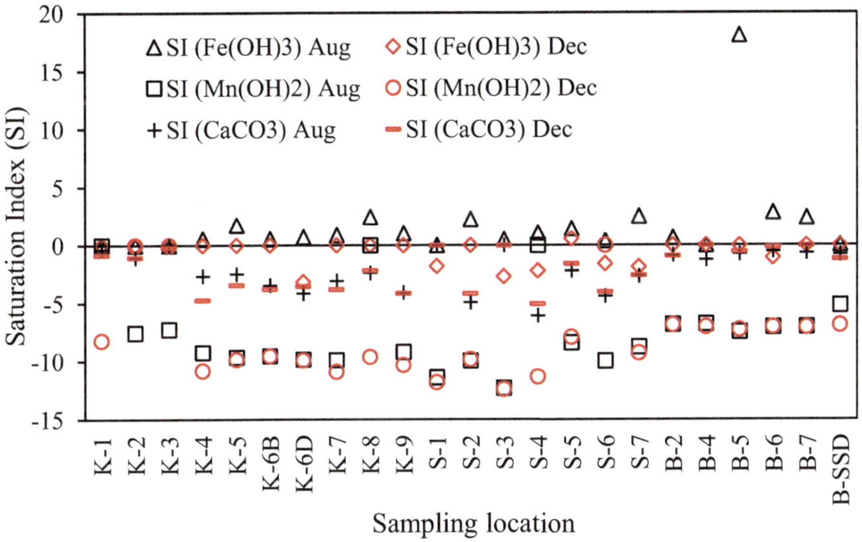

Fig. 26.6 Saturation index (SI) of $CaCO_3$, $Mn(OH)_2$, and $Fe(OH)_3$ of well water in August and December 2019

26.4 Pollution Level and Suitability for Water Use

Almost all shallow groundwater samples in Kampong Thom and Siem Reap during both rainy and dry periods showed pH values out of the DWS desirable range, 6.5–8.5 by the Ministry of Industry, Mines and Energy (MIME 2004), except K-1, K-2, and K-3. In contrast, most of the groundwater in Battambang had higher pH values within the standard range. Samples K-3 and S-7 were observed to have TDS higher than 800 mg/L. With TDS values higher than 900 mg/L, the taste of water would become slightly salty or unacceptable (Bruvold and Ongerth 1969; WHO 1996). The turbidity for drinking purposes was recommended by Cambodian DWS and the WHO to be <5 NTU (Nephelometric Turbidity Unit)/FNU; therefore, around half number of total samples in August and December exceeded the permissible value. The ORP values were not specified in the WHO, but the desirable value of ORP < 650 mV was generally recommended.

For the major ions, it was observed that their concentrations were approximate or under the maximum/desirable value of DWS and the WHO, as presented in Table 26.1, and there were no clear guidelines mentioned related to some major ions. The concentrations of Cr, Cd, Pb, and Cu were considered far below the high limit of WHO drinking water standard (typically around 0.01 mg/L) since they are under the detection limit of AAS. However, the upper range of Mn (2.7 mg/L) and Fe (24 mg/L) concentrations of all water samples from three provinces in Fig. 26.4 was significantly higher than the drinking water standard for Mn (desire <0.1 mg/L) and Fe (desire <0.3 mg/L), particularly most wells from Kampong Thom and Siem

Reap in the rainy season. These higher concentrations are harmful to human health in case of long-term intake. For a safe drink and use, the possible removal techniques should be carried out by injecting oxygen gas (O_2) to generate the oxidation process that can release carbon dioxide (CO_2) from the groundwater and change Fe^{2+} and Mn^{2+} into the insoluble Fe^{3+} and Mn^{4+} compounds. Alternatively, utilizing a boiling process, the present Fe and Mn could turn to precipitate in a hydroxide form by releasing CO_2 and increasing pH; thus, their insoluble compounds/precipitates could be removed by filtration (Ellis et al. 2000; Ahmad 2012).

Key Points

- Most of well waters satisfied the water drinking standard of Cambodia and the WHO, except for some well waters that had low pH, high TDS, and alarming concentrations of Fe and Mn.
- The hydrochemical facies were dominated by $CaHCO_3$ in both seasons, including CaCl in the rainy season, suggesting that the geochemistry of well waters was mainly controlled by carbonate minerals.
- The low pH of well waters in Kampong Thom and Siem Reap from the shallow groundwater source itself passed through an unsaturated zone containing high CO_2 or from dissolution of the existing sulfur minerals that resulted in more acidic water.
- Mn and Fe concentrations of groundwater were up to 2.7 and 24 mg/L, respectively; this relatively high concentration could also conform to their natural co-occurrence. However, Fe concentrations showed a considerable seasonal difference, and this might result from the excessive iron that is washed out from the top and subsoil by rainwater or flooding water from the TSL in August and gathered into the well.
- Possible Fe and Mn removal techniques such as oxidation process by O_2 injection or boiling process including microfiltration process should be applied for a safe drink and use.
- Soil/rock profile in the well and its mineralogical properties should be further studied for a better understanding.

References

Ahmad, M. (2012). Iron and manganese removal from groundwater: geochemical modeling of the Vyredox method (Master's thesis). University of Oslo

An TD, Tsujimura M, Le Phu V, Kawachi A, Ha DT. Chemical characteristics of surface water and groundwater in coastal watershed, Mekong Delta, Vietnam. Procedia Environ Sci. 2014;20:712–21.

Appelo CAJ, Postma D. Geochemistry, groundwater and pollution. 2nd ed. London: A.A. Balkema; 2005.

Bruvold WH, Ongerth HJ. Taste quality of mineralized water. J Am Water Works Assoc. 1969;61:170.

Eang KE, Igarashi T, Fujinaga R, Kondo M, Tabelin CB. Groundwater monitoring of an open-pit limestone quarry: groundwater characteristics, evolution and their connections to rock slopes. Environ Monit Assess. 2018a;190(193):1–15.

Eang KE, Igarashi T, Kondo M, Nakatani T, Tabelin CB, Fujinaga R. Groundwater monitoring of an open-pit limestone quarry: water-rock inter-action and mixing estimation within the rock layers by geochemical and statistical analyses. Int J Min Sci Technol. 2018b;28(6):849–57.

Ellis D, Bouchard C, Lantagne G. Removal of iron and manganese from groundwater by oxidation and microfiltration. Desalination. 2000;130(3):255–64.

Fitts CR. Groundwater science. San Diego: Academic Press (Elsevier Science); 2002.

Homoncik SC, MacDonald AM, Heal KV, Dochartaigh BÉÓ, Ngwenya BT. Manganese concentrations in Scottish groundwater. Sci Total Environ. 2010;408:2467–73.

Kou Y, Li Z, Hua K, Li Z. Hydrochemical characteristics, controlling factors, and solute sources of streamflow and groundwater in the Hei River catchment, China. Water. 2019;11(11):2293.

Ministry of Industry, Mining and Energy (2004). Kingdom of Cambodia drinking water quality standards. Drinking Water Quality Standards, 1–20.

Shahin, K.K., Ahmed, A., ZeinEldin, A., Ismail Shahin, S., Innovation, K.K., Zeineldin, A., Ismail, S. (2019). Sustainable treatment for high iron concentration in groundwater for irrigation purposes. 2019 ASABE annual international meeting. Vol. 1, p. 10–16.

World Health Organization (1996). WHO guidelines for drinking-water quality: total dissolved solids in drinking-water. Health criteria and other supporting information, 2nd ed. Vol. 2.

World Health Organization (2017). WHO guideline for drinking-water quality. 4th ed.

Part VI
Microbial Community

Chapter 27
Bacterial Communities: Their Dynamics and Interactions with Physicochemical Factors

Vannak Ann, Porsry Ung, Chanthol Peng, Manabu Fujii, Yasunori Tanji, and Kazuhiko Miyanaga

27.1 Bacterial Communities in Lakes

In both soil and aquatic systems, bacterial communities perform key functions in ecosystem processes, such as material cycle, the decomposition of organic matter, and the breakdown of pollutants (Pesce et al. 2008; Gonzalez et al. 2012). Lakes are excellent systems for investigating bacterial community dynamics because lakes have not only a sharp boundary definition but also strong environmental gradients (Linz et al. 2017). Thus, lakes are biologically complex freshwater ecosystems that create habitats for a wide range of macro- and microorganisms (Carpenter et al. 2011; see also Chaps. 31, 32, 33 and 34). Lake water and sediments of different environmental habitats with their unique characteristics result in their unique bacterial community composition (BCC), and such difference accounts for different microbial biogeographies in lake environments (Lindström and Langenheder 2012; Yang et al. 2016). On the other hand, the establishment of BCC in environmental water may vary from location to location depending on community adaptation, community interactions, environmental factors, hydrology, and human activities (Newton et al. 2011; Wang et al. 2018). For instance, BCC varies with different physicochemical characteristics (Zwart et al. 2002; Chang et al., 2020) among lakes (Van der Gucht et al. 2005; Yang et al. 2016) and across space and time that represents habitat heterogeneity (Shade et al. 2008).

V. Ann (✉) · C. Peng
Institute of Technology of Cambodia, Phnom Penh, Cambodia
e-mail: ann.v@itc.edu.kh

P. Ung
Ministry of Industry, Science, Technology and Innovation, Phnom Penh, Cambodia

M. Fujii · Y. Tanji · K. Miyanaga
Tokyo Institute of Technology, Tokyo, Japan
e-mail: miyanaga.kazuhiko@jichi.ac.jp

C. Yoshimura et al. (eds.), *Water and Life in Tonle Sap Lake*,
https://doi.org/10.1007/978-981-16-6632-2_27

Tonle Sap Lake (TSL) has a unique and dynamic flood-pulse ecosystem driven by seasonal flooding from the Mekong River (Holtgrieve et al. 2013; see also Chap. 1), and TSL is arguably known as one of the world's most productive freshwater lake ecosystems (Daly et al. 2020; see also Chaps. 31, 32, 33 and 34). As a result, the water volume change, flow direction, mixing of water and sediments, and others affecting BCC must be considered. A better understanding of the drivers and controls of bacterial communities (on spatial and longtime scales) would improve both our knowledge of the fundamental properties of those communities and our ability to predict community states (Linz et al. 2017). However, compared with that in temperate environmental water, the shifting of BCC in tropical environmental water as found in lakes has not been well understood (Ung et al. 2019). Furthermore, the comprehensive dataset of the shifting of BCC in water column and sediment in aquatic ecosystems is limited since studies have focused either on surface waters or sediments (Liu et al. 2013; Li et al. 2015; Chen et al. 2016; Hengy et al. 2017; Li et al. 2017; Wan et al. 2017).

27.2 Dominant Genera and their Spatiotemporal Distribution

The water and sediment samples in the TSL were collected at cross sections (CS) 1, 2, 3, and 4 (northwestern part); CS 5, 6, and 7 (southeastern part); and at three floating villages (Kampong Plouk (KP), Kampong Luong (KL), and Chhnok Tru (CT)). The surface water samples were collected using a water sampler (General Oceanics, Miami, FL, USA), while sediment samples were collected using a core sampler (Rigosha & Co., Ltd., Saitama, Japan). This sampling was conducted four times in December 2016 (Dec-16, dry season), March 2017 (Mar-17, the transition from dry to rainy season), and June and August 2017 (Jun-17 and Aug-17, rainy season) (see map of the study area (Fig. 1, Ung et al. 2019), published in Science of the Total Environment 664: 414-423. doi.org/10.1016/j.scitotenv.2019.01.351, as well as the detailed methods to measure physicochemical parameters and microbial communities).

The range (i.e., maximum–minimum) for the water depth of TSL was approximately 9.3 m (mean ± standard deviation, 3.9 ± 2.2), while, respectively, varied temperature of the surface water, 8.7 m (29.7 ± 1.8); electrical conductivity (EC), 21.5 mS/m (12.8 ± 2.3); dissolved oxygen (DO), 8.5 mg/L (6.8 ± 1.7); pH, 2.4 (7.9 ± 0.5); oxidation–reduction potential (ORP), 454 mV (213 ± 86); total dissolved solids (TDS), 95.3 mg/L (36.8 ± 33.1); and turbidity, 337 Nephelometric Turbidity Unit (NTU) (31 ± 58.5).

Based on the 16S rRNA gene sequence results from next-generation sequencing analysis, microbial communities in the lake water and sediment were chosen at the genus level for the major lineages set greater than 5% (see Figs. 27.1 and 27.2, respectively). Most of the genera were widely distributed in the lake water and

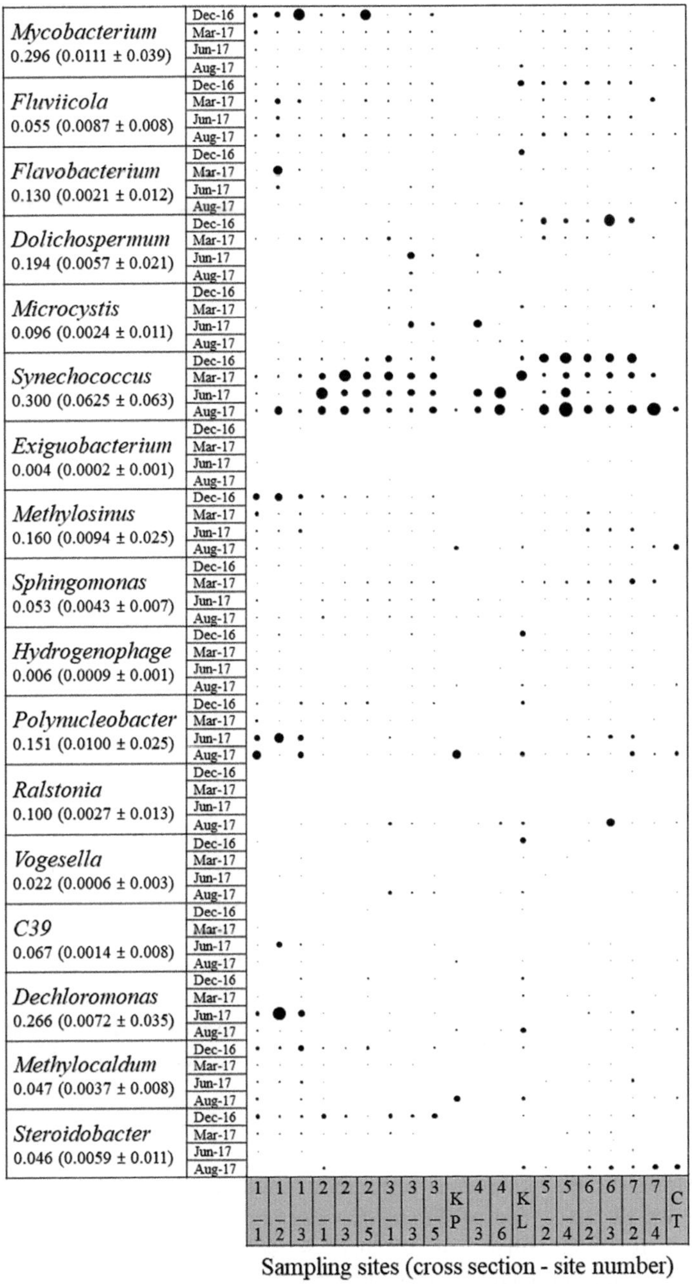

Fig. 27.1 Spatiotemporal patterns of assigned major genera on bubble charts, with the range value (mean ± standard deviation), showing their relative abundance in the surface water of the lake. Blank spaces represent low detection or no data available. Kampong Plouk (KP), Kampong Luong (KL), and Chhnok Tru (CT) are the sampling sites at the floating villages

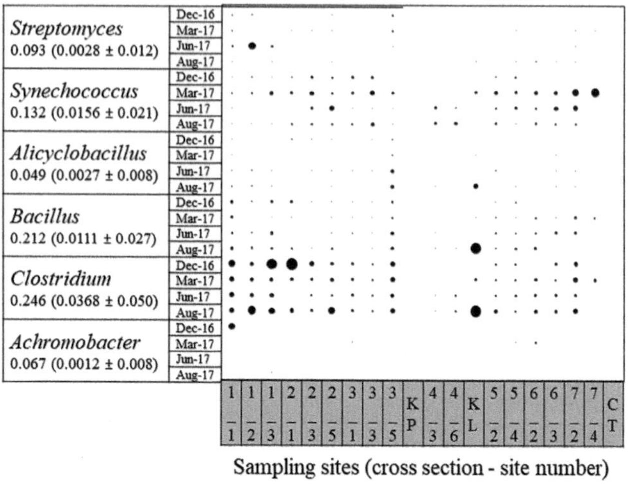

Fig. 27.2 Spatiotemporal patterns of assigned major genera on bubble charts, with the range value (mean ± standard deviation), showing their relative abundance in the sediment of the lake. Blank spaces represent low detection or no data available. Kampong Plouk (KP), Kampong Luong (KL), and Chhnok Tru (CT) are the sampling sites at the floating villages

sediment; however, their abundance depended on location and season. With relative abundance about 30%, *Synechococcus* was the most abundant and widely distributed genus in the lake water (Fig. 27.1). Considered as an important contributor to primary production in many freshwater lakes (Ruber et al. 2018), *Synechococcus* (also called *picocyanobacteria*; diameter, 0.2–2 μm; cell wall, Gram-negative type; see Hoiczyk and Hansel 2000; Bell and Kalff 2001; and Callieri et al. 2007) belongs to *Cyanobacteria* at the phylum level. In the study of Ung et al. (2019), the abundance of *Cyanobacteria* was highly measured during the high-water-level season (Dec-16 and Aug-17) in the southeastern part (CS 4–3 to 7–4) of the lake. Conversely, its high abundance was measured during the low-water-level season (Mar-17 and Jun-17) in the northwestern part (CS 1–1 to 3–5). These results seem to suggest that the water flow from the Tonle Sap River (TSR) to the TSL during rainy season and from the TSL to the TSR during dry season strongly affects the abundance of *Cyanobacteria*. Similarly, *Cyanobacteria* was also dominantly found in some freshwater lakes in the world, such as Lake Loosdrecht (the Netherlands), Lake Ijssel (the Netherlands), and Lake Soyang (South Korea) (Yadav et al. 2019). Moreover, the biodiversity of bacteria found in other freshwater lake ecosystems shows the existence of *Synechococcus* and *Microcystis*. In addition, they appear to mix with other genera, such as *Aphanizomenon*, *Planktothrix*, and *Polynucleobacter*, in lakes Loosdrecht (the Netherlands), Baikal (Russia), and Tanganyika (Africa).

Synechococcus was not only dominant in this lake water but also in its sediment (Fig. 27.2). *Synechococcus* in the sediment was highly observed during dry season (Mar-17), especially in the northwestern part of the lake, probably due to the water flow exchanges at the junction of the TSR and TSL. Those exchanges might play a role in the mixing process of the surface water and sediment in other parts of the lake. Another process was previously reported by Weisse (1993), in which a great interannual variability of *Synechococcus* was associated with the differences in weather conditions causing different spring mixing regimes and water column stabilization timing. Further characterization of these dynamics is necessary to truly understand the role of *Synechococcus* in environmental change, as well as that of other related microorganisms in this river–lake system.

On the other hand, *Mycobacterium*, *Flavobacterium*, *Polynucleobacter*, *Methylosinus*, *Dechloromonas*, and *Methylocaldum* were relatively dominant in the lake water of the northwestern part. *Methylosinus* and *Methlocaldum*, as methane-oxidizing bacteria, were highly detected in the northwestern part; this may be caused by the high concentration of methane, which is due to the methane fermentation of a high concentration of organic compounds and a low concentration of DO. It means that the northwestern part is the dead end of the lake, although a little water flow exists during the transition from dry to rainy season, and the area may tend to be a reductive environment. *Dolichospermum* and *Sphingomonas* were present in high abundance in the lake water of the southeastern part. *Microcystis* was also found in lake water, and its abundance increased during the transition from dry to rainy season in Jun-17. As shown in a recent study, during the bloom of *Cyanobacteria* in Lake Chaohu, *Microcystis* and *Dolichospermum* were the two major genera (Guan et al. 2020); specifically, the ranges of the cell diameter and colony size of *Microcystis* were small throughout the whole year, but those of *Dolichospermum* significantly changed over time and in different parts of the lake. It seems to be true that both genera coexist with the abundance of *Cyanobacteria*, which was also abundantly found in the water column of this river–lake system (Ung et al. 2019). Meanwhile, the most abundant genus in the sediment in Aug-17 is the *Clostridium*, which appeared in the northeastern part of the lake and is close to the floating village of KL. Generally, *Clostridium* sp. is known as a facultative anaerobic bacterium that can survive not only in the surface water but also in the sediment as fermentative microbes.

27.3 Relationships between Dominant Genera and their Physicochemical Factors

Most of the dominant genera significantly correlated with the physicochemical factors of lake water, especially with DO and pH (Table 27.1). In addition to both variables, *Mycobacterium* (individually, still determined by ORP), *Methylosinus* (by ORP and EC), and *Methylocaldum* were determined by temperature, TDS, and turbidity; *Steroidobacter* was significantly correlated with water depth, temperature, EC, and TDS; and *Hydrogenophaga* with turbidity. It is possible that these microbes are adsorbed to the surface of minute solids such as inorganic and organic substances

Table 27.1 Pearson's r correlation coefficients between the relative abundance of the dominant genera and physicochemical factors

	Water depth	Temperature	EC	DO	pH	ORP	TDS	Turbidity
Mycobacterium		−0.34**		−0.60**	−0.32**	0.46**	−0.40**	−0.38**
		(107)		(107)	(107)	(107)	(50)	(50)
Fluviicola					−0.28**			
					(107)			
Dolichospermum				0.21*				
				(107)				
Microcystis		0.30**			0.34**			
		(107)			(107)			
Synechococcus		0.35**		0.43**	0.30**	−0.24*		
		(107)		(107)	(107)	(107)		
Methylosinus		−0.26**	−0.40**	−0.56**	−0.42**	0.33**	−0.41**	−0.41**
		(107)	(107)	(107)	(107)	(107)	(50)	(50)
Sphingomonas	−0.28**				0.23*			
	(114)				(107)			
Hydrogenophaga				−0.29**	−0.34**			−0.29*
				(107)	(107)			(50)
Polynucleobacter			0.20*	−0.36**	−0.36**			
			(107)	(107)	(107)			
Dechloromonas			0.22*	−0.21*	−0.26**	0.43**	−0.45**	
			(107)	(107)	(107)	(107)	(50)	
Methylocaldum		−0.34**		−0.74**	−0.50**		−0.45**	−0.40**
		(107)		(107)	(107)		(50)	(50)
Steroidobacter	0.27**	−0.48**	−0.45**	−0.26**	−0.32**		−0.38**	
	(114)	(107)	(107)	(107)	(107)		(50)	

The significant levels (two-tailed) of the correlations are 0.01 (**) and 0.05 (*). Sample sizes are in bracket. *EC* electrical conductivity, *DO* dissolved oxygen, *ORP* oxidation–reduction potential; and *TDS* total dissolved solids. Blank spaces are not statistically significant

present in lake water. Therefore, as the concentration of TDS and/or turbidity increases, the abundance of those genera in lake water decreases. Besides DO and pH, *Dechloromonas* was also significantly related to EC, ORP, and TDS. As the physiological function of *Dechloromonas* is thought to be a reductive dichlorination from chlorine-containing materials, the abundance of this genus was related to ORP. In our further study, the relationships between the abundance of *Dechloromonas* and the original source or the concentration of chlorinated compounds in lake water must be verified.

Dolichospermum showed a significant correlation with DO in the water of TSL. This genus was also observed, in Missisquoi Bay (Lake Champlain), following low chlorine and mid-$KMnO_4$ (287.7 mg·min/L) exposure (Moradinejad et al. 2020). *Microcystis* significantly correlated with temperature and pH, and in addition to these variables, *Synechococcus* significantly correlated with DO and ORP (Table 27.1). It means that their photosynthesis is significantly affected by these physicochemical factors. However, although pH showed a significant correlation with *Microcystis* and *Synechococcus*, it is difficult to directly relate them because of the complexity of pH. However, the consistency of trends remains seen; for example, in Western Lake Erie, the shifts of cyanobacterial community composition dominated by OTUs of *Microcystis* and *Synechococcus*, which dynamically fluctuated during the bloom, could be partitioned into components predicted by pH, temperature, chlorophyll a, and water mass movements (Berry et al. 2017).

Key Points

- *Synechococcus* was the most detected genus in both lake water and sediment, while *Microcystis* increased its dominance in lake water during the transition from dry to rainy season.
- Most of the genera in lake water were statistically related to physicochemical factors, especially DO and pH.
- The succession of dominant species and their abundances might be regulated by the hydrodynamics of this river–lake system.
- The microbial communities in lake water are sensitive to environmental fluctuations.

References

Bell T, Kalff J. The contribution of picophytoplankton in marine and freshwater Systems of Different Trophic Status and Depth. Limnol Oceanogr. 2001;46:1243–8.

Berry MA, Davis TW, Cory RM, Duhaime MB, Johengen TH, Kling GW, Marino JA, Den Uyl PA, Gossiaux D, Dick GJ, Denef VJ. Cyanobacterial harmful algal blooms are a biological disturbance to Western Lake Erie bacterial communities. Environ Microbiol. 2017;19:1149–62.

Callieri C, Modenutti B, Queimaliños C, Bertoni R, Balseiro E. Production and biomass of picophytoplankton and larger autotrophs in Andean Ultraoligotrophic Lakes: differences in light harvesting efficiency in deep layers. Aquat Ecol. 2007;41:511–23.

Carpenter SR, Stanley EH, Vander Zanden MJ. State of the World's freshwater ecosystems: physical, chemical, and biological changes. Annu Rev Environ Resour. 2011;36:75–99.

Chang W, Sun J, Pang Y, Zhang S, Gong L, Lu J, Feng B, Xu R. Effects of different habitats on the bacterial community composition in the water and sediments of Lake Taihu. China Environ Sci Pollut Res. 2020;27:44983–94.

Chen Y, Dai Y, Wang Y, Wu Z, Xie S, Liu Y. Distribution of bacterial communities across plateau freshwater Lake and upslope soils. J Environ Sci. 2016;43:61–9.

Daly K, Ahmad SK, Bonnema M, Beveridge C, Hossain F, Nijssen B, Holtgrieve G. Recent warming of Tonle Sap Lake, Cambodia: implications for one of the World's most productive inland fisheries. Lakes Reserv Res Manag. 2020;25:133–42.

Gonzalez A, King A, Robeson MS II, Song S, Shade A, Metcalf JL, Knight R. Characterizing microbial communities through space and time. Curr Opin Biotechnol. 2012;23:431–6.

Guan Y, Zhang M, Yang Z, Shi X, Zhao X. Intra-annual variation and correlations of functional traits in Microcystis and Dolichospermum in Lake Chaohu. Ecol Indic. 2020;111:106052.

Hengy MH, Horton DJ, Uzarski DG, Learman DR. Microbial community diversity patterns are related to physical and chemical differences among Temperate Lakes near Beaver Island, MI. PeerJ. 2017;5:e3937.

Hoiczyk E, Hansel A. Cyanobacterial cell walls: news from an unusual prokaryotic envelope. J Bacteriol. 2000;182:1191–9.

Holtgrieve GW, Arias ME, Irvine KN, Lamberts D, Ward EJ, Kummu M, Koponen J, Sarkkula J, Richey JE. Patterns of ecosystem metabolism in the Tonle Sap Lake, Cambodia with links to capture fisheries. PLoS One. 2013;8:e71395.

Li D, Jiang X, Wang J, Wang K, Zheng B. Effect of sewage and industrial effluents on bacterial and archaeal communities of creek sediments in the Taihu Basin. Water. 2017;9:373.

Li J, Zhang J, Liu L, Fan Y, Li L, Yang Y, Lu Z, Zhang X. Annual periodicity in planktonic bacterial and archaeal community composition of eutrophic Lake Taihu. Sci Rep. 2015;5:15488.

Lindström ES, Langenheder S. Local and regional factors influencing bacterial community assembly. Environ Microbiol Rep. 2012;4:1–9.

Linz AM, Crary BC, Shade A, Owens S, Gilbert JA, Knight R, McMahon KD. Bacterial community composition and dynamics spanning five years in freshwater Bog Lakes. mSphere. 2017;2: e00169–17.

Liu Y, Yao T, Jiao N, Liu X, Kang S, Luo T. Seasonal dynamics of the bacterial Community in Lake Namco, the largest Tibetan Lake. Geomicrobiol J. 2013;30:17–28.

Moradinejad S, Trigui H, Guerra Maldonado JF, Shapiro J, Terrat Y, Zamyadi A, Dorner S, Prévost M. Diversity assessment of toxic cyanobacterial blooms during oxidation. Toxins. 2020;12:728.

Newton RJ, Jones SE, Eiler A, McMahon KD, Bertilsson S. A guide to the natural history of freshwater lake bacteria. Microbiol Mol Biol Rev. 2011;75:14–49.

Pesce S, Bardot C, Lehours AC, Batisson I, Bohatier J, Fajon C. Effects of diuron in microcosms on natural riverine bacterial community composition: new insight into phylogenetic approaches using PCR-TTGE analysis. Aquat Sci. 2008;70:410–8.

Ruber J, Geist J, Hartmann M, Millard A, Raeder U, Zubkov M, Zwirglmaier K. Spatio-temporal distribution pattern of the picocyanobacterium *Synechococcus* in lakes of different trophic states: a comparison of flow cytometry and sequencing approaches. Hydrobiologia. 2018;811: 77–92.

Shade A, Jones SE, McMahon KD. The influence of habitat heterogeneity on freshwater bacterial community composition and dynamics. Environ Microbiol. 2008;10:1057–67.

Ung P, Peng C, Yuk S, Tan R, Ann V, Miyanaga K, Tanji Y. Dynamics of bacterial Community in Tonle sap Lake, a large tropical flood-pulse system in Southeast Asia. Sci Total Environ. 2019;664:414–23.

Van der Gucht K, Vandekerckhove T, Vloemans N, Cousin S, Muylaert K, Sabbe K, Gillis M, Declerk S, De Meester L, Vyverman W. Characterization of bacterial communities in four Freshwater Lakes differing in nutrient load and food web structure. FEMS Microbiol Ecol. 2005;53:205–20.

Wan Y, Ruan X, Zhang Y, Li R. Illumina sequencing-based analysis of sediment bacteria community in different trophic status freshwater lakes. Microbiologyopen. 2017;6:e00450.

Wang L, Zhang J, Li H, Yang H, Peng C, Peng Z, Lu L. Shift in the microbial community composition of surface water and sediment along an Urban River. Sci Total Environ. 2018;627:600–12.

Weisse T. Dynamics of autotrophic picoplankton in marine and freshwater ecosystems. In: Jones JG, editor. Advances in microbial ecology. New York: Plenum Press; 1993. p. 327–70. https://doi.org/10.1007/978-1-4615-2858-6_8.

Yadav AN, Yadav N, Kour D, Kumar A, Yadav K, Kumar A, Rastegari AA, Sachan SG, Singh B, Chauhan VS, Saxena AK. Bacterial community composition in lakes. In: Freshwater microbiology. London: Academic; 2019. p. 1–71. https://doi.org/10.1016/B978-0-12-817495-1.00001-3.

Yang J, Jiang H, Wu G, Liu W, Zhang G. Distinct factors shape aquatic and sedimentary microbial community structures in the lakes of Western China. Front Microbiol. 2016;7:1782.

Zwart G, Crump BC, Kamst-van Agterveld MP, Hagen F, Han SK. Typical freshwater bacteria: an analysis of available 16S rRNA gene sequences from plankton of lakes and Rivers. Aquat Microb Ecol. 2002;28:141–55.

Chapter 28
Microcystin Production and Oxidative Stress

Kota Nakatani, Kohei Nasukawa, Tetsuro Kikuchi, Manabu Fujii, Kazuhiko Miyanaga, Yasunori Tanji, and Vannak Ann

28.1 Cyanobacteria and Toxin Production

Algal bloom is a phenomenon in which the color of the surface water turns into blue–green because of cyanobacterial phytoplankton overgrowth (the color originates from chlorophyll a and phycocyanin) in eutrophic lakes and ponds (Sengco and Anderson 2004). Cyanobacterial bloom outbreak poses a range of environmental and social issues, such as deficiency of dissolved oxygen and associated death of fishes and animals, deterioration of landscape, filtration fouling during the water treatment, and mold odor of the treated water (Hallegraeff 1992). Cyanobacteria, such as *Microcystis aeruginosa*, *Anabaena flos-aquae*, and *Planktothrix agardhii*, also often synthesize cyanotoxin (e.g., hepatotoxins and neurotoxins) as secondary metabolites (Pearson et al. 2010), causing disease and even death of wild and livestock animals (Hallegraeff 1992). Among cyanobacterial species, *M. aeruginosa* is the most commonly observed species in eutrophic closed lakes all over the world.

Microcystin (MC) is a frequently observed cyanobacterial hepatotoxin. MC consists of cyclic peptides (six amino acids) plus a side amino acid chain called 3-amino-9-methoxy-2,6,8-trimethyl-10-phenyl-4,6-decadienoic acid. There are a number of MC variants (over 200, including MC-LR, MC-RR, and MC-YR), depending on the type of amino acid resident in the cyclic peptides. MC is highly toxic, e.g., the amount of lethal toxicity for MC (to rats) is from 1/10 to 1/200 compared with that for potassium cyanide (Schmidt et al. 2014). The MC concentration in drinking water set in the World Health Organization (WHO) guidelines is

K. Nakatani · K. Nasukawa · T. Kikuchi · M. Fujii (✉) · K. Miyanaga · Y. Tanji
Tokyo Institute of Technology, Tokyo, Japan
e-mail: fujii.m.ah@m.titech.ac.jp; miyanaga.kazuhiko@jichi.ac.jp

V. Ann
Institute of Technology of Cambodia, Phnom Penh, Cambodia

C. Yoshimura et al. (eds.), *Water and Life in Tonle Sap Lake*,
https://doi.org/10.1007/978-981-16-6632-2_28

1 μg/L (WHO 1998). Thus, it is necessary to mitigate the occurrence of toxic cyanobacteria to reduce human health risks and develop the sustainable water environment.

However, although the genes involved in MC synthesis (e.g., *mcyA-J*) have been well known, the mechanisms behind the cellular MC synthesis and resultant spatio-temporal occurrence in lakes remain inconclusive (Kaebernick et al. 2000). For example, it is well recognized that there are toxin-producing and toxin-nonproducing strains of cyanobacterial species (e.g., *M. aeruginosa* PCC7806 and 7005, respectively), regardless of their higher gene similarities (Neilan et al. 1997). The presence of nontoxic strain indicates that MC is not necessarily an inevitable biomolecule for basal cellular metabolisms, although it is expected to play a role in some cellular functions, such as intracellular oxidative stress regulation, defense against predators, nutrient uptake, and quorum sensing (Kearns and Hunter 2001; Kurmayer and Juttner 1999; Zhai et al. 2012). Another previous study also reported that the abundance of toxin-producing genes, such as *mcyA* and *mcyE*, was not always positively correlated with the concentrations of MC in aquatic environment and a significant correlation may not be obtained in the absence of a single or multiple *mcy* genes required for MC synthesis (via gene deletion, recombination, etc.) (Beversdorf et al. 2015). Therefore, comprehensively understanding the environmental factors, water quality, and cellular variables related to MC synthesis in the environment is important.

Previous studies have suggested that MC synthesis is associated with intracellular oxidative stress (Zilliges et al. 2011). Oxidative stress can be caused by the generation of intracellular reactive oxygen species (ROS: superoxide, hydrogen peroxide, hydroxyl radical, etc.) under nutrient-limiting environments and high light conditions (Latifi et al. 2009). Cyanobacteria such as *M. aeruginosa* may mitigate ROS-mediated cellular damage (e.g., photosynthetic protein and DNA) via the generation of MC. For example, it has been reported that the N-methyldehydroalanine site of MC binds to the cysteine residue of photosynthetic protein under high light intensity, thereby suppressing protein degradation initiated by oxidative stress (Zilliges et al. 2011). In addition to light condition, deficiency of some important nutrients may also cause oxidative stress. Iron is a cofactor for antioxidant enzymes, such as catalase, peroxidase, and superoxide dismutase, and is an important trace metal nutrients for regulating oxidative stress in cells (Lundrigan et al. 1997). Iron deficiency in aqueous environment has been well investigated, and iron limitation is known to increase oxidative stress and consequently induce MC production (Alexova et al. 2011). Furthermore, while reports are limited, the presence of excess heavy metals is likely to increase oxidative stress in cyanobacterial cells (Choudhary et al. 2007). For example, redox-reactive trace metals (such as Fe and Cu) can be related to the intracellular generation of ROS via electron transfer from metal ions (e.g., Haber–Weiss reaction) (Dang et al. 2012), whereas excess toxic heavy metals (such as Zn) may inhibit cellular metabolism via the inactivation of antioxidant enzymes and thylakoid membranes (Tripathi and Gaur 2004).

Although the investigation on the mechanism of cyanobacterial toxin production in aquatic systems, such as freshwater lakes, has been addressed in many previous studies, the occurrence of cyanobacterial toxin and its relation to environmental factors remain controversial. In this chapter, we introduce our field survey and laboratory experiment on the mechanisms of MC generation by summarizing the results mainly reported by Nakatani et al. (2019). In particular, we focus on the impact of divalent metals (e.g., Zn and Cu) on oxidative stress and MC synthesis.

28.2 Microcystin (MC) Concentration in Tonle Sap Lake and Eutrophic Lakes in Japan

We conducted the field survey on MC contamination in Tonle Sap Lake, Cambodia (Fig. 28.1; see Chap. 1 for details regarding this lake). The shipboard field survey was conducted in 2017 (March for the dry season and August for the wet season). Surface water was collected at several designated sampling stations, and water qualities, including MC concentrations, were measured across the Tonle Sap Lake. During the shipboard survey, we observed the occurrence of phytoplankton bloom during the dry season (e.g., area near the Kampong Phluk Port, as shown in Fig. 28.1a; see also Fig. 31.3a in Chap. 31). The spatial distribution of MC concentration during the dry season ranged from 8.2 to 70,692 ng/L (Fig. 28.1b). The water near the Kampong Phluk Port showed the highest MC concentration, whereas the middle of the lake tended to have lower concentration (e.g., order of 10 ng/L) most likely because of the low nutrient load. Since the WHO guideline suggests MC concentration lower than 1000 ng/L for drinking purpose, it is highly recommended that water with MC concentration over this guideline should not be directly used for domestic purpose. Compared with the dry season, the MC concentrations observed during the shipboard survey in the wet season were low and, in many cases, undetectable (Fig. 28.1c).

Field survey was also conducted in lakes and dam reservoirs (Lake S, Lake R, and Lake K) near the Kanto Region, Japan, in July 2016. During the field campaign, the occurrence of cyanobacterial bloom was visually observed in all water bodies (Fig. 28.2a). In Lake R, for example, parts of the water surface were blue–green in color, whereas the surface water in Lake K was covered with floating algal scums. Visual inspection using optical microscope identified some cyanobacterial genera, including *Anabaena*, whereas *Microcystis* was observed to be the dominant species in the lakes investigated, showing the highest cell density of 1.4×10^7 cell/mL for the bloom water samples. Comprehensive dataset for the water quality variables (such as pH, electronic conductivity, water temperature, MC, dissolved oxygen, chlorophyll a, trace metals, nutrients, and cell concentrations) were reported in Nakatani et al. (2019). Briefly, MC concentrations ranged from 0.6 to 124 μg/L among ten samples. For fundamental water quality, the water temperature, pH, and dissolved oxygen concentration for bloom waters ranged from 27 °C to 36 °C, 8.8 to 10.8, and 7.5 to over 22 mg/L, respectively, with a high pH indicating alkalization due to the consumption of carbon dioxide by photosynthesis. Cellular MC

(A) Algal bloom at Kampong Phluk Port in Tonlé Sap Lake

(B) Special distribution of microcystin (ng/L)

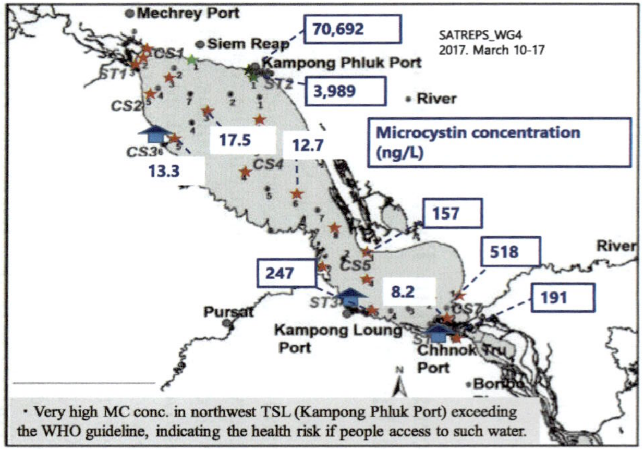

(C) Microcystin concentrations in the dry and wet seasons

Sampling site code and name	Dry seaon (March 2017)		Wet seaon (August 2017)	
	MC[ng/L]	Chl-a [μg/L]	MC[ng/L]	Chl-a [μg/L]
CS3_1: Near Kampong Phluk Port	3989	179	B.D.	3.3
CS5_4: Near Kampong Loung Port	249	24	B.D.	4.9
CS6_1: Southeast shore	516	69	18.4	11
CS7_4: Near Chhnok Tru Port	193	11	B.D.	7.3

Fig. 28.1 (**a**) Algal bloom at Kampong Phluk Port in Tonle Sap Lake (picture taken from sampling boat). (**b**) Special distribution of microcystin (MC) across the Tonle Sap Lake (dry season [March] in 2017). (**c**) Comparison of microcystin (MC) concentrations in the dry (March 2017) and wet (August 2017) seasons

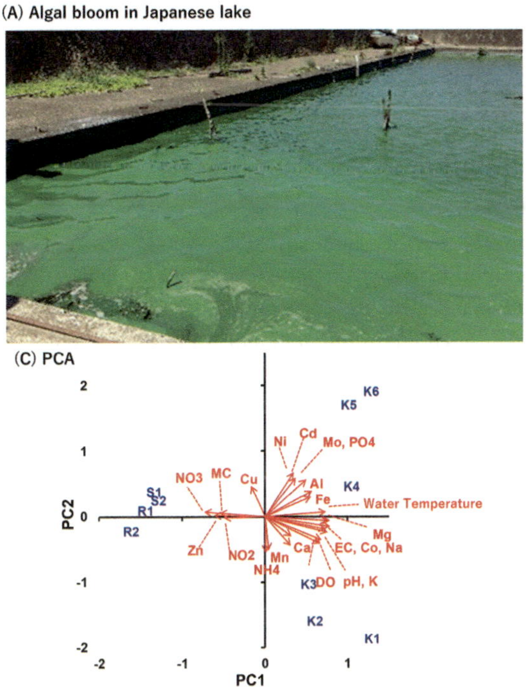

(B) Simple correlation analysis

Water quality variables	Correlation coefficient[a]	Significance[b]
pH	−0.70	*
EC (mS.m^{-1})	−0.68	*
Water temperature (°C)	−0.67	*
Dissolved oxygen (mg.L^{-1})	−0.69	*
Cd (nM)	−0.16	
Co (nM)	−0.67	*
Cu (nM)	0.22	
Fe (nM)	−0.42	
Mn (nM) [c]	−0.39	
Mo (nM)	−0.34	
Ni (nM)	−0.11	
Zn (nM)	0.29	
Ca (μM)	−0.03	
Mg (μM)	−0.60	
Na (μM)	−0.69	*
Al (μM)	−0.49	
K (μM)	−0.67	*
NO_2 (μM)	0.51	
NO_3 (μM)	0.87	**
NH_4 (μM)	0.24	
PO_4 (μM)	−0.39	

[a] Italics are Spearman's rank correlation coefficients. The rest are Pearson's product rate correlation coefficients.

[b] *: $p < 0.05$; **: $p < 0.01$

Fig. 28.2 (**a**) Algal bloom in Japanese lake. (**b**) Principal component analysis (PCA) for microcystin (MC) concentration per cell (pg/cell) and water quality parameters. The water quality data was collected from the field survey in Japanese (Kanto Region) lakes. (**c**) Simple correlation analysis between microcystin (MC) concentration per cell (pg/cell) and water quality parameters in the Japanese lakes (adapted from Nakatani et al. (2019) with partial modification)

concentration (i.e., MC amount per phytoplankton cell in unit of pg/cell) was also calculated to represent cellular MC production.

Principal component analysis (PCA) was performed for the water quality parameters examined in the Japanese field campaign (where enough dataset of water quality is available), as shown in Fig. 28.1b. The first principal component (54% contribution) was associated with pH, water temperature, dissolved oxygen, electronic conductivity, major cations (Ca, K, Mg, and Na), and others (e.g., Co, NO_3, and Zn), whereas the second principal component (23% contribution) was associated with PO_4, Cd, Mo, and Ni. The result from the PCA suggested that lakes S and R had different characteristics with respect to cellular MC concentration from Lake K. To obtain further insights into the relation of cellular MC concentration with water quality variables, simple correlation analyses were conducted between MC concentration and other water quality parameters (Fig. 28.1c). As a result, cellular MC concentration showed a strong positive correlation ($p < 0.01$, $R = 0.87$) with NO_3, whereas significant negative correlations ($p < 0.05$, R values ranging from −0.70 to

−0.67) were observed with some parameters such as pH, dissolved oxygen, K, Na, and Co.

28.3 MC Production in Relation to Divalent Metals

It is well recognized that the growth and metabolism of cyanobacteria algae in aquatic environments can be largely affected by the temporal and spatial variations of a range of environmental and water quality variables and competition with other microorganisms for nutrient uptake. Such environmental and water quality changes may cause different levels of cellular stress and therefore produce MC. In particular, here, given the expected importance of dissolved divalent metal impact on the cellular oxidative stress, we focus on the relation of dissolved divalent metal concentration with MC production via laboratory-based culturing experiment. As illustrated in Fig. 28.3a, b, laboratory incubation of *M. aeruginosa* PCC7806 at different metal concentrations for Cu and Zn indicated that cellular oxidative stress and MC concentrations per cell were affected by the Cu and Zn concentrations in the culture medium. A multiple comparison test (Tukey's HSD) for oxidative stress and MC concentrations per cell (measured at different metal concentrations) showed that oxidative stress and MC concentrations increased with increasing concentrations of Cu and Zn in the medium (e.g., in cases where metal concentrations were 2–10 times higher than those for original Fraquil* medium, $p < 0.05$). Overall, the increase in cell oxidative stress resulted in an increase in MC concentration per cell. Therefore, we conducted a linear regression analysis of oxidative stress and MC production in a laboratory culture test and found a positive and significant correlation ($p < 0.05$, Fig. 28.3c), implying that the increase in MC production is due to the increase in oxidative stress. The data presented here only showed that increasing heavy metal ions caused MC production and oxidative stress response. However, it is also recognized that deficiency of trace nutrient metal (iron) also induces oxidative stress and consequently MC synthesis (Alexova et al. 2011).

Redox-active metals, such as Cu, are involved in the intracellular production of hydroxyl radicals that initiate oxidative stress via the Haber–Weiss reaction. Zn is also known to be a trace metal essential for the growth of blue–green algae; however, it is also a nonredox-active heavy metal that exerts negative effects on cells as an oxidative stress factor (Zeng et al. 2009). Nonredox-active heavy metals, such as Zn, Pb, and Cd, are not directly involved in the production of ROS but may increase cellular oxidative stress by inactivating intracellular antioxidants and disrupting metabolic balance (Zeng et al. 2009). For example, in chloroplasts, divalent metal ions, such as Zn^{2+} and Pb^{2+}, have been suggested to increase oxidative stress by inhibiting electron transfer in thylakoid membranes involved in photosynthesis (Pinto et al. 2003).

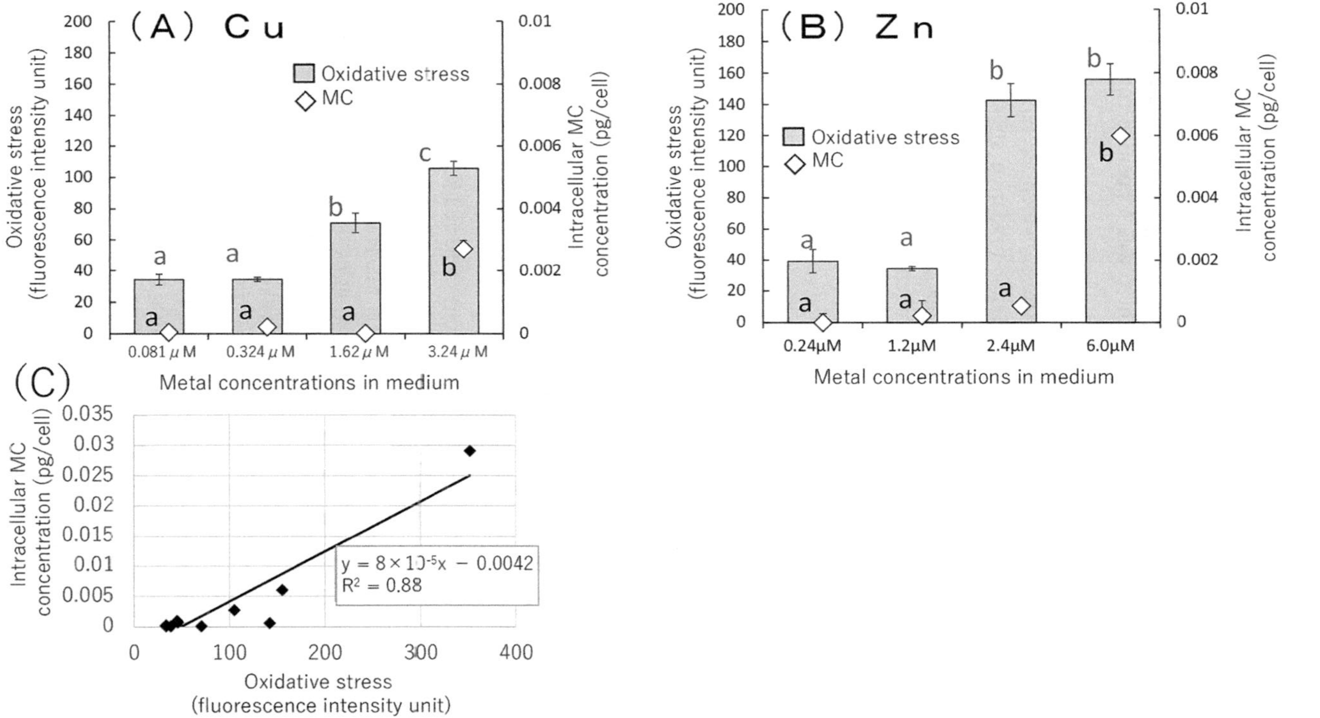

Fig. 28.3 (**a**, **b**) The effect of metals (Cu and Zn) on oxidative stress and microcystin (MC) production. The results from laboratory culture tests with varying metal concentrations are shown as error bars representing standard errors ($n = 2$). *Microcystis aeruginosa* PCC7806 was used as the culture strain. The results of multiple comparison tests (Tukey's HSD) for oxidative stress values and microcystin (MC) concentrations per cell obtained under each metal concentration are shown as alphabets, where different alphabets indicate a significant difference between the treatments ($p < 0.05$). Oxidative stress values and microcystin (MC) concentrations are indicated by gray and black color alphabets, respectively. (**c**) Relationship between oxidative stress and microcystin (MC) production from the laboratory culture test using *M. aeruginosa* with different metal concentrations. Solid lines and equation represent the results from linear regression analysis (adapted from Nakatani et al. (2019) with partial modification)

Key Points

- Algal bloom water in the field campaign in Cambodia and Japan contained MC with statistical analyses (PCA and correlation analyses) of water quality data showing positive correlations between intracellular MC and water quality, such as metals and nutrients.
- Laboratory-based incubation experiment indicated that the metal-induced oxidative stress and intracellular MC concentration of *M. aeruginosa* increased in the presence of excess heavy metals (Cu and Zn). There was a positive correlation between oxidative stress and MC concentration.
- The findings of this study generally suggest that metal ions, such as heavy metals, present in the water cause cellular oxidative stress, which in turn induces cyanobacterial toxicity. On the other hand, cellular oxidative stress may be affected by a variety of environmental conditions, including nutrients, major and trace metals, their chemical species, water temperature, and light conditions, and studies are needed to develop a more comprehensive understanding of this issue.

References

Alexova R, Fujii M, Birch D, Cheng J, Waite TD, Ferrari BC, Neilan BA. Iron uptake and toxin synthesis in the bloom-forming Microcystis aeruginosa under iron limitation. Environ Microbiol. 2011;13:1064–77.

Beversdorf LJ, Chaston SD, Miller TR, McMahon KD. Microcystin mcyA and mcyE gene abundances are not appropriate indicators of microcystin concentrations in lakes. PLoS One. 2015;10(5):e0125353.

Choudhary M, Jetley UK, Khan MA, Zutshi S, Fatma T. Effect of heavy metal stress on proline, malondialdehyde, and superoxide dismutase activity in the cyanobacterium Spirulina platensis-S5. Ecotoxicol Environ Saf. 2007;66:204–9.

Dang TC, Fujii M, Rose AL, Bligh M, Waite TD. Characteristics of the freshwater cyanobacterium Microcystis aeruginosa grown in iron-limited continuous culture. Appl Environ Microbiol. 2012;78:1574–83.

Hallegraeff GM. Harmful algal blooms in the Australian region. Mar Pollut Bull. 1992;25:186–90.

Schmidt JR, Wilhelm SW, Boyer GL. The fate of microcystins in the environment and challenges for monitoring. Toxins. 2014;6(12):3354–87.

Kaebernick M, Neilan BA, Borner T, Dittmann E. Light and the transcriptional response of the microcystin biosynthesis gene cluster. Appl Environ Microbiol. 2000;66:3387–92.

Kearns KD, Hunter MD. Toxin-producing Anabaena flos-aquae induces settling of Chlamydomonas reinhardtii, a competing motile alga. Microb Ecol. 2001;42:80–6.

Kurmayer R, Juttner F. Strategies for the co-existence of zooplankton with the toxic cyanobacterium Planktothrix rubescens in Lake Zurich. J Plankton Res. 1999;21:659–83.

Latifi A, Ruiz M, Zhang CC. Oxidative stress in cyanobacteria. FEMS Microbiol Rev. 2009;33: 258–78.

Lundrigan MD, Arceneaux JEL, Zhu WM, Byers BR. Enhanced hydrogen peroxide sensitivity and altered stress protein expression in iron-starved mycobacterium smegmatis. Biometals. 1997;10: 215–25.

Nakatani K, Nasukawa K, Matsumae H, Wei W, Kikuchi T, Fujii M. Effect of dissolved metal ion concentration on toxin production of freshwater cyanobacteria. J Jpn Soc Civil Eng Ser G (Environ Res). 2019;75:III_385–93.

Neilan BA, Jacobs D, DelDot T, Blackall LL, Hawkins PR, Cox PT, Goodman AE. rRNA Sequences and evolutionary relationships among toxic and nontoxic cyanobacteria of the genus Microcystis. Int J Syst Bacteriol. 1997;47:693–7.

Pearson L, Mihali T, Moffitt M, Kellmann R, Neilan B. On the chemistry, toxicology and genetics of the cyanobacterial toxins, microcystin, nodularin, saxitoxin and Cylindrospermopsin. Mar Drugs. 2010;8:1650–80.

Pinto E, Sigaud Kutner TCS, Leitao MAS, Okamoto OK, Morse D, Colepicolo P. Heavy metal-induced oxidative stress in algae. J Phycol. 2003;39:1008–18.

Sengco MR, Anderson DM. Controlling harmful algal blooms through clay flocculation. J Eukaryot Microbiol. 2004;51:169–72.

Tripathi BN, Gaur JP. Relationship between copper- and zinc-induced oxidative stress and proline accumulation in Scenedesmus sp. Planta. 2004;219:397–404.

WHO. 1998. Cyanobacterial toxins: microcystin LR in drinking water. Background document for development of WHO Guidelines for drinking water quality.

Zeng J, Yang LY, Wang WX. Cadmium and zinc uptake and toxicity in two strains of Microcystis aeruginosa predicted by metal free ion activity and intracellular concentration. Aquat Toxicol. 2009;91:212–20.

Zhai CM, Zhang P, Shen F, Zhou CX, Liu CH. Does Microcystis aeruginosa have quorum sensing? FEMS Microbiol Lett. 2012;336:38–44.

Zilliges Y, Kehr JC, Meissner S, Ishida K, Mikkat S, Hagemann M, Kaplan A, Borner T, Dittmann E. The cyanobacterial hepatotoxin microcystin binds to proteins and increases the fitness of Microcystis under oxidative stress conditions. PLoS One. 2011;6(3):e17615.

Chapter 29
Multidrug-Resistant Bacteria

Reasmey Tan, Chanthol Peng, Sophea Chheun, Monychot Tepy Chanto, Porsry Ung, Kazuhiko Miyanaga, and Yasunori Tanji

29.1 Floating Villages and Antibiotic-Resistant Bacteria (ARB)

The Tonle Sap Lake (TSL), the largest freshwater body in Southeast Asia, is home to an estimated 300,000 people. Some of these people are living in about 170 floating villages, whereas some are living in stile-supported houses located along the flood-plain (Matsui et al. 2006). These floating villages are located at different regions in five provinces around the TSL, including Kampong Chhnang, Pursat, Battambang, Siem Reap, and Kampong Thom. Normally, the lifestyle of people living in these floating villages, including animal farming and having stores, police stations, and school, is similar in many ways to that on land.

Antibiotics are the most potent drugs in the treatment of infections caused by bacteria. As a consequence, the rapid emergence of antibiotic-resistant bacteria (ARB) occurs globally, and ARB has become a threat to public health. Particularly in the floating villages in the TSL, pathogens can cause serious problems to the residents of the floating villages if pathogens associated with waterborne infection show resistance against antibiotics. Om et al. (2015) reported that antibiotic resistance in Cambodia may be due to the improper use of antibiotics, including self-medication by the community, uncontrolled sale of antibiotics, and unregulated antibiotic use in food animals. Furthermore, Levy and Marshall (2004) showed

R. Tan (✉) · C. Peng · S. Chheun · M. T. Chanto
Institute of Technology of Cambodia, Phnom Penh, Cambodia
e-mail: rtan@itc.edu.kh

P. Ung
Ministry of Industry, Science, Technology and Innovation, Phnom Penh, Cambodia

K. Miyanaga · Y. Tanji
Tokyo Institute of Technology, Tokyo, Japan
e-mail: miyanaga.kazuhiko@jichi.ac.jp

C. Yoshimura et al. (eds.), *Water and Life in Tonle Sap Lake*,
https://doi.org/10.1007/978-981-16-6632-2_29

that doctors who cannot distinguish between bacterial and viral infections inappropriately prescribe antibiotics. Although ARB is among the largest risks to global health, little is known about their concentration in the TSL, specifically in floating villages where human activities (aquaculture farming, discharging human waste, bathing, washing, etc.) are high.

In this chapter, we will present the concentrations of ARB in eight different floating villages around the TSL. We will first present the concentrations of ARB in these floating villages and then compare the concentrations of ARB and multidrug-resistant (MDR) bacteria in these floating villages with those in the Tonle Sap River (TSR), Mekong River (MR), and domestic wastewater (WW) in Phnom Penh (PP) in March 2018.

29.2 Concentration of Antibiotic-Resistant Bacteria (ARB) in Floating Villages

For the TSL, the samples from eight different floating villages, including Chhnok Tru (Kampong Chhnang), Phat Sanday (Battambang), Prek Khsach (Battambang), Kampong Luong (Pursat), Kampong Phluk (Siem Reap), Chong Khneas (Siem Reap), Kbal Tor (Battambang), and Kaoh Chiveang (Battambang), were collected in March 2018 (Fig. 29.1).

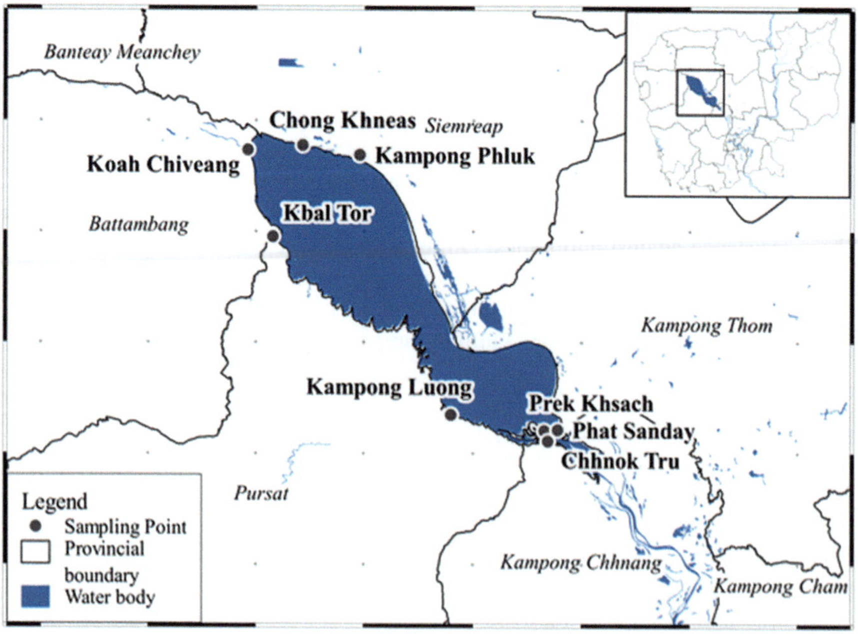

Fig. 29.1 Sampling sites in eight floating villages around the Tonle Sap Lake (TSL)

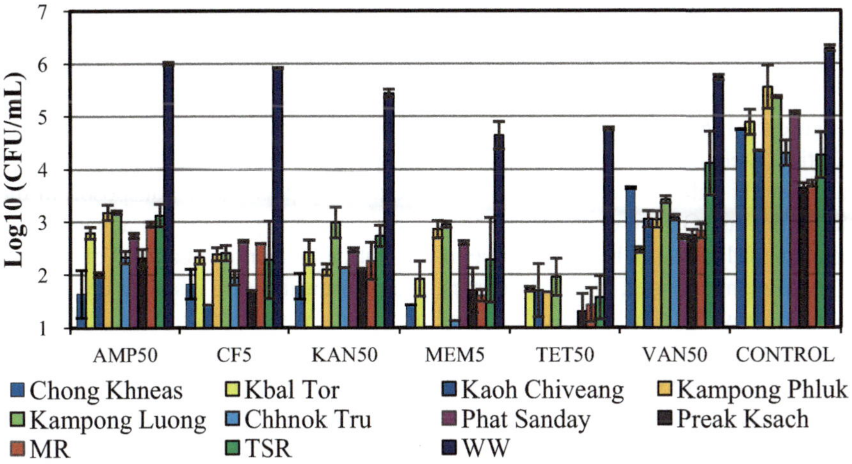

Fig. 29.2 Concentration of antibiotic-resistant bacteria (ARB) in the floating villages of the Tonle Sap Lake (TSL) and Phnom Penh (PP) in March 2018

To detect ARB, different concentrations of six antibiotics were used with Reasoner's 2A (R2A) agar (Difco proteose peptone, 0.5 g/L; casamino acid, 0.5 g/L; yeast extract, 0.5 g/L; glucose, 0.5 g/L; soluble starch, 0.5 g/L; dipotassium phosphate, 0.3 g/L; magnesium sulfate heptahydrate, 0.05 g/L; sodium pyruvate, 0.3 g/L; and agar powder, 15 g/L): ampicillin (50 μg/mL) (AMP50), ciprofloxacin (5 μg/mL) (CF5), kanamycin (50 μg/mL) (KAN50), meropenem (5 μg/mL) (MEM50), tetracycline (50 μg/mL) (TET50), and vancomycin (30 μg/mL) (VAN50). Different effective concentrations are used for each antibiotic, and these are common concentrations in the study of antibiotic resistance. In addition, cycloheximide was also used as antifungal agent to suppress the growth of fungi. For the control, only cycloheximide was added into the R2A agar after cooling down to 50 °C. ARB was then counted using the spread plate technique with 100 μl of inoculation volume and incubation at 30 °C for 48 h.

Among the six antibiotics used in this study, in March 2018, the concentration of vancomycin-resistant bacteria was found to be the highest, followed by that of ampicillin-resistant bacteria (Fig. 29.2). In contrast, the concentration of tetracycline-resistant bacteria was found to be the lowest. Moreover, the concentrations of ARB in Kampong Luong and Kampong Phluk were higher than those in other villages, while the lowest concentration of ARB was in the Kaoh Chiveang floating village. Furthermore, the concentration of ARB in the Chong Khneas floating village showed the highest for vancomycin resistance and the lowest for tetracycline resistance. The result also showed that meropenem-resistant bacteria were not detected in Kaoh Chiveang, but vancomycin-resistant bacteria were the highest. However, ARB was high for all types of antibiotics for Kampong Phluk floating village in Siem Reap.

According to Fig. 29.2, the concentrations of ARB were high in four floating villages, namely, Kampong Luong, Kampong Phluk, Kbal Tor, and Phat Sanday. Many factors lead to the fluctuation of ARB concentration in each floating village. The first factor is the horizontal gene transfer of antibiotic resistance genes (ARGs), because the high amount of antibiotic pollution facilitates the receiving of ARGs. In addition, keeping the antibiotic concentration at an incomplete inhibitory level significantly enhances conjugal transfer mediated by plasmid or transposon in the environments (Ohlsen et al. 2003). Moreover, antibiotic pollution in aquatic environment is also caused by the improper use of antibiotics for the treatment of disease. In the floating villages, hospitals, clinics, and pharmacies were installed on or close to the lake water; thus, most of the used antibiotics were discarded into the water. Furthermore, antibiotics used in aquaculture and fish farms are also a major source of antibiotic pollution. People in the floating villages around the TSL had their own aquaculture during the dry season and frequently used antibiotics not only to treat diseases in aquaculture but also to promote the growth of aquatic organisms. Importantly, Cabello et al. (2013) reported that 80% of antibiotics used in aquaculture would enter the water environment. Secondly, water pollution is also caused by the increased development of ARB by horizontal gene transfer. In a large community, pollution is also higher because of human activities. In this sense, high bacterial density and nutrients are appropriate conditions to accelerate gene recombination and transfer (Schluter et al. 2007). In addition, Joffre et al. (2016) showed that, among the aquaculture in Cambodia, the cage culture of freshwater fishes dominated around 53% of the total production, whereas the number of cages around the TSL increased in most provinces. The cage culture of fishes uses the existing water resources but encloses the fish in a cage, thus permitting water exchange and waste.

29.3 Comparison of the Tonle Sap Lake (TSL) and Water Bodies Around PP

To compare the ARB in other sites, additional surface waters were sampled in the TSR, MR, and WW in PP. The TSR sample was collected under the Cambodia–Japan Friendship (Chroy Changva) Bridge, Sangkat Tuol Sangke, Khan Russey Keo, PP; the MR sample was collected near Svay Chrum Ferry Dock, Sangkat Chroy Changva, and Khan Chroy Changva; and the WW sample was collected near the boat rental area near Wat Phnom, Sangkat Wat Phnom, and Khan Daun Penh. The sources of WW were mainly from Preah Ket Mealea Hospital, Kantha Bopha IV Children's Hospital, and National Maternal and Child Health Center; domestic WW are directly discharged without treatment to the TSR. The surface water samples from PP sites were monthly collected during the dry season from November 2017 to April 2018, as shown in Fig. 29.3.

For the MR, the concentrations of ARB in March and April substantially increased compared with other months, except for tetracycline-resistant bacteria,

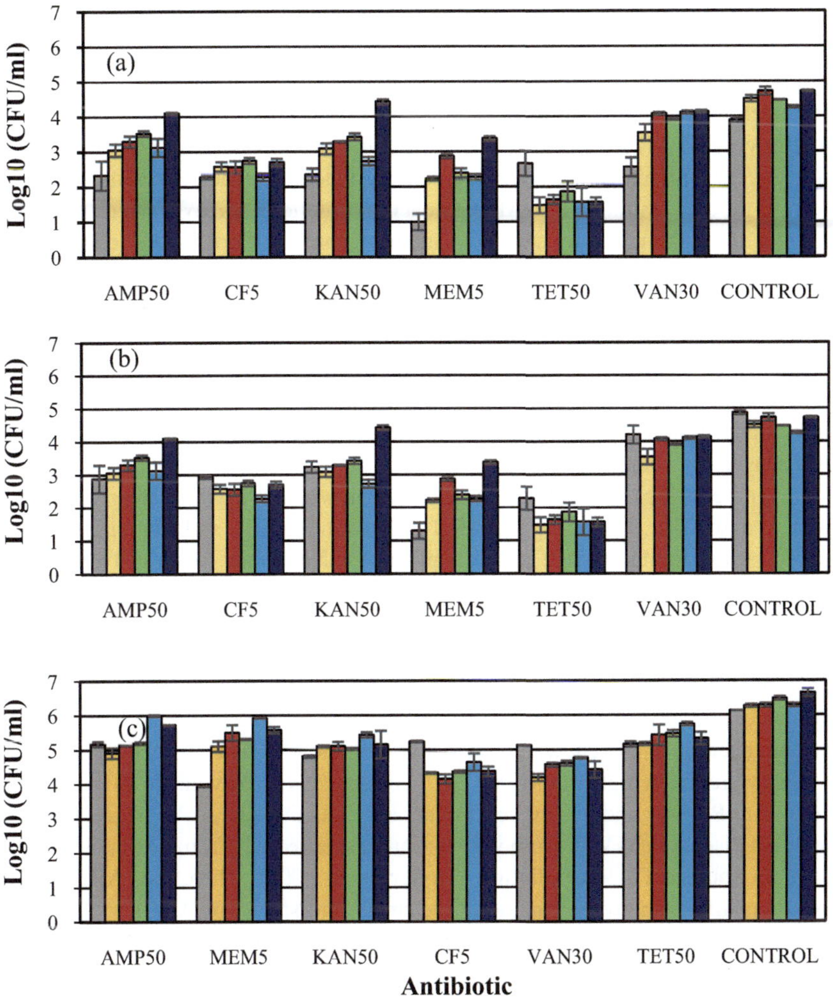

Fig. 29.3 Concentration of antibiotic-resistant bacteria (ARB) in Phnom Penh (PP) sites from November 2017 (Nov. 2017) to April 2018 (Apr. 2018): (**a**) Mekong River (MR), (**b**) Tonle Sap River (TSR), and (**c**) domestic wastewater (WW)

which were the highest in November. In this case, the decrease of water level might be because these 2 months were the complement of dry season, when the water level was at the lowest. Meanwhile, the pollution of antibiotic and other pollution activities seemed to be the same. For the TSR, the concentrations of vancomycin-, ampicillin-, and kanamycin-resistant bacteria in each month were higher than those of ciprofloxacin-, meropenem-, and tetracycline-resistant bacteria. Interestingly, the concentrations of ARB in April were the highest for most of antibiotics.

For the WW, the concentrations of ARB were high for all types of antibiotics. The concentrations of ampicillin-, ciprofloxacin-, kanamycin-, and vancomycin-resistant bacteria were higher than those of meropenem- and tetracycline-resistant bacteria. Moreover, the pattern of ARB concentrations for each antibiotic was similar for each month, whereas ARB concentrations in March and April were higher for all antibiotics, except for tetracycline-resistant bacteria, which were the highest in November. Overall, the concentrations of ARB to each antibiotic in the WW were not totally different from month to month. Two main reasons may contribute to this steady concentration. Firstly, domestic WW come from different sources, whereas hospital WW, the main source of ARB, was directly discharged to the TSR through WW pipeline without any storage for the long term, and hence, bacterial communities do not change much. Secondly, the sample was directly collected from sewage drainage system; therefore, the level of water seems to be the same from month to month or without any dilution, which is similar to that in the TSR and MR, where the water level seasonally changed.

November is the first month of the dry season, whereas March is nearly the end of the dry season. By comparing the ARB in November 2017 and March 2018, ARB concentrations in March were higher than those in November for almost all of the six antibiotics and sites, especially in the MR and WW. This might be due to the low level of water, allowing the concentration of ARB to be more concentrated in March. Moreover, ARB in the WW was much higher than those in the TSR and MR. In addition, ARB in the TSR was higher than those in the MR for most of antibiotic resistance, except for tetracycline-resistant bacteria in November and ciprofloxacin-resistant bacteria in March, which were higher in the MR. Therefore, the percentages of ARB were much higher in March 2017 compared with November 2018. The TSR is closer to PP City and has higher human activities than the MR and is polluted by the direct discharge of WW from various sources, especially from hospitals, which is the main source of ARB pollution (Taylor et al. 2007). Therefore, ARB play an important role in the recombination, exchange, and spread of ARGs to the indigenous or pathogen in the environmental water (Zhang et al. 2009). Moreover, inappropriate antibiotic use, such as inappropriate dosage, over prescription, omission prescription, incorrect selection, incorrect route, and incorrect duration, is also counted as one of the reasons of antibiotic pollution.

As shown in Fig. 29.2, the concentrations of ARB in eight floating villages around the TSL were compared with those in the MR, TSR, and WW in PP in March 2018. The results showed that ARB in the WW were the highest, whereas ARB in Kampong Luong and Kampong Phluk were higher than those in the TSR and MR. Nevertheless, the concentration of the total bacteria in the floating villages was higher than in PP sites (TSR and MR), except for the WW. Therefore, it illustrated that the percentage of ARB in the floating villages was lower than in PP sites.

The pollution load in PP is higher than in the floating villages since the number of population and hospitals is also much higher. In this case, municipal WW including hospital, domestic WW, and industry WW are the main factors to the increase in the percentage of ARB in PP sites. Nevertheless, people in the floating villages directly use lake water for drinking, washing, cooking, and other activities in daily life and therefore are facing serious health risks more than those in PP sites (see Chaps. 40 and 41). Evidently, Iversen et al. (2004) stated that new resistance characteristics could be horizontally transferred to pathogenic bacteria in human, thus causing the difficulties in the treatment of infectious diseases and a threat to public health.

29.4 Proportion of MDR Bacteria in Floating Villages

In this study, Müeller–Hinton agar was used as the medium for the disc diffusion method, which is a common method for MDR bacteria detection (Hudzicki 2009). The inhibiting zone was recorded for interpreting the result as whether susceptible (S) or resistant (R). When a single bacterium is resistant to more than one antibiotic, it is said to be MDR. The results of MDR bacteria were interpreted as the percentage of MDR bacteria for each sample site; a total of 24 isolated strains (100%) were chosen (four stains for each antibiotic). The maximum percentage of MDR bacteria was 52.4% for Prek Ksach, and the minimum percentage of MDR bacteria was 40% for Kaoh Chiveang (Fig. 29.4). Overall, the MDR bacteria were not absolutely different among these floating villages and slightly decreased from 50% in Kbal Tor to 47.4% in Chhnok Tru, 44% in Phat Sanday, 44.4% in Chong Khneas, and 41.7% in both Kampong Luong and Kampong Phluk.

The MDR bacteria in the floating villages might depend on the level of pollution and sampling site. It is similar to the study on bacterial community structure and

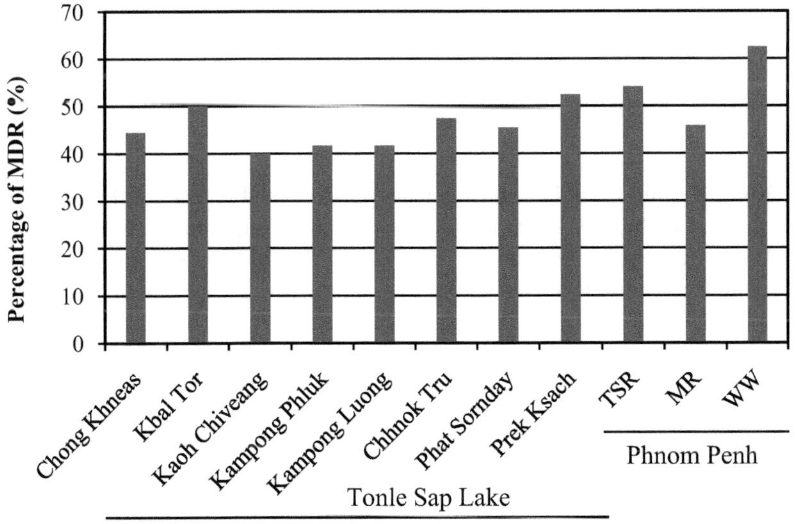

Fig. 29.4 Percentage of multidrug-resistant (MDR) bacteria against at least four different antibiotics in the floating villages and Phnom Penh (PP) in March 2018

cyanotoxin in Chaps. 27 and 28. The highest amount of microcystin was detected in Kampong Phluk and the lowest amount in Kampong Luong. Some floating villages are near the fish farm, e.g., the Prek Ksach floating village. Thus, in floating villages, the pollution of antibiotics and ARB were not only from human waste but also from the aquaculture. In the dry season, the water level in the TSL sharply decreased; thus, people can make their own small-scale fish farm. Fish farming is a major factor in the spread of antibiotics caused by antibiotic overuse in feed for growth promotion and disease prevention. According to Om and Mclaws (2016), Cambodian farmers focused on the benefits of food animal production rather than concerns about the consequences of antibiotic use. In addition, they made a cocktail of several antibiotics for treating the diseases of their animals. They do not understand the actions for and indications of antibiotic use and chose antibiotics in treating sick animals based only on trial and error (Om and Mclaws 2016). Therefore, this may contribute to the antibiotic pollution in the floating villages around the TSL.

The MDR bacteria in PP sites were higher than those in the floating villages around the TSL, especially the WW and TSR, although the MDR bacteria in Prek Ksach, Kbal Tor, and Chhnok Tru showed to be higher than those in the MR in PP (Fig. 29.4). WW samples might contain antibiotics that flowed from the nearby hospitals. According to Li et al. (2010), WW contained an abundance of bacteria with a long-term exposure to high concentration of various types of antibiotic, resulting in it having the highest amount of MDR bacteria. Although both the TSR and MR are the receiving rivers, the TSR received more WW than did the MR. Moreover, although the amount of pollutants was diluted in much water flow, the level of pollution in the TSR remains higher than that in the TSL because there are many human activities around the TSR. In contrast, the main sources of pollution in the floating villages around the TSL were both from human and animal wastes, especially antibiotics used in aquaculture.

29.5 Bacterial Strains Showing Multidrug Resistance

One isolated strain from the lake water sample of Kbal Tor that was first resistant to TET showed multidrug resistance to all six types of antibiotics. According to 16S rRNA gene sequencing with universal primers 27F-1492R, the strain was identified as *E. coli*. These species have many strains, some of which were nonpathogenic, but some can cause serious infection, e.g., *E. coli* O157:H7 (Al-Mohanna 2011). There are three bacterial strains from Kampong Phluk that were resistant to AMP, MEM, VAN, and TET. Two of them were pathogenic and were identified as *Serratia marcescens*, whereas the other strain was nonpathogenic and was identified as *Caulobacter* sp. (Table 29.1). Indeed, *S. marcescens* is a Gram-negative bacillus that naturally inhabits in soil and water. It is mainly associated with urinary and respiratory tract infections, endocarditis, osteomyelitis, septicemia, wound infections, eye infections, and meningitis. Moreover, infections caused by them may be difficult to treat because of their resistance to a variety of antibiotics (Herra and

Table 29.1 MDR bacteria present in water collected from the floating villages around the TSL and PP sites

MDR bacterial strain	Antibiotic resistance	Related species	Identity (%)
TSL site			
KT-TET	All six types	*Escherichia coli* strain UFV 478	1341/1341 (100)
KP-MEM	All six types	*Sphingomonas desiccabilis* strain OLM58	1315/1316 (99)
KP-VAN	AMP, CF, KAN, MEM, VAN	*Serratia marcescens* strain FY	1344/1344 (100)
KP-TET	AMP, CF, KAN, TET	*Serratia marcescens* strain FZSF02	1355/1355 (100)
PP site			
TSR-VAN	All six types	*Klebsiella pneumoniae* strain Kpn1693	1353/1353 (100)
MR-MEM	All six types	*Stenotrophomonas maltophilia* strain S-3	1351/1362 (99)
WW-CF	All six types	*Escherichia coli* strain J-8	1347/1347 (100)

[a]*KT* Kbal Tor, *KP* Kampong Phluk

Falkiner 2015). Importantly, these MDR bacteria may cause severe effect on the residents of the floating villages since they directly use the lake water for many activities in daily life. If these waterborne pathogens infect the residents of the floating villages, treating the disease will be difficult since these bacteria already showed resistance to more than four types of antibiotics.

Regarding PP sites, one of the four isolated strains with vancomycin resistance in the TSR showed multidrug resistance to all of the antibiotics used in this study, and the strain was identified as *Klebsiella pneumoniae* (Table 29.1). *Klebsiella pneumoniae* is a dangerous pathogenic bacteria that can cause fever and kidney, wound, and bloodstream infection (Tsay et al. 2002). In addition, Patel et al. (2008) showed that carbapenem-resistant *K. pneumoniae* infection is associated with many healthcare-related risk factors and high mortality because of limited antimicrobial options for treatment. Moreover, one isolated strain from the MR that was also resistant to all of six antibiotics was identified as *Stenotrophomonas maltophilia*, which is also a pathogenic bacterium. *Stenotrophomonas maltophilia* is an aerobic, Gram-negative bacillus living in natural and hospital environments. Moreover, it is considered a global MDR human opportunistic pathogen. It is most commonly associated with respiratory and other serious infections (Zhao et al. 2015). The MDR bacteria isolated from the CF agar plate of WW was identified also as *E. coli*.

Key Points

- The percentage of MDR bacteria in the target areas ranged from 40% to 62.5%, and the percentage depends on the water source.

- The ARB concentration was the highest in the polluted water and during the end of the dry season.
- The concentrations of ampicillin- and vancomycin-resistant bacteria were higher than other types of ARB in the TSL.
- Of the eight floating villages, Kampong Luong showed the highest concentrations of ARB against all antibiotics, whereas kanamycin- and meropenem-resistant bacteria were not detected in Kaoh Chiveang. However, Prek Ksach showed the highest percentage of MDR bacteria.

References

Al-Mohanna MT. Escherichia coli and klebsiella. Microbiology. 2011:216–26.

Cabello FC, Godfrey HP, Tomova A, Ivanova L, Dölz H, Millanao A, Buschmann AH. Antimicrobial use in aquaculture re-examined: its relevance to antimicrobial resistance and to animal and human health. Environ Microbiol. 2013;15:1917–42.

Herra C, Falkiner F. Serratia marcescens. Microbe. 2015;46:903–12.

Hudzicki J. Kirby-Bauer disk diffusion susceptibility test protocol. Am Soc Microbiol. 2009:1–14.

Iversen A, Kühn I, Rahman M, Franklin A, Burman LG, Olsson B, Torell E, Möllby R. Evidence for transmission between humans and the environment of a *nosocomial strain* of *Enterococcus faecium*. Environ Microbiol. 2004;6:55–9.

Levy SB, Marshall B. Antibacterial resistance worldwide : causes, challenges and responses. Nat Med. 2004;10(12):122–9.

Li D, Yu T, Zhang Y, Yang M, Li Z, Liu M, Qi R. Antibiotic resistance characteristics of environmental bacteria from an oxytetracycline production wastewater treatment plant and the receiving river. Appl Environ Microbiol. 2010;76:3444–51.

Matsui S, Keskinen M, Sokhem P, Nakamasura M. Tonle Sap, experience and lessons learned brief. Shiga: International Lake Environment Committee Foundation; 2006. p. 407–19.

Ohlsen K, Ternes T, Werner G, Wallner U, Löffler D, Ziebuhr W, Witte W, Hacker J. Impact of antibiotics on conjugational resistance gene transfer in *Staphylococcus aureus* in sewage. Environ Microbiol. 2003;5:711–6.

Om C, Mclaws M. Antibiotics : practice and opinions of Cambodian commercial farmers, animal feed retailers and veterinarians. Antimicrob Resist Infect Control. 2016:1–8.

Joffre, O., So N., Chheng P., 2016. Aquaculture production in Cambodia: trends and patterns in recent years. Inland Fisheries Research and Development Institute (Fisheries Administration) and WorldFish. Phnom Penh, Cambodia. p. 1–26.

Om C, McLaws M-L, Vlieghe E, Daily F, McLaughlin J. Cambodia: the first national study of antibiotic prescribing and resistance using mixed methods approach. Antimicrob Resist Infect Control. 2015;4:183.

Patel G, Huprikar S, Factor SH, Jenkins SG, Calfee DP. Outcomes of carbapenem-resistant *Klebsiella pneumoniae* infection and the impact of antimicrobial and adjunctive therapies. Infect Control Hosp Epidemiol. 2008;29:1099–106.

Schluter A, Szczepanowski R, Puhler RA, Top EM. Genomics of IncP-1 antibiotic resistance plasmids isolated from wastewater treatment plants provides evidence for a widely accessible drug resistance gene pool. FEMS Microbiol Rev. 2007;31:449–77.

Taylor P, Kim S, Aga DS. Potential ecological and human health impacts of antibiotics and antibiotic-resistant bacteria from wastewater treatment plants. J Toxicol Environ Health B Crit Rev. 2007:37–41.

Tsay R-W, Siu LK, Fung C-P, Chang F-Y. Characteristics of bacteremia between community-acquired and nosocomial *Klebsiella pneumoniae* infection. Arch Intern Med. 2002;162:1021.

Zhang X, Zhang T, Fang HHP. Antibiotic resistance genes in water environment. Appl Microbiol Biotechnol. 2009;82(3):397–414.

Zhao Y, Niu W, Sun Y, Hao H, Yu D, Xu G, Shang X, Tang X, Lu S, Yue J, Li Y. Identification and characterization of a serious multidrug resistant *Stenotrophomonas maltophilia* strain in China. Biomed Res Int. 2015;2015:1–10.

Joffre O, So N, Chheng P. Aquaculture production in Cambodia: trends and patterns in recent years. Technical report. 2016:1–14.

Chapter 30
Antibiotic Resistance of Intestinal Bacteria

Masateru Nishiyama, Mith Hasika, Jian Pu, In Sokneang, and Toru Watanabe

30.1 Prevalence of Antibiotic-Resistant Bacteria (ARB) in Southeast Asia

The increased use of antibiotics and growth-promoting agents in humans and animals has led to an increase in antibiotic-resistant bacteria (ARB). The prevalence of antibiotic resistance genes associated with ARB and the resulting ARB-induced disease is a threat to human health in the twenty-first century. Based on the analysis of the review on antimicrobial resistance report, the number of resulting deaths could increase to ten million people per year by 2050, unless action is taken to decrease ARB (O'Neil 2014). Antibiotic consumption in humans and animals is increasing globally at low-/middle-income countries. The capita antimicrobial consumption increased in Thailand, Malaysia, and Vietnam between 2005 and 2010 but decreased in Indonesia and the Philippines; the data of Cambodia and Myanmar were not available (CDDEP 2015). Asian countries, particularly Southeast (SE) Asia, are regarded as hotspots of ARB compared to Western countries due to antimicrobial usage, different public hygiene systems such as wastewater treatment plants and drainage systems, and a lack of policies related to antibiotic use (Zellweger et al. 2017).

Upon surveillance of hospitals in seven Asian countries, a number of clinically important ARB, which were resistant to fluoroquinolones, third-generation cephalosporins, and aminoglycosides, were identified (Jean and Hsueh 2011). Based on

M. Nishiyama (✉) · T. Watanabe
Yamagata University, Tsuruoka, Japan
e-mail: m-nishiyama@tds1.tr.yamagata-u.ac.jp

M. Hasika · I. Sokneang
Institute of Technology of Cambodia, Phnom Penh, Cambodia

J. Pu
University of Tokyo, Tokyo, Japan

C. Yoshimura et al. (eds.), *Water and Life in Tonle Sap Lake*,
https://doi.org/10.1007/978-981-16-6632-2_30

integrating routine data from a database, each year in Thailand, an extra 20,000 people die due to multidrug-resistant bacteria, a number 3–5 times higher than those for the United States and European countries (Lim et al. 2016). In Cambodia, the rate of ARB isolated from hospitalized children infected with bacteria increased from 2007 to 2016 (Fox-Lewis et al. 2018). The study showed that *Escherichia coli* and *Klebsiella pneumoniae* isolated from the blood cultures collected from hospitalized patients were resistant to ampicillin (ABPC)–gentamicin (47.2% and 62.1%, respectively) and third-generation cephalosporins (49.5% and 78.8%, respectively). The monitoring of animal farms, including pigs and broiler chickens, and meat products in Cambodia and Thailand revealed the resistance of *E. coli* isolates to antibiotics (i.e., ABPC (break point (BP), 32 μg/mL), tetracycline (TC) (BP, 16 μg/mL), and sulfamethoxazole (BP, 512 μg/mL)), with the percentage of resistant strains isolated from Cambodia being lower than that in Thailand (Trongjit et al. 2016). Although most of ARB researches in SE Asia focus on clinical settings and livestock, ARB are also present in rivers, soils, and coastal areas. However, there is little information on ARB in Cambodia, especially for water sources.

Approximately 1.7 million people live in the villages in the Tonle Sap Lake (TSL) and the surrounding floodplains (see also Chap. 1). Floating village communities depend on the lake in their day-to-day activities, such as drinking, cooking, washing, and disposing waste. The population growth around the TSL has degraded water quality in the lake and has resulted in an increase in waterborne illnesses for those consuming the water (the current state of health and hygiene in the TSL is summarized in Chap. 40; see also Chaps. 22 and 23 regarding water quality). Although pathogenic microorganisms cause infection for the people living around the TSL, there is no information on the presence of ARB in the aquatic environment in Cambodia. Therefore, this chapter introduces the prevalence of fecal indictor bacteria *Enterococci* spp. and *E. coli* in water collected from the TSL, Cambodia. To evaluate antibiotic resistance, *Enterococci* spp. and *E. coli* were isolated from each of the water samples and evaluated for antimicrobial resistance.

30.2 Antibiotic Resistance of *Enterococcus* Spp. in Drinking Water Resources

Enterococci are Gram-positive bacteria that form part of the intestinal microflora of the human and animal gastrointestinal tracts. Because of their abundance in the feces of animals and long-term survival in the environment, they have been used as indicators of fecal contamination in food, drinking water, and aquatic environments in developed countries (U.S.EPA 1986). The genus *Enterococcus* has 35 recognized species, and *Enterococcus faecalis* and *Enterococcus faecium* are the two species most frequently associated with enterococcal disease in clinical settings and cause and are responsible for as much as one-third of the nosocomial infections worldwide (Werner et al. 2008).

The drinking water samples were collected from 16 villages in four provinces (Kampong Chhnang, Battambang, Kampong Thom, and Siem Reap) around the TSL in July and August 2018 and April 2019. *Enterococcus* spp. were isolated by the membrane filtration method using m-*Enterococcus* agar (Difco, BD). The isolates most frequently associated with a range of enterococcal diseases in clinical settings, *E. faecalis* and *E. faecium*, were identified using polymerase chain reaction (PCR) analysis targeted for enterococcal superoxide dismutase (*sodA*) gene (Jackson et al. 2004). The disc diffusion method was used to evaluate the antibiotic resistance of *Enterococcus* spp. according to the Clinical and Laboratory Standards Institution (CLSI 2017; Dolinsky 2017). Based on the recommendations of the CLSI guidelines, the following antibiotic discs were used: ABPC (10 μg), imipenem (IPM) (10 μg), levofloxacin (LVFX) (5 μg), vancomycin (VCM) (30 μg), ciprofloxacin (CPFX) (5 μg), TC (30 μg), tigecycline (TGC) (15 μg), minocycline (MINO) (30 μg), doxycycline (DOXY) (30 μg), and EM (15 μg). The levels of antibiotic susceptibility (susceptible, intermediately resistant, or resistant) were determined by measuring the sizes of the growth-inhibited zones around the discs and comparing the values with those of the standards reported in the CLSI guidelines and the European Committee on Antimicrobial Susceptibility Testing (EUCAST) (CLSI 2017; EUCAST 2019).

As summarized in Table 30.1, among the 281 water samples (i.e., rain, well, lake, river, and filtered water), 69% (193/281 samples) had enterococci. Two of the samples from Kampong Preah and Ses Slab villages were classified as too numerous to count (TNTC). *Enterococcus* sp. was detected in all drinking water samples from Kampong Preah village. The water quality standards in developed countries do not accept any level of enterococci; thus, the drinking water consumed by inhabitants in these villages is considered contaminated.

All enterococci isolated from each sample were identified as *E. faecalis* or *E. faecium* using PCR analysis. Of the 416 isolates collected from the drinking water samples, 320 (77%) belonged to the species *E. faecalis* and 27 (6.5%) to *E. faecium*. The 347 isolates identified as *E. faecalis* or *E. faecium* were evaluated for antimicrobial susceptibility to clinically important antibiotics using the disc diffusion method. Table 30.2 shows the number of isolates collected from drinking water in all villages, with percentages classified as susceptible, intermediately resistant, and resistant to each of antibiotics following the CLSI and EUCAST guidelines.

Of the 347 enterococcal isolates from the drinking water samples, 2 (0.6%) of *E. faecalis* were found to be resistant to VCM, which is the most effective antibiotic against Gram-positive bacteria. Eight *E. faecalis* and one *E. faecium* isolates showed resistance to ABPC. There were 329 (92%) and 290 (81%) enterococcal isolates resistant or intermediately resistant to CPFX and EM, respectively. These percentages of resistance were similar to the finding of Nishiyama et al. (2017), who reported the resistance of enterococci isolated from urban river where human influenced in Japan. The percentage of isolates being resistant or intermediately resistant to TC groups (TC, TGC, MINO, and DOXY) showed no significant differences, except for TGC (2.0%), and this might be due to the fact that TC and MINO, which are categorized as the TC group, are commonly used for not only the

Table 30.1 Prevalence of enterococci in drinking water collected from villages in the Tonle Sap Lake (TSL)

Province	Village	No. of water samples	No. of detected enterococci from samples	No. of enterococci (CFU/100 mL)			No. of isolates
				Min	Max	Average	
Battambang	Anlong ta-Uor	20	10	ND	3.2×10^2	2.2×10^1	15
	Kbal Tol	16	9	ND	3.0×10^3	2.4×10^2	11
	Praek Norin	20	11	ND	9.4×10^4	4.7×10^3	25
	Praek Tal	20	11	ND	9.0×10^4	4.5×10^3	11
	Rohal Suong	20	12	ND	9.1×10^4	9.0×10^3	25
	Sdei	23	9	ND	3.1×10^1	5.0	43
Kampong Chhnang	Chnok Tru	17	13	ND	4.5×10^3	2.9×10^2	44
	Kampong Preah	15	15	2.5	TNTC	2.3×10^1	27
	Kan Yuo	15	14	ND	5.3×10^2	1.1×10^2	49
	Me Lom	15	9	ND	1.2×10^2	2.2×10^1	34
	Ses slab	16	13	ND	TNTC	2.1×10^1	33
	Sras Keo	17	13	ND	1.1×10^2	1.8×10^1	16
	Toul Rourka	19	16	ND	1.0×103	1.5×10^2	19
	Tuol Thlok	13	9	ND	1.3×10^2	1.6×10^1	17
Kampong Thom	Phat Sanday	17	15	ND	3.0×10^3	2.2×10^2	25
Siem reap	Tnot Kombot	18	14	ND	TNTC	5.3×10^1	22
Total	16	281	193				416

ND not detected, *TNTC* too numerous to count

treatment of humans but also the treatment of or prevention for animals and fish (CDDEP 2015). Multidrug resistance (resistant to ≥ 3 groups of antimicrobial agents) was observed in 5.3% (17/320 isolates) and 3.7% (1/27 isolates) of *E. faecalis* and *E. faecium*, respectively. One isolate of *E. faecalis* isolated from Kampong Chhnang was resistant to eight antimicrobial agents (LVFX, VCM, CPFX, TC, TGC, MINO, DOXY, and EM), representing four chemical classes. This indicates that the antimicrobial agents for the effective treatment of humans reflected by *Enterococcus* are decreased. The pollution of aquatic environments with antibiotic-resistant *Enterococcus* spp. is a significant concern. Accurate information on the prevalence of antibiotic resistance of *Enterococcus* spp. in aquatic environments is important for the improvement of public health in the TSL. More and

Table 30.2 Antibiotic susceptibilities of enterococci isolated from drinking water

	Enterococcus faecalis (329 isolates)			*Enterococcus faecium* (27 isolates)		
	Susceptible	Intermediately resistant	Resistant	Susceptible	Intermediately resistant	Resistant
Antimicrobial agent	No. of isolates			No. of isolates		
ABPC	320	1	8	26	0	1
CPFX	326	3	0	26	1	0
DOXY	236	71	22	14	9	4
EM	201	126	2	25	2	0
IPM	23	276	30	4	15	8
LVFX	216	6	107	16	0	11
MINO	322	4	3	27	0	0
TC	220	56	53	16	7	4
TGC	230	27	72	16	6	5
VCM	58	234	37	8	16	3

ABPC ampicillin, *IPM* imipenem, *LVFX* levofloxacin, *VCM* vancomycin, *CPFX* ciprofloxacin, *TC* tetracycline, *TGC* tigecycline, *MINO* minocycline, *DOXY* doxycycline; *EM* erythromycin

continuous investigations on the antibiotic resistance of *Enterococcus* spp. in the TSL should be conducted.

30.3 Antibiotic Resistance of *E. Coli* Isolated from the Tonle Sap Lake (TSL)

Escherichia coli, a Gram-negative, rod-shaped bacterium from the family *Enterobacteriaceae*, is another fecal indicator bacterium in water resources. *Escherichia coli* is classified as commensal and harmless because its natural habitat is in the human and animal gastrointestinal tracts (Croxen et al. 2013). However, several highly adapted *E. coli*-specific virulence genes can confer an increased ability to adapt to new niches, allowing them to cause disease widely (Kaper et al. 2004). Pathogenic *E. coli* can cause a broad spectrum of human diseases spanning from the gastrointestinal tract to extraintestinal sites such as the urinary tract, bloodstream, and central nervous system (Croxen et al. 2013). *Escherichia coli* strains are some of the most important etiological agents of diarrhea. These strains are associated with intestinal diseases, which are well known as intestinal pathogenic or diarrheagenic *E. coli*, although not all of the subtypes of this group necessarily cause diarrhea. Antibiotic resistance has also led to problems for the treatment of nosocomial infections.

Sampling was conducted during the low (dry) and high (wet) water-level seasons in June and October 2019 for the investigation of *E. coli*. The samples were collected from 38 and 27 locations from the lake (including three floating villages), as well as 11 tributaries, during the dry and wet seasons, respectively. The membrane filtration method, with Chromocult Coliform Agar ES (Merck, Darmstadt, Germany) used as the selective media detection, was used for the detection of *E. coli* from each sample, and all isolates were identified as *E. coli* by the detection of specific genes (*uidA*). The antibiotic resistance of the identified isolates was evaluated using the same method in Sect. 30.2.

The number and prevalence of *E. coli* in the water samples from the TSL and its tributaries are shown in Table 30.3. Of the 92 samples from both seasons, 79 formed at least 1 *E. coli* colony on the selective media, and the percent positive for *E. coli* contamination was higher in the dry season (96% or 49/51 samples) than in the wet season (72% or 30/42 samples). The decrease in *E. coli* concentration in the wet season might be due to the dilution in the lake by seasonal flow from the Mekong River. *Escherichia coli* was detected in all lake water samples near the floating villages and tributaries in both seasons. The highest *E. coli* concentrations in the lake water were observed near the floating villages, suggesting that fecal materials were discharged from the floating villages. These data show that *E. coli* concentration is largely affected by human activity, which is remarkable in the dry season.

Escherichia coli isolates collected from both dry and wet seasons were then evaluated for antimicrobial susceptibility and used for antibiotic resistance testing

Table 30.3 Enumeration of *E. coli* in the Tonle Sap Lake (TSL) and its tributaries in the dry and wet seasons

		Dry season			Wet season		
		Floating villages	Lake	Tributaries	Floating villages	Lake	Tributaries
No. of samples		3	37	11	4	27	11
No. of detected *E. coli* from samples		3	35	11	4	15	11
No. of *E. coli* (CFU/100 mL)	Min.	2.0×10^2	N.D.	4.0×10^1	5.0×10^3	N.D.	6.2×10^1
	Ave.	3.6×10^3	4.0×10^1	3.3×10^2	5.4×10^3	6	8.9×10^2
	Max.	8.2×10^3	5.2×10^2	1.5×10^3	6.0×10^3	4.4×10^1	4.1×10^3

ND not detected

Table 30.4 Antibiotic susceptibility of *E. coli* isolated from drinking water

	Dry season ($n = 69$)			Wet season ($n = 43$)		
	Susceptible	Intermediately resistant	Resistant	Susceptible	Intermediately resistant	Resistant
Antimicrobial agents	No. of isolates			No. of isolates		
ABPC	24	2	43	13	2	28
IPM	64	5	0	42	0	1
GM	66	0	3	37	0	6
CTX	53	4	12	35	0	8
CFX	48	1	20	32	1	10
CPFX	65	0	4	35	4	4
TP	61	3	5	36	3	4
C/A	44	2	23	31	4	8
ST	43	0	26	28	0	15
AZT	61	1	7	35	0	8
CP	48	6	15	24	5	14
FOM	57	5	7	33	1	9
TC	29	0	40	25	0	18

ABPC ampicillin (10 μg), *IPM* imipenem (10 μg), *GM* gentamycin (10 μg), *CTX* cefotaxime (30 μg), *CFX* cefoxitin (30 μg), *CPFX* ciprofloxacin (5 μg), *TP* tazobactam/piperacillin (100/10 μg), *C/A* amoxicillin/clavulanate (20/10 μg), *ST* sulfamethoxazole/trimethoprim (23.75/1.25 μg), *AZT* aztreonam (30 μg), *CP* chloramphenicol (30 μg), *FOM* fosfomycin (200 μg), *TC* tetracycline (30 μg)

(Table 30.4). The percentage of resistance and intermediate resistance to ABPC and TC among the 112 isolates in the lake water were 67% (75 isolates) and 52% (58 isolates), respectively. Many *E. coli* isolates were resistant to both antibiotics. Approximately 30% of the isolates in both seasons were resistant to cefotaxime (CTX) and cefoxitin (CFX), which are categorized as a third-generation cephem, and antimicrobial resistance to these antibiotics has increased worldwide. Notably, some isolates were resistant to tazobactam/piperacillin (TP) and amoxicillin/clavulanate (C/A), which were used to evaluate extended-spectrum β-lactamase (ESBL)-producing bacteria. Additionally, one and seven isolates were resistant or intermediately resistant to imipenem (IMP), which is classified as carbapenem. Carbapenems are the most effective antimicrobial agents for diseases caused by ESBL-producing bacteria, and ARB detected in the TSL were observed to be also resistant to it.

Several water samples exceeded the allowable amounts of both indicator bacteria for water standard of Cambodia. In particular, water pollution by microorganisms is noticeable at points associated with human activity. However, there was no correlation between the number of enterococci and *E. coli* and the percentage of resistance of both bacterial isolates in this study. The existence of both resistance bacteria is difficult to be predicted from indicator bacteria; thus, directly measuring them is necessary. As mentioned, most of enterococci and *E. coli* are nonpathogenic. It should be noted that these investigations did not identify the pathogenicity of both bacteria and that these resistant isolates do not necessarily cause diseases to humans. Analyzing the presence or absence of pathogenicity in the future work is necessary to evaluate the infectivity of ARB.

Key Points

- The drinking and lake water collected from points associated with human activities showed a higher concentration of enterococci and *E. coli* than the water standard of Cambodia.
- As the results of antimicrobial resistance test against *Enterococcus* spp. and *E. coli* identified in each water resource, many of the enterococcal isolates and *E. coli* in the TSL were resistant to several important antibiotics.
- The risk of infection due to ARB in the TSL was identified. To reduce the risk caused by ARB, accurate information concerning the prevalence of ARB in the TSL should be compiled in a comprehensive manner.

References

Clinical Laboratory Standards Institute. Performance standards for antimicrobial susceptibility testing. CLSI supplement M100. 27th ed. Wayne, PA: Clinical Laboratory Standards Institute; 2017. isbn:1-56238-805-3.

Center for Disease Dynamics, Economics & Policy (CDDP). The state of the world's antibiotics 2015.

Croxen MA, Law RJ, Scholz R, Keeney KM, Wlodarska M, Finlay BB. Recent advances in understanding enteric pathogenic *Escherichia coli*. Clin Microbiol Rev. 2013;26(4):822–80.

Dolinsky, A. L. (2017). M100 Performance standards or antimicrobial susceptibility testing J Serv Mark (vol. 8, issue 3).

European Committee on Antimicrobial Susceptibility Testing (2019). Breakpoint tables for interpretation of MICs and zone diameters version 9.0. http://www.eucast.org.

Fox-Lewis A, Takata J, Miliya T, Lubell Y, Soeng S, Sar P, Rith K, McKellar G, Wuthiekanun V, McGonagle E, Stoesser N, Moore CE, Parry CM, Turner C, Day NPJ, Cooper BS, Turner P. Antimicrobial resistance in invasive bacterial infections in hospitalized children, Cambodia, 2007–2016. Emerg Infect Dis. 2018;24(5):841–51.

Jackson CR, Fedorka-Cray PJ, Barrett JB. Use of a genus- and species-specific multiplex PCR for identification of enterococci. J Clin Microbiol. 2004;42(8):3558–65.

Jean SS, Hsueh PR. High burden of antimicrobial resistance in Asia. Int J Antimicrob Agents. 2011;37(4):291–5.

Kaper JB, Nataro JP, Mobley HLT. Pathogenic Escherichia coli. Nat Rev Microbiol. 2004;2(2): 123–40.

Lim C, Takahashi E, Hongsuwan M, Wuthiekanun V, Thamlikitkul V, Hinjoy S, Day NPJ, Peacock SJ, Limmathurotsakul D. Epidemiology and burden of multidrug-resistant bacterial infection in a developing country. elife. 2016;5:1–18.

Nishiyama M, Ogura Y, Hayashi T, Suzuki Y. Antibiotic resistance profiling and genotyping of vancomycin-resistant enterococci collected from an urban river basin in the Provincial City of Miyazaki, Japan. Water. 2017;9(2)

O'Neil J. (2014). Review on antibiotic resistance. Antimicrobial Resistance:Tackling a crisis for the health and wealth of nations. Health and Wealth Nations, 2014.

Trongjit S, Angkittitrakul S, Chuanchuen R. Occurrence and molecular characteristics of antimicrobial resistance of Escherichia coli from broilers, pigs and meat products in Thailand and Cambodia provinces. Microbiol Immunol. 2016;60(9):575–85.

U. S. Environmental Protection Agency. Ambient water-quality criteria for bacteria-1986. Washington, DC: Office of Water Regulations and Standards, U. S. Environmental Protection Agency; 1986.

Werner G, Coque TM, Hammerum AM, Hope R, et al. Emergence and spread of vancomycin resistance among enterococci in Europe. Euro Surveill. 2008;13(47) pii: 19046

Zellweger RM, Carrique-Mas J, Limmathurotsakul D, Day NPJ, Thwaites GE, Baker S, Ashley E, de Balogh K, Baird K, Basnyat B, Benigno C, Bodhidatta L, Chantratita N, Cooper B, Dance D, Dhorda M, van Doorn R, Dougan G, Hoa NT, et al. A current perspective on antimicrobial resistance in Southeast Asia. J Antimicrob Chemother. 2017;72(11):2963–72.

Part VII
Flora and Fauna

Chapter 31
Primary Production

Say Samal, Uk Sovannara, Ly Sophanna, Rajendra Khanal, Dilini Kodikara, Sok Ty, Oeurng Chantha, Manabu Fujii, and Chihiro Yoshimura

31.1 Primary Producers and Their Productivity

Primary production (PP) refers to the light-mediated assimilation of inorganic carbon in the process of photosynthesis (Koponen et al. 2010). PP and its controlling factors (e.g., nutrients, light energy, water temperature, dissolved oxygen, and flow) in aquatic ecosystems may vary greatly according to the environmental conditions and system dynamics. Shallow and deep lakes function in distinct ways regarding the ecosystem processes (e.g., nutrient cycling and light penetration). For instance, shallow lakes are generally prone to frequent mixing and sediment resuspension (Padisák and Reynolds 2003; see also Chap. 19), whereas deep lakes tend to hold water for a long time, owing to their larger volume and thermal stratification. Tonle Sap Lake (TSL) is strongly influenced by a seasonal flood pulse and thus highly seasonally variable (e.g., hydrodynamics, water quality, and sediment and nutrient dynamics; refer also to Chaps. 1, 19, 24, and 27), also influencing PP. This chapter

S. Samal
Ministry of Environment, Phnom Penh, Cambodia

U. Sovannara (✉) · D. Kodikara · M. Fujii · C. Yoshimura
Tokyo Institute of Technology, Tokyo, Japan

L. Sophanna
Ministry of Environment, Phnom Penh, Cambodia

Tokyo Institute of Technology, Tokyo, Japan

R. Khanal
Tokyo Institute of Technology, Tokyo, Japan

Policy Research Institute, Kathmandu, Nepal

S. Ty · O. Chantha
Institute of Technology of Cambodia, Phnom Penh, Cambodia

C. Yoshimura et al. (eds.), *Water and Life in Tonle Sap Lake*,
https://doi.org/10.1007/978-981-16-6632-2_31

describes PP in TSL and its seasonal variability in relation to the flood pulse and its significant role in supporting the secondary production (e.g., fisheries).

The phytoplankton and macrophytes, including floating and rooted types, are the major primary producers in TSL and its floodplain (Chea et al. 2016; Koponen et al. 2010). At least 197 phytoplankton species have been identified to inhabit TSL's ecosystem (MRC 2010). The phytoplankton communities were mainly composed of five phyla, viz., Cyanophyta (blue-green algae), Chlorophyta (green algae), Heterokontophyta (stramenopiles and diatoms), Euglenophyta, and Dinophyta (Ohtaka et al. 2010). Chea et al. (2016) estimated the biomass of the phytoplankton and macrophytes in the TSL system at 5.48 ton/km^2 and 1489 ton/km^2, with the ratio of production to the biomass of the phytoplankton and macrophytes at 185/year and 1.67/year, respectively.

In the open area of the lake, sedge communities such as *Cyperus malaccensis*, *Cyperus halpan*, *Rhynchospora corymbosa*, and *Scirpus grossus* are also dominant, followed by a free-floating herbs water hyacinth *Eichhornia crassipes* and Asian watermoss *Salvinia cucullata* (MRC 2015). Dense mats of tall herbaceous vegetation are found on the shallower shoreline of the lake (McDonald et al. 1997), which typically becomes dislodged and forms floating mats during the high-water period. These floating mats consist of various grasses and sedges, along with several larger plants (i.e., *Polygonaceae*). The submerged parts of these floating vegetations form a complicatedly intertwined root and rhizome system, which creates a special biotope that harbors diverse aquatic organisms (e.g., sessile algae, zooplankton, phytoplankton, and invertebrates; Junk 1973, 1977; Heckman 1998; Tanaka and Ohtaka 2010).

The floodplain's vegetation is patchy (Lamberts 2001), with large differences in density, height, and other aspects, and is distinguished into three major types of short-tree and shrubland vegetation with grasslands, forest, and aquatic herbaceous vegetation, totally comprising over 190 species (Koponen et al. 2010). Submerged primarily aquatic vegetation is virtually absent, and rooted-submerged species are uncommon in the lake, except for the non-rooting species (e.g., bladderwort *Utricularia aurea*; Campbell et al. 2006). Likewise, floating-leaved plants are rarely found within the lake, except in occasional pools and ponds around the lake's periphery.

TSL is renowned for its high productivity and fisheries production. The productivity of TSL is considered to be among the highest in the world, owing to a combination of factors (ADB 2000). The factors include high water temperature, annual flooding, and the supporting role of the flooded forests in stimulating the development of microorganisms and the phytoplankton, as well as the zooplankton. The average phytoplankton primary productivity in TSL and its floodplain has been estimated to be aNPP = 2.0 g-C/m^2/day (Holtgrieve et al. 2013), which falls within the range of phytoplankton productivities of other large lakes (>500 km^2, Fig. 31.1). Recently, Chea et al. (2016) calculated the total system throughput of the TSL ecosystem to be 38.5 g/m^2/day in a holistic food web model using Ecopath with Ecosim. Total NPP and the net system production were 9.6 g/m^2/day and 1.8 g/m^2/day, respectively.

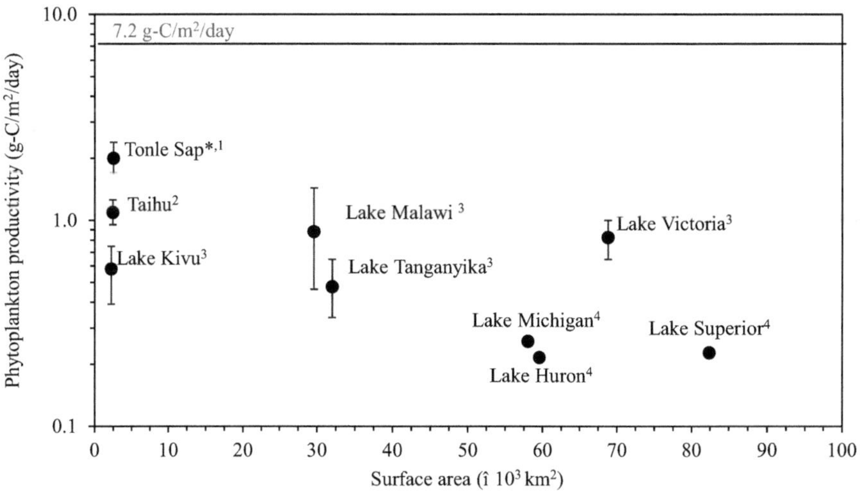

Fig. 31.1 Phytoplankton primary productivity in TSL compared to those in eight other large lakes (>500 km^2) around the world. The gray line denotes the highest phytoplankton productivity in other large lakes (range, 0.01–7.2 g-C/m^2/day; Alin and Johnson 2007). *TSL was plotted in the graph using the surface area of the lake during the dry season (~2500 km^2); its surface area increases to ~15,732 km^2 during the wet season (Chap. 1). [1]Holtgrieve et al. (2013), [2]Deng et al. (2017), [3]Darchambeau et al. (2014), [4]Fahnenstiel et al. (2016)

Seasonal flooding benefits both primary and secondary production (Holtgrieve et al. 2013), perhaps as a result of biological adaptation and migration to the unique physical habitat condition. Together with the annual flood cycle, inundated forests are among the driving factors in maintaining high productivity (e.g., fish production; Say 2009, and citation therein). Recent evidence also confirmed that the productivity of TSL is supported by not only autochthonous PP but also macrophytes and allochthonous inputs from terrestrial plants and seasonally inundated floodplains (Hiroki et al. 2020). The seasonally inundated forests in the floodplains play a critically important role as habitats, nursing ground, and food sources for aquatic fauna (Chaps. 32 and 33), and they also offer submerged surfaces for the proliferation of periphyton (MRC 2010). In the TSL ecosystem, the periphyton is either epiphytic (using plants as a substrate) or epipelic (growing on sediments).

31.2 Phytoplankton Community and Biomass

Characterized by the remarkable monotonal flood pulse, TSL cyclically oscillates between the high- and low-water conditions, whereas the extensive area of floodplains around the lake perpetually shifts between the aquatic and terrestrial phases, i.e., seasonal inundation (see Chap. 1 for details). Such a unique hydrological condition underpins not only the complexity, dynamics, and spatiotemporal variability of the system's biogeochemical processes but also the heterogeneous habitats and breeding colonies for aquatic and terrestrial biotic communities (Chaps. 32–34). There exist also many seasonal interrelations among those components in the system (Uk et al. 2018, and citations therein).

Although the composition has not changed significantly, at least since the 1950s, the major taxa of the phytoplankton communities vary seasonally (Blache 1951; Ohtaka et al. 2010). For instance, blue-green algae (*Anabaena* and *Microcystis*) have been reported to dominate the communities during the low-water period (i.e., the dry season), whereas diatoms (mostly *Aulacoseira granulata*) are predominant throughout the lake during the high-water period (i.e., the wet season; Ohtaka et al. 2010).

Chlorophyll a (Chl-*a*, a common photosynthetic pigment indicating algae biomass) and blue-green algae–phycocyanin (BGA–PC, a pigment specific to cyanobacteria) were measured during our recent field survey in 2017–2019. The results indicated that Chl-a concentration in the surface water of TSL varied seasonally and spatially (Fig. 31.2), with a higher level in the low-water period (March and June) taken as an overall pattern. The whole-lake averaged ($\pm$ standard deviation) concentration of Chl-*a* in the surface water (0–30 cm) of TSL ranged from as low as 2.6 ± 3.3 μg/L in September 2017 to as high as 82.4 ± 224.5 μg/L in March 2017. BGA-PC varied from below the detection limit (<0.01 μg/L) to 21.26 μg/L over the investigated period, showing a comparatively higher concentration during the low-water period (whole-lake average of 2.38 ± 2.64 μg/L) relative to the high-water period (0.14 ± 0.14 μg/L). During our field survey, we also visually observed the prevalence of algal bloom on the water surface during the low-water period, especially at the northern part of TSL and the area around the floating villages (Fig. 31.3a).

This result is consistent with the previous studies. During 2001–2002, Sarkkula et al. (2003) reported that the phytoplankton biomass and Chl-a concentration in TSL peaked at 2.5 mg-C/L and ca. 60 μg/L, respectively, during the low-water period in April. Based on satellite images and in situ measurement between 2004 and 2006, Say (2009) also reported that the lowest level of Chl-a was recorded during the high-water period (<10 μg/L), increasing to around 30 μg/L during the falling and rising period, and generally peaked during the low-water period (often >60 μg/L). The spatial distribution of Chl-*a* was patchy during the low-water period, with its horizontal heterogeneity being reduced once the flooding phase began, and finally became homogeneous with the lowest level of Chl-a during the high-water period. Ohtaka et al. (2010) and Mizuno and Mori (1970) also described that a bloom of floating blue-green algae was seen dominantly during the low-water period, whereas the floating algal scum was hardly observed during the high-water period.

Factors limiting PP in TSL somehow have still not been clearly determined (Holtgrieve et al. 2013; Hiroki et al. 2020), although P is possibly a limiting factor among the macronutrients (Chap. 24), and dissolved inorganic carbon could also limit PP in TSL (Hiroki et al. 2020). Despite the biomass peak during the low-water period, a majority of the annual biological production occurs during the high-water period, which is attributable to the immense volume of the floodwater and large water surface area, as well as high sediment input (Sarkkula et al. 2003). Hiroki et al.

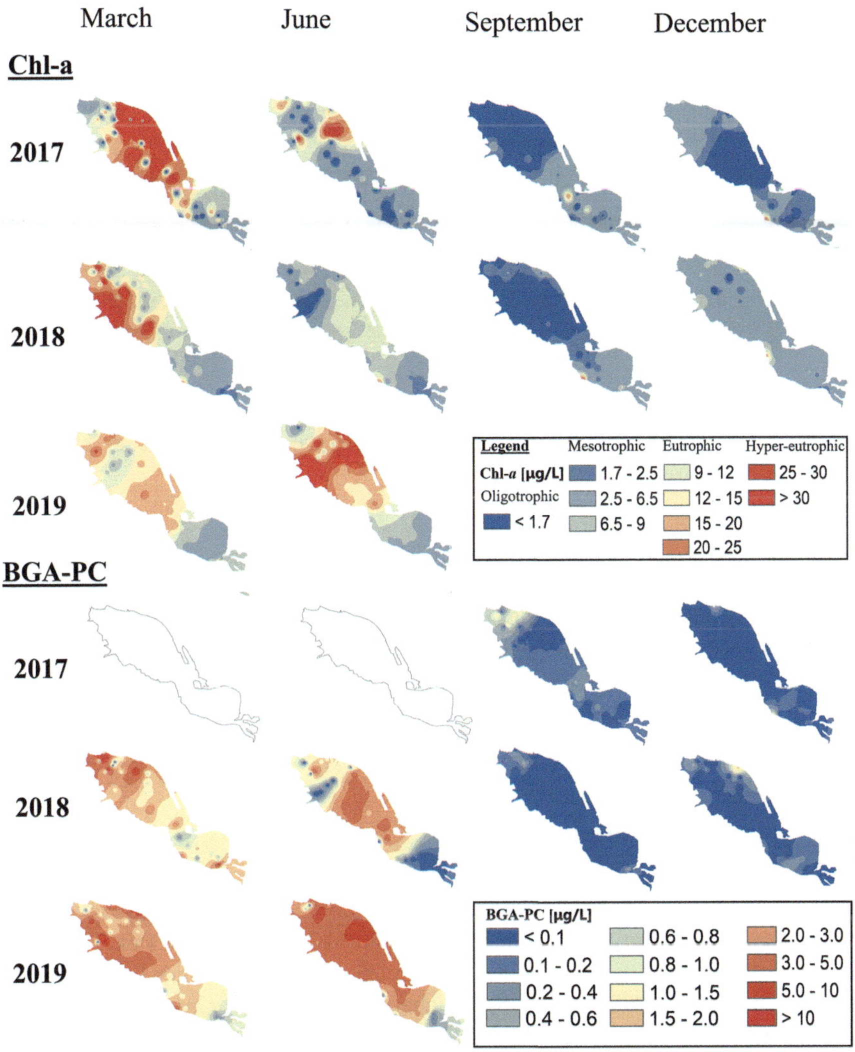

Fig. 31.2 Distribution of Chl-*a* concentrations based on our field survey (2017–2019) and spatial interpolation (inversed distance weighted). Oligotrophic[1], Chl-*a* < 1.7 μg/L; mesotrophic[1], 1.7 < Chl-*a* < 9 μg/L; eutrophic, Chl-*a* > 9 μg/L; hyper-eutrophic[2], Chl-*a* > 25 μg/L. [1]Cunha et al. (2012), [2]UESPA (1988). Note: The white maps denote no available data

(2020) also demonstrated higher PP during the high-water period compared to that during the low-water period because PP occurred mostly near the water surface (low PP below the surface) during the low-water period due to high turbidity. In addition, no sign was found for the photoinhibition effect on the seasonality in PP in TSL, although such a phenomenon was likely the cause for seasonality in PP in other water bodies in the lower Mekong River basin chosen for their study. Rather, a light

Fig. 31.3 Pictures of (**a**) the presence of algae bloom taken during our field survey at Kampong Luong floating village (February 26, 2020) and (**b**) the presence of water hyacinth taken during our pre-survey at the northern upstream of the lake (December 8, 2015). (**a**) Algae bloom at Kampong Luong. (**b**) Water hyacinth

condition is closely related to the seasonal flood pulse-induced turbidity variation, and it was probably among the factors responsible for PP seasonality in TSL. Holtgrieve et al. (2013) also suggested that reduced suspended sediment and stronger irradiation for autotrophs might increase phytoplankton productivity per unit area in TSL or might conversely decrease because of reduced nutrient inputs with flooding.

31.3 Trophic Status

Based on Chl-*a* concentration from our survey (2016–2019), TSL was classified as being reciprocated between being oligotrophic (Chl-*a* < 1.7 μg/L) and eutrophic (Chl-*a* > 9 μg/L) or hyper-eutrophic (Chl-*a* > 25 μg/L) (United State Environmental Protection Agency (USEPA 1988); Cunha et al. 2012). Our results are in good accordance with previous studies. Sarkkula et al. (2003) reported that TSL is naturally mesotrophic, whereas Junk et al. (2006) claimed that the lake is mesotrophic to eutrophic. More recently, Burnette et al. (2017) also concluded that TSL's trophic state generally fluctuates between eutrophic and hyper-eutrophic, i.e., Chl-a >25 μg/L.

According to its spatiotemporal distribution, the trophic status of TSL varied both seasonally and spatially, and the lake typically has a relatively high trophic status in the dry season (e.g., March) and gradually changes to a lower trophic state as the rainy season and reversal flow start (e.g., June). By the late rainy season (e.g., September), the lake shifts to the lowest trophic state (oligotrophic/mesotrophic, Fig. 31.1). This variation of Chl-a is consistent with the trend of total phosphorus (TP) concentration depicted in Fig. 24.2 (Chap. 24). Higher nutrient concentrations (e.g., phosphorus and nitrogen) were generally observed during the low-water period (the dry season) than during the high-water period (the wet season) in TSL. With the beginning of seasonal rainfall, the southeastern part of the lake first drops its trophic state, with decreases in concentrations of Chl-a, total suspended solids (TSS), and TP (see Chap. 24 for seasonal variations of TSS and TP). This can be because of a large mass of water starting to enter the lake from MR through the reversal flow in TSR and with increasing rainfall.

Secchi disk transparency remarkably differs between the high-water period (0.8–2.7 m) and the low-water period (<2 cm), according to Ohtaka et al. (2010). If we calculate the photic zone depth (E), the uppermost layers receiving sunlight for phytoplankton's photosynthesis, as $E = 2.5 \times$ Secchi depth, and then the photic zone during the low-water period would be as shallow as <5 cm, in comparison to the high-water period when E ranges from 2.0 to 6.8 m. Limitation of light due to turbidity could severely inhibit the growth of submerged plants in TSL, although such type of plants (e.g., Hydrocharitaceae, Najadaceae, Nymphaeaceae, and Potamogetonaceae) is pervasive throughout Southeast Asia (Campbell et al. 2006). Instead, algae and floating mats of herbaceous vegetation are common in TSL, as both primary producers outcompete submerged vegetation for light (Campbell et al. 2006; Day et al. 2011).

Owing to socioeconomic development around the lake, water pollution by excessive nutrient loading (e.g., phosphorus) has led to the explosive development of algae bloom (Fig. 31.3a; refer also to Fig. 28.1a in Chap. 28) and invasive plant species such as water hyacinth (Fig. 31.3b; see also Kuenzer 2013), which could have various negative environmental and social impacts. The facts that the occurrence of algae blooms and cyanobacteria in higher abundance during the low-water period was typically dominated by *Anabaena* and *Microcystis* species (refer to Sect. 31.2) and that these species can produce toxins (e.g., microcystin; see Chap. 28 for details) also suggest a potential threat to the aquatic organisms (e.g., fish) and a health risk to humans.

31.4 Trophic Transfer to Secondary Production

A major part of PP is estimated to channel through the food web, which ends up in remarkably high fish production (Sarkkula et al. 2003; Koponen et al. 2010; Holtgrieve et al. 2013). Indeed, the fish community could influence the overall structure of an ecosystem through trophic interactions (Carpenter and Kitchell 1993), nutrient cycling (Vanni 2002), and primary productivity (Schindler et al. 1997). The fish production and other secondary production in TSL depend on and are limited by the PP of the ecosystem and processes of the flood pulse (Koponen et al. 2010).

The seasonal flood pulse appeared to be the key factor affecting the fish community in TSL (Lamberts 2006; Say 2009; Chan et al. 2017; see also Chap. 33). Previous studies have demonstrated a close relationship between the annual extent of flooded forest and fish production of the industrial fisheries along the Tonle Sap River, which connects MR to TSL (Baran et al. 2001). The level and duration of flood influenced not only the breeding, growth, and survival of fish (Welcomme 1999) but also the food sources for fish (Campbell et al. 2006; MRC 2010; Pool et al. 2017). During the high-water period, the surrounding floodplains are inundated, submerging a substantial amount of terrestrial-derived organic plant matters, thereby providing food sources for fish (e.g., nutrient-rich detritus, seeds, and fruits). The periphyton, which is among the major food sources for fish, depends on the presence of a suitable substrate, nutrients, water quality, exposure time to light, and time since colonization.

Using stable isotope signatures of carbon and nitrogen, Say (2009) investigated the trophic linkages of six commercially and biologically distinctive fish species (i.e., *Barbonymus gonionotus*, *Channa striata*, *Cirrhinus microlepis*, *Cyclocheilichthys enoplos*, *Henicorcycus lobatus*, and *Pangasius larnaudii*; refer to Rainboth 1996; Lim et al. 1999; and Chap. 33 for taxonomy and ecology of these species), suggesting that algae carbon from the phytoplankton or periphyton and other food items (e.g., algae scums, zooplankton, benthic fauna, small invertebrates, and detritus) are the primary energy source of the studied fish species. Similarly, other fish species such as *Catlocarpio siamensis*, *Cirrhinus siamensis*, *Osteochilus hasseltii*, *Pangasianodon gigas*, *Pangasius krempfi*, and *Pangasius macronema* also feed on algae, periphyton, phytoplankton, detritus, and plant materials (e.g., leaves, fruits, roots, debris, and plant fragments), whereas other carnivorous species such as *Cyclocheilichthys enoplos*, *Lycothrissa crocodilus*, *Micronema apogon*, and *M. bleekeri* feed on a variety of food items, including snails, algae, earthworms, detritus, insect larvae, crustaceans, bivalves, shrimps, and even fish (Poulsen et al. 2004). These migratory species longitudinally swim from MR to the TSL system either to seek spawning, feeding, or refuge habitats during the high-water period (the wet season) and return to MR when the water in the lake recedes during the low-water period (the dry season).

In Cambodia, it is estimated that approximately two million people are directly involved in the TSL fishery (Nam and Song 2011). The annual catch from TSL was within the range of 150,000–253,000 ton/year (average value, 216,303 ton/year), representing 60% of the nation's total catch in the inland fisheries. A hydrodynamic productivity model (Holtgrieve et al. 2013) revealed that the removal of aNPP by the secondary production and fisheries harvest, considering a wide range of possible fisheries harvests, in TSL is equivalent to 7–69%, which is certainly higher than the global averages of 8% for marine and 24% for freshwater systems (Pauly and Christensen 1995). They also mentioned that this fact is attributable to the efficient carbon transfer through the food web and fish production supported also by terrestrial-derived materials.

Interestingly, the above estimate is probably low as it is based solely on the hypothetical fishery yield in TSL over a range of 150,000–300,000 ton/year. These estimates exclude the spawning stock of fish that remains in the lake, fish migrating out from the lake, non-fish predation (i.e., birds), and natural mortality. Many species of birds (e.g., water birds), zoobenthos, mammals, reptiles (e.g., snakes), and invertebrates (e.g., shrimps) are also important components of the food web

(MRC 2010; Uk et al. 2018; Chaps. 32, 33, and 48). The contribution of the microbial loop in nutrient and carbon cycles is also likely substantially important to the maintenance of fisheries, albeit unexplored for this ecosystem (Holtgrieve et al. 2013). Adding the aforementioned missing paths and processes might increase the required amount of PP for fish.

Key Points

- The planktonic Chl-*a* varied spatially and temporally, with the whole-lake averaged (±standard deviation) concentration ranging from 2.6 ± 3.3 μg/L during the low-water period to as high as 82.4 ± 224.5 μg/L during the high-water period.
- Cyanobacteria (blue-green algae) occurred in higher abundance and peaked (21.3 μg/L) during the low-water period compared to those during the high-water period (below detection limit of <0.01 μg/L most of the time).
- The productivity of TSL is considered among the highest in the world. The average net primary productivity in TSL and its floodplain between 1997 and 2008 was estimated at 2.4 million ton-C/year (average areal rates of 2.0 ± 0.2 g-C/m^2/day).
- PP in the TSL ecosystem appears to be effectively transferred to fish productivity and other secondary production. Based on the hypothetical fisheries harvest from TSL, fisheries harvest in TSL is roughly estimated to be equivalent to 7–69% of the total aNPP of the lake.

Acknowledgments The authors would like to thank Dr. Srey Sunleang, Deputy Director-General, General Directorate of Administration for Nature Conservation and Protection, Ministry of Environment, Cambodia, for his coordination, and my (corresponding author's) best friend Dr. Try Sophal, a postdoctoral fellow at Disaster Prevention Research Institute, Kyoto University, for his comments and technical support.

References

ADB. Environment in transition. Cambodia, LAO PDR, Thailand, Vietnam. Manila: Asian Development Bank; 2000.

Alin SR, Johnson TC. Carbon cycling in large lakes of the world: a synthesis of production, burial, and lake-atmosphere exchange estimates. Glob Biogeochem Cycle. 2007;21:GB3002.

Baran E, Van ZN, Ngor PB. Floods, floodplains and fish production in the Mekong Basin: present and past trends. pp. 920–932. In: Ali A, et al., editors. Proceedings of the Second Asian Wetlands Symposium, 27–30 August 2001, Penang, Malaysia. Pulau Pinang, Malaysia: Penerbit Universiti Sains Malaysia; 2001. p. 1116.

Blache J. Apercu sur le plancton des eaux douces du Cambodge. Cybium. 1951;6:62–94.

Burnette WC, Wattayakorn G, Supcharoen R, Sioudom K, Kum V, Chanyotha S, Kritsananuwat R. Groundwater discharge and phosphorus dynamics in a flood-pulse system: Tonle Sap Lake, Cambodia. J Hydrol. 2017;549:79–91.

Campbell IC, Poole C, Giesen W, Valbo-Jorgensen J. Species diversity and ecology of Tonle Sap Great Lake, Cambodia Aquat. Science. 2006;68:355–73.

Carpenter SR, Kitchell JF. The trophic cascade in lakes. Cambridge: Cambridge University Press; 1993.

Chan B, Ngor PB, So N, Lek S. Spatial and temporal changes in fish yields and fish communities in the largest tropical floodplain lake in Asia. Ann Limnol Int J Lim. 2017;53:485–93.

Chea R, Guo C, Grenouillet G, Lek S. Toward an ecological understanding of a flood-pulse system lake in a tropical ecosystem: food web structure and ecosystem health. Ecol Model. 2016;323:1–11.

Cunha DGF, Oguira AP, Calijuri MC. Nutrient reference concentrations and trophic state boundaries in subtropical reservoirs. Water Sci Technol. 2012;65:1461–7.

Darchambeau F, Sarmento H, Descy JP. Primary production in a tropical large lake: the role of phytoplankton composition. Sci Total Environ. 2014;473–474:178–88.

Day MB, Hodell DA, Brenner M, Curtis JH, Kamenov GD, Guilderson TP, Peterson CL, Kenney FW, Kolata AL. Middle to late Holocene initiation of the annual flood pulse in Tonle Sap Lake, Cambodia. J Paleolimnol. 2011;45(1):85–99.

Deng Y, Zhang Y, Li D, Shi K, Zhang Y. Temporal and spatial dynamics of phytoplankton primary production in Lake Taihu Derived from MODIS Data. Remote Sens. 2017;9(3):195.

Fahnenstiel GL, Sayers MJ, Shuchman RA, Yousef F, Pothoven SA. Lake-wide phytoplankton production and abundance in the Upper Great Lakes: 2010–2013. J Great Lakes Res. 2016;42 (3):619–29.

Heckman CW. The seasonal succession of biotic communities in wetlands of the tropical wet-and-dry climatic zone: V. Aquatic invertebrate communities in the Pantanal Mato Grosso, Brazil. Int Rev Gesamten Hydrobiol. 1998;83:31–63.

Hiroki M, Tomioka N, Murata T, Imai A, Jutagate T, Preecha C, Avakul P, Fukushima M. Primary production estimated for large lakes and reservoirs in the Mekong River Basin. Sci Total Environ. 2020;747:141133. ISSN 0048-9697

Holtgrieve GW, Arias ME, Irvine KN, Lamberts D, Ward EJ, Kummu M, Koponen J, Sarkkula J, Richey EJ. Patterns of ecosystem metabolism in the Tonle Sap Lake, Cambodia with links to capture fisheries. PLoS One. 2013;8(8):e71395.

Junk WJ. Investigations on the ecology and production-biology of the “floating meadows” (Paspalo-Echinochloetum) on the Middle Amazon. Part II. The aquatic fauna in the root zone of floating vegetation. Amazoniana: Limnologia et Oecologia Regionalis Systematis Fluminis Amazonas. 1973;4(1):9–102.

Junk WJ. The invertebrate fauna of the floating vegetation of Bung Borapet, a reservoir in central Thailand. Hydrobiologia. 1977;53:229–38.

Junk WJ, Brown M, Campbell IC, Finlayson M, Gopal B, Ramberg L, Warner BG. The comparative biodiversity of seven globally important wetlands: a synthesis. Aquat Sci. 2006;68:400–14.

Koponen J, Lamberts D, Sarkkula J, Inkala A, Junk W, Halls A, Kshatriya M. Primary and fish production report. Vientiane, Laos PDR; 2010.

Kuenzer C. Threatening Tonle Sap: challenges for Southeast Asia’s largest Freshwater Lake. Pac Geogr. 2013;40(August):29–31.

Lamberts D. Tonle Sap fisheries: a case study on floodplain gillnet fisheries. Bangkok: RAP Publication; 2001.

Lamberts D. The Tonle Sap Lake as a productive ecosystem. Int J Water Resour Dev. 2006;22:481–95.

Lim P, Lek S, Touch TS, Mao S, Chhouk B. Diversity and spatial distribution of freshwater fish in Great Lake and Tonle Sap river (Cambodia, Southeast Asia). Aquat Living Resour. 1999;12(6): 379–86.

McDonald JA, Pech B, Phauk V, Leeu B. Plant communities of the Tonle Sap Floodplain. Final Report in Contribution to the Nomination of Tonle Sap as a Biosphere Reserve for UNESCO’s Man in the Biosphere Program; 1997.

Mizuno T, Mori S. Preliminary hydrobiological survey of some Southeast Asian inland waters. Biol J Linn Soc. 1970;2:77–117.

MRC. Assessment of Basin-wide development scenarios: impacts on the Tonle Sap Ecosystem. Basin Development Plan Program, Phase 2.10. Vientiane, Lao PDR; 2010.

MRC. The Council Study: study on the sustainable management and development of the Mekong River, including impacts of mainstream hydropower projects. BioRA PROGRESS REPORT 1: Appendix D: field trip part I–specialists' notes. The Mekong River Commission; 2015.

Nam S, Song SL. Fisheries management and development in Tonle Sap Great Lake, Cambodia. Paper presented to the Consultation on Development Trends in Fisheries and Aquaculture in Asian Lakes and Reservoirs, 20–23 September 2011, Wuhan, China; 2011.

Ohtaka A, Watanabe R, Im S, Chhay R, Tsukawaki S. Spatial and seasonal changes of net plankton and zoobenthos in Lake Tonle Sap, Cambodia. Limnology. 2010;11:85–94.

Padisák J, Reynolds CS. Shallow lakes: the absolute, the relative, the functional and the pragmatic. Hydrobiologia. 2003;506(1–3):1–11.

Pauly D, Christensen V. Primary production required to sustain global fisheries. Nature. 1995;374: 255–7.

Pool T, Holtgrieve G, Elliott V, McCann K, McMeans B, Rooney N, Smits A, Phanara T, Cooperman M, Clark S, Phen C, Chhuoy S. Seasonal increases in fish trophic niche plasticity within a flood-pulse river ecosystem (Tonle Sap Lake, Cambodia). Ecosphere. 2017;8(7) https://doi.org/10.1002/ecs2.1881.

Poulsen AF, Hortle KG, Valbo-Jorgensen J, Chan S, Chhuon CK, Viravong S, Bouakhamvongsa U, Suntornratana K, Yoorong N, Nguyen TT, Tran BQ. Distribution and Ecology of Some Important Riverine Fish Species of the Mekong River Basin. MRC Technical Paper No. 10. ISSN: 1683-1489; 2004.

Rainboth WJ. FAO species identification field guide for fisheries purposes. Fishes of the Cambodian Mekong. Rome: Food and Agriculture Organisation of the United Nations; 1996. p. 265.

Sarkkula J, Kiirikki M, Koponen J, Kummu M. Ecosystem processes of the Tonle Sap Lake. In: 1st Workshop of Ecotone Phase II. Phnom Penh, Cambodia; 2003. pp. 1–14.

Say S. Trophic linkage: the importance of micro-algae to the fisheries of Boeung Tonle Sap, Cambodia. Dissertation. Monash University, Clayton, Victoria, Australia; 2009.

Schindler DE, Carpenter SR, Cole JJ, Kitchell JF, Pace ML. Influence of food web structure on carbon exchange between lakes and the atmosphere. Science. 1997;277:248–51.

Tanaka S, Ohtaka A. Freshwater Cladocera (Crustacea, Branchiopoda) in Lake Tonle Sap and its adjacent waters in Cambodia. Limnology. 2010;11:171–8.

Uk S, Yoshimura C, Siev S, Try S, Yang H, Oeurng C, Li S, Hul S. Tonle Sap Lake: Current status and important research directions for environmental management. Lakes Reserv Res Manag. 2018;23(3):1–13.

United States Environmental Protection Agency (USEPA). The lake and reservoir restoration guidance manual, EPA 440/5–88-002. Criteria and Standard Division Nonpoint Source Branch, Washington, DC; 1988.

Vanni MJ. Nutrient cycling by animals in freshwater ecosystems. Annu Rev Ecol Syst. 2002;33: 341–70.

Welcomme RL. A review of a model for qualitative evaluation of exploitation levels in multispecies fisheries. Fish Manag Ecol. 1999;6:1–19.

Chapter 32
Flooded Forests

Ly Sophanna, Uk Sovannara, Theng Vouchlay, Sun Visal, Lim Puy, Rajendra Khanal, Srey Sunleang, and Pham Ngoc Bao

Supplementary Information The online version of this chapter (https://doi.org/10.1007/978-981-16-6632-2_32) contains supplementary material, which is available to authorized users.

L. Sophanna (✉)
Ministry of Environment, Phnom Penh, Cambodia

Tokyo Institute of Technology, Tokyo, Japan
e-mail: ly.s.aa@m.titech.ac.jp

U. Sovannara
Tokyo Institute of Technology, Tokyo, Japan

T. Vouchlay
Tokyo Institute of Technology, Tokyo, Japan

Institute of Technology of Cambodia, Phnom Penh, Cambodia

S. Visal
Ministry of Environment, Phnom Penh, Cambodia

Wildlife Conservation Society, Phnom Penh, Cambodia

L. Puy
Tonle Sap Authority, Phnom Penh, Cambodia

R. Khanal
Tokyo Institute of Technology, Tokyo, Japan

Policy Research Institute, Kathmandu, Nepal

S. Sunleang
Ministry of Environment, Phnom Penh, Cambodia

P. N. Bao
Institute for Global Environmental Strategies, Hayama, Japan

C. Yoshimura et al. (eds.), *Water and Life in Tonle Sap Lake*,
https://doi.org/10.1007/978-981-16-6632-2_32

32.1 Distribution of Flooded Forests

In general, the wetland ecosystem is characterized by swamps and flooded forests, and this combination accounts for the higher portion of the global wetlands than marsh, bog, and fen, where herbs are the most dominant vegetation (Scarano 2009). Flooded forests are freshwater swamp forests that are wholly or seasonally inundated. In Cambodia flooded forests cover around 4778 km^2, and the majority of them is located in the wetland around Tonle Sap Lake (TSL), largely the three Ramsar sites (MoE 2018), covering approximately 3660 km^2 (including shrublands and gallery forests) (Rundel 2000). The remaining flooded forest is along the Mekong River (MR) and the Tonle Sap River and at other relatively small wetlands across the country (Vathana 2003).

Wetlands cover over 30% of the total area of Cambodia (Vathana 2003). Five of the country's wetlands have been designated as wetlands of international importance, also known as Ramsar sites, covering a total area of 852.35 km^2 (Ramsar 2021). Those registered wetlands are (1) Boeung Chhmar and its associated river system and floodplain (280 km^2); (2) Prek Toal (213 km^2); (3) Stung Sen (93 km^2); (4) middle stretches of the MR, North of Stung Treng (146 km^2); and (5) Koh Kapik and associated islets (120 km^2). Of them, the first three sites are situated in the floodplain of TSL.

Only limited flooded forest cover remains pristine in TSL because of human activities (Parolin and Wittmann 2010) (see Chap. 48 for the loss of flooded forests). Most of the flooded forests are mosaic (MoE 2018) and often patchy (Campbell et al. 2006). Bonheur and Lane (2002) identified 190 flooded vascular plants in TSL, and more recently, a survey conducted by the Tonle Sap Authority (TSA) registered 182 flooded plant species around the lake (TSA 2017). The flooded plants are classified into three distinct types: gallery forests with a height up to 15–20 m, shrubs that are 2–4 m tall, and a herbaceous community (Fig. 32.1, Supplementary Video 32.1) (Bonhuer 2001).

The gallery forests originally dominated along the dry season's shoreline of TSL, covering around 10% of the floodplain (~15,000 km^2, including waterbodies) (Rundel 2000), whereas Koponen et al. (2003) found that they covered around 1.4% of the floodplain (~14,839 km^2, including waterbodies). The gallery forests were indicated around 5.6% of the floodplain (14,128 km^2, excluding the permanent waterbody) (see Chap. 48 for the map, indicated mainly by the dense forest in Fig. 48.1). The short-tree shrublands consist of shrubs and stunted trees with occasional scattered taller trees, covering a relatively larger part (~34%) of the TSL floodplain (Koponen et al. 2003). The floodplain of TSL nurtures herbaceous vegetation of flooded grasslands and dense mats of floating and submerged aquatic vegetation, accounting for around 10% of the floodplain (Koponen et al. 2003). However, submerged aquatic species are completely absent, because most of the plants are emergent. Only some ceratophyllids such as *Utricularia aurea*, charids (*Nitella* species), and leminds (*Lemna minor*) were recorded.

The flooded forests in TSL are ideal and important protective habitats as feeding and/or breeding grounds for aquatic and terrestrial biota such as fishes, globally threatened mammals, and waterbirds (Mekong River Commission (MRC), 2010;

Fig. 32.1 Typical vegetations in TSL. (**a**) Gallery forests (*Barringtonia acutangula* and *Coccoceras anisopodum*) in June 2020, (**b**) gallery forests (*Barringtonia acutangula*) in October 2019, (**c**) shrubs (*Vitex holpadenon* and *Schoutenia godefroyana*) in February 2019, and (**d**) the herbaceous community in Prek Toal Ramsar site in July 2019. (**a**) Gallery forests in Stung Sen Ramsar site. (**b**) Gallery forests in Prek Toal Ramsar site. (**c**) Shrubland in Prek Toal Ramsar site. (**d**) Herbaceous community in Prek Toal

Davidson 2006; Bonheur and Lane 2002; see Chaps. 31 and 33–34 for details). TSL annually provides 150,000–253,000 ton/year of fish catch (Nam and Song 2011), a major protein in the Cambodian diet, owing to active primary and secondary productions in the lake, which are supported by flooded forests in its floodplain (MRC 2010).

32.2 Inundation Pattern and Flooded Forests

The distribution, structure, and variation of flooded vegetation in TSL depend primarily on the water regimes (i.e., fluctuation of the water level and flood duration) (Bonheur and Lane 2002). The gallery forests are located on the shoreline of the permanent lake, where the land is seasonally flooded for around 9 months every year. The shrub communities farther from the shoreline (mostly in areas where soils are non-saturated during the dry season) are flooded from 5 to 8 months. Shrubs can be found in the understory of gallery forests or in isolation. The flooded grasslands of herbaceous habitats are flooded for 3 to 5 months (Arias et al. 2012; Bonhuer 2001), in the middle-outer floodplain (Davidson 2006). Other herbaceous vegetations of 1–3 m tall may be emergent from water, whereas the free-floating macrophytes are more typical on the lake and located in clearings within the gallery forests.

The distribution, structure, and composition of gallery forests and shrubs in TSL are also mainly a function of seasonal flood dynamics and the heterogeneity of soil moisture conditions (Arias et al. 2013; Parolin and Wittmann 2010). First, this showcases the zonation of flooded plants concerning their distribution as a function of the flooding gradient, which is consistent with other large floodplains (e.g., the Central Amazonia in South America and the Kakadu Region in Northern Australia) (Parolin and Wittmann 2010). For instance, the unique difference of TSL from the Central Amazonian floodplain is that TSL receives a monomodal flood pulse resulting from seasonal reversal flow from MR during the rainy season. This system makes the gallery forests and shrubs in TSL distinct from typical swamp forests in a number of respects (Rundel 2000). The gallery forests and shrubs are low in stature of different floristic plants containing a number of narrowly distributed endemics. In addition, when the flood recedes at the start of the dry season, the flooded grasslands are exposed to the drier condition for at least 6 months. This grassland serves as a favorable habitat for some bird species, in particular for the critically endangered Bengal Florican (*Houbaropsis bengalensis*) (see also Chap. 34). By contrast, in the Central Amazonian floodplain of Varzea, grasses can be found in longer-flooded environments (Parolin and Wittmann 2010).

Second, the distribution of forest cover in TSL may also be closely related to the content of clay and silt in the soil (Daskin et al. 2019). The high soil moisture condition also contributes to taller trees; this explains the trees' growth closer to open waters (Arias et al. 2013; Parolin and Wittmann 2010; Veneklaas et al. 2005). Several species that would normally be classified as shrubs in shrubland can also reach the size of trees within the gallery forests under such conditions. The sandier soil content decreases the tree cover due to reduced nutrient availability. In fact, extensive sparse short grasslands occur on sandy soils on the outer edge of the TSL floodplain (Arias et al. 2013; Davidson 2006).

Trees are fundamentally terrestrial specialists that tend to die more readily in response to flooding than to desiccation (Parolin and Wittmann 2010). To survive under the flooding stress, flooded trees need to be flood-deciduous. Most gallery forests and shrubs in TSL are deciduous and may thus be associated with an adaptive response during the periodic flood pulse. Leaf shedding by those trees when partially or completely submerged during the monsoonal wet season (rather than in the dry season) has been highlighted in previous studies (Campbell et al. 2006; Rundel 2000). However, some tree species are classified as evergreen (e.g., *Barringtonia acutangula* and *Combretum trifoliatum*), although they flush new leaves contemporaneously with the flood-deciduous species as the water recedes (in November–December). Flowering and fruit production of the gallery forests and shrubs are active after several months of new leaf flush. This occurs during the late dry and early wet seasons. Fruits then turn mature at the time of submergence, suggesting that fish may be a key seed dispersal agent (Davidson 2006), in addition to flood waters (Rundel 2000) and waterbirds (Green et al. 2016).

Regarding the distribution of flooded vegetation in TSL, there are two noticeable knowledge gaps. First, to date, there is no validated distribution map of each individual vegetation type, namely, gallery forest, shrubland, and flooded grassland. Thus, there is a need to create a detailed map of each vegetation type, using up-to-date data and technology. Second, there is a need to reinvestigate the influence of seasonal flood dynamics and heterogeneity of soil moisture and properties on their potential distribution. In addition to the literature, further research may be required to understand how flooded vegetation is spatially distributed under heterogeneous physical and chemical conditions (e.g., inundation regime and soil moisture/property). This will enhance our capacity to deal with the degradation of the individual types of flooded vegetation due to forest fires (see Chap. 48 for forest fires) and agricultural intensification and expansion (Mahood et al. 2020). Such study will enable to produce an actual map of vegetation distribution, which can be compared with the reinvestigated map of potential vegetation distribution.

In the permanent freshwater of the lake, there is an absence of flooded forest stands, meaning that the flooded forests cannot persist in permanently inundated conditions. Based on the land use and land cover (LULC) map developed by the Japan International Cooperation Agency (JICA) in 1999 and the aerial imagery captured in 2004–2005 for the Tonle Sap Environmental Management Project, Arias et al. (2012) predicted hydrological alteration caused by climate change and infrastructure development. Their result indicated a significant impact on the flooded forest ecosystem, which decreases the size of gallery forests (−75% to −83%) while increasing the area of open water (+18% to +21%). However, this study does not include how well-flooded forests adapt to such hydrological alteration. In general, an organism adapts and evolves itself to changes in the habitat, such as the flood-deciduous forests, to survive. In addition, groundwater might have a potential influence on flooded forests as they actually play a key role in the recharge of shallow groundwater (Hughes et al. 2003; Daskin et al. 2019). In this respect, the adaptation and evolution of flooded forests to hydrological change deserve further research considering groundwater and LULC in the floodplain and the inflow river basins. In case there will be an impact on the flooded forests, future scenarios on their ecosystem functions and services require to be investigated to understand their potential changes.

32.3 Common and Dominant Species

The common species of gallery forests in TSL are *Barringtonia acutangula* (Lecythidaceae), *Diospyros cambodiana* (Ebenaceae), *Terminalia cambodiana* (Combretaceae), *Coccoceras anisopodum* (Euphorbiaceae), *Crudia chrysantha* (Caesalpiniaceae), *Croton caudatus* (Euphorbiaceae), *Ficus* sp. (Moraceae), *Elaeocarpus griffithii*, and *Garcinia loureirin* (Guttiferae) (see Campbell et al. 2006; Davidson 2006, for other gallery forest species). Among them, the dominant species are *Barringtonia acutangula*, *Diospyros cambodiana*, and *Terminalia cambodiana*. As these three species show affinities with coastal mangrove habitats, they may be called freshwater mangrove (Davidson 2006).

Barringtonia acutangula (local name: Daem raing teuk) stands 4–10 m tall, with a fissured gray bark, and generally grows in moist places along rivers in Southeast Asia (Dy 2000). This species forms a dense canopy and stands in some protected sites such as Kampong Phluk in Siem Reap province. As one of the few species with perennial leaves, even when submerged, its wood is digested from the inside by aquatic bacteria, gradually transforming the tree into a ring of treelets and giving an eerie aspect to older trees. *Diospyros cambodiana* (local name: Daem phtoul) stands 15–20 m tall and is found mostly along the riverbanks. This tree species is often preferred by large waterbirds and can accommodate over 50 nests of mixed waterbird species per tree. *Terminalia cambodiana* (local name: Daem tá uë) stands 8–15 m tall and is rare. The species is of high quality for firewood.

The common species of shrubs are *Barringtonia acutangula* (Lecythidaceae), *Bridelia cambodiana* (Euphorbiaceae), *Brownlowia paludosa* (Tiliaceae), *Combretum trifoliatum* (Combretaceae), *Croton krabas* (Euphorbiaceae), *Gardenia cambodiana* (Rubiaceae), *Gmelina asiatica* (Verbenaceae), *Phyllanthus emblica* (Euphorbiaceae), and *Vitex holoadenon* (Verbenaceae) (see more shrub species in Davidson 2006; Rundel 2000).

Vitex holoadenon (local name: Tien Prey) is a predominant shrub species in TSL. Its leaves are round-based with pointed tips, whereas yellow flowers are clustered with broadened lower tips. In October and November, its emerging stems become a favorable communal roost for whiskered tern. *Combretum trifoliatum* (local name: Trâhs) is a climbing shrub of secondary formation of Southeast Asia and is abundant in wetlands. The species generates short shoots during the dry season or climbs on trees during the elevated water season. *Schoutenia godefroyana* (local name: Lo: nhië) stands 2–3 m tall. It is the secondary forest of Cambodia, which is abundant on the shore of TSL. Its fragrant smelling flowers are of interest to bees.

Herbaceous communities include grasslands, vines, and aquatic plants. The communities include marsh sesbania (*Sesbania javanica*), cogongrass (*Imperata cylindrica*), morning glory (*Ipomoea aquatica*), water hyacinth (*Eichhornia crassipes*), *Ipomoea chrysioides*, water lily (*Nymphaea nouchali*), and other species (Bonhuer 2001; Davidson 2006). *Sesbania javanica* (local name: Snaô) is a woody herbaceous plant, which stands 2–4 m tall in swampy places of tropical Asia and is abundant on TSL. Its yellow pea-like flowers hang under the stem that generally blossoms in November and December. *Ipomoea aquatica* (local name: Tra Kuon) is a sprawling vine, creeping on mud, or floating on water. It is naturally found within the flooded forest but is also guarded with pegs in the waterbodies of TSL close to the villages during the dry season. *Eichhornia crassipes* (local name: Kâmphlôk) is generally found in stagnant waters or slow-flowing rivers. This free-floating perennial aquatic plant has a high reproduction rate with runners that eventually form daughter plants, potentially colonizing new areas and becoming invasive. The species also produces long-lived viable seeds.

32.4 Ecosystem Services of Flooded Forests

Approximately 1.7 million people live in TSL and its floodplain (Salmivaara et al. 2016) and receive multiple benefits from this ecosystem. The flooded forests provide ecosystem services, which may be classified into four major categories: provisioning, cultural, regulating, and supporting services. The economic valuation of ecosystem services of flooded forests highlights the relationship between the services of forests and human well-being. Thus, it will provide fundamental information that supports decision-making and contributes to better conservation and management of flooded forests in TSL. The following sections outline those services of flooded forests based on the relevant literature and our survey conducted in February and March 2020 in villages in TSL. We randomly selected 22 villages from nine communes and then selected 97 households (hh) from those villages. All houses were located near flooded forests within an average distance of 500 m.

32.4.1 Provisioning and Cultural Services

Provisioning services of flooded forests in TSL include fishery/aquaculture (Nam and Song 2011); timber and firewood (Vathana 2003); non-timber forest products, wild animals/birds, and traditional medicines (Rab et al. 2006); and domestic water use and irrigation (Uk et al. 2018). Our survey also revealed that the provisioning services of flooded forests include fish, fuelwood, wild food, traditional medicine, and honey. These services provide an average economic value of US$914/person/year or US$5021/hh/year. The economic value from fishing alone holds the highest value at US$892/person/year. Fuelwood and wild food also contribute a considerable annual economic benefit of US$12/person/year and US$8/person/year, respectively. In addition, traditional medicine and honey collection contribute the same average economic value of US$1/person/year. This highlights the importance of these provisioning services for local people, given that the gross domestic product per capita of Cambodia was US$1643 in 2019 (World Bank 2021).

Cultural services of flooded forests in TSL stem from the close relationship that local communities have with the flooded forests (Keskinen 2006), especially in the floating communities. Based on our survey, cultural services of flooded forests that the local communities value are cultural heritage, religious belief, aesthetics, inspiration, social relation, education, and ecotourism value. Ecotourism could directly bring benefits to some respondents in villages where ecotourism sites are in place.

Many respondents are aware of social values, with communication and solidarity among villagers generating momentum for the conservation of flooded forests through community activities. With such interrelationships, flooded forests are better protected. Respondents value the aesthetics of flooded forests, providing beautiful landscapes, exhibiting the greenery of nature, bringing lively feelings, and attracting tourists, although a respondent negatively emphasizes that flooded forests create a dark environment, which makes catching fish difficult.

Religious beliefs linked to flooded forests have long been practiced, with links to praying to the forests for high fish catches, happiness, safety and security, and prosperity. Educational values are placed on the importance of the future generations to have profound knowledge of different species of flooded forests and how to get optimal benefits from flooded forests and love and protect for sustained flooded forest conservation. Cultural heritage mainly addresses the cultural uses of flooded forests when fishing (i.e., resting, cooking, and tightening fishing nets). Inspirational value is associated with songs (e.g., the beauty of Prek Toal), articles, and myths of flooded forests, which become well-known as a result of the description of the beautiful landscape of flooded forests. This can mainstream the idea of loving flooded forests and help in their protection.

32.4.2 Regulating and Supporting Services

Regulating services of flooded forests in TSL include flood regulation in the MR and Delta (Pokhrel et al. 2018), water regulation, water supply, aquifer recharge (Vathana 2003), carbon sequestration (JICA 2017), and water quality regulation (Hoshikawa et al. 2019; Siev et al. 2018). From our survey, the majority of local communities (94% of respondents) acknowledge that flooded forests in the TSL help regulate local air temperature, water purification (69% of the respondents), storm regulation (90%), and flood regulation (60%). Local communities move their houses close to gallery forests where their houses can be protected from storms in the rainy season.

Supporting services of flooded forests in TSL comprise habitats for biodiversity, namely, fish-habitat seasonal breeding, nursery grounds, and forage areas (Bonheur and Lane 2002; Chan et al. 2020; Keskinen 2006; Lamberts 2001) (see also Chap. 33). During the inundated periods, flooded forests are important food sources, such as nutrient-rich detritus, seeds, and fruits, for fish (Campbell et al. 2006; Saint-paul et al. 2000). Importantly, previous studies in TSL (Chan et al. 2020) and the floodplain lake of the lower Amazon River (Castello et al. 2018) confirmed that fish yield has a significant positive correlation with the cover of flooded forests. Other supporting services are primary production (Lamberts 2001; Uk et al. 2018; see also Chap. 31), deposition of sediments (Hoshikawa et al. 2019), rich deposited sediments, endangered waterbird habitats, and breeding grounds (Sun and Mahood 2015) (see also Chap. 34).

Key Points

- Flooded forests are mainly distributed directly around TSL, with their coverage being approximately 3660 km^2, largely within the lake's three Ramsar sites.
- Flooded vegetation in the wetland consists of 190 vascular plant species and is divided into three major types: gallery forests, shrubs, and herbaceous

communities. The gallery forests and shrubs are ideal and important protective habitats as feeding and/or breeding grounds for aquatic and terrestrial biota.

- The distribution, structure, and variation of flooded forests in TSL depend primarily on seasonal flood dynamics (i.e., fluctuation of water levels and flood duration) and heterogeneity of soil moisture and properties.
- Flooded forests provide a myriad of benefits to humans and the environment through their services, which constitute provisioning, cultural, regulating, and supporting services. Local communities in TSL place substantial values on these services. They benefited from the economic value of provisioning services at US $914/person/year in 2020.

Acknowledgments We are thankful to Mr. Sive Thea, the technical assistant of the SATREPS project in Cambodia, for his dedicated assistance during the interview survey.

References

Arias ME, Cochrane TA, Piman T, Kummu M, Caruso BS, Killeen TJ. Quantifying changes in flooding and habitats in the Tonle Sap Lake (Cambodia) caused by water infrastructure development and climate change in the Mekong Basin. J Environ Manag. 2012;112:53–66.

Arias ME, Cochrane TA, Norton D, Killeen TJ, Khon P. The flood pulse as the underlying driver of vegetation in the largest wetland and fishery of the mekong basin. Ambio. 2013;42(7):864–76.

Bonheur N, Lane BD. Natural resources management for human security in Cambodia's Tonle Sap biosphere reserve. Environ Sci Policy. 2002;5(1):33–41.

Bonhuer N. Tonle Sap Ecosystem and Value. Kingdom of Cambodia Ministry of Environment; 2001. pp. 1–12.

Campbell IC, Poole C, Giesen W, Valbo-Jorgensen J. Species diversity and ecology of Tonle Sap Great Lake, Cambodia. Aquat Sci. 2006;68(3):355–73.

Castello L, Hess LL, Thapa R, McGrath DG, Arantes CC, Renó VF, Isaac VJ. Fishery yields vary with land cover on the Amazon River floodplain. Fish Fish. 2018;19(3):431–40.

Chan B, Brosse S, Hogan ZS, Ngor PB, Lek S. Influence of local habitat and climatic factors on the distribution of fish species in the tonle sap lake. Water (Switzerland). 2020;12(3) https://doi.org/10.3390/w12030786.

Daskin JH, Aires F, Staver AC. Determinants of tree cover in tropical floodplains. Proc R Soc B Biol Sci. 2019;286(1914)

Davidson PJA. The Biodiversity of the Tonle Sap Biosphere Reserve 2005 status review. (March); 2006.

Dy P. Dictionary of Plants Used in Cambodia. Imprimerie Olympic, Phnom Penh (published by the author); 2000.

Green AJ, Soons M, Brochet A-L, Kleyheeg E. Dispersal of plants by waterbirds. In: Sekercioglu ÇH, Wenny DG, Whelan CJ, editors. Why birds matter. Avian ecological function and ecosystem services. Chicago: The University of Chicago Press; 2016. p. 147–95.

Hoshikawa K, Fujihara Y, Siev S, Arai S, Nakamura T, Fujii H, Sok T, Yoshimura C. Characterization of total suspended solid dynamics in a large shallow lake using long-term daily satellite images. Hydrol Process. 2019;33(21):2745–58.

Hughes F, Richards K, Girel J, Moss T, Muller E, Nilsson C, Rood S. The Flooded Forest: Guidance for policy makers and river managers in Europe on the restoration of floodplain forests. Europe. 2003;96.

JICA. (2017). Project for Facilitating the Implementation of REDD+ Strategy and Policy for Cambodia. *Jica*, (November).
Keskinen M. The lake with floating villages: socio-economic analysis of the Tonle Sap Lake. Int J Water Resour Dev. 2006;22(3):463–80.
Koponen J, Sarkkula J, Keskinen M, Varis O, Hellsten S, Elise Järvenpää TD. Draft MRCS/WUP-FIN report: Tonle Sap development scenario impacts and guidelines. Tonle Sap development scenario Water Utilization Program - Modelling of the Flow Regime and Water Quality of the Tonle Sap; 2003
Lamberts D. Tonle Sap fisheries: a case study on floodplain gillnet fisheries in SIEM Reap, Cambodia; 2001.
Mahood SP, Poole CM, Watson JEM, Sharma S, Garnett ST, Mackenzie RA. Agricultural intensification is causing rapid habitat change in the Tonle Sap Floodplain, Cambodia. Wetl Ecol Manag. 2020;28(5):713–26.
MoE. Cambodia Forest Cover 2016. (March), 1–22; 2018.
MRC. Assessment of Basin-wide development scenarios: Impacts on the Tonle Sap Ecosystem. Basin Development Plan Program, Phase 2. 10. Vientien, Lao PDR; 2010.
Nam S, Song SL. Fisheries management and development in Tonle Sap Great Lake, Cambodia. Consultation on Development Trends in Fisheries and Aquaculture in Asian Lakes and Reservoirs. 2011;2011(4).
Parolin P, Wittmann F. Struggle in the flood: tree responses to flooding stress in four tropical floodplain systems. AoB PLANTS. 2010;2010:1–19.
Pokhrel Y, Shin S, Lin Z, Yamazaki D, Qi J. Potential disruption of flood dynamics in the lower MEKONG river basin due to upstream flow regulation. Sci Rep. 2018;8(1):1–13.
Rab MA, Navy H, Ahmed M, KeAng S, Viner K. Socioeconomics and values of resources in Great Lake-Tonle Sap and Mekong-Bassac area: results from a sample survey in Kampong Chhnang, Siem Reap and Kandal Provinces, Cambodia. WorldFish Center Discussion Series. 2006;(4), iv +98 pp.
Ramsar. Ramsar Sites Information Service. Retrieved from: https://rsis.ramsar.org/ris-search/?f[0]=regionCountry_en_ss%3ACambodia&pagetab=1, Accessed on 08 Feb 2021; 2021.
Rundel PW. Forest habitats and flora in Lao PDR, Cambodia and Vietnam: Conservation Priorities in Indochina. Hanoi, Vietnam; 2000.
Saint-paul U, Zuanon J, Correa MAV, Berger U, Junk WJ. Fish communities in central Amazonian white- and blackwater oodplains. Environ Biol Fish. 2000:235–50.
Salmivaara A, Kummu M, Varis O, Keskinen M. Socio-economic changes in Cambodia's unique Tonle Sap Lake area: a spatial approach. Appl Spat Anal Policy. 2016;9(3):413–32.
Scarano FR. Flooded forests. In: Del Claro K, Oliveira PS, Rico-Gray V, editors. Tropical biology and conservaion management: Vol. IV. Abu Dhabi: EOLSS Publications; 2009.
Siev S, Yang H, Sok T, Uk S, Song L, Kodikara D, Oeurng C, Hul S, Yoshimura C. Sediment dynamics in a large shallow lake characterized by seasonal flood pulse in Southeast Asia. Sci Total Environ. 2018;631–632:597–607.
Sun V, Mahood S. Wildlife monitoring at Prek Toal Ramsar Site, Tonle Sap Great Lake, 2013 and 2014 (Vol. 1); 2015.
TSA. Technical report and research on socio-economics, natural resources, and managment of Tonle Sap Lake in Khmer. Phnom Penh; 2017.
Uk S, Yoshimura C, Siev S, Try S, Yang H, Oeurng C, Li S, Hul S. Tonle Sap Lake: Current status and important research directions for environmental management. Lake Reserv Res Manag. 2018;23(3):177–89.
Vathana K. Review of Wetland and Aquatic Ecosystem in the Lower Mekong River Basin of Cambodia. The Cambodian National Mekong Committee Secretariat (CNMCS) and The Mekong River Commission Secretariat (MRCS), (August), 100; 2003.
Veneklaas EJ, Fajardo A, Obregon S, Lozano J. Gallery forest types and their environmental correlates in a Colombian savanna landscape. Ecography. 2005;28(2):236–52.

World Bank. GDP per capita (current US$) in 2019 – Cambodia. World Bank national accounts data, and OECD National Accounts data files Retrieved from: https://data.worldbank.org/indicator/NY.GDP.PCAP.CD?locations=KH, Accessed on 08 Feb 2021; 2021.

Chapter 33
Aquatic Fauna and Aquaculture

Lim Puy

33.1 The Floodplain and Its Habitats

The ecosystem of Tonle Sap Lake (TSL) serves as a complex flood regulator that is fundamental for wetlands and biodiversity of the lake itself and the wider Mekong Delta. The flood-pulse-induced bi-directional flow in the TSL system seasonally inundates the lake floodplains during the rainy season, creating heterogeneous habitats and diverse ecosystems for aquatic fauna communities (Figs. 33.1 and 33.2; see Chaps. 1–4, 31, 32, and 34). This chapter describes major aquatic fauna with their habitats and present status.

The wetland around TSL including the grassland covers approximately 2520 km^2 (Legris and Blasco 1972). We conducted a study during the dry and the rainy seasons between 2018 and 2019; the results indicated that the inundated area of the seasonally flooded habitats (Fig. 33.1) varied from 1314 to 5343 km^2 when the water level at the hydrological stations varied from 2 to 5 m at the beginning of the rainy season (July–August), respectively. As the seasonally flooded habitats become inundated when the water level rises, reaching 8–9 m in October, many plant species produce fruit and seeds (Arias et al. 2013).

The junction between the lake and its outlet to Tonle Sap River (TSR), called "bottle-neck," plays a prominent role in maintaining the minimum water level in the lake necessary for the survival of the biotic communities by constituting a natural sediment trap for the inflow from Mekong River (MR, Fig. 33.2). The 11 major tributaries of TSL and their sub-basins also play a secondary role in maintaining the water level in the back swamp of TSL during the dry season (see also Chap. 8).

L. Puy (✉)
Tonle Sap Authority, Phnom Penh, Cambodia
e-mail: puy.lim@toulouse-inp.fr

C. Yoshimura et al. (eds.), *Water and Life in Tonle Sap Lake*,
https://doi.org/10.1007/978-981-16-6632-2_33

Fig. 33.1 Habitats for spawning and growth for the fish community. (**a**) *Barringtonia acutangula*, *Eichhornia crassipes*. (**b**) *Phragmites* spp. (**c**) *Cynodon dactylon*. (**d**) *Cyperus elatus*

33.2 Fish

33.2.1 Inventory of Fish Species

Fish species in the TSL system have been well reported compared to other faunas. Campbell et al. (2006) claimed 149 fish species in TSL, whereas Nam et al. (2006) highlighted around 200 fish species, encompassing 35 families and 90 genera. Based on the inventory in 2010–2019, the Tonle Sap Authority (TSA) has listed 167 species, belonging to 12 orders, 35 families, and 94 genera. Among all the identified samples, 73% belonged to several genera including Cypriniformes (genus: *Paralaubuca*, *Puntioplites*, *Labiobarbus*, and *Henicorhynchus*), Siluriformes (genus: *Mystus* and *Pangasius*), and Perciformes (genus: *Parambassis*). The inventoried 167 species were classified into four categories: high consumption (commonly caught/processed by population for food), high market value, endangered species, and introduced species (Tables 33.1 and 33.2).

The knowledge of the relationship between species and habitats is necessary for better understanding of the environmental requirements of fish fauna, which are critically important for the conservation of fish fauna diversity and productivity (Chan et al. 2020), especially in a complex hydroecological system such as TSL. During the early tropical monsoon rainfalls, many fish species in MR start to

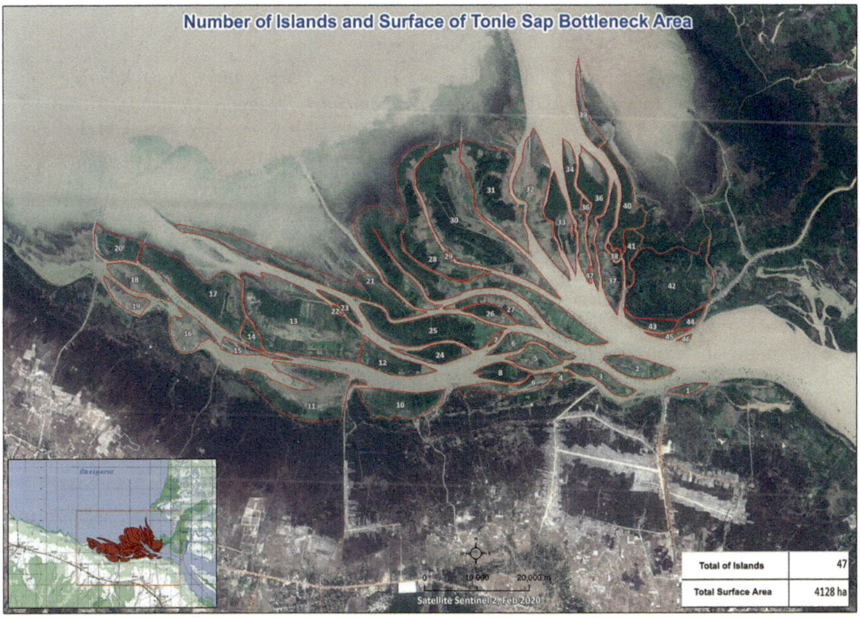

Fig. 33.2 Chhnok Trou and Tonle Sap River branches forming a complex of 47 islands at the bottle-neck junction with Tonle Sap Lake, at a water level <0.5 m

Fig. 33.3 *Catlocarpio siamensis* (local name: Trey Kol Reang, left panel) and *Probarbus jullieni* (Trey Trasark, right panel)

disperse for spawning and longitudinally migrate to the lower floodplain habitats (e.g., TSL and flooded area) for feeding and growth. These migratory fish species access the seasonally flooded habitats to feed during the wet season (Campbell et al. 2006). During the dry season, when the water level recedes, the longitudinal migratory fishes migrate from TSL back to MR to escape from the contracting condition in the floodplain and the lake, for better water quality and deeper pools as a refuge in the dry season (Poulsen et al. 2004). By surveying the dai fishery lot (bagnet fishery, mainly located along TSR; see Chap. 48) from 2011 to 2020, we observed that genera *Clupeichthys*, *Paralaubuca*, *Puntioplites*, *Thynnichthys*,

Table 33.1 Fish species classification

Fish species classification	Fish
High consumption and processing comprising 43 species belonging to 3 orders and 13 genera	*Clupeidea*, *Clupeichthys*, *Paralaubuca*, *Thynnichthys*, *Dangila*, *Henichorhynchus*, *Botia*, *Mystus*, *Pangasius*, *Parambassis*, *Pseudambassis*, *Anabas*, and *Trichogaster*
High market value include 24 species belonging to 12 genera	*Chilata*, *Notopterus*, *Osteocheilus*, *Micronema*, *Ompok*, *Wallago*, *Helicophagus*, *Macrognathus*, *Boesemania*, *Oxyelectis*, *Channa*, and *Cynoglossus*
Endangered: 11 vulnerable, near-threatened, endangered, or critically endangered species on the International Union for Conservation of Nature's Red List (IUCN 2021)	*Tenualosa thibaudeaui*, *Lycothrissa crocodilus*, *Probarbus jullieni*, *Probarbus labeamajor*, *Probarbus labeaminor*, *Catlocarpio siamensis*, *Pangasianodon gigas*, *Bagarius yarrelli*, *Glyptothorax fuscus*, *Glyptothorax lampris*, and *Datnioides microlepis* (Fig. 33.3)
Introduced species inhabit the lake and breed in cages or ponds around TSL	*Cyprinus carpio*, *Hypophthalmichthys molitrix*, *Piaractus brachypomus*, and *Oreochromis mossambicus* (Table 33.2; see species numbers 29–32)

Dangila, *Henichorhynchus*, and *Botia* are the most abundant for the longitudinal migration. As these species are typically intolerant to low oxygenated water, they need the open water area of the lake where wind-induced water current oxygenates (Ngor et al. 2018).

The floodplain-resident or non-migratory fish species, by contrast, spend most of their lifespans in the flooded forest (Campbell et al. 2006). The floodplain residents mostly live in the lakes and marshes or swamps on the floodplains near the permanent waterbodies and migrate to flooded areas during the flooded season. The floodplain residents such as *Channa striata* and *Cerberus microlepis* have developed their external organs to breathe the air and thus adapted to the low oxygen and harsh environmental conditions in swamps and small floodplain lakes during the dry season (Poulsen et al. 2002). *C. striata* and all the genera of *Trichogaster* occur predominantly in the northern part of the lake, where the area is predominantly covered by a flooded forest, swampy and shrubland, aquatic vegetation, and rice fields (Chan et al. 2020). The flooded forest areas are also likely important habitats to rear and forage for *Pangasianodon hypophthalmus*, *C. striata*, and *Cyclocheilichthys enoplos* (Chan et al. 2020) because the flooded forests typically constitute an appropriate feeding ground providing a variety of terrestrial prey (e.g., insects, frogs, and small mammals) for opportunistic predators (e.g., *Phalacronotus* spp. and *C. striata*; Poulsen et al. 2004). *C. striata* needs flooded vegetations for breeding and hatching (Rainboth 1996).

Aquaculture is mainly practiced on floating cages attached to houses in TSL and the ponds on non-flooded land. During 2018–2019, TSA inventoried fish farms in 12 large floating villages, i.e., Psar Chhnang, Chhnok Trou, Kampong Luong, Reang

Table 33.2 The fish species of high market value for fresh consumption and exotic species in TSA inventory on fish biodiversity in TSL and its dependency during 2010–2019

	Species	Name[a]	Class	Order	Family	Genus	Habitat[b]	Size[b]	Trophic level[b]	Status	IUCN Red list status[a]	
												Last assessed
1	*Chitala ornata*	Clown featherback	Actinopterygii	Osteoglossiformes	*Notopteridae*	*Notopterus*	pelagic	100.0 SL	3.7	Native	LC	08/30/19
2	*Notopterus notopterus*	Bronze featherback	Actinopterygii	Osteoglossiformes	*Notopteridae*	*Notopterus*	demersal	60.0 SL	3.6	Native	LC	02/09/11
3	*Cyclocheilichthys enoplos*	–	Actinopterygii	Cypriniformes	*Cyprinidae*	*Cyclocheilichthys*	benthopelagic	74.0 SL	3.2	Native	LC	02/09/11
4	*Cirrhinus microlepis*	Small Scaled Mud Carp	Actinopterygii	Cypriniformes	*Cyprinidae*	*Cirrhinus*	benthopelagic	65.0 SL	2.4	Native	VU	02/17/11
5	*Hemibagrus nemurus*	Yellow Catfish	Actinopterygii	Siluriformes	*Bagridae*	*Hemibagrus*	benthopelagic	65.0 SL	3.6	Native	LC	09/01/18
6	*Belodontichthys dinema*	–	Actinopterygii	Siluriformes	*Siluridae*	*Belodontichthys*	Wetlands (inland)	–	–	–	LC	05/30/19
7	*Cryptopterus microneme*	–	–	–	–	–	–	–	–	–	–	–
8	*Wallago attu*	Wallago	Actinopterygii	Siluriformes	*Siluridae*	*Wallago*	demersal	240.0 TL	3.7	Native	VU	08/12/19
9	Pangasius djambal	–	Actinopterygii	Siluriformes	*Pangasiidae*	*Pangasius*	Wetlands (inland)	–	–	–	LC	09/01/18
10	Pangasius larnaudii	Black-spotted catfish	Actinopterygii	Siluriformes	*Pangasiidae*	*Pangasius*	benthopelagic	130.0 SL	3.3	Native	LC	02/26/11
11	*Macrognathus taeniagaster/ Macrognathus semiocellatus**	–	Actinopterygii	Synbranchiformes	*Mastacembelidae*	*Macrognathus*	benthopelagic	16.0 SL	3.3	Native	LC	12/11/19
12	*Macrognathus siamensis*	Spotfin Spiny Eel	Actinopterygii	Synbranchiformes	*Mastacembelidae*	*Macrognathus*	benthopelagic	30.0 SL	3.3	Native	LC	02/24/11

(continued)

Table 33.2 (continued)

	Species	Name[a]	Class	Order	Family	Genus	Habitat[b]	Size[b]	Trophic level[b]	Status	IUCN Red list status[a]	
												Last assessed
13	*Boesemania microlepis*	Smallscale Croaker	Actinopterygii	Perciformes	*Sciaenidae*	*Boesemania*	benthopelagic	100.0 SL	3.7	Native	NT	02/26/11
14	*Oxyeleotris marmorata*	Marble goby	Actinopterygii	Perciformes	*Eleotridae*	*Oxyeleotris*	demersal	65.0 SL	3.9	Native	LC	08/24/18
15	*Channa micropeltes*	Giant Snakehead	Perciformes	Channidae	*Channidae*	*Channa*	benthopelagic	130.0 SL	3.8	Native	LC	12/03/19
16	*Channa striata*	Striped snakehead	Actinopterygii	Perciformes	*Channidae*	*Channa*	benthopelagic	100.0 SL	3.6	Native	LC	01/19/11
17	*Cynoglossus microlepis*	Smallscale tonguesole	Actinopterygii	Pleuronectiformes	*Cynoglossidae*	*Cynoglossus*	demersal	32.5 SL	3.5	Native	LC	01/19/11
18	*Tenualosa thibaudeaui*	Mekong herring	Actinopterygii	Clupeiformes	*Clupeidae*	*Tenualosa*	pelagic	30.0 SL	2	Native	VU	02/22/11
19	*Lycothrissa crocodilus*	Sabretoothed Thryssa	Actinopterygii	Clupeiformes	*Engraulidae*	*Lycothrissa*	pelagic	30.0 SL	3.7	Native	LC	12/02/19
20	*Probarbus jullieni*	Jullien's Golden Carp	Actinopterygii	Cypriniformes	*Cyprinidae*	*Probarbus*	demersal	150.0 SL	3.2	Native	CR	01/29/19
21	*Probarbus labeamajor*	Thicklipped Barb	Actinopterygii	Cypriniformes	*Cyprinidae*	*Probarbus*	benthopelagic	150.0 SL	2.5	Native	EN	02/25/11
22	*Probarbus labeaminor*	Thinlip barb	Actinopterygii	Cypriniformes	*Cyprinidae*	*Probarbus*	Wetlands (inland)	–	–	–	NT	02/25/11
23	*Catlocarpio siamensis*	Giant barb	Actinopterygii	Cypriniformes	*Cyprinidae*	*Catlocarpio*	benthopelagic	300.0 TL	2.9	Native	CR	04/05/11
24	*Pangasianodon gigas*	Mekong giant catfish	Actinopterygii	Siluriformes	*Pangasiidae*	*Pangasianodon*	benthopelagic	300.0 TL	2.3	Native	CR	04/13/11
25	*Bagarius yarrelli*	–	Actinopterygii	Siluriformes	*Sisoridae*	*Bagarius*	Wetlands (inland)	–	–	–	VU	06/06/19
26	*Glyptothorax fuscus*	–	Actinopterygii	Siluriformes	*Sisoridae*	*Glyptothorax*	Wetlands (inland)	–	–	–	LC	06/04/19

27	*Glyptothorax lampris*	Torrent catfish	Actinopterygii	Siluriformes	*Sisoridae*	*Glyptothorax*	benthopelagic	12.1 SL	3.2	Native	LC	05/06/11
28	*Datnioides microlepis*	Finescale tigerfish	Actinopterygii	Perciformes	*Datnioididae*	*Datnioides*	benthopelagic	45.0 SL	3.6	Native	LC	08/09/19
29	*Cyprinus carpio*	Common carp	Actinopterygii	Cypriniformes	*Cyprinidae*	*Cyprinus*	benthopelagic	120.0 TL	3.4	Exotic	VU	01/01/08
30	*Hypophthalmichthys molitrix*	–	Actinopterygii	Cypriniformes	*Cyprinidae*	*Hypophthalmichthys*	Wetlands (inland),Arti-ficial/Aquatic	–	–	Exotic	NT	01/20/11
31	*Piaractus brachypomus*	–	Actinopterygii	Characiformes	–	–	–	–	–	Exotic	–	–
32	*Oreochromis mossambicus*	Mozambique Tilapia	Actinopterygii	Perciformes	*Cichlidae*	*Oreochromis*	Wetlands (inland),Arti-ficial/Aquatic	–	–	Exotic	VU	10/02/17

Note: LC, VU, NT, EN, and CR denote less concern, vulnerable, near-threatened, endangered, and critically endangered, respectively. The sign "–" denotes not available; SL, standard length (cm); TL, total length (cm); trophic level, the trophic position of a fish in the food web. Source [a] ICUN Red List, [b] FishBase (Froese and Pauly 2021), * Synonyms

Til, Kampong Khlaeng, Kampong Phluk, Chong Khneas, Me Chrey, Prek Toal, Prey Chas, Peam Bang, and Phat Sanday. Approximately 6520 cages and 210 ponds have been designed for fish aquaculture, and 650 cages are for crocodile farms. The main fish species reared include *C. striata*, *Channa micropeltes*, *Clarias* spp., *Pangasius* spp., and *Hemibagrus wyckioides*. Fish aquaculture around TSL produces around 14,000 tons/year and can be sold fresh for consumption.

33.3 Reptiles

It has been reported that 46 species of reptiles inhabit TSL's ecosystem (Bonheur and Lane 2002). The Tonle Sap Biosphere Reserve supports internationally significant populations of at least eight globally threatened reptile species, including the critically endangered Siamese crocodile *Crocodylus siamensis*; at least six species of freshwater turtle, including the endangered yellow-headed temple turtle *Heosemys annandalii*; and the near-threatened Burmese python *Python molurus* (Davidson 2006).

The majority of snakes are semi-aquatic homalopsid water snakes, previously known as *homalopsines* (Brooks et al. 2007; Lawson et al. 2005; Saint Girons and Pfeffer 1972). Brooks et al. (2007) reported 11 species of snake from 5 families, viz., Homalopsidae (*Enhydris enhydris*, *Enhydris longicauda* (Fig. 33.4a), *Homalopsis buccata*, *Erpeton tentaculatum*, *Enhydris bocourti*, and *Enhydris plumbea*), Colubridae (*Xenochrophis piscator* and *Cylindrophis ruffus*), Elapidae (*Naja kaouthia/siamensis*), Boidae (*Python molurus*), and Acrochordidae (*Acrochordus granulatus*). Most of the species are restricted to vegetated areas, with their catches higher in the flooded grassland than in other habitats, except for *Erpeton tentaculatum* and *E. plumbea*. Species such as *E. enhydris* and *Cylindrophis ruffus* were very occasionally caught in the open water of the lake. Only three individuals of *Acrochordus granulatus* were recorded in the open lake (Brooks et al. 2007). *E. plumbea* showed the highest catch per unit effort in deep water. All of the other species showed significant negative relationships between catch per unit effort and water depth.

Crocodile (*Crocodylus siamensis* or local name Kra peu trey) is present in both the crocodile farm and the TSL floodplain. Within the TSL basin, there has been a substantial boom in the local crocodile farm industry, and approximately 600 crocodile farms exist around TSL. Local people breed and rear the native Siamese crocodile (*Crocodylus siamensis*, Fig. 33.4c) and the non-native Cuban crocodile (*Crocodylus rhombifer*) for commercial purposes (Campbell et al. 2006).

Varanus salvator (local name: An sorng, Fig. 33.4d) lives in the flooded forests around TSL, spawning in the hole of big trees during the rainy season. Around TSL, this species has become very rare and is caught by the fisherman only in the Prek Toal Biosphere Reserve. This species has been classed in CITES Appendix II of the Convention in International Trade in Endangered Species of Wild Fauna and Flora (CITES), for trade control to avoid utilization incompatible to its survival.

Fig. 33.4 (**a**)[1] Tonle Sap water snake *Enhydris longicauda* (local name: Pous ph'ab), (**b**)[1] Yellow-Headed Temple Turtle *Heosemys annandalii* (local name: Song Kal), (**c**)[1] Crocodile *Crocodylus siamensis (*local name*:* Kra peu trey), and (**d**)[2] *Varanus salvator* from a camera trap in 2015 (local name: An sorng). Photo credit ([1]author; [2]Sun Visal)

At least six species of turtles have been confirmed in TSL (Davidson 2006). In 2018–2019, TSA inventoried different turtles in TSL through surveys in floating villages with people living there since the 1980s. As a result, seven turtle species were identified, viz., yellow-headed temple turtle *Heosemys annandalii*, rice-field terrapin *Malayemys subtrijuga* and Asiatic soft-shell turtle, Asian box turtle *Cuora amboinensis*, black mash turtle *Siebenrockiella crassicollis*, Asiatic soft-shell turtle *Amyda cartilaginea*, and P'dao kramoun (no scientific name available yet). The spawning period of most turtles starts from November to January when the water level in TSL recedes and in the rainy season from July to August for the Asian soft-shell turtle.

33.4 Mollusca and Arthropod

After fish, Mollusca are the second most important contributor to food provisioning in TSL (Rainboth 1996). According to MRC (2003), bivalves (e.g., Asian clam *Corbicula* spp., Fig. 33.5a) are abundant in the dry season, but people start harvesting it at the end of the rainy season, whereas gastropods (e.g., apple snails *Pila* spp.) are abundant in the beginning of the rainy season and in the beginning of the dry season, when floodwaters recede and the water temperature reaches its

Fig. 33.5 (**a**) Asian clam *Corbicula moretiana* (local name, leas; class, Bivalvia; family, Corbiculidae), (**b**) snail *Pila polita* (local name: k'chorng boeng), (**c**) freshwater shrimp *Neocaridina* spp. (local name, Kam peus touch/angkam; family, Atyidae), and (**d**) crab *Siamthelphusa improvisa* (local name, Kdam mouk yeak; class, Malacostraca; family, Gecarcinucidae)

minimum in the year. The report from Ting et al. (2020) recorded 33 Crustacea and Mollusca species in TSL.

Our surveys in TSL from 2018 to 2019 found 31 species, 15 bivalves (5 families), and 16 gastropods (8 families), including three new records for Cambodia (i.e., *Scaphula minuta*, *Novaculina siamensis*, and *Wattebledia siamensis*), a globally invasive species such as *Pomacea maculata* and *Limnoperna fortune*, which modifies the presence and abundance of native macro-invertebrate fauna and fish diets and changes the ecological condition (Boltovskoy 2015).

Until now, the estimation of catch and trade of Gastropoda in TSL has not been assessed. The preliminary inventory by TSA on the caught snail for sale at the local markets and for export from the six provinces around the lake indicated that snails were caught by rudiment and a landing net. The period of snail collection starts in the beginning of the rainy season when the water levels rise. Five species from two families were identified, including Viviparidae [*Melongirapongeasis aeruginasesaiy* (local name: Ka chav), *Filopaludina martensi cambodjensis* (local name: Ka chav Ka'ek/Ka chav kut srouch), *Mekongina phaericula* (local name: Ka chav doung)] and Pilidae [*Pila polita* (local name: k'chorng boeng, Fig. 33.5b) and *Pila scutata* (k'chorng boeng krang/kut teal)].

Freshwater shrimps are dominant in the upper and the middle of TSL and near the organic substrate-rich shoreline. Their exploitation is typically practiced from December to June when the water level of TSL is between 2 and 3 m. Four main species were recorded: *Macrobrachium nipponense* (Kam peus), *Macrobrachium lotidachylus* (Kam peus), *Macrobrachium ohione* (Kam peus dai thom), and *Neocaridina* spp. (Kam peus touch/angkam) (Fig. 33.5c). The freshwater shrimp is sold fresh for consumption and processed as dried shrimp. In 2019, the estimated total annual catch of Crustacea (e.g., Crab *Siamthelphusa improvisa*, Fig. 33.5d) in TSL was approximately 59 tons. Different fishing gears and traps are mostly brush bundle for shrimp.

Key Points

- TSL and its surrounding floodplains form heterogeneous habitats, including flooded forests, shrubs, grasslands, and aquatic vegetations. These provide the variety of habitat for aquatic fauna to accomplish a large part of their life cycle and support high biodiversity and productivity.
- For fish, 167 species, belonging to 12 orders, 35 families, and 94 genera, were identified in the inventory on fish biodiversity in TSL and its floodplains in 2010–2019, approx. 75% of which belong to the order Cypriniformes, Siluriformes, and Perciformes.
- The migratory fish species access the seasonally flooded habitats to feed and grow during the wet season, whereas the floodplain-resident or non-migratory fish species spend most of their lifespans in the flooded forests.
- TSL is also home to at least 15 species of bivalves, 16 gastropods, 11 snake species, 7 turtle species, and other arthropod species.

References

Arias ME, Cochrane TA, Kummu M, Lauri H, Koponen J, Holtgrieve GW, Piman T. Impacts of hydropower and climate change on drivers of ecological productivity of Southeast Asia's most important wetland. Ecol Model. 2013;272:252–63.

Boltovskoy D. Ecology and environmental impact of *Limnoperna fortunei*: introduction. In: Boltovskoy D, editor. Limnoperna Fortunei. Invading nature, Springer series in invasion ecology, vol. 10. Cham: Springer; 2015.

Bonheur N, Lane DB. Natural resource management for human security in Cambodia's Tonle Sap Biosphere Reserve. Environ Sci Pol. 2002;5:33–41.

Brooks ES, Allison HE, Reynold DJ. Vulnerability of Cambodian water snakes: Initial assessment of the impact of hunting at Tonle Sap Lake. Biol Conserv. 2007;139:401–14.

Campbell I, Poole C, Giesen W, Valbo-Jorgensen J. Species diversity and ecology of Tonle Sap Great Lake, Cambodia. Aquat Sci. 2006;68(3):355–73.

Chan B, Brosse S, Hogan ZS, Ngor BP. Influence of local habitat and climatic factors on the distribution of fish species in the Tonle Sap Lake. Water. 2020;12(3):786.

Davidson PJA. Biodiversity of the Tonle Sap Biosphere Reserve, 2005 status review; 2006.

Froese R, Pauly D, eds. FishBase. World Wide Web electronic publication. www.fishbase.org, version (06/2021); 2021.

IUCN. The IUCN Red List of Threatened Species. Version 2021-1.https://www.iucnredlist.org. ISSN 2307-8235; 2021.

Lawson R, Slowinski JB, Crother BI, Burbrink FT. Phylogeny of the Colubroidea (Serpentes): new evidence from mitochondrial and nuclear genes molecular. Phylogenet Evol. 2005;37:581–601.

Legris P, Blasco F. Notice de la carte: Cambodge International du Tapis Vegetal. Extrait des travaux de la Section Scientifique et Technique de l'Institute Francaise de Pondichery, Toulous: hors serie no 1, Toulouse, France; 1972.

MRC. State of the Basin Report 2003. Phnom Penh: Mekong River Commission; 2003.

Nam S, Vann LS, Baran E, Authur R.. An Evaluation of Fish Species and Genetic Diversity of the Tonle Sap Great Lake, Cambodia. Keynote speech at the International Workshop and Training on Fish Diversity of the Mekong River. Tohoku University, Sendai, Japan; 2006.

Ngor PB, Grenouillet G, Phem S, So N, Lek S. Spatial and temporal variation in fish community structure and diversity in the largest tropical flood-pulse system of South-East Asia. Ecol Freshw Fish. 2018;27:1087–100.

Poulsen AF, Ouch P, Viravong S, Suntornratana U, Nguyen TT. Fish Migrations of the Lower Mekong River Basin: Implications for Development, Planning and Environmental Management; Mekong River Commission: Phnom Penh, Cambodia, 2002; p. 62. ISSN 1683-1489.

Poulsen AF, Hortle KG, Chan S, Chhuon CK, Viravong S, Bouakhamvongsa K, Suntornratana U, Yoorong N, Nguyen TT, Tran BQ. Distribution and Ecology of Some Important Riverine Fish Species of the Mekong River Basin; Mekong River Commission: Phnom Penh, Cambodia; 2004. Vol. 89. ISSN 1683-1489.

Rainboth WJ. FAO species identification field guide for fisheries purposes. Fishes of the Cambodian Mekong. Room, Food and Agriculture Organisation of the United Nations; 1996. p. 265.

Saint Girons H, Pfeffer P. Notes sur l'ecologie des serpents du Cambodge Zoologische Mededelingen. Leiden: Rijkmuseum van Natiirlijke Historie; 1972. p. 65–86.

Ting HN, Ekgachai J, Chirasak S, Chhuoy S, Pin K, Arthit P, Warut S, Ruttapon S, Hogan ZS, Ngor PB. Annotated checklist of freshwater molluscs from the largest freshwater lake in Southeast Asia. Zookeys. 2020;958:107–41.

Chapter 34
Waterbirds

Ly Sophanna, Uk Sovannara, Sun Visal, Son Virak, Hong Chamnan, Seng Bunthoeun, Taing Porchhay, Pham Ngoc Bao, and Srey Sunleang

34.1 Waterbirds and Their Habitat

Tonle Sap Lake (TSL) is one of the largest waterbird conservation sites in the region, owing to its unique habitat, both in structure and floristic composition, driven largely by the seasonal flooding regime (Rundel 2000) (see Chaps. 1, 31, and 32). The productivity of the TSL floodplain and the increased accessibility of this enormous and bountiful feeding ground in the dry season (see also Chap. 33) are crucial for a wide range of waterbirds. When the water recedes during the dry season, fish and

---

Supplementary Information The online version of this chapter (https://doi.org/10.1007/978-981-16-6632-2_34) contains supplementary material, which is available to authorized users.

L. Sophanna (✉)
Ministry of Environment, Phnom Penh, Cambodia

Tokyo Institute of Technology, Tokyo, Japan
e-mail: ly.s.aa@m.titech.ac.jp

U. Sovannara
Tokyo Institute of Technology, Tokyo, Japan

S. Visal · S. Virak · H. Chamnan
Ministry of Environment, Phnom Penh, Cambodia

Wildlife Conservation Society, Phnom Penh, Cambodia

S. Bunthoeun
Ministry of Environment, Phnom Penh, Cambodia

Asian Institute of Technology, Khlong Nueng, Thailand

T. Porchhay · S. Sunleang
Ministry of Environment, Phnom Penh, Cambodia

P. N. Bao
Institute for Global Environmental Strategies, Hayama, Japan

C. Yoshimura et al. (eds.), *Water and Life in Tonle Sap Lake*,
https://doi.org/10.1007/978-981-16-6632-2_34

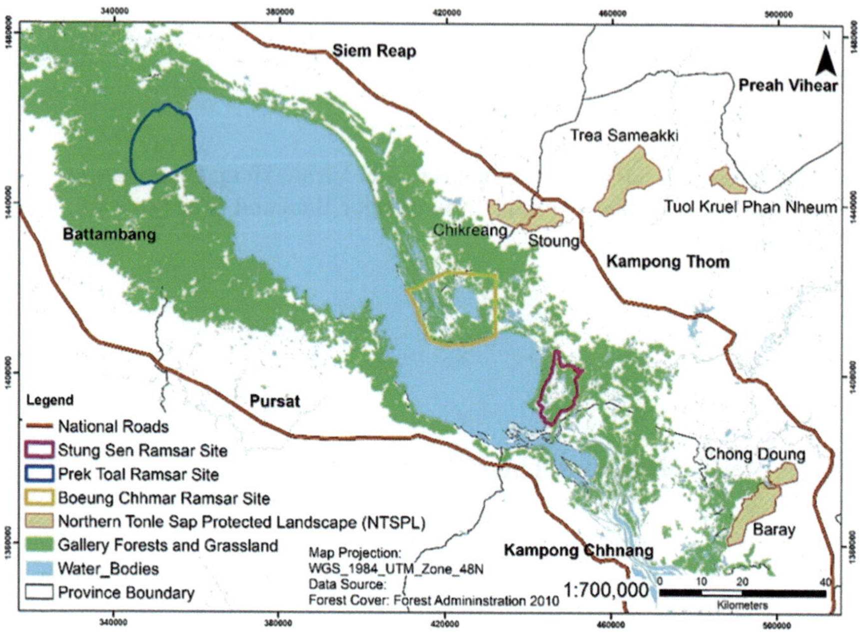

Fig. 34.1 Locations of the protected areas of the three Ramsar sites and the Northern Tonle Sap Protected Landscape

many other aquatic organisms are increasingly confined to isolated ponds scattered throughout the floodplain, and it becomes easier for waterbirds to catch them as the water levels decline.

Prek Toal, also known as Prek Toal Ramsar site, harbors dense flooded forests, which are optimal habitats for various waterbirds (Figs. 34.1 and 34.2, Supplementary Videos 34.1 and 34.2). The main compositions of these habitats include different gallery forest species, shrubs, herbaceous communities, and exposed muds along the shoreline during the dry season (see Chap. 32). According to our continuous monitoring effort between 2004 and 2018, 135 species of birds have been found in Prek Toal. Among them, five are globally threatened, and another five are near-threatened species according to the International Union for Conservation of Nature. The five globally threatened waterbird species in Prek Toal are the milky stork *Mycteria cinerea*, greater adjutant *Leptoptilos dubius*, lesser adjutant *Leptoptilos javanicus*, spot-billed pelican *Pelecanus philippensis*, and masked finfoot *Heliopais personatus*. The five near-threatened species are the oriental darter *Anhinga melanogaster*, painted stork *Mycteria leucocephala*, black-necked stork *Ephippiorhynchus asiaticus*, black-headed ibis *Threskiornis melanocephalus*, and grey-headed fish eagle *Ichthyophaga ichthyaetus*. The Prek Toal waterbird colonies are the only remaining breeding stronghold for the globally threatened spot-billed pelican and milky stork in mainland Southeast Asia (Sun and Mahood 2015). In addition, it is the largest remaining breeding site in the region for the oriental darter,

Fig. 34.2 A flock of oriental darter on gallery forests in November 2019 in the Prek Toal Ramsar site

lesser adjutant, greater adjutant, black-headed ibis, painted stork, and grey-headed fish eagle. These waterbird species are short-distance migrants, migrating within Cambodia and neighboring countries (see Sect. 34.2 for their distribution).

Both Stung Sen and Boeung Chhmar Ramsar sites (Fig. 34.1) are also home to various waterbird species including globally threatened and near-threatened species. The sites share similar ecosystem characteristics. Interestingly, the habitat of Stung Sen is different from the other two Ramsar sites in TSL. It is actually characterized by the presence of distinctive species of tall gallery forests, namely, *Garcinia loureirin*, *Crudia chrysantha*, and *Samandura harmandii* (see Chap. 32, Fig. 32.1a). Those gallery forests and surrounding ecosystems accommodate various key waterbird species including the grey-headed fish eagle, oriental darter, black-headed ibis, spot-billed pelican, lesser adjutant, and many others. Boeung Chhmar, by contrast, along the landlocked waterways, developed large and dense gallery forests between Peam Bang and Moat Khla, most of which are mature and closed canopy forests.

The ecosystem of the lowland Northern Tonle Sap Protected Landscape (NTSPL) (31,159 ha) (Fig. 34.1) is highly diverse, including dominated vast inundated grasslands (predominantly *Saccharum spicatum*, *Paspalum scrobiculatum*, and *Imperata cylindrica*) and densely distributed scrub (predominantly *Gmelina asiatica*) mixed with grasslands, rice fields, natural ponds, impounded open water, small artificial canals, and fishery-based systems. Such diverse ecosystems are favorable feeding grounds for bird species, including critically endangered Bengal

Florican *Houbaropsis bengalensis*, white-shouldered ibis *Pseudibis davisoni*, globally vulnerable sarus crane, greater adjutant, lesser adjutant, and black-necked stork.

34.2 Characteristics and Distribution of Waterbirds

Most waterbirds inhabit the colonies in Prek Toal during their breeding season, whereas some other bird species whose breeding grounds are grasslands (e.g., Bengal Florican) appear in the lowland NTSPL. The pattern of movement after the breeding season varies strongly depending on the species and seasonal flooding. The characteristics and distribution of major large bird species in TSL are summarized in this section.

With the exception of the Bengal Florican and sarus crane, the home ranges of many important waterbirds are poorly understood after the breeding season or during the elevated water period. A certain proportion of some species stay at TSL (e.g., the oriental darter and the Asian openbill). Those leaving TSL were concluded to migrate across Cambodia to neighboring countries such as Vietnam and Thailand. There remains a gap in knowledge of their actual distribution and seasonal movement. The lifespan of those important waterbirds is also unknown in Cambodia. Therefore, further studies are required to understand their home range and lifespan to better manage their population both in breeding and non-breeding seasons.

Bengal Florican (*Houbaropsis bengalensis*), a "critically endangered" bustard. Its existence is restricted to the dry season. The main conservation area of the species is the floodplain grasslands of TSL (Gray et al. 2009), particularly NTSPL. During the water elevated period in TSL, this species migrates to the higher lowland in the upper part of NTSPL, namely, Trea Sameakki and Tuol Kruel Phan Nheum (Fig. 34.1). Its breeding season is from March to August, with 1–2 eggs per nest on grassland. Florican typically feeds on insects such as grasshoppers and beetles, crabs, snakes, shoots, and greases (see also Sect. 34.4).

Sarus crane (*Grus Antigone*), "vulnerable." NTSPL of TSL's floodplain is one of the feeding sites for cranes in the non-breeding season. During the dry season, the species migrates to various feeding sites across Cambodia, namely, NTSPL, Ang Trapeang Thmor, Boeung Prek Lapouv (Sophanna et al. 2019), Anlung Pring-protected landscapes, and Phu My Grassland Nature Reserve in Vietnam. Its breeding nest is made of grasses deposited within very shallow water from June to September, with 1–4 eggs per nest. Cranes feed on the tubers of water chestnut (*Eleocharis dulcis*), insects, small fish, fallen rice grains, and other tubers.

Milky stork (*Mycteria cinerea*), "endangered." This species is primarily a coastal bird, but it is also found in small numbers inland. In Cambodia, the habitats of this bird are mudflats of tidal areas, mangrove forests, and flooded forests, especially in TSL. This species is present all year-round in Cambodia, particularly in Prek Toal, where it breeds during the breeding season. Its breeding season starts from December to April, and it lays 1–4 eggs per nest. This species feeds on fish, frogs, crustaceans, and amphibians.

Masked finfoot (*Heliopais personatus*), "endangered." Cambodia supports one of the most important worldwide populations of this species (Chowdhury et al. 2020). It inhabits forests along rivers, streams, lakes, flooded forests, and mangrove forests. Occurring all year in Cambodia, this species is recorded in forested creeks in the semi-evergreen forests in the northern plains of the country, the floodplains of TSL (Chowdhury et al. 2020). Its breeding season starts from June–October, having 4–6 eggs per nest. Its diet includes small fish, invertebrates, crustaceans, and amphibians.

Greater adjutant (*Leptoptilos dubius*), an "endangered" large stork. Its habitats are wet grasslands, flooded forests, ponds, and rice fields. This species resides year-round across Cambodia, occurring in the highest numbers in Prek Toal in the dry season, with peak nest counts mostly in March–April (Sun and Mahood 2015). Its breeding season is between October and March, with 2–4 eggs per nest. This species feeds on frogs, reptiles, fish, large insects, small mammals (e.g., rats), and birds.

Lesser adjutant (*Leptoptilos javanicus*), a "vulnerable" large stork. The habitats of this bird are mangrove forests, muddy fields, and flooded forests, specifically in TSL. The species lives year-round in Cambodia and is largely distributed in low-lands across the county, mostly occurring in Prek Toal in the dry season, with peak nest counts between February and April. Its diet is largely similar to that of greater adjutant. Its breeding season starts from September to June, with 2–4 eggs per nest.

34.3 Waterbird Community in the Prek Toal Ramsar Site

During the 15-year period (2004–2018) of our record, the total number of breeding pairs of most of the important waterbird species in Prek Toal indicated a significant increase (Fig. 34.3a). Contradictory to their global status either as threatened or near-threatened (Sun and Mahood 2015), such species increased in Prek Toal as a result of better monitoring at TSL compared to 2004. This highlights the importance and success of conservation acts such as the establishment of TSL as UNESCO Biosphere Reserve in 2001 by the Royal Decree of the Royal Government of Cambodia, the collaboration between the Ministry of Environment and the Wildlife Conservation Society since 2001, and the designation of Prek Toal as the Ramsar site in 2015. These, through monitoring, patrols, and law enforcement, have positively contributed to conservation of the species' breeding grounds and habitats but are also likely a result of the degradation of other sites in the region and displacement of populations (Hughes 2017), possibly coming to TSL.

Numbers of greater adjutant, lesser adjutant, oriental darter, and Asian openbill have substantially fluctuated over the 11-year period (2008–2018). This fluctuation might have been from their out-migration. Not all populations of these species stay in Prek Toal, as some of them move out after the breeding season or when the water levels rise. Staying at or moving out from Prek Toal or TSL does not ensure a high survival rate for the species. Factors related to human activities, such as hunting, egg theft from the nests, foraging ground loss due to land conversion, and poisoning in

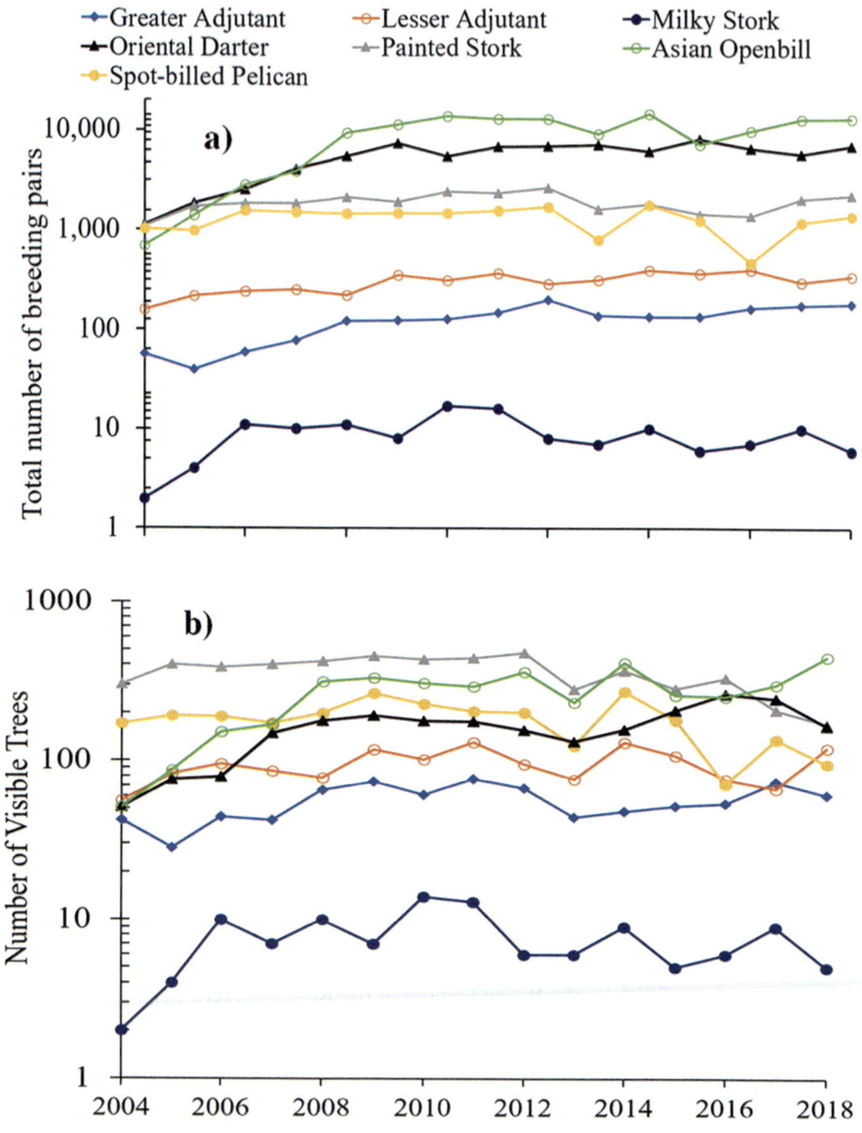

Fig. 34.3 (**a**) The total number of breeding pairs of key waterbirds in log scale in Prek Toal during 2004–2018 and (**b**) the number of visible trees (nesting trees) in log scale in Prek Toal during 2004–2018

some areas within the TSL floodplain (Bonheur and Lane 2002; Mahood et al. 2020a), may have declined their population. We noticed that some of the waterbirds had been illegally hunted at the edge of the TSL floodplain when they temporarily inhabited it during the elevated water period.

The number of visible trees (nesting trees that allow recording of waterbirds from a distance using a telescope from the counting platform) used between 2004 and 2018 has increased for the nesting species of the Asian openbill, oriental darter, lesser adjutant, milky stork, and greater adjutant (Fig. 34.3b). As such, these key waterbird species used more visible trees as their total number of breeding pairs also significantly elevated (Fig. 34.3a). The number of visible trees used for nesting only decreased during this period for painted stork and spot-billed pelican. However, the total number of breeding pairs of these two key waterbirds increased (Fig. 34.3a), meaning that the number of nests per visible tree for these species has also increased, respectively, from 3.6 and 6 nests per visible tree in 2004 to 13 and 14 nests per visible tree in 2018.

There is a significant positive relationship between the number of visible trees and the total number of breeding pairs of those seven waterbird species ($p < 0.001$; $r = 0.57$; 95% confidence interval (CI), 0.42–0.69). It is likely that when there are more visible trees, there is a greater number of breeding pairs coming in TSL for nesting. Preserving the habitats for tree-nesting waterbirds will contribute to a suitable conservation of nesting habitat and may increase the population of nesting birds, thereby catalyzing the possibility of having new colony sites (Baker et al. 2015). This showcases the importance of the flooded forests (i.e., gallery forests) as the ideal habitats for those key waterbird species and, perhaps, other species.

34.4 Bengal Florican

TSL floodplain supports the critically endangered Bengal Florican *Houbaropsis bengalensis*, which is the last extant population of the Indochinese subspecies *blandini* (Packman et al. 2014). Bonheur and Lane (2002) categorized Bengal Florican as a waterbird; however, this species is in fact not included in the Asian waterbird census regardless of its status (Mundkur et al. 2017). This means that it is not a waterbird, but only a grassland bird inhabiting the grassland of a wetland area. Its population was estimated to be between 350 and 1500 individuals, residing only in three countries, namely, Cambodia, India, and Nepal (Birdlife International 2021). Of those estimated individuals, 432 individuals were estimated to be in Cambodia in 2012 (Packman et al. 2014), almost one-third of the global population size of this species. However, its latest estimated population number in Cambodia was 138 individuals recorded, with a 95% CI of 119–156, in 2018, and a 3:1 sex ratio for males to females for the 104 displaying males (Mahood et al. 2020b) (Fig. 34.4).

The population of the total displaying male (capable to reproduce) Bengal Florican in Cambodia was found to decline by between 44% and 64%, down from 376 (95% CI, 293–462) in 2006 to 216 (95% CI, 156–275) in 2012 (Packman et al. 2014) (Fig. 34.4).

Based on the recent estimate in 2018, its rate of decline was 52%, a decline compared to that recorded in 2012. The number of displaying male Bengal Floricans in NTSPL has fluctuated since recording began in 2009 (Fig. 34.4), starting at below

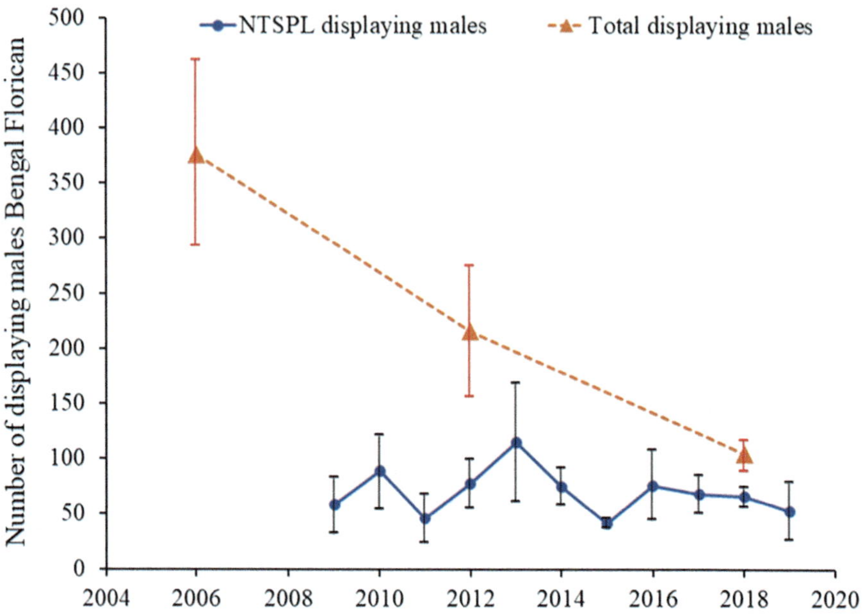

Fig. 34.4 The number of the total displaying males in Cambodia and displaying males Bengal Florican recorded in NTSPL in 2009–2019; error bars denote 95% confident interval of the number of displaying male Bengal Floricans

60 in 2009 to a sharp increase at 113 in 2013, followed by a dramatic drop to 41 individuals in 2015. Of note, NTSPL supported around 63% of the total population number of Bengal Florican estimated in 2018 in Cambodia (Fig. 34.4).

The last four recorded years (2016–2019) still, nonetheless, indicated a sustained decreasing trend, which could be a significant concern for the conservation of this species. Such concern can be exacerbated when the habitats and surrounding ecosystems of TSL have been degraded as a result of flooded forest fires (see also Chap. 48 for the reasons of forest fires) and the intensification and expansion of rice cultivation (Mahood et al. 2020a). Between 1993 and 2018, the scrubland and grassland cover remarkably decreased from ~8660 to ~6776 km^2 and from ~3160 to ~519 km^2, respectively. Based on the survey data of previous studies conducted across Cambodia in 2006 (Gray et al. 2009), 2012 (Packman et al. 2014), and 2018 (Mahood et al. 2020b), statistically, the number of displaying male Bengal Floricans has a significant positive relation with the grassland cover they inhabit ($p < 0.001$; $r = 0.48$; 95% CI, 0.24–0.67). Preserving more grassland for this species will contribute to the recovery of its population; otherwise, the rate of decline will continue to drop, and that may lead to the extinction of this species in Cambodia in the near future.

We observed an effective conservation act in Prek Toal regarding important waterbird species recorded from 2004. This ensures that those key species continue to use the wetland. However, their habitats in the TSL floodplain, especially the grasslands of NTSPL, are under threat from direct human activities (i.e., forest fires and land conversion). The indirect threats, such as climate change, transboundary issues, and pollutant load from the tributaries and rice fields, may also be perceived as threats to the aforementioned waterbird species. The hydrological alteration caused by climate change and infrastructure development will also significantly affect the flooded forest ecosystem (Arias et al. 2013, 2014) (see also Chap. 32). It will decrease the size of gallery forests and flooded grasslands in the TSL floodplain. Changes to habitat will have wider implications on the food web and interactions of species, so there are also indirect implications beyond the waterbirds being discussed here. These interactions in TSL are poorly understood at present (Arias et al. 2014) but required for maintaining the biodiversity and the intact ecosystem.

The intensification of rice cultivation in the TSL floodplain is elevating the use of chemical fertilizers and pesticide (Chap. 37), which could harm various ecosystems in and around TSL through the food chain. Studied fish species in TSL have a bioaccumulated higher concentration of heavy metals than the maximum permissible limit set by the World Health Organization (Sokneang et al. 2021). The studied fish at a higher trophic level tend to show a higher concentration of heavy metals, particularly *Anabas testudineus* (see also Chaps. 36 and 38). In addition, Chanvorleak et al. (2021) found fungicides, of either biphenyl or fluquinconazole, in the four most dominant fish species of the Cyprinidae family in TSL. Because large waterbirds in Prek Toal feed on fish, they may be exposed to those pollutants. A higher heavy metal content in birds has been identified to have caused reproductive impairment in birds, and they affect fledgling success, resulting in a slower growth rate for chicks, nestling mortality, and a stress response (Baos et al. 2006; Liu et al. 2015; Scheuhammer et al. 2001; Spahn and Sherry 1999).

These direct and indirect threats potentially put these species under severe threats if there are no countermeasures. Therefore, further studies are required to understand the multi-species interactions in TSL in the context of the habitat change and the chemical pollution on large waterbirds and other bird species.

Key Points

- Totally, 135 bird species were recorded in Prek Toal between 2004 and 2018.
- Gallery forests, shrublands, grasslands, and muds along the shoreline and isolated ponds scattered in the floodplain of TSL are favorable feeding, roosting, and breeding grounds for large waterbirds, particularly in the dry season.
- The increased population in key large waterbirds in TSL indicates effective conservation acts since 2001. However, it is also a possible result of the degradation of other sites in the region and displacement of populations.
- TSL is home to the critically endangered Bengal Florican *Houbaropsis bengalensis*, whose population in Cambodia was one-third of the global population size in 2012. However, the population of displaying males has encountered a continuous decline.
- Both direct and indirect impacts on TSL's ecosystem will potentially expose large waterbird and other bird species in TSL to severe threats.

Acknowledgments The authors are thankful to Eang Phallis and Neab Samneang, officers of the Department of Freshwater Wetlands Conservation of the Ministry of Environment, Cambodia, for providing the supporting documents. The authors also would like to thank the Cambridge University Press for the permission on employing the figure on the estimate of population of displaying male Bengal Floricans in Cambodia, cited in this chapter as Mahood et al. (2020b).

References

Arias ME, Cochrane TA, Norton D, Killeen TJ, Khon P. The flood pulse as the underlying driver of vegetation in the largest wetland and fishery of the mekong basin. Ambio. 2013;42(7):864–76.

Arias ME, Cochrane TA, Elliott V. Modelling future changes of habitat and fauna in the Tonle Sap wetland of the Mekong. Environ Conserv. 2014;41(2):165–75.

Baker NJ, Charles D, Dieter B, K. K. Reproductive Success of Colonial Tree-nesting Waterbirds in Prairie Pothole Wetlands and Rivers throughout Northeastern South Dakota. Am Midl Nat. 2015;174(1):132–49.

Baos R, Blas J, Bortolotti GR, Marchant TA, Hiraldo F. Adrenocortical response to stress and thyroid hormone status in free-living nestling white storks (Ciconia ciconia) exposed to heavy metal and arsenic contamination. Environ Health Perspect. 2006;114(10):1497–501.

BirdLife International. Species factsheet: *Houbaropsis bengalensis*. Downloaded from http://www.birdlife.org on 08/02/2021; 2021.

Bonheur N, Lane BD. Natural resources management for human security in Cambodia's Tonle Sap biosphere reserve. Environ Sci Policy. 2002;5(1):33–41.

Chanvorleak P, Kearakvattey K, Vorleak P, Sereyvath Y, Eden GM, Winarto K, Hirofumi H. Assessment of Pesticide Residues in Surface Water and Fish from Chhnok Tru, Kampong Chhnang. Insights and Challenges toward Achievings SDGs, 2020. Phnom Penh, Cambodia; 2021.

Chowdhury SU, Yong DLI, Round PD, Mahood S, Tizard R, Eames JC. The status and distribution of the Masked Finfoot Heliopais personatus — Asia's next avian extinction? (December); 2020.

Gray TNE, Collar NJ, Davidson PJA, Dolman PM, Evans TD, Fox HN, Chmanan H, Borey R, Kimhout S, van Zalinge RN. Distribution, status and conservation of the Bengal Florican Houbaropsis bengalensis in Cambodia. Bird Conserv Int. 2009;19(1):1–14.

Hughes AC. Understanding the drivers of Southeast Asian biodiversity loss. Ecosphere. 2017;8(1): e01624.

Liu J, Liang J, Yuan X, Zeng G, Yuan Y, Wu H, Huang X, Liu J, Hua S, Li F, Li X. An integrated model for assessing heavy metal exposure risk to migratory birds in wetland ecosystem: a case study in Dongting Lake Wetland, China. Chemosphere. 2015;135:14–9.

Mahood SP, Poole CM, Watson JEM, Sharma S, Garnett ST, Mackenzie RA. Agricultural intensification is causing rapid habitat change in the Tonle Sap Floodplain, Cambodia. Wetl Ecol Manag. 2020a;28(5):713–26.

Mahood SP, Hong C, Virak S, Sum P, Garnett ST. Catastrophic ongoing decline in Cambodia's Bengal Florican Houbaropsis bengalensis population. Bird Conserv Int. 2020b;30(2):308–22.

Mundkur T, Langendoen T, Watkins D, editors. The Asian Waterbird Census 2008–2015 - results of coordinated counts in Asia and Australasia. Ede: Wetlands International; 2017.

Packman CE, Showler DA, Collar NJ, Virak S, Mahood SP, Handschuh M, Evans TD, Chamnan H, DOLMAN PM. Rapid decline of the largest remaining population of Bengal Florican Houbaropsis bengalensis and recommendations for its conservation. Bird Conserv Int. 2014;24(4):429–37.

Rundel PW. Forest habitats and flora in Lao PDR, Cambodia and Vietnam: Conservation Priorities in Indochina. Hanoi, Vietnam; 2000.

Scheuhammer AM, Perrault JA, Bond DE. Concentrations in Eggs of Common Loons (Gavia Immer). Environ Monit Assess. 2001;72:79–94.

Sokneang I, Sovannmony N, Soukim H, Dung VP, Masateru N, Hasika M, Toru W. Bioaccumulation of heavy metals and trace elements in six fish species from Tonle Sap Lake, Cambodia. Insights and Challenges toward Achievings SDGs, 2020. Phnom Penh, Cambodia; 2021.

Sophanna L, Pok H, Avent T. Climate Change Vulnerability Assessment Boeung Prek Lapouv Protected Landscape, Cambodia. Bangkok, Thailand; 2019.

Spahn SA, Sherry TW. E nvironmental Contamination and Toxicology Cadmium and Lead Exposure Associated with Reduced Growth Rates, Poorer Fledging Success of Little Blue Heron Chicks (Egretta caerulea) in South Louisiana Wetlands. Nat Wildlife. 1999;384:377–84.

Sun V, Mahood S. Wildlife Monitoring at Prek Toal Ramsar Site, Tonle Sap Great Lake, 2013 and 2014 (Vol. 1); 2015.

Part VIII
Chemical Pollution

Chapter 35
Chemical Analysis of Heavy Metals and Pesticides: Pretreatment

Winarto Kurniawan, Phat Chanvorleak, Ty Boreborey, Eden M. Andrews, Kuok Fidero, and Hirofumi Hinode

35.1 Importance of Pretreatment

A suitable analysis method is required to achieve accurate and reliable measurements when analyzing samples. Depending on the properties of the samples, additional pretreatment of the samples is needed before subjecting them to an appropriate analytical procedure to guarantee the accuracy and reliability of results. Without pretreatment, the samples might contain other components that will produce noises during analysis, thus decreasing the signal-to-noise (S/N) ratio and measurement accuracy. These components might also produce additional false signals, which may lead to false-positive results. Moreover, some of the target elements might not be readily detectable by the equipment chosen; thus, it is necessary to increase their detectability through pretreatment.

There are various pretreatment methods available, and based on the purposes of pretreatment, these methods can be categorized as follows:

- Methods to protect instruments.
- Methods to extract target components.
- Methods to remove impurities from the samples.
- Methods to improve S/N ratio (to improve accuracy).

W. Kurniawan (✉) · E. M. Andrews · H. Hinode
Tokyo Institute of Technology, Tokyo, Japan
e-mail: kurniawan.w.ab@m.titech.ac.jp

P. Chanvorleak · T. Boreborey
Institute of Technology of Cambodia, Phnom Penh, Cambodia

K. Fidero
Institute of Technology of Cambodia, Phnom Penh, Cambodia

Ministry of Industry, Science, Technology and Innovation, Phnom Penh, Cambodia

C. Yoshimura et al. (eds.), *Water and Life in Tonle Sap Lake*,
https://doi.org/10.1007/978-981-16-6632-2_35

In the next chapters in this part of the book (Chap. 36), we will discuss the analytical methods used for the measurement of the concentration of pollutants in Tonle Sap Lake, namely, atomic absorbance spectroscopy (AAS) for heavy metals and gas chromatography-mass spectrophotometry (GC-MS) for pesticides. These methods were selected on the basis of the factors affecting their overall performance such as cost, speed, and level of accuracy. Since the basic theory and operation of both AAS and GC-MS are already well established, this chapter will focus only on the pretreatment method used to prepare samples before being subjected to an analytical procedure. The reasons for the selection of the methods are also discussed briefly below.

35.2 Pretreatment for Measurement of Heavy Metals

AAS is widely used to analyze ions in an aqueous solution owing to its versatility, reliability, and simplicity of operation. Furthermore, in the field of environmental water sample analysis, AAS is already an established basic technique in various analytical laboratories (Fishman and Downs 1966). This is because AAS offers sufficient sensitivity for many applications and is relatively interference-free for the determination of most metals and metalloids (Hill and Fisher 2017). Owing to these reasons, the pretreatment methods of water samples for AAS analysis are also well developed and established.

In this research, the focus of AAS analysis was for the measurement of heavy metals in both lake water samples and lake sediment samples. The analytical methodology and results are further presented in Chap. 36. The preparation of lake water samples for measurement was simple and straightforward because the sample required by the equipment is in the form of liquid, and the obtained sample is already in the form of the aqueous solution, thus eliminating the need to perform phase conversion operation. An additional step was required to prepare measurement samples from the lake sediment, that is, to digest the targeted elements before proceeding with the measurements.

35.2.1 Pretreatment of Water Samples

As mentioned before, the water samples came from Tonle Sap Lake, a freshwater lake, and thus the pretreatment procedure to determine metal concentration is straightforward. The matrices that might exist in the lake water samples also pose no or little influence on the detection signal, such as fine particles that can adsorb target heavy metals. Moreover, since the range of concentrations of heavy metals detected was above the limit of detection of AAS (in the order of parts per billion (ppb)), it was not necessary to concentrate the samples. Therefore, the main pretreatment procedure used here was conducted to obtain a total concentration of each metal and to protect the equipment. The pretreatment is slightly different depending on which size fraction or speciation you wish to determine. In what follows, we present the pretreatment procedure for the case of determining the total concentration of dissolved and acid-soluble fractions of heavy metals. In case you wish to quantify only the dissolved fraction, you should apply filtration first and then acidify the sample.

First, nitric acid was added until pH became 2 to keep heavy metals soluble in the solution because there is a possibility of heavy metals being adsorbed on the surface of suspended particles in the sample. The dissolution of solid heavy metals in particles due to oxidation by nitric acid would also occur; thus, the measurement results also included those of the heavy metals from those particles (mostly clay in the case of Tonle Sap Lake).

Second, water samples were filtered using analytical grade filter paper (Whatman GMF150, pore size of 2 μm) to remove suspended solids. After filtration, water samples might still contain fine particles that can clog the nebulizer of the AAS equipment; thus, further filtration is required. Water samples were then filtered using a cellulose acetate filter with a pore size of 0.45 μm. This kind of procedure is a basic procedure for analysis using AAS (Sharma and Tyagi 2013).

Heavy metals targeted in this research were cadmium (Cd), chromium (Cr), copper (Cu), iron (Fe), manganese (Mn), lead (Pb), and zinc (Zn). These metals were selected according to the result of elemental scanning of lake water samples using the inductively coupled plasma–atomic emission spectrophotometry method. Note that for the analysis involving a mixture of elements that precipitate at different pH ranges, multiple analyses using different samples with various pH values might be required.

35.2.2 *Pretreatment of Sediment Samples*

The preparation of measurement samples from the lake sediment requires pretreatment that aims to protect instruments and extract/digest target compounds. This is because lake sediments are in solid form, which cannot be directly used as measurement samples for AAS. The target analysis compounds (heavy metals and, in some cases, silicon) need to be extracted beforehand. A simple method such as the X-ray fluorescence method or energy-dispersive X-ray spectrophotometry can be used as an alternative option for elemental analysis of solid samples; however, their high limit of detection might restrict their application for detection of trace elements.

Extra care should be taken in measuring the weight of samples used for extraction or digestion because an error in weight measurement can contribute significantly to an error in the subsequent calculation of concentration. Samples used for extraction or digestion need to be oven-dried overnight, at a temperature slightly higher than 100 °C to remove water from the sediment. If the dried samples are kept in a desiccator, it is not necessary to dry the samples 1 day prior to the extraction or digestion. However, it is also not a good practice to keep the samples for a long period of time (more than 1 week) before analysis because water moisture absorbed on the samples might accumulate over time and affect the measurement results.

Various digestion methods have been developed to extract target components from solid samples (Sneddon et al. 2006), such as using mineralizers ranging from acids (hydrochloric acid to hydrofluoric acid) to piranha solutions (a mixture of sulfuric acid and 30% hydrogen peroxide with the common ratio of 3:1), heating in conditions ranging from mild (low temperature and atmospheric pressure) to severe (hydrothermal condition), and using heating equipment such as a conventional static oven or microwave. There is no single best method that will work all the time. The selection of the most suitable method is heavily influenced by the type of target components, while factors such as safety hazards and environmental impact also need to be considered. In the next chapters, because the target compound was heavy metals (Cd, Cr, Cu, Fe, Mn, Pb, and Zn), which are soluble in inorganic acids, a combination of hydrochloric acid and nitric acid was used for digestion based on the US EPA 3050B method.

Similar to the analysis of water samples, the lake water samples for analysis were added with nitric acid to reach a pH value of 2 in order to prevent the precipitation of heavy metals. The samples were then also filtered using a cellulose acetate filter with 0.45-μm pore size.

35.3 Pretreatment for Measurement of Pesticides

Most pesticides are found in relatively low concentrations in water samples in the order of ppb or less (Choquette and Kroening 2009; Székács et al. 2015), because most pesticides are organic compounds with high molecular weights and are non-polar, thus possessing low solubility in water. This fact is not to undermine the importance of pesticide monitoring and control in the environment because their bioaccumulation properties can pose a high risk even at very low concentrations of exposure (Chopra et al. 2011). Caution should be taken in analyzing pesticides using GC-MS since the limit of detection for pesticides using GC-MS is usually close to or higher than the concentration value found in lake water samples. Furthermore, the presence of other components in water samples can complicate the detection of pesticides, owing to the additional noise or false signal generation.

Considering these concerns, analysis of pesticides by GC-MS needs to be preceded with pretreatment procedures, mainly to extract target components and improve the S/N ratio, while also protecting the instruments from fouling. Target components are extracted by isolating the pesticides. Many extraction methods have been developed for this purpose (Table 35.1). Among those methods, liquid-liquid extraction (LLE) and solid-phase extraction (SPE) are the commonly used methods for the pretreatment of water samples containing pesticides. LLE is used owing to its wide adoption as a standardized method, whereas SPE is gaining more attention because of its low health hazard and environmental impact (Jinya 2013). Column chromatography (open column) is not often used because of its low flexibility. It

Table 35.1 Comparison of various pesticide extraction methods

Method	Principle	Advantage	Disadvantage
Liquid-liquid extraction (LLE)	Extraction using a liquid solvent	Wide applicability Wide suitability of target elements Low cost Fewer adsorption problems	A large amount of solvent usage Possibility of emulsion Complexity (requires a skilled operator)
Column chromatography (open column)	Extraction based on the difference in elution time through a specific stationary phase	Wide applicability Good reproducibility High loading capacity	A large amount of solvent usage Complexity (requires a skilled operator) Narrow suitability of target elements for simultaneous operation
Solid-phase extraction (SPE)	Extraction using a solid-phase adsorbent	Less solvent usage The flexibility of elution solvent Wide suitability of target elements	High running cost Adsorption with material
QuEChERS	Extraction using a commercial dispersive SPE pack prepared for food samples	Less solvent usage Easy handling Time- and cost-effective	Limited purification capability Need additional procedure to increase performance

Adapted from GL Sciences Inc. training handout on the SPE method (n.d.)

requires more preparation and time; thus, it is not suitable for regular analysis. Meanwhile, the QuEChERS (quick, easy, cheap, effective, rugged, safe) method has low purification ability, which can lead to contamination problems and reduce the accuracy of measurement (Bursić et al. 2016). Procedures have been developed to improve the QuEChERS method; however, these procedures need additional chemicals and pretreatment steps (Perestrelo et al. 2019). Therefore, justification on the overall cost performance is necessary for this method to be used in the future.

To improve the S/N ratio of analysis, the concentration of samples is necessary. It can be done by direct evaporation of samples (in such case, extraction of components was not performed) or evaporation of component solvent after extraction. For the case of the SPE method, the extraction step also serves the purpose of concentration, because the process reduces the volume of samples (typically from 2 L of water to 1 mL of solvent) while keeping the number of target compounds almost constant.

The SPE method was used for analyzing the samples from the lake, owing to its advantages listed in Table 35.1, as well as its additional purpose that also helps in the concentration of samples, thus reducing the time required for water or solvent evaporation. This method was also recommended in a technical report by Jinya (2013) and also had been used in an analysis of pesticides in water (Johnson et al. 1991; Kapsi et al. 2020). Compared to the LLE method, which is highly manual and whose results are sometimes influenced by the skill of operators (such as the operator ability to manually use a separation funnel), the SPE method is relatively simple. It is less influenced by the skill of operators and increases the reliability and the reproducibility of results. Furthermore, it can be performed in parallel for several samples at once, thus reducing the total time required for analysis compared to that with the LLE method. Another reason for selecting the SPE method is the fact that its disadvantage can be minimized. The influence of adsorption with material on the analysis result can be minimized by spiking samples with a known amount of surrogate(s) to measure the recoverability and then applying it to the measurement result. The problem concerning the running cost can be addressed in the future by developing a low-cost solid phase. For example, activated carbon, which is used in commercial solid-phase extraction, can be replaced by self-developed activated carbon after proper investigation.

Furthermore, according to a technical report by Jinya (2013), the LLE method and SPE method showed comparable results for the analysis of pesticides in water samples. This guarantees the reliability of the SPE method. From the environmental and chemical waste management perspectives, the SPE method also showed significant advantages, which are its low organic solvent consumption (up to several hundred times lower than that of LLE) and low amount of organic solvent vapor released. However, even though the amount of organic solvent vapor released is low, the analysis should be done under a fume hood to reduce exposure. One of the drawbacks of the SPE method is the handling of chemical waste, due to the generation of a used solid phase. Nonetheless, the handling of this waste is relatively simple; that is, it can be disposed of in a capped waste bin. The advantages the of SPE method outweigh its drawbacks.

The outline of this pretreatment procedure is shown in Fig. 35.1, which was adapted from the method developed by Kadokami (2013). The aqueous sample needs to be filtered using Whatman GMF150 filter paper to remove fine particles that might clog a GC column. For the extraction, PLS3 (average particle size, 60 μm; specific surface area, 600 m^2/g; pore size, 7 nm) and AC2 (average particle size, 60/150 mesh; specific surface area, 800–1200 m^2/g) (GL Sciences) were used as solid phases. PLS3 is a solid phase based on non-polar styrene-divinylbenzene and methacrylate copolymer, which can extract the non-polar target components in water samples. AC2 is an activated carbon-based solid phase, which can extract both polar and non-polar target components. PLS3 was used as the main extraction phase, whereas the role of AC2 was to extract target components that cannot be extracted by PLS3 owing to the overcapacity of PLS3 or the polarity of the target components.

Target components from the solid phase were eluted using acetone and dichloromethane for PLS3 and acetone only for AC2. Acetone is suitable to elute the target component from the solid phase, and dichloromethane was used to ensure that no target component was left in the PLS3. Elution from AC2 was done using acetone only because most of the target components were extracted in PLS3,

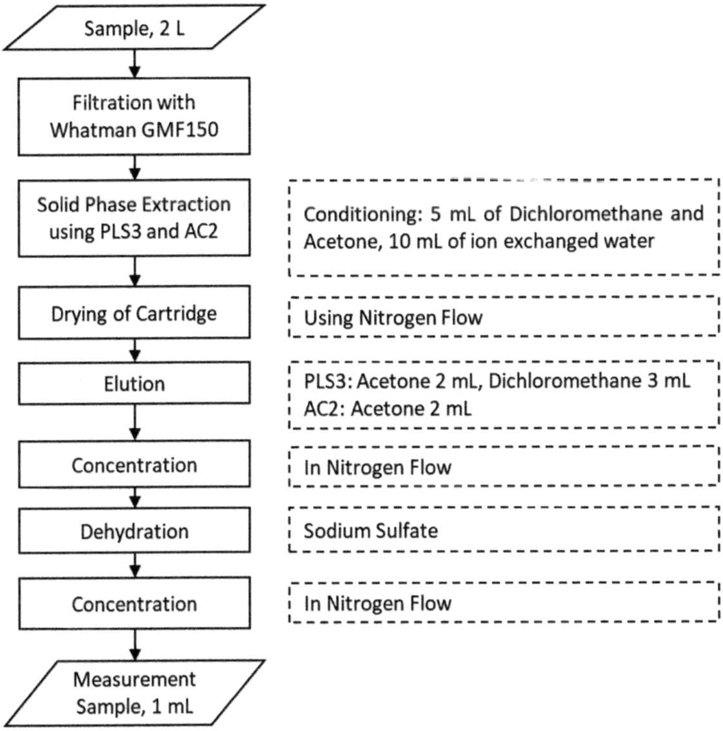

Fig. 35.1 Procedure for solid-phase extraction of pesticides from water samples (adapted from Jinya 2013)

whereas AC2 extracts only the small remnants. Thus, it is not necessary to use dichloromethane in AC2. Sodium sulfate was added and later removed from the extracted samples to remove the water content that remained in the samples. The addition of sodium sulfate was continued until salting-out occurs to confirm that all the water was removed. This step protects the GC column from possible clogging. Furthermore, all drying and evaporation steps in this procedure were performed using nitrogen gas, to avoid the possible oxidation or degradation of target components, which would decrease the accuracy of the analysis.

To confirm the necessity to perform extraction of target components from lake water samples, the measurements of samples prepared following the concentration step only and samples prepared following the concentration and extraction steps were compared (Table 35.2). It can be observed that samples prepared following the concentration step only showed no detection of target components, while samples prepared following the concentration and extraction steps showed detection of target components. This result shows that extraction is necessary to prepare the measurement samples from lake water. The sample prepared without extraction was concentrated by evaporating water from the sample in a vacuum by heating and at a temperature below at about 80 °C or lower, which led to evaporation or degradation

Table 35.2 Comparison of measurement results of samples prepared with extraction and without extraction

Pesticides	With extraction (ppb)	Without extraction (ppb)
Malathion	1.354	ND
Mefenoxam	0.140	ND
Metalaxyl	0.059	ND

The concentration process was applied to both methods. ND, not detected (i.e., below the detection limit)

of target components, and thus, no pesticide was detected. By contrast, evaporation of samples prepared using the extraction method was done by evaporation of the organic solvent (acetone, dichloromethane, or hexane) at low temperatures and a shorter processing time. This led to small evaporation and degradation of target components, resulting in better accuracy.

The SPE method is applicable also for the analysis of residual pesticides in lake sediments and food products such as vegetables (see Chap. 39 for vegetables). For such a purpose, an additional extraction step is necessary to isolate pesticides from solid samples. Organic solvents such as acetone and dichloromethane can be used, and the procedure can be done using Soxhlet or microwave extraction (Onuska and Terry 1993).

In conclusion, suitable pretreatments are necessary to obtain reliable measurements from samples. The selection of the pretreatment procedure should be based on the characteristics of the samples to be analyzed. The pretreatment methods described in this chapter were used for the analysis of samples from Tonle Sap Lake, and the results are presented in Chaps. 36 and 37.

Key Points

- Pretreatment of samples before an analytical procedure is necessary because it helps remove other components in the sample that possibly produce noises and false detection signals, and it also promotes the detection of the target elements.
- The selection of the pretreatment procedure should be based on the characteristics of the samples to be analyzed.
- For heavy metal analysis using AAS, conventional pretreatment of nitric acid addition and filtration was recommended. Acid addition aims to prevent adsorption and precipitation of target heavy metals, while filtration aims to remove particles and prevent equipment from fouling.
- For pesticide analysis using GC-MS, the use of the solid-phase extraction method was recommended. The solid-phase extraction method showed good reproducibility of results and requires less amount of organic solvent, a feature that is superior from the points of view of performance and environmental aspects.

Acknowledgments The authors would like to express their sincere gratitude to Shimadzu Corp. for the training on GC-MS operation and the technical support for the selection of a suitable analysis method for pesticides.

References

Bursić V, Vuković G, Zeremski T, Dušan M, Gvozdenac S, Popović A, Petrović A. Advantages and disadvantages of active carbon in QuEChERS sample preparation method. Sci Bull Ser Biotechnol. 2016;XX(February 2017):6–9.

Chopra AK, Sharma MK, Chamoli S. Bioaccumulation of organochlorine pesticides in aquatic system - an overview. Environ Monit Assess. 2011;173(1–4):905–16. https://doi.org/10.1007/s10661-010-1433-4.

Choquette AF, Kroening SE. Water Quality and Evaluation of Pesticides in Lakes in the Ridge Citrus Region of Central Florida. U.S. Geological Survey Scientific Investigation Report 2008–5178; 2009. p. 55.

Fishman, M. J., & Downs, S. C. (1966). Methods for Analysis of Selected Metals in Water by Atomic Absorption. Geological Survey Water-Supply Paper 1540-C,. 45 pp.

GL Science Inc., Training Handout on SPE Method; n.d.

Hill SJ, Fisher AS. Atomic absorption, method and instrumentation. In: Lindon J, Tranter GE, Koppenaal D, editors. Encyclopedia of spectroscopy and spectrometry. 3rd ed. Hoboken: Academic Press; 2017.

Jinya D. Development of Solid-Phase Extraction Method for Simultaneous Analysis of Semi-Volatile Organics Compounds Using a GC-MS Database System. Shimadzu Technical Report C146-E202; 2013. p. 8.

Johnson WE, Fendinger NJ, Plimmer JR. Solid-phase extraction of pesticides from water: possible interferences from dissolved organic material. Anal Chem. 1991;63(15):1510–3. https://doi.org/10.1021/ac00015a003.

Kadokami K. Development of a Novel Automated Identification and Quantification System with a Database for GC-MS. Shimadzu Technical Report C146-E203; 2013. p. 6.

Kapsi M, Tsoutsi C, Albanis T. Simple analytical methodology based on solid phase extraction for monitoring pesticide residues in natural waters. MethodsX. 2020;7(April 2019):101011. https://doi.org/10.1016/j.mex.2020.101011.

Onuska FI, Terry KA. Extraction of pesticides from sediments using a microwave technique. Chromatographia. 1993;36(1):191–4. https://doi.org/10.1007/BF02263861.

Perestrelo R, Silva P, Porto-Figueira P, Pereira JAM, Silva C, Medina S, Câmara JS. QuEChERS - Fundamentals, relevant improvements, applications and future trends. Anal Chim Acta. 2019;1070:1–28. https://doi.org/10.1016/j.aca.2019.02.036.

Sharma B, Tyagi S. Simplification of metal ion analysis in fresh water samples by atomic absorption spectroscopy for laboratory students. J Lab Chem Educ. 2013;2013(3):54–8. https://doi.org/10.5923/j.jlce.20130103.04.

Sneddon J, Hardaway C, Bobbadi KK, Reddy AK. Sample preparation of solid samples for metal determination by atomic spectroscopy - an overview and selected recent applications. Appl Spectrosc Rev. 2006;41(1):1–14.

Székács A, Mörtl M, Darvas B. Monitoring pesticide residues in surface and ground water in hungary: surveys in 1990–2015. J Chem. 2015;2015:1–15.

Chapter 36
Heavy Metals

Boreborey Ty, Chanvorleak Phat, Kuok Fidero, Eden M. Andrews, Winarto Kurniawan, and Hirofumi Hinode

36.1 Heavy Metals in the Lake Water

"Heavy metals" is the collective definition given to metallic chemical elements and some metalloids that have high atomic weight and high density and induce toxicity at certain levels of exposure to the environment and humans (Briffa et al. 2020; Tchounwou et al. 2012). They also cause a serious concern because they accumulate into the food chain and they cannot be degraded by natural processes in soil and sediments (Paul et al. 2015). The occurrence of heavy metals in waters is from natural or anthropogenic sources (Mohiuddin et al. 2015). They occur naturally in soils and are usually derived from the parent material in the bedrock. They are also natural constituents of lake waters, and some of the heavy metals are present at low concentrations. Some heavy metals such as iron (Fe), manganese (Mn), magnesium (Mg), copper (Cu), nickel (Ni), and zinc (Zn) are essential nutrients that are required for different biochemical and physiological functions. At trace amounts, heavy metals play an important role in biological systems and the ecosystem (Tchounwou et al. 2012) but are considered toxic at higher levels.

During the last few decades, the rapid expansion of human activity has also accelerated the risk of environmental pollution with heavy metals (Wu et al. 2007). Aside from natural sources, their presence in the environment is increased by agricultural runoff, industrial wastes, improper waste disposal, leaking landfills,

B. Ty (✉) · C. Phat
Institute of Technology of Cambodia, Phnom Penh, Cambodia
e-mail: b.ty@itc.edu.kh; mariquit.e.ab@m.titech.ac.jp

K. Fidero
Institute of Technology of Cambodia, Phnom Penh, Cambodia

Ministry of Industry Science, Technology and Innovation, Phnom Penh, Cambodia

E. M. Andrews · W. Kurniawan · H. Hinode
Tokyo Institute of Technology, Tokyo, Japan

C. Yoshimura et al. (eds.), *Water and Life in Tonle Sap Lake*,
https://doi.org/10.1007/978-981-16-6632-2_36

Table 36.1 Some common heavy metals including metalloids in everyday life

List of common heavy metals		
Titanium (Ti)	Copper (Cu)	Platinum (Pt)
Vanadium (Vn)	Zinc (Zn)	Gold (Au)
Chromium (Cr)	Arsenic (As)	Mercury (Hg)
Manganese (Mn)	Molybdenum (Mo)	Lead (Pb)
Iron (Fe)	Silver (Ag)	Selenium (Se)
Cobalt (Co)	Cadmium (Cd)	Arsenic (As)
Nickel (Ni)	Tin (Sn)	Aluminum (Al)

and mining (Edokpayi et al. 2014). Table 36.1 shows a list of heavy metals that are common in everyday life. Heavy metals that are discharged into the lake from both natural and anthropogenic sources are distributed between bed sediments and aqueous phases. This is why it is recommended that in the evaluation of heavy metal contamination in a water body, both sediments and water properties should be considered (Lee et al. 2003).

To date, studies assessing the presence of heavy metals in the water in Tonle Sap Lake (TSL) are scarce. Our research group conducted a survey of heavy metal pollution in TSL. We investigated the heavy metal pollution in both the surface water and lake sediments. The first study focused on the determination of heavy metal levels in the water. Twenty-eight surface water samples were collected from TSL during the dry season (March 2017) and the wet season (June 2017). The locations of the sampling points are shown on the map in Fig. 36.1.

The analysis follows the procedure outlined in Chap. 35. The heavy metal concentrations were analyzed using the BGC-D2 method and an atomic absorption spectrometer (Shimadzu AA-7000, Japan). We detected the presence of heavy metals in the TSL water, but most of them were generally within the Cambodian water quality standards for public water areas, as stipulated in the Sub Decree (Anukret) on Water Pollution Control 1999 (No. 27 ANRK. BK, 1999) by the Royal Government of Cambodia except for some points that showed levels of lead (Pb) in the water exceeding the limits. The relative abundance of metals in terms of mean concentration (mg/l) in water samples during the whole survey period was found in the following order: Fe > Mn > Pb > Cu > Cd > Cr. It was also observed that most of the heavy metals that were measured in the lake water were seasonally varied; the recorded levels of the heavy metals in the rainy season were higher than those in the dry season, which is possibly caused by surface runoff.

Among the heavy metals that we tested in the TSL surface water, Fe, Mn, and Pb, were detected at high levels compared to other metals analyzed (Fig. 36.2). The average detected levels of total dissolved Fe in the water samples from the lake were 0.97 ± 2.24 mg/l for the dry season and 3.76 ± 2.9 mg/l for the rainy season. For Mn, the concentration is found to be at 0.15 ± 0.22 mg/l (dry season) and 0.20 ± 0.1 mg/l (rainy season). Pb was measured at 0.10 ± 0.09 mg/l (dry season) and 0.15 ± 0.07 mg/l (rainy season), which is over the national standards for Pb public water, <0.01 mg/l. Pb at these levels in the water renders it unfit for public consumption. Chromium was not detected in both seasons. Cu was found at very low concentrations with a mean value of 0.03 ± 0.01 mg/l during the rainy season. Zn

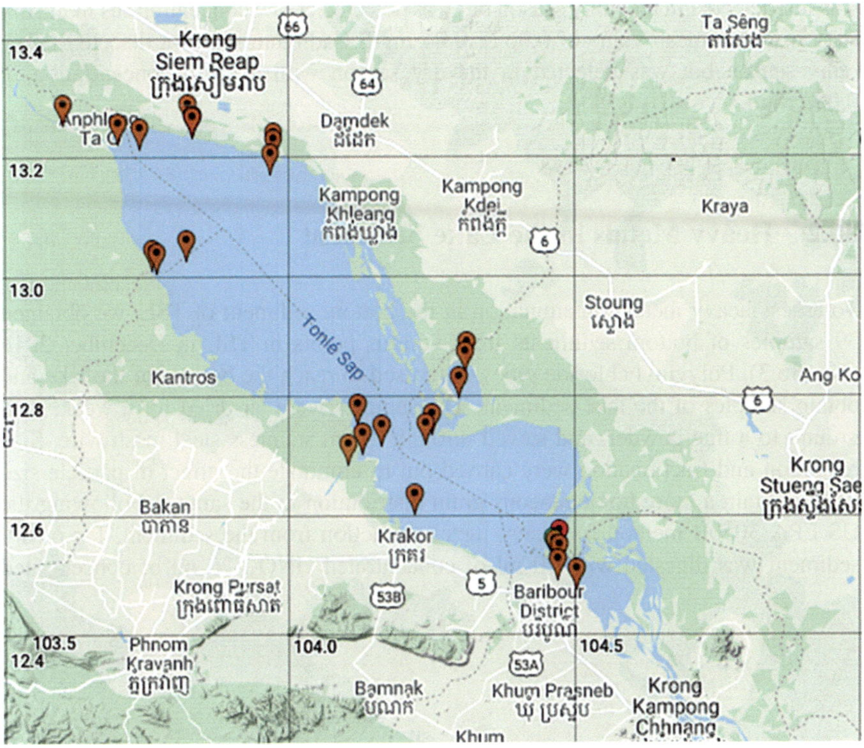

Fig. 36.1 Sampling points of surface water in TSL for heavy metal analysis

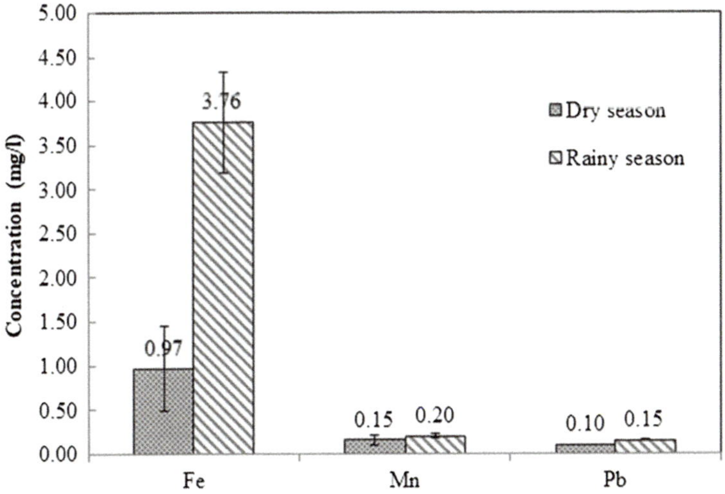

Fig. 36.2 Mean concentrations of Fe, Mn, and Pb measured in the dry (March 2017) and rainy seasons (June 2017) from 28 sampling points in TSL

was not detected in the rainy season but was detected at low concentrations in the dry season with a mean value of 0.06 ± 0.04 mg/l. Cadmium was not detected in the rainy season but was detected in the dry season with a mean concentration of 0.003 mg/l.

36.2 Heavy Metals in the Lake Sediment

To assess heavy metal contamination in the bottom sediment of TSL, we obtained 39 samples of bottom sediments from various points in TSL in December 2016 (Fig. 36.3). Polyvinyl chloride pipes were used to reach the bottom of the lake and obtain samples of the lake sediment. The samples were air-dried for a week, then ground to a fine powder, and sieved through 2-mm stainless steel mesh wire. Size reduction and fractionation were carried out to eliminate the effect of particle size and to obtain a more homogeneous grain distribution in the sample. Following the US EPA 3050B method for heavy metal extraction from the sediment, 1 g of dry sediment was digested with 20 ml of concentrated HNO_3, 20 ml of concentrated

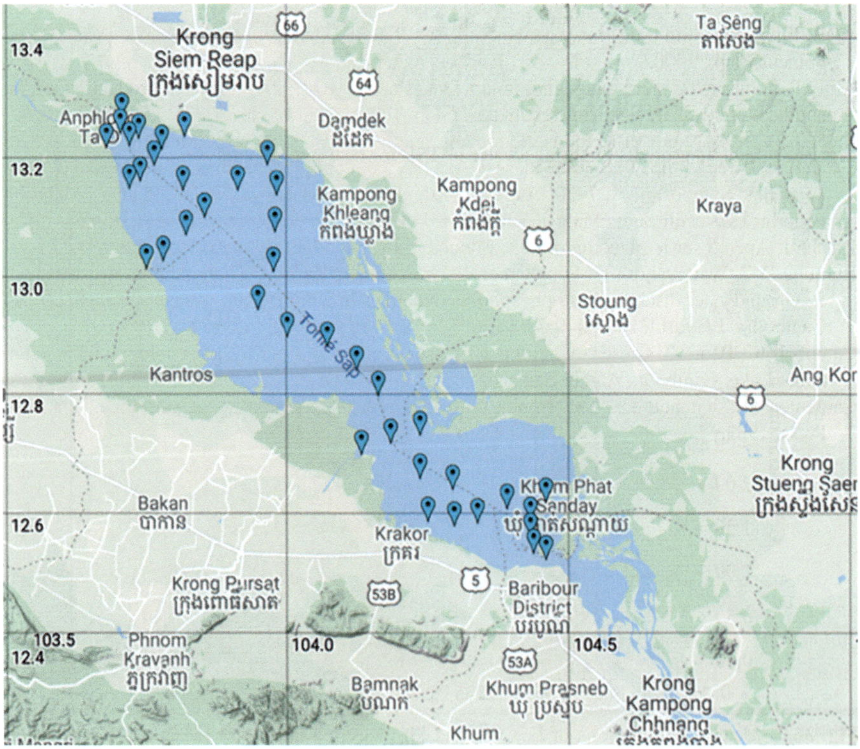

Fig. 36.3 Sampling points of sediment in TSL for heavy metal analysis

HCl, and 10 ml of H_2O_2 (30%) and then heated at 95 °C using a hot plate (Edgell 1989). The digested solution was cooled and filtered through a syringe filter (nominal pore size 45 μm). The filtered digested solution was then diluted to 50 ml with distilled water. Finally, the selected heavy metals in the digested sample were analyzed using the procedure that was outlined in Chap. 35, using an atomic adsorption spectrophotometer (AA-7000, Shimadzu Japan).

Based on our findings, the relative abundance of metals in sediment samples in terms of their mean values follows the order Fe > Mn > Pb > Zn > Cu > Cr > Cd. The concentration of Pb varied from 22.9 to 159.9 μg/g, with the mean value being 43.9 μg/g.

Additionally, the extent of heavy metal pollution in the sediment was also determined using the following quantitative indices: pollution load index (PLI), contamination factor (CF), and geo-accumulation index (I_{geo}). The PLI gives an indication of the overall level of heavy metal pollution by comparing the number of times by which the heavy metal concentrations in the sediment exceed the background concentration (Priju and Narayana 2006). It can be calculated using the CF:

$$\mathrm{PLI} = [\mathrm{CF}_1 \times \mathrm{CF}_2 \times \mathrm{CF}_3 \ldots \times \mathrm{CF}_n]^{1/n}$$

where CF is the quotient obtained by dividing the concentration of each heavy metal with its background values or its natural abundance in the sediment:

$$\mathrm{CF}_i = \frac{C_{\mathrm{metal}\ i(\mathrm{sample})}}{C_{\mathrm{metal}\ i(\mathrm{background})}}$$

PLI values of 0 or less than 1 indicate that there is no pollution, whereas PLI values greater than 1 indicate there is pollution or deterioration of site quality (Tomlinson et al. 1980). The PLI values for heavy metals in TSL sediments in Fig. 36.4 show that all the computed values are generally below 1 for all sampling points, suggesting no significant pollution at all the studied sites.

Another index that we used to assess heavy metal contamination is I_{geo}, which is commonly used to measure the pollution of specific heavy metals in freshwater sediments (Singh et al. 1997). I_{geo} is calculated using the following formula:

$$I_{geo} = \log_2\left[\frac{C_n}{1.5\,B_n}\right]$$

where C_n is the measured concentration (mg/kg) and B_n is the geochemical background value (mg/kg) of the n metal. The world surface rock average values (Martin and Meybeck 1979) were used as background values (B_n) to compute for the I_{geo} values, which are presented in Table 36.2. After obtaining the I_{geo} values, the level of pollution based on the value with respect to the I_{geo} index is determined, as shown in Table 36.3.

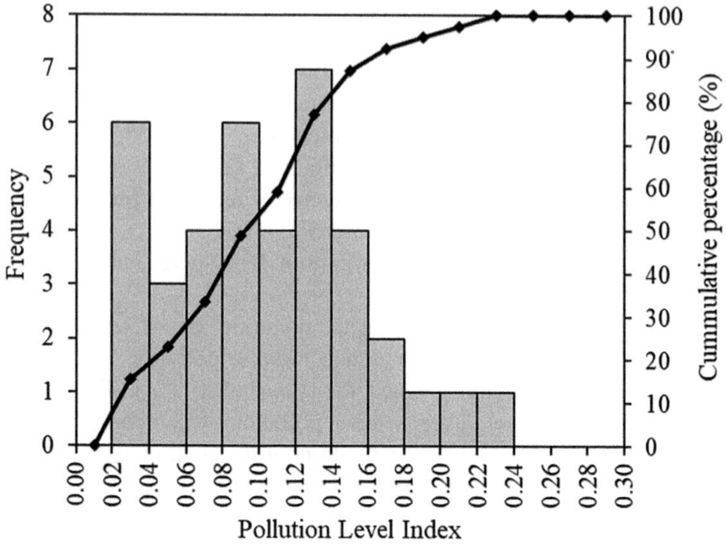

Fig. 36.4 Histogram of pollution load index of sediments in Tonle Sap Lake

Table 36.2 Geochemical background value of world surface rock average concentration (Martin and Meybeck 1979)

Heavy metal	World surface rock average (mg/kg)
Cu	32
Mn	750
Cd	0.2
Pb	16
Cr	71
Fe	35,900
Zn	127

Table 36.3 Classification of sediments based on I_{geo} (Muller 1969)

Class	Geo-accumulation index, I_{geo}	Sediment contamination
6	>5	Very strong
5	4–5	Strong to very strong
4	3–4	Strong
3	2–3	Moderate to strong
2	1–2	Moderate
1	0–1	Uncontaminated to moderate
0	<0	Practically uncontaminated

Based on the calculated I_{geo} values, most of the sediments in the sites are uncontaminated except for those in two sites (CS3-5 and CS3-6), which showed moderate to strong contamination of Pb (Fig. 36.5). These sampling points are located near the floating village of Koh Krek, and the high level of Pb at these points may be due to anthropogenic activities from the area where chemicals such as

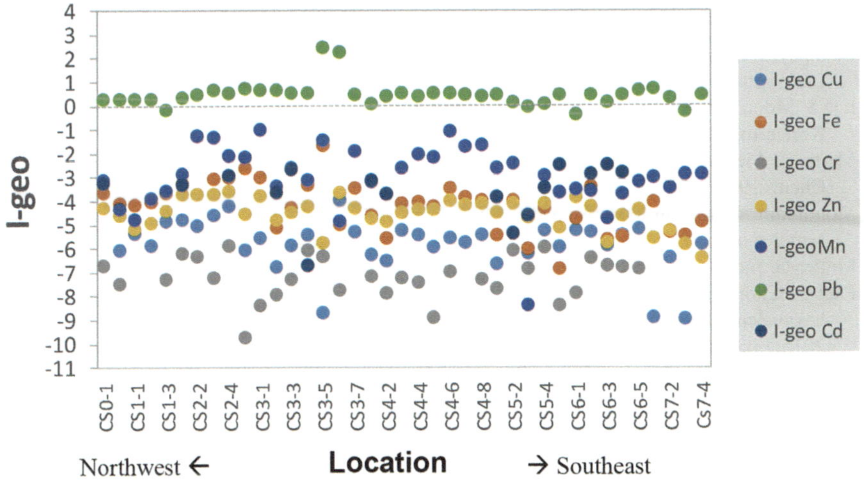

Fig. 36.5 Geo-accumulation index values (I_{geo}) of Tonle Sap Lake sediments

gasoline, oil, and pesticides, among others, are used. Further analysis of physicochemical properties of the sediments in TSL can be found in Chap. 21.

Key Points

- Heavy metals cause serious concern for the health and environment because they accumulate into the food chain and cannot be degraded by natural processes in soil and sediment.
- For the water in TSL, the heavy metal levels detected were within allowable limits based on Cambodian water quality standards for lake water, except for the Pb levels in the water, which exceeded the set standard value.
- The observed levels of heavy metals in the surface water of TSL are reportedly higher in the rainy season than in the dry season, which can be due to the difference in the direction of the flow of the water in the two seasons; in the rainy season, runoff pollutants are flushed out from the tributaries of TSL, where mining areas, agricultural fields, and small industrial areas are located nearby.
- The sediments in TSL were found to be uncontaminated with heavy metals except for Pb, which was found at moderate to strong pollution levels in two sites near Koh Krek floating village.
- We highly recommend continuing the monitoring of heavy metal levels in the waters and sediment of TSL to ensure that the water and sediment quality in TSL will not deteriorate.

References

Briffa J, Sinagra E, Blundell R. Heavy metal pollution in the environment and their toxicological effects on humans. Heliyon. 2020;6(9):e04691.

Edgell K. USEPA Method Study 37-SW-846 Method 3050 Acid Digestion of Sediments, Sludges and Soils-EPA Contract No 68-03-3254, November 1988; 1989.
Edokpayi JN, Odiyo JO, Olasoji SO. Assessment of heavy metal contamination of Dzindi River, in Limpopo Province, South Africa. Int J Nat Sci Res. 2014;2:185–94.
Lee S, Moon JIW, Moon HIS. Heavy metals in the bed and suspended sediments of Anyang River, Korea: Implications for water quality. Environ Geochem Health. 2003;25:433–52.
Martin J, Meybeck M. Elemental mass-balance of material carried by major world rivers. Mar Chem. 1979;7(3):178–206.
Mohiuddin KM, Alam MM, Ahmed I, Chowdhury AK. Heavy metal pollution load in sediment samples of the Buriganga river in Bangladesh. J Bangaladesh Agril Univ. 2015;13(2):229–38.
Muller G. Index of geo-accumulation in sediments of the Rhine. Geol J. 1969;2:109–18.
Paul D, Choudhary B, Gupta T, Jose MT. Spatial distribution and the extent of heavy metal and hexavalent chromium pollution in agricultural soils from Jajmau, India. Environ Earth Sci. 2015;73:3565–77. https://doi.org/10.1007/s12665-014-3642-6.
Priju CP, Narayana AC. Spatial and Temporal Variability of Trace Element Concentrations in a Tropical Lagoon, Southwest Coast of India: Environmental Implications. J Coast Res. 2006;39: 1053–7.
Singh M, Ansari A, Muller G, Singh IB. Heavy metals in freshly deposited sediments of the Gomati river a tributary of the Ganga River: Effects of human activities. Environ Geol. 1997;29:246–52.
Tchounwou PB, Yedjou CG, Patlolla AK, Sutton DJ. Heavy metal toxicity and the environment. Exp Suppl. 2012;101:133–64.
Tomlinson D, Wilson J, Harris C, Jeffrey D. Problems in the assessment of heavy-metal levels in estuaries and the formation of a pollution index. Helgol Mar Res. 1980;33:566–75.
Wu Y, Hou X, Cheng X, Yao S, Xia W, Wang S. Combining geochemical and statistical methods to distinguish anthropogenic source of metals in lacustrine sediment: a case study in Dongjiu Lake, Taihu Lake catchment, China. Environ Geol. 2007;52:1467–74.

Chapter 37
Residual Pesticides

Chanvorleak Phat, Boreborey Ty, Fidero Kuok, Eden M. Andrews, Winarto Kurniawan, and Hirofumi Hinode

37.1 Pesticide Usage Around Tonle Sap Lake

Tonle Sap Lake (TSL) functions as the main freshwater resource for various food and agricultural activities in Cambodia (Campbell et al. 2009; Tangdamrongsub et al. 2016; see also Chaps. 1–4 and 31–34 regarding the lake ecosystem and its ecological and socioeconomic settings). Fisheries and crop cultivation in the TSL basin benefited from the abundant freshwater, nutrients, and enriched soils supplied by seasonal flood pulse and rainfall; together, these ecosystem services supported livelihoods in the region for generations (Lin and Qi 2017). Within this basin, the main floodplain area for agriculture production in Cambodia covers over 1.4 million ha in total and extends to the six provinces, Kampong Chhnang, Pursat, Battambang, Banteay Meanchey, Siem Reap, and Kampong Thom (Song et al. 2011). Agricultural activities in the TSL basin has grown with the facilitation of irrigation, and now, more than half of Cambodian rice fields are located within this area (Tangdamrongsub et al. 2016). In line with the sustainable development goals, particularly SDG2 (Zero Hunger), achieving food security through sustainable agriculture is needed. The challenge for sustainable intensification of agricultural cultivation, however, lies in enhancing productivity while lowering production costs or finding less costly alternative resources (Flor et al. 2019). In this aspect, the rice-growing regions of Asia including Cambodia face a particular challenge with the high usage of agro-chemicals. Pesticide use in Cambodia is exponentially increasing

C. Phat (✉) · B. Ty
Institute of Technology of Cambodia, Phnom Penh, Cambodia
e-mail: phatchanvorleak@itc.edu.kh

F. Kuok
Ministry of Industry, Science, Technology and Innovation, Phnom Penh, Cambodia

E. M. Andrews · W. Kurniawan · H. Hinode
Tokyo Institute of Technology, Tokyo, Japan

C. Yoshimura et al. (eds.), *Water and Life in Tonle Sap Lake*,
https://doi.org/10.1007/978-981-16-6632-2_37

(Matsukawa et al. 2016). For instance, in 2011, the amount of imported pesticides was documented at 5598 tons and then rose to 41,648 tons in 2016 (Matsukawa et al. 2016; Flor et al. 2019).

With this trend, inappropriate pesticide applications together with their widespread mismanagement can bring notable concerns related to human health, environmental pollution, pesticide resistance, and negative effects on non-target yet beneficial organisms (Van Toan et al. 2013). Several monitoring studies worldwide have illustrated the potential of pesticides to contaminate surface and ground waters due to runoff, leaching into groundwater and spray drift (Jensen et al. 2011; Papadakis et al. 2015; Kapsi et al. 2019). The residual pesticides present in water bodies are closely related to the agricultural activities of surrounding areas (Papadakis et al. 2015). In this regard, the investigation of pesticide residues losses to watercourses, and their propagation through biological chains from agricultural fields is critically important (Lamers et al. 2011). Therefore, monitoring and understanding the potential residual pesticides not only in the areas where they are applied but also in proximal areas across vulnerable landscapes like TSL are very crucial (Lamers et al. 2011; Kapsi et al. 2019). Although numerous studies have been conducted, there is still limited information about the status of pesticide pollution in Cambodia. This chapter focuses on the investigation of levels of pesticide residues in the surface water of TSL to provide scientific information and a better understanding of the extent of pesticide pollution in TSL.

37.2 Study Area and Pesticide Analysis

In our survey, water samples were collected four times during the rainy season (June and September 2018) and the dry season (December and March 2018) at 12 different sites, which include seven junctions of tributaries of TSL and five floating communities (Fig. 37.1). Those sites included the junction of the Sen River (SR), Boribo River (BRR), Stoung River (STR), Chmar River (CR), Pursat River (PSR), Sangker River (SKR), and Dauntry River (DTR) and floating communities of Chhnok Tru (CT), Kampong Luong (KL), Kos Khaek (KK), Kampong Phluk (KP), and Chong Kneas (CK).

To date, several methods have been used for the determination of pesticides. Pesticide analysis involves two main steps—pesticide extraction and detection of extracted pesticides (Samsidar et al. 2018). Various extraction methods have been conducted in different studies, namely, liquid–liquid extraction, solid-phase extraction (SPE), matrix solid-phase dispersion, QuEChERS (quick, easy, cheap, effective, rugged, and safe) extraction, and solid-phase microextraction, among others (Van Toan et al. 2013; Farina et al. 2017; Machado et al. 2017; Samsidar et al. 2018). For analytical detection, pesticides are usually detected by classical analytical methods such as gas chromatography–mass spectrometry (GC-MS), liquid chromatography–mass spectrometry, enzyme-linked immunosorbent assay, and capillary electrophoresis (Machado et al. 2017; Samsidar et al. 2018).

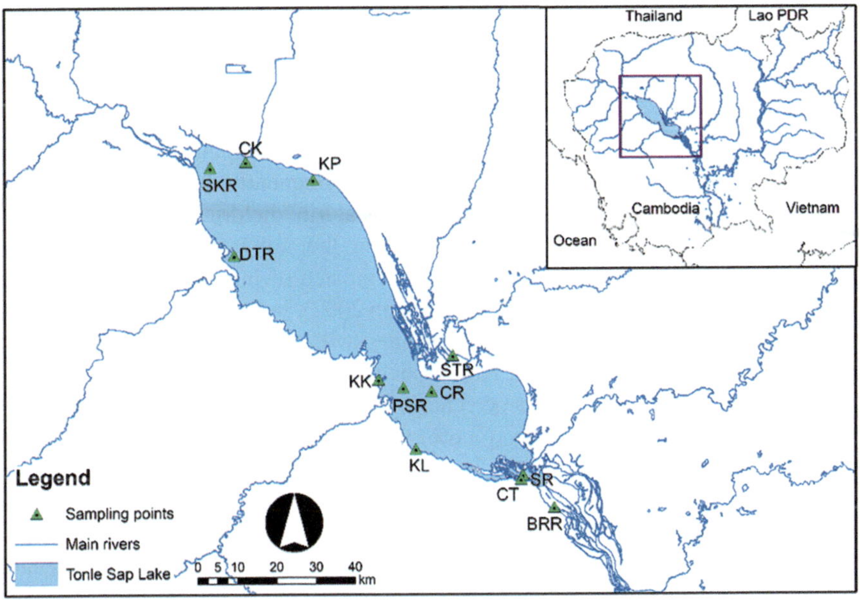

Fig. 37.1 Locations of water sampling points in the lake

Pesticide extraction from the water samples was done by SPE according to the method of Jinya (2013) with some modifications (refer to Chap. 35 for more details). One liter of water sample was passed through InertSep PLS3 (GL Sciences), a copolymer-based sorbent composed of nitrogen-containing methacrylate and styrene divinylbenzene, in combination with AC2, a low ash content of activated carbon cartridge, to extract the pesticides. The pesticide compounds were analyzed using the GC-MS model TQ8040 series (Shimadzu, Japan) with a DB-5 ms column. The analytes were identified by comparing the retention time and mass spectra of the sample analyte with those of the external reference standard solutions that were running at the same condition as the samples. Quantification of the pesticide concentration was carried out using linear integration of the standards and sample data based on peak areas.

37.3 Residual Pesticides

Pesticide use in Cambodia has significantly increased throughout the years. A survey on pesticide use among farmers revealed that some farmers are spraying their crops as often as 20 times in a season (Bell et al. 2016). There is a high possibility that these compounds can contaminate water sources, especially during the rainy season due to surface runoff. In the rural part of Cambodia, particularly within the TSL basin, the surface water still serves as one of the main drinking water sources (Irvine et al. 2006). Direct intake of pesticides via drinking water is one of the main exposure routes in humans. Moreover, surface water is also used for cooking and personal hygiene, thus adding another exposure pathway that potentially threatens human health (Environmental and Agency 2004; Van Toan et al. 2013).

In our survey in 2018, a total of 20 pesticide compounds were detected (Tables 37.1 and 37.2). Fungicides were the most frequently detected, followed by insecticides and herbicides. Insecticides such as aldrin, dieldrin, endrin, heptachlor, isazofos, and isoxathion were not detected in the rainy season. Recently, the applications of these insecticides have been regulated owing to their persistence in the environmental system (Hung and Thiemann 2002). In September, chloroneb and mefenoxam were found in all sampling sites with maximum concentrations of 4.19 μg/L at CT and 3.84 μg/L at KP, respectively. Metalaxyl and methamidophos were frequently detected with a percentage of detection of 90.90%, followed by atrazine (72.7%) and pyroquilon (63.6%). All of these compounds were also the predominant compounds detected in June. The highest pesticide concentration detected in June was 3.88 μg/L for mefenoxam in KL. Usually, we detect five different pesticide compounds that coexisted in the water samples that were taken throughout the year, but there were also instances where water samples from the lake contain ten different pesticides, and these samples were taken from BRR, CK, and CT during the rainy season (Fig. 37.2).

Because most agricultural cultivations begin in the rainy season, the application of pesticides also intensifies during this time of the year. Coupled with heavy rainfalls, this leads to the discharge of residual pesticides into the nearby streams, lakes, and rivers. High levels and percentage of pesticide detection in this study showed the severity of pesticide contamination in TSL owing to the increase in pesticide use, incorrect applications, and improper disposal of pesticide containers. Cambodia does not manufacture pesticides locally, so a large proportion of the pesticides that are for sale in Cambodia are imported from neighboring countries such as Vietnam and Thailand, and usually, the pesticides are labeled in a language incomprehensible to Cambodian farmers (Environment Justice Foundation (EJF) 2002; Irvine et al. 2006). This further aggravates the previously mentioned problems, leaving the farmers confused on what chemical they are using, how to use it properly, and eventually how to dispose of the used containers (EJF 2002).

More pesticide residues were detected in the dry season but at lower concentrations compared to those detected in the rainy season. Interestingly, o,p′-DDT (dichloro-diphenyl-trichloroethane) was detected only in September and December at a very high concentration of 16.59 μg/L at KP. DDT is a predominant contaminant throughout Asia, particularly in developing countries, viz., Cambodia,

Vietnam, China, Thailand, and India (Hong et al. 2008). In the 1990s, DDT was used to control malaria and dengue fever's mosquito vectors, but because of its carcinogenic effect, the policy on DDT application was modified in 1993 that led to the banning of the use of DDT in Cambodia (EJF 2002; Ministry of Agriculture 2019). The occurrence of this compound in various study sites might be due to its persistence in the environment or illegal application (Environment Justice Foundation (EJF) 2002; Hong et al. 2008).

Table 37.1 Pesticide residues in water samples during the rainy season (2018)

Group	Compounds	Jun. 2018				Sep. 2018			
		Min (μg/L)	Max (μg/L)	LM	PD (%)	Min (μg/L)	Max (μg/L)	LM	PD (%)
Fungicides	Azaconazole	0.80 ± 0.001	0.80 ± 0.001	PSR	8.33	–	–	–	0
	Chloroneb	0.52 ± 0.03	2.21 ± 0.05	KL	91.66	0.48 ± 0.01	4.59 ± 0.24	CT	100
	Hexachlorobenzene	–	–	–	0	–	–	–	0
	Mefenoxam	1.19 ± 0.01	3.88 ± 0.01	KL	91.66	0.82 ± 0.03	3.84 ± 0.01	KP	100
	Metalaxyl	0.49 ± 0.01	2.03 ± 0.03	SR	91.66	0.34 ± 0.01	2.90 ± 0.03	CT	90.90
	Pyroquilon	1.14 ± 0.01	3.31 ± 0.01	SKR	83.33	0.49 ± 0.01	1.94 ± 0.01	SR	63.63
	Triadimefon	0.70 ± 0.02	0.70 ± 0.02	PSR	8.33	1.00 ± 0.00	1.00 ± 0.00	CT	9.09
Herbicides	Anilofos	1.07 ± 0.00	1.07 ± 0.00	CR	8.33	0.54 ± 0.01	0.67 ± 0.01	STR	45.45
	Atrazine	0.26 ± 0.00	0.44 ± 0.03	SKR	41.66	0.11 ± 0.01	0.41 ± 0.01	CT	72.72
	Terbacil	1.06 ± 0.01	1.08 ± 0.02	KP	16.66	–	–	–	0
Insecticides	Aldrin	–	–	–	0	–	–	–	0
	Dieldrin	–	–	–	0	–	–	–	0
	Endrin	–	–	–	0	–	–	–	0
	Heptachlor	–	–	–	0	–	–	–	0
	Isazofos	–	–	–	0	–	–	–	0
	Isoxathion	–	–	–	0	–	–	–	0
	Malathion	1.64 ± 0.01	1.64 ± 0.01	PSR	8.33	–	–	–	0
	Methamidophos	0.67 ± 0.02	1.92 ± 0.01	KL	75.00	0.66 ± 0.02	1.92 ± 0.02	KP	90.90
	Methyl parathion	1.23 ± 0.01	1.27 ± 0.01	KP	66.66	0.80 ± 0.01	0.81 ± 0.01	CR	18.18
	o,p′-DDT	–	–	–	0	1.68 ± 0.01	2.17 ± 0.01	KL	27.27
	Parathion	–	–	–	0	–	–	–	0

Values are expressed as mean ± standard deviation of triplicated analyses. Water samples from CK in March and September and from SR in December were not available for analysis. Min, minimum concentration detected; max, maximum concentration detected; "–," not detected; LM, location of max; PD, percentage of detection.

Table 37.2 Pesticide residues in water samples during the dry season (2018)

Group	Compounds	Dec. 2018				Mar. 2018			
		Min (μg/L)	Max (μg/L)	LM	PD (%)	Min (μg/L)	Max (μg/L)	LM	PD (%)
Fungicides	Azaconazole	0.2 ± 0.00	0.24 ± 0.05	BRR	36.36	–	–	–	0
	Chloroneb	0.41 ± 0.01	2.78 ± 0.18	DTR	54.54	0.42	0.98 ± 0.12	PSR	27.27
	Hexachlorobenzene	0.16 ± 0.00	1.26 ± 0.01	SKR	81.81	–	–	–	0
	Mefenoxam	0.47 ± 0.00	0.88 ± 0.51	KL	72.72	0.47	0.47	SR	18.18
	Metalaxyl	0.19 ± 0.00	0.37 ± 0.25	KL	63.63	0.19	0.19	SR	18.18
	Pyroquilon	1.32 ± 0.00	1.32 ± 0.00	KK	9.09	–	–	–	0
	Triadimefon	–	–	–	0	–	–	–	0
Herbicides	Anilofos	0.27 ± 0.00	0.42 ± 0.18	CK	54.54	0.27	0.27	SR	18.18
	Atrazine	0.07 ± 0.01	0.15 ± 0.00	DTR	45.45	–	–	–	0
	Terbacil	0.63 ± 0.00	0.67 ± 0.00	CK, DTR	54.54	–	–	–	0
Insecticides	Aldrin	–	–	–	0	–	–	–	0
	Dieldrin	0.18 ± 0.01	0.27 ± 0.07	CK	63.63	1.79	1.79	PSR	9.09
	Endrin	0.55	0.68 ± 0.15	CK	18.18	0.55	0.55	SR	9.09
	Heptachlor	0.16 ± 0.01	0.19 ± 0.04	CK, DTR	63.63	0.16	0.16	SR	18.18
	Isazofos	–	–	–	0	2.03 ± 0.01	2.09 ± 0.01	PSR	9.09
	Isoxathion	0.26 ± 0.00	0.26 ± 0.00	BRR	9.09	0.54	0.54	SR	9.09
	Malathion	–	–	–	0	3.6 ± 0.00	3.63 ± 0.01	PSR	18.18
	Methamidophos	0.41 ± 0.20	0.41 ± 0.20	CR	9.09	0.41	0.41	SR	18.08
	Methyl parathion	0.47 ± 0.00	0.61 ± 0.20	CT	27.27	–	–	–	0
	o,p′-DDT	7.64 ± 0.00	16.59 ± 1.69	KP	18.18	–	–	–	0
	Parathion	0.78 ± 0.00	0.78 ± 0.00	KK	9.09	–	–	–	0

Values are expressed as mean ± standard deviation of triplicated analyses. Water samples from CK in March and September and from SR in December were not available for analysis. Min, minimum concentration detected; max, maximum concentration detected; "–," not detected; LM, location of max; PD, percentage of detection

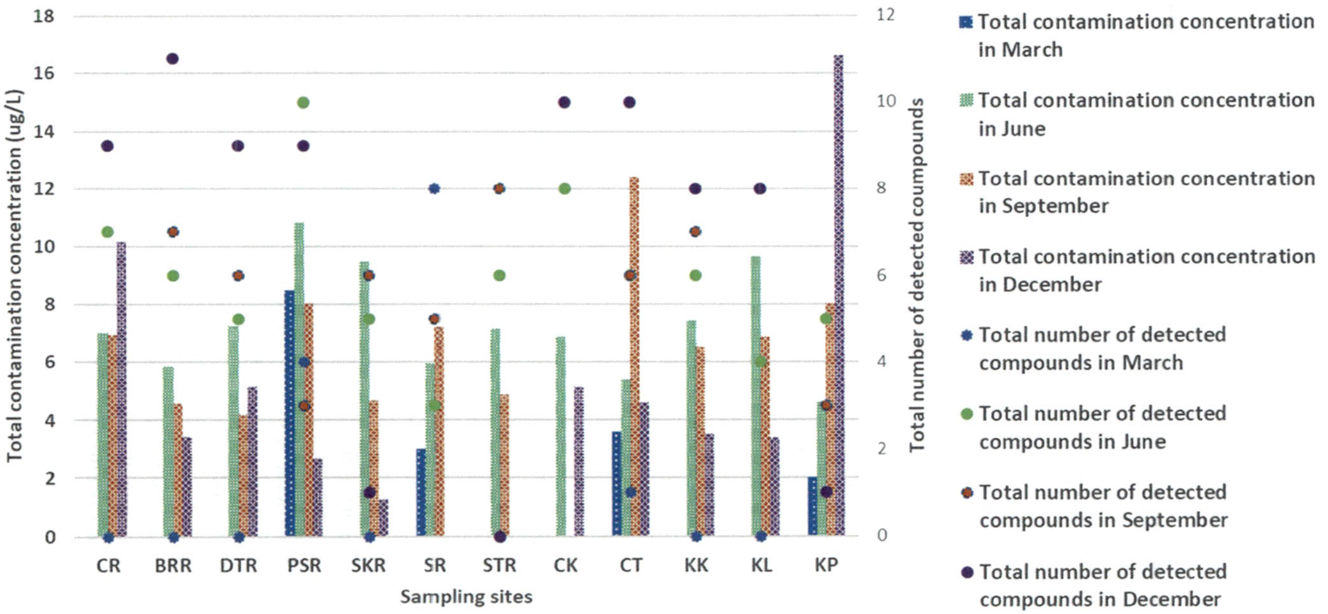

Fig. 37.2 Total concentration and number of pesticide residues detected in water samples from TSL in 2018. Note that water samples from CK in March and September and from SR in December were not available for analysis

The predominant pesticides detected in December were hexachlorobenzene (81.81% of total sites), mefenoxam (72.72%), metalaxyl (63.63%), dieldrin (63.63%), and heptachlor (63.63%). The levels of dieldrin and heptachlor detected in December were much higher than the regulatory limit in drinking water of 0.03 μg/L (EPA 2014). In Cambodia, the standard limit for hexachlorobenzene and dieldrin in surface water are <0.03 and <0.01 μg/L, respectively (Ministry of Environment Cambodia 1999). All detected levels of these two compounds in this study were over the set standard limits, which indicate their intensive use in agricultural areas around TSL. Endrin, a banned insecticide in Cambodia, was also detected in two sites (CR and CK) in December and one site (SR) in March with concentrations of 0.55, 0.68, and 0.55 μg/L, respectively. These levels were over the standard limits for surface water set by the Ministry of Environment of <0.01 μg/L (Ministry of Environment Cambodia 1999). Occurrences of these organochlorine pesticides in TSL in high levels can lead to serious problems due to their harmful effects on reproduction, development, and immunological function in humans and wildlife (Hong et al. 2008).

Only 11 residual pesticides were detected in March 2018 with a low percentage of detection (9.09–27.27%). Among these compounds, malathion was detected with the highest concentration of 3.63 μg/L at PSR. Maximum levels of pesticide residue were mostly detected at SR and PSR. The maximum number of compounds detected in one single sample was 11 at BRR, whereas five or more compounds co-occurred in 80% of the samples in December, and only one or two compounds were detected in most sites in March. The low levels of pesticide contamination in March might be due to the absence of or limited rainfall and runoff during the dry season, thereby reducing the incidents of pesticide leaching from agricultural areas.

Intensified agricultural activities within the lower Mekong basin and the TSL basin lead to the great volume of pesticide discharge into the water body of TSL (72 L or kg per hectare per year for vegetables and 1 L or kg per hectare per year for rice) (Environment Justice Foundation (EJF) 2002). The total contamination levels of pesticide residues at all sampling points raise a big concern with levels ranging from 2.03 to 8.49 μg/L in March, 4.59 to 10.83 μg/L in June, 4.19 to 12.42 μg/L in September, and 1.26 to 16.59 μg/L in December (Fig. 37.2). Among the seven junctions of TSL tributary sites, PSR was the most contaminated site, with the highest total contamination levels for all tested periods except in December. The maximum number of pesticide residues detected in the water sample from PSR was ten active compounds. This might be due to the upstream location of PSR in the heavy agricultural landscape with high usage of pesticides (Kimkhuy and Chhay 2014). In addition, data from a previous survey disclosed that farmers frequently mix several types of pesticides in one spray tank and then apply them all at the same time together without considering the types of pesticides used (Matsukawa et al. 2016). This contributed to the co-occurrence of various active compounds in water samples.

For floating communities, KP and CT were the most polluted areas. It is worth noting that CT is one of the floating communities of TSL that has a special characteristic; during the rainy season, most of the village is underwater, whereas large areas of land for cultivation are available during the dry season. This contributes to the abundance of rice and other crop cultivation activities in this area, making it prone to pesticide pollution. Kimkhuy and Chhay (2014) reported that there is a high pesticide usage in the area of KP. The highest total contamination level was found at KP in December, with a concentration of 16.59 μg/L mainly from o,p′-DDT. According to the Environment Protection Agency (EPA 2014), the sum of all individual pesticides detected and quantified or the total contamination levels of pesticides in drinking water should not be over 0.50 μg/L. Some people living on TSL source their drinking water from the lake. However, the total contamination levels of pesticides discovered in our study were much higher than the limit values, which indicate the problem of residual pesticide pollution in the waters of TSL.

The use of pesticides in Asia, Africa, Latin America, the Middle East, and Eastern Europe has been reported to bring serious environmental problems (Elibariki and Maguta 2017), especially with the pollution caused by organochlorine pesticides such as DDTs, aldrin, dieldrin, endrin, heptachlor, and hexachlorobenzene. These pesticides have been banned in Cambodia (Ministry of Agriculture 2019); however, the occurrences of these compounds in different sites in TSL can be linked to their persistence in the environment, illegal use, improper disposal, or emissions from certain point sources (Elibariki and Maguta 2017). Pesticide contamination in TSL poses significant health issues for the people that may have consumed the contaminated drinking water, vegetables, and fish. Monitoring the levels of pesticide residues in TSL is being continued to provide more information on the pesticide pollution in TSL. This will further contribute to the information available on residual pesticides found in the surface waters of Cambodia, and collectively, this valuable information can serve as scientific evidence for policymakers in Cambodia to reflect on the effectiveness of the current environmental policies.

Key Points

- Many residual pesticides were detected in TSL in our survey in 2018. Fungicides and insecticides such as chloroneb, mefenoxam, metalaxyl, and methamidophos were the predominant residues in the rainy season, whereas hexachlorobenzene and mefenoxam compounds were mostly detected in the dry season.
- According to the standard for surface water of the Cambodian Ministry of Environment (1999) and US EPA drinking water standards (2014), the water in TSL is contaminated with residual pesticides and is not safe for human consumption.
- DDTs, dieldrin, endrin, heptachlor, and hexachlorobenzene have been detected in the waters in TSL even though they have been already banned in Cambodia. This can be due to their persistence in the environment, or, more problematic, some of these chemicals are still being illegally used in agriculture in Cambodia.
- The problem in residual pesticide pollution can be highly related to unregulated selling of pesticides, incorrect application and use of pesticides, and improper disposal of pesticide containers in Cambodia.

References

Bell A, Zhang W, Nou K. Pesticide use and cooperative management of natural enemy habitat in a framed field experiment. Agric Syst. 2016;143:1–13. https://doi.org/10.1016/j.agsy.2015.11.012.

Campbell IC, Say S, Beardall J. Tonle Sap Lake, the Heart of the Lower Mekong. In: The Mekong. Amsterdam: Elsevier; 2009. p. 251–72.

Elibariki R, Maguta MM. Status of pesticides pollution in Tanzania – a review. Chemosphere. 2017;178:154–64. https://doi.org/10.1016/j.chemosphere.2017.03.036.

Environment Justice Foundation (EJF). Death in Small Doses: Cambodia's Pesticides Problems and Solutions. Environ Justice Found London, UK 37; 2002.

Environmental AUS, Agency P. National Center for Environmental Assessment. In: Exposure; 2004. https://cfpub.epa.gov/ncea/risk/recordisplay.cfm?deid=85843

EPA. Drinking Water Parameters. In: Environ. Prot. Agency; 2014. www.epa.ie

Farina Y, Abdullah MP, Bibi N, Khalik WMAWM. Determination of pesticide residues in leafy vegetables at parts per billion levels by a chemometric study using GC-ECD in Cameron Highlands, Malaysia. Food Chem. 2017;224:186–92. https://doi.org/10.1016/j.foodchem.2016.11.113.

Flor RJ, Maat H, Hadi BAR, Then R, Kraus E, Chhay K. How do stakeholder interactions in Cambodian rice farming villages contribute to a pesticide lock-in? Crop Prot. 2019;104799 https://doi.org/10.1016/j.cropro.2019.04.023.

Hong SH, Yim UH, Shim WJ, Oh JR, Viet PH, Park PS. Persistent organochlorine residues in estuarine and marine sediments from Ha Long Bay, Hai Phong Bay, and Ba Lat Estuary, Vietnam. Chemosphere. 2008;72:1193–202. https://doi.org/10.1016/j.chemosphere.2008.02.051.

Hung DQ, Thiemann W. Contamination by selected chlorinated pesticides in surface waters in Hanoi, Vietnam. Chemosphere. 2002;47:357–67. https://doi.org/10.1016/S0045-6535(01)00342-3.

Irvine K, Murphy TP, Sampson M, Dany V, Vermette S, Tang T. An overview of water quality issues in Cambodia. J Water Manag Model. 2006:6062. https://doi.org/10.14796/JWMM.R225-02.

Jensen HK, Konradsen F, Jørs E, Petersen JH. Dalsgaard A (2011) Pesticide use and self-reported symptoms of acute pesticide poisoning among aquatic farmers in Phnom Penh, Cambodia. J Toxicol. 2011; https://doi.org/10.1155/2011/639814.

Jinya D. Report development of solid-phase extraction method for simultaneous analysis of semi-volatile organic compounds Using a GC-MS Database. System. 2013:1–8.

Kapsi M, Tsoutsi C, Paschalidou A, Albanis T. Environmental monitoring and risk assessment of pesticide residues in surface waters of the Louros River (N.W. Greece). Sci Total Environ. 2019;650:2188–98. https://doi.org/10.1016/j.scitotenv.2018.09.185.

Kimkhuy K, Chhay N. Does Cambodia need integrated pest management? Past experience, present knowledge and future prospects; 2014.

Lamers M, Anyusheva M, La N, Nguyen VV, Streck T. Pesticide pollution in surface- and groundwater by paddy rice cultivation: a case study from northern vietnam. Clean (Weinh). 2011;39:356–61. https://doi.org/10.1002/clen.201000268.

Lin Z, Qi J. Hydro-dam – A nature-based solution or an ecological problem: The fate of the Tonlé Sap Lake. Environ Res. 2017;158:24–32. https://doi.org/10.1016/j.envres.2017.05.016.

Machado I, Gérez N, Pistón M, Heinzen H, Cesio MV. Determination of pesticide residues in globe artichoke leaves and fruits by GC–MS and LC–MS/MS using the same QuEChERS procedure. Food Chem. 2017;227:227–36. https://doi.org/10.1016/j.foodchem.2017.01.025.

Matsukawa M, Ito K, Kawakita K, Tanaka T. Current status of pesticide use among rice farmers in Cambodia. Appl Entomol Zool. 2016;51:571–9. https://doi.org/10.1007/s13355-016-0432-5.

Ministry of Agriculture F and F. Prakas on list of pesticides in Cambodia.pdf. Ministry of Agriculture, Forestry and Fisheries, Phnom Penh; 2019.

Ministry of Environment Cambodia. Sub-decree No 27 on the Control of Water Pollution 1999; 1999.

Papadakis EN, Vryzas Z, Kotopoulou A, Kintzikoglou K, Makris KC, Papadopoulou-Mourkidou E. A pesticide monitoring survey in rivers and lakes of northern Greece and its human and ecotoxicological risk assessment. Ecotoxicol Environ Saf. 2015;116:1–9. https://doi.org/10.1016/j.ecoenv.2015.02.033.

Samsidar A, Siddiquee S, Shaarani SM. A review of extraction, analytical and advanced methods for determination of pesticides in environment and foodstuffs. Trends Food Sci Technol. 2018;71:188–201. https://doi.org/10.1016/j.tifs.2017.11.011.

Song S, Lim P, Meas O, Mao N. The agricultural land use situation on the periphery of the Tonle Sap Lake. Int J Environ Rural Dev. 2011;2:66–71.

Tangdamrongsub N, Ditmar PG, Steele-Dunne SC, Gunter BC, Sutanudjaja EH. Assessing total water storage and identifying flood events over Tonlé Sap basin in Cambodia using GRACE and MODIS satellite observations combined with hydrological models. Remote Sens Environ. 2016;181:162–73. https://doi.org/10.1016/j.rse.2016.03.030.

Van Toan P, Sebesvari Z, Bläsing M, Rosendahl I, Renaud FG. Pesticide management and their residues in sediments and surface and drinking water in the Mekong Delta, Vietnam. Sci Total Environ. 2013;452–453:28–39. https://doi.org/10.1016/j.scitotenv.2013.02.026.

Chapter 38
Heavy Metal Accumulation in Fish

In Sokneang, Sengly Sroy, Molin Soeung, Masateru Nishiyama, Viet-Dung Pham, Soukim Heng, Hasika Mith, Sovannmony Nget, and Toru Watanabe

38.1 Fishes and Heavy Metals in Tonle Sap Lake

Tonle Sap Lake (TSL), the largest freshwater lake in Southeast Asia, is the main source of supply of fish and fish products for local consumption and export (see also Chaps. 2–4, 31–33, and 48 for details regarding fish and fish production). This lake has two directional flows every year, in and out of the Mekong stream with 4500 km^2 of flooded forest. It creates a unique ecosystem favorable to the existence of the fish, providing a breeding ground for around 149 fish species (see Chaps. 31–33 for details regarding fish habitats and ecology). In 2002, TSL produces 60% of 400,000 tons of total fish production for fishery business that is worth over US$ 1.4 billion (Boran 2005) (refer also to Chap. 48 regarding fish stock). It supplies up to 80% of animal protein requirements to approximately two million Cambodian people (Baran and Gallego 2015).

Heavy metals (HMs) pertain to elements with high atomic weights and densities at least five times greater than water (Tchounwou et al. 2012). HMs such as arsenic (As), chromium (Cr), cadmium (Cd), mercury (Hg), lead (Pb), manganese (Mn), nickel (Ni), and zinc (Zn) are potentially hazardous in combined or elemental form. They are highly soluble in aquatic environments and thus can be easily absorbed by living organisms. Once HMs enter the food chain, they may end up accumulating in the human body (Kinuthia et al. 2020). In recent years, the growing concern about pollution caused by HMs is brought about by the increasing agricultural, domestic, industrial, and technological activities (He et al. 2005). Their sources in the

I. Sokneang (✉) · S. Sroy · M. Soeung · S. Heng · H. Mith · S. Nget
Institute of Technology of Cambodia, Phnom Penh, Cambodia
e-mail: in@itc.edu.kh

M. Nishiyama · V.-D. Pham · T. Watanabe (✉)
Yamagata University, Tsuruoka, Japan
e-mail: to-ru@tds1.tr.yamagata-u.ac.jp

C. Yoshimura et al. (eds.), *Water and Life in Tonle Sap Lake*,
https://doi.org/10.1007/978-981-16-6632-2_38

environment include agricultural waste, pharmaceutical wastes, domestic effluents, and waste effluent from industrial activities (Jan et al. 2015). Over time, HMs penetrate the water reservoirs via different pathways such as the atmosphere, waste drainage, soil erosion, and human activities (Staniskiene et al. 2006).

In lakes, HMs are continuously accumulated into aquatic organisms, which are also eventually consumed by the fish. Alternatively, HMs can also directly penetrate the fish through its skin and gill (Staniskiene et al. 2006). Because of the increasing development of areas around TSL and poor wastewater management, there is a serious concern of HM contamination in the lake ecosystem. More than 900,000 people living in floating villages dump their domestic wastes, including plastic and human waste, directly into the lake. This lack of a proper waste management system also raises serious concerns (Johnstone et al. 2013). This chapter provides the result of our study about HM contamination in fishes in TSL, which possibly causes consumers health problems.

38.2 Monitoring of Heavy Metals in Fishes

Six local fish species, namely, Trey Brorma (*Boesemania microlepis*), Trey Deab (*Channa micropeltes*), Trey Khman (*Hampala dispar*), Trey Kranh (*Anabas testudineus*), Trey Ptouk (*Channa striata*), and Trey Slat (*Notopterus notopterus*), which are often found in TSL, were selected for our investigation (Fig. 38.1). The fishes were randomly collected at the ports in Chhnoktrou commune (Kampong Chhnang province), Phat Sanday commune (Kampong Thom province), Kbal Tor commune (Battambang province), and Kampong Plouk commune (Siem Reap province) from September 2017 to December 2018 (Fig. 38.2).

The fish samples were cleaned and washed with tap water. The fish muscle (edible part) was separated with the use of a knife and then washed with distilled water. Then, it was freeze-dried at a temperature of −50 °C and a pressure of 4 Pa for 24 h, broken into small pieces, and then passed through a nylon sieve (0.5 mm). Two grams of the freeze-dried sample was digested using a mixture of concentrated HCl and HNO_3 acids at a high temperature (170 °C). The digested sample was analyzed in triplicate for Zn, Pb, Cr, Cu, Cd, Mn, and Ni using the ICP-MS model Elan DRCII of PerkinElmer Japan Co., Ltd., Japan. Simultaneously, As in the samples was analyzed with an atomic absorption spectrophotometer machine (model AA-7000) with a hydride vapor generator (HVG-1; Shimadzu Corp., Japan) after the digestion using HNO_3 and H_2SO_4 acids at a different temperature (200 °C).

Fig. 38.1 Fish species monitored for heavy metal contamination in Tonle Sap Lake. The ruler in the photos is in centimeters. The photos were taken by an author (S. H.) and her colleagues

38.3 Occurrence and Concentration of Heavy Metals in Fishes

The mean concentrations of HMs in the muscles of all six fish species are summarized in Table 38.1. The HM concentrations varied widely among species. The levels of Zn, Pb, Cr, Cu, Cd, and As in all species were below their maximum permissible limits (MPLs). By contrast, the mean concentrations of Mn and Ni in all species exceeded the limits. The highest concentration of Mn was found in *Anabas testudineus* (10.11 mg/kg) followed by *Boesemania microlepis* (4.07 mg/kg),

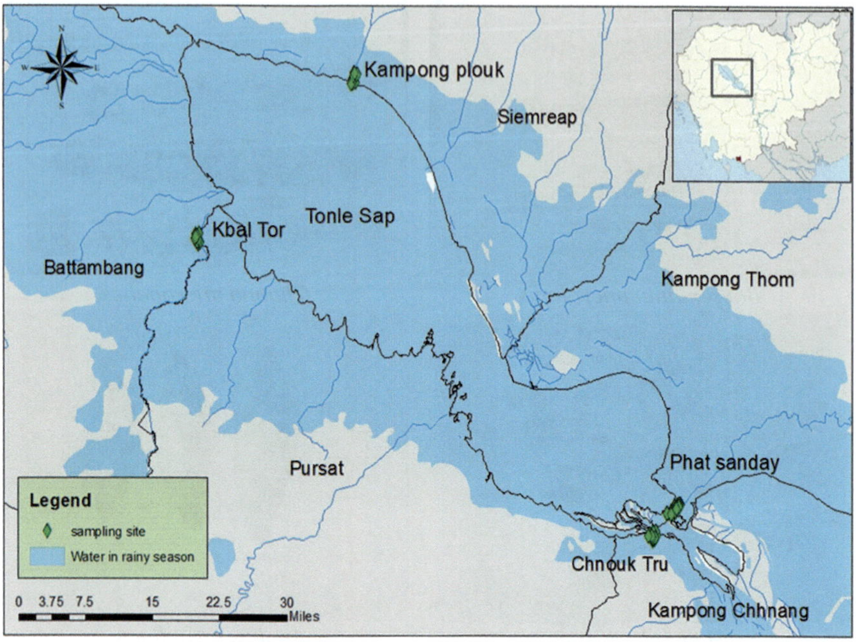

Fig. 38.2 Locations of fish sampling (the ports where the fishes were collected)

whereas the highest concentration of Ni was found in *Anabas testudineus* (0.79 mg/kg) followed by *Channa micropeltes* (0.41 mg/kg) and *Boesemania microlepis* (0.36 mg/kg). Except for Pb and Cd, there was a significant difference in HM concentrations among different fish species ($p < 0.05$). *Anabas testudineus* tended to accumulate Zn, Cu, Cr, As, Mn, and Ni more effectively than did the other species. Such differences were not observed for Zn in the other four species except *Boesemania microlepis* and *Anabas testudineus*; the same can be said for Cu, As, Mn, and Ni among the five species except *Anabas testudineus*. Although a variety of findings on the bioaccumulation of HMs and trace elements in fishes have been reported, some studies mentioned higher bioaccumulation of HMs in predatory and semi-predatory fishes than in non-predatory fishes. A higher bioaccumulation was also reported in demersal than in neritic and pelagic fishes (Julián et al. 2015; Chen et al. 2004; Velusamy et al. 2014).

Aside from biological factors that are dependent on fish species, environmental factors in TSL may have also contributed to the high accumulation of HMs in *Anabas testudineus* found in the lake. Another research group investigated the HM contamination in the surface water and lake sediment in TSL at the same period and found out that Mn and Pb concentrations in the surface water and the Pb level in the lake sediment were relatively high (refer to Chap. 36). This finding was consistent with a higher concentration of Mn in the observed fishes, but the contribution of sediment to Pb accumulation in the fishes was unknown. However, when all the fish

Table 38.1 Heavy metal concentration (mg/kg-wet) of fishes collected from Tonle Sap Lake

Fish species	Local name	Zn	Pb	Cr	Cu	Cd	As	Mn	Ni
Boesemania microlepis	Trey Brorma	14.3[a, b]	0.02	0.12[a, b]	0.22[b]	ND	0.006[b]	4.07[b]	0.36[b]
Channa micropeltes	Trey Deab	7.77[b]	0.06	0.07[a, b]	0.17[b]	ND	0.005[b]	1.19[b]	0.41[b]
Hampala dispar	Trey Khman	11.5[b]	0.05	0.04[b]	0.30[b]	0.004	0.003[b]	0.85[b]	0.18[b]
Anabas testudineus	Trey Kranh	20.3[a]	0.06	0.16[a]	0.44[a]	0.006	0.017[a]	10.1[a]	0.79[a]
Channa striata	Trey Ptouk	10.9[b]	0.10	0.08[a, b]	0.22[b]	0.006	0.002[b]	1.10[b]	0.22[b]
Notopterus	Trey Slat	6.68[b]	0.07	0.08[a, b]	0.21[b]	ND	0.009[b]	0.80[b]	0.21[b]
MPL		100[1]	1[1]	2[2]	20[1]	0.5[1]	1[2]	0.5[2]	0.05[3]

ND, not detected; MPL, maximum permissible limit. Different letters (a > b) indicate a significant difference of heavy metals between fish species at $p < 0.05$.

[1] Thailand. Ministry of Public Health. Notification of Ministry of Public Health, No. 98, B.-E. 2529, Re: Prescribing Standards of Contaminated Substances. R. Thai Gov. Gazette 1986, 103, 98 (in Thai).

[2] Food and Agricultural Organization (FAO). Compilation of Legal Limits for Hazardous Substances in Fish and Fishery Products; Fisheries circular No. 764; Food and Agricultural Organization: Rome, Italy, 1983. http://www.fao.org/3/q5114e/q5114e.pdf (accessed on 7 July 2020).

[3] U.S. Department of Health and Human Services. ATSDR Toxicology Profile for Nickel. 2005. Available online: http://www.atsdr.cdc.gov/toxprofiles/tp15.pdf (accessed on 7 July 2020).

species were pooled, the HM concentrations varied among the collection sites, but a particularly high accumulation of Zn, Cr, Cu, and Mn was found at certain sites near local ports (data not shown).

Although not analyzed here, the seasonal effect on HM accumulation in fishes should be taken into account. Barry et al. (2018) found that some fishes they have analyzed had significantly higher concentrations of HMs in the dry season ($p < 0.05$) than in the rainy season. In the rainy season, a large amount of water is delivered from the Mekong River to the lake (Chaps. 11, 12, and 15; Matsui et al. 2005; Mekong River Commission (MRC) 2008, 2010) as well as the water from 12 tributaries. These waters can reduce the HM contamination in the lake water by dilution but may also bring industrial and/or agricultural waste to the lake. The rainy season (June to October) is also the rice-growing period in Cambodia, and pesticides are used in the areas around the lake. As reported in past studies (Thiyagarajan et al. 2012; Colin et al. 2016; Vallod and Sarrazin 2010), the bioaccumulation of HMs in fishes should be further investigated considering the link among several factors influencing the fish aquatic ecosystem and the socio-environmental conditions in the lake basin, including land cultivation; trends in preference of pesticides used; seasonal differences; changes in industrial, municipal, and agricultural wastes released into the water; hydrodynamics; and other water quality characteristics, among others.

Key Points

- Six species of fish that are commonly caught in Tonle Sap Lake were tested for HM, and all of them were found to have Mn and Ni concentrations exceeding the MPLs set by the FAO/WHO. These HMs are essential elements for many living organisms including humans, but chronic exposure to a high level of these metals causes adverse effects, for instance, neurological effects (Mn) and allergy and kidney diseases (Ni).
- The highest concentrations of Mn and Ni were found in *Anabas testudineus* (known as Trey Kranh locally) probably because it is a predatory fish (carnivorous fish), which has been reported to show bioaccumulation of HMs through the food web.
- The concentrations of Pb in surface water and sediments were relatively high. However, it was not significantly accumulated in any investigated species of fish.

References

Baran E, Gallego G. Cambodia's fisheries: a decade of changes and evolution. Catch Cult. 2015;21: 28–31.

Barry CK, Aung NM, Na P, Stéphane B, Phoeung CL, Mondarin C, Boon HT. Human exposure to trace elements in central Cambodia: Influence of seasonal hydrology and food-chain bioaccumulation behavior. Ecotoxicol Environ Saf. 2018;162:112–20.

Boran E. Cambodian inland fisheries: fact, figures and context, vol. 1751. Penang: Published by WorldFish Center; 2005. p. 16.

Chen YC, Chen CY, Hwang HJ, Chang WB, Yeh WJ, Chen MH. Comparison of the metal concentrations in muscle and liver tissues of fishes from the Erren River, Southwestern Taiwan, after the restoration in 2000. J Food Drug Anal. 2004;12(4):358–66.

Colin N, Porte C, Fernandes D, Barata C, Padrós F, Carrassón M, MacedaVeiga A. Ecological relevance of biomarkers in monitoring studies of macro-invertebrates and fish in Mediterranean rivers. Sci Total Environ. 2016;540:307–23.

He ZL, Yang XE, Toffella PJ. Trace elements in agroecosystems and impacts on the environment. J Trace Elem Med Biol. 2005;19(2–3):125–40.

Jan AT, Azam M, Siddiqui K, Ali A, Choi I, Haq QM. Heavy metals and human health: mechanistic insight into toxicity and counter defense system of antioxidants. Int J Mol Sci. 2015;16(12): 29592–630.

Johnstone G, Puskur R, Declerck F, Mam K, Il O, Mak S, Pech B, Seak S, Chan S, Hak S, Lov S, Suon S, Proum K, Rest S. Tonle Sap scoping report. CGIAR Research Program on Aquatic Agricultural Systems. Penang, Malaysia.Project Report: AAS-2013-28; 2013. Available from URL: http://pubs.iclarm.net/resource_centre/AAS-2013-28.pdf. Accessed 27 Apr 2021.

Julián ZR, Sara EGR, Claudia MRB. Content of Hg, Cd, Pb and As in fish species: a review. Vitae. 2015;22(2):148–59.

Kinuthia GK, Ngure V, Beti D, Lugalia R, Wangila A, Kamau L. Levels of heavy metals in wastewater and soil samples from open drainage channels in Nairobi, Kenya: community health implication. Sci Rep. 2020;10:8434.

Matsui S, Keskinen M, Sokhem P, Nakamura M. Tonle Sap: Experience and lessons learned brief. International Waters: Science Database; 2005. Available from URL: http://www.worldlakes.org/uploads/25_Lake_Tonle_Sap_27February2006.pdf. Accessed 27 Apr 2021.

Mekong River Commission (MRC). An assessment of water quality in the Lower Mekong Basin. MRC Technical Paper No. 19; 2008.

Mekong River Commission (MRC). Impacts on the Tonle Sap ecosystem-Assessment of basin-wide development Scenarios. Technical Note 10; 2010.

Staniskiene B, Matusevicius P, Budreckiene R, Skibniewska KA. Distribution of heavy metals in tissues of freshwater fish in Lithuania. Polish J of Environ Stud. 2006;15:585–91.

Tchounwou PB, Yedjou CG, Patlolla AK, Sutton DJ. Heavy metals toxicity and the environment. EXS. 2012;101:133–64.

Thiyagarajan D, Dhaneesh KV, Kumar TTA, Kumaresan S, Balasubramanian T. Metals in fish along the southeast coast of India. Bull Environ Contam Toxicol. 2012;88(4):582–8.

Vallod D, Sarrazin B. Water quality characteristics for draining an extensive fish farming pond. Hydrol Sci J. 2010;55(3):394–402.

Velusamy A, Satheesh KP, Ram A, Chinnadurai S. Bioaccumulation of heavy metals in commercially important marine fishes from Mumbai Harbor, India. Mar Pollut Bull. 2014;81:218–24.

Chapter 39
Pesticide Residues in Vegetables from Provinces Around Tonle Sap Lake

Chanvorleak Phat, Yoeun Sereyvath, Fidero Kuok, Eden M. Andrews, Winarto Kuriniawan, and Hirofumi Hinode

39.1 Pesticide Applications in Vegetable Fields

The awareness of the importance of vegetables in a balanced and healthy diet has led consequently to the increased consumption of vegetables. Vegetables are considered an essential part of a healthy diet because they provide the majority of micronutrients, vitamins, and minerals (Abang et al. 2014; Leong et al. 2020). However, growing fresh vegetables on a large scale faces various constraints, which can be attributed to biotic and abiotic factors. Biotic problems, such as insects, pathogens, weeds, nematodes, and parasitic plants, are more pronounced in tropical regions like Cambodia, where high rainfall and humidity (see Chap. 6) enhance disease incidence on vegetables. Regular rainfalls and heavy dews coinciding with warm temperatures and dry climates are favorable conditions for fungi growth, insect infestation, weed conquest, and plague infection (Adjrah et al. 2013; Abang et al. 2014).

Pesticides play a crucial role in agricultural practices to control pests, insects, weeds, and diseases, thus minimize crop losses, and increase the yield (Bhandari et al. 2019; Leong et al. 2020). This is the reason the use of pesticides per hectare also increased by 11% annually from 1997 to 2010 (Schreinemachers et al. 2012). Particularly, fungicides and insecticides are applied frequently during vegetable cultivation especially shortly before the harvest to protect the crops (Jankowska et al. 2019). But the misuse of chemical pesticides has created alarming problems

C. Phat (✉) · Y. Sereyvath
Institute of Technology of Cambodia, Phnom Penh, Cambodia
e-mail: phatchanvorleak@itc.edu.kh

F. Kuok
Ministry of Industry, Science, Technology and Innovation, Phnom Penh, Cambodia

E. M. Andrews · W. Kuriniawan · H. Hinode
Tokyo Institute of Technology, Tokyo, Japan

C. Yoshimura et al. (eds.), *Water and Life in Tonle Sap Lake*,
https://doi.org/10.1007/978-981-16-6632-2_39

including pest resistance, impairment of beneficial organisms, and environmental pollution (Abang et al. 2014). In countries located in Southeast Asia, including Cambodia, chronic overuse and misuse of agricultural pesticides in crop production expose farmers/consumers and ecosystems to risks of disease and deterioration, respectively (Schreinemachers et al. 2012; Adjrah et al. 2013). The use and application of pesticides without proper management result in environmental and public health problems due to water and soil pollution, damage non-target organisms, and can cause pest resistance. Despite compliance with local agricultural policies regarding the use of pesticides, a considerable amount of pesticide residues and metabolites are still found in the harvested vegetables and crops.

When these pesticide-contaminated foodstuffs are consumed, the pesticides can accumulate in the body, and even in small quantities, pesticides may cause cancer or severe health problems in humans, particularly affecting the nervous reproductive and endocrine systems. The potential effect of pesticide residues on public health is a global concern. The levels of pesticide residues and types of pesticides used might vary in different regions owing to the differences in geographic and dietary habits (Pang et al. 2020). The Codex Alimentarius Commission in collaboration with the Environmental Protection Agency has set maximal residual limits (MRLs) for particular pesticides in food including vegetables (FAO/WHO 2015) to ensure food safety. Monitoring and assessment of pesticides and proper regulation should be carried out continuously to make sure that concentrations of pesticides accumulated in food are within the MRLs (Leong et al. 2020).

39.2 Status of Pesticide Use in Cambodia

The misuse of pesticides in developing countries involves the use of pesticides banned by the local government, excessive use of pesticides, lack of protective gear when handling the pesticides, inappropriate storage of pesticides, improper handling of pesticide containers, and, in extreme cases, reuse of empty pesticide containers as kitchen tools (Fan et al. 2015). Each year, three million farmers suffer from serious pesticide poisoning in the rural areas of developing countries, 25 million farmers suffer from mild poisoning, and approximately 180,000 deaths among farmworkers are reported. All of these are due to misperceptions and lack of awareness, regulation, and education among pesticide users (Fan et al. 2015).

Cambodia does not have a pesticide manufacturing facility within the country (Preap and Sareth 2015). Cambodian farmers use imported pesticides that are usually labeled in the language of the source country and are therefore incomprehensible to them. This is very problematic because the instructions on usage, information on possible hazards, and proper safety practices have not been made available (Jensen et al. 2011; Swain et al. 2011). It is estimated that 33% of pesticides in the Cambodian market have been prohibited by Cambodian law, and there is a widespread trade of unauthorized pesticides across borders (Jensen et al. 2011; Adjrah et al. 2013).

To address this situation, the Cambodian government has introduced a new law limiting the use of dangerous pesticides and has committed to updating the available instructions (Matsukawa et al. 2016). The Cambodian government reinforces laws and numerous other sub-decrees to control the selling and use of pesticides, particularly the Class I compounds, which are categorized as extremely and highly hazardous compounds. Cambodia is also a signatory country in the Stockholm Convention on Persistent Organic Pollutants and is active in the interim proceedings of the Rotterdam Convention on the Prior Informed Consent for Certain Hazardous Chemicals and Pesticides in International Trade (Swain et al. 2011). Cambodia's plan and policies have promoted eco-friendly measures of agro-production, integrated pest management (IPM) practices, and organic farming that explicitly or indirectly support the principle of reduction of the risk of pesticides for food safety (Preap and Sareth 2015).

39.3 Collection and Analysis of Vegetable Samples

Analysis of pesticide residues on fresh produce is an essential requirement from human health security and international trade points of view. Because of the diversity of the molecular characteristic of pesticides and the complexity of the food matrix, effective analytical tools are needed to analyze them. Chromatography combined with mass spectrometry (MS) is commonly used for pesticide residue analysis (Rajski et al. 2016). The selectivity of accurate MS can be even higher when full-scan MS is combined with tandem MS (MS/MS). A full scan offers results based on elemental composition (accurate mass and isotopic pattern), whereas fragmentation of MS/MS relies on molecular structures. Two molecules with identical elemental composition but the distinct structure (e.g., structural isomers) are almost the same in full-scan MS; however, they may show different fragment ions. Depending on the applied method, MS/MS information can be used for both qualitative and quantitative purposes (Rajski et al. 2016).

In our study, we determined the levels of pesticide residues in vegetables including bitter melon, cabbage, chili, corn, cucumber, eggplant, melon, sesame, spinach, and yardlong bean collected from provinces around Tonle Sap Lake (TSL) (Fig. 39.1). The vegetable samples were collected on August 26–31, 2019; slightly washed with tap water to remove dirt; and stored at −20 °C before further analysis. Pesticide residues in vegetables were extracted and analyzed at the laboratory of the Institute of Technology of Cambodia and scanned for the 451 pesticide compounds using gas chromatography (GC)-MS/MS (GC-MSTQ8040, Shimadzu) equipped with a pesticide database. After the scan, 23 different types of pesticide residues were quantified on the vegetables, and their toxicity and status are given in Table 39.1.

The pesticides from the vegetable samples were extracted using solid-phase extraction. Ten grams of sample was required for the analysis, and 20 mL of acetonitrile was used as the extraction solvent. The sample and extraction solvent

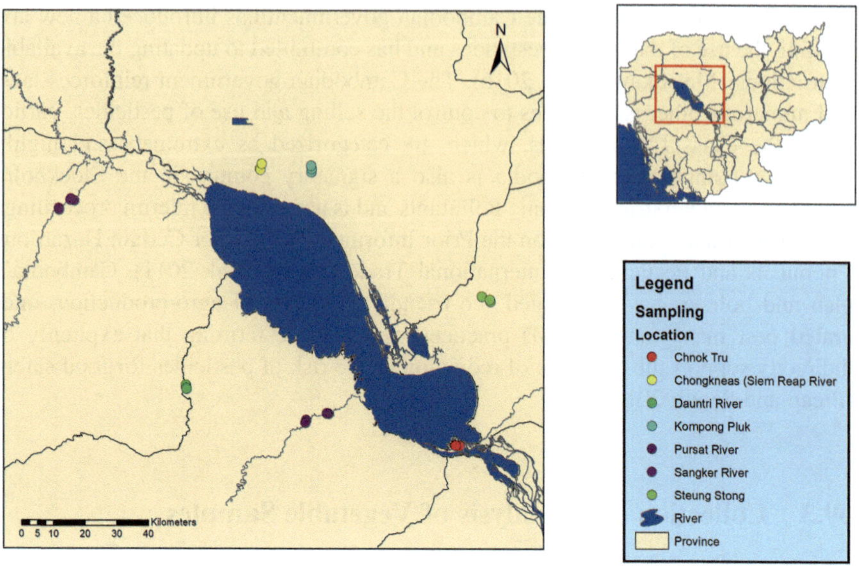

Fig. 39.1 Locations of the farms where we sampled vegetables around Tonle Sap Lake (yellow rhombuses)

were mixed, and then the solvent was passed through an InertSep PLS-3 sorbent (GL Science), combined with AC2, a low ash content of activated carbon cartridge, to recover the pesticide. The analytical methods were the same as described in Chap. 37.

39.4 Pesticide Residues in Vegetables Farmed Around TSL

Thirty-three residual compounds were detected in vegetables collected from small farms in different provinces around TSL (Table 39.2). Yardlong bean and cabbage farmed along the Sangker River, Battambang, were found to be contaminated with many pesticide residues (eight and seven compounds, respectively). Fluquinconazole (fungicide) and dimethametryn (herbicide) were most frequently detected in 27% of tested samples. Those samples were from Chhnok Tru, Chong Kneas, Kampong Phluk, the Pursat River, the Sangker River, and the Stoung River.

Mevinphos (insecticide) was found in 15% of samples including cabbage, mung bean, spinach, and yardlong bean from Battambang and Pursat province. Other residual compounds such as carbaryl, chlopyrifos, and cypermethrin found in this study were also the main active compounds used by local farmers in vegetable production in many countries in sub-Saharan Africa (de Bon et al. 2014). Cypermethrin, detected in spinach from the Dountri River, was one of the ten most commonly reported pesticides with chronic hazards to human health, and it

Table 39.1 Target pesticide compounds for the quantification and their toxicity and status in Cambodia (Ministry of Agriculture 2019)

Group	Compounds	Status in Cambodia	Toxicity class (WHO)	
Fungicides	Azaconazole		II	Moderately hazardous
	Chloroneb		O	Obsolete or discontinued for use as pesticides
	Hexachlorobenzene	Banned	Ia	Extremely hazardous
	Mefenoxam		II	Moderately hazardous
	Metalaxyl		II	Moderately hazardous
	Pyroquilon		II	Moderately hazardous
	Triadimefon	Permitted	II	Moderately hazardous
Herbicides	Anilofos	Permitted	II	Moderately hazardous
	Atrazine	Permitted	III	Slightly hazardous
	Terbacil		U	Unlikely to present acute hazard
Insecticides	Aldrin	Banned	O	Obsolete or discontinued for use as pesticides
	Chlordane	Banned	II	Moderately hazardous
	Dieldrin	Banned	O	Obsolete or discontinued for use as pesticides
	Endrin	Banned	O	Obsolete or discontinued for use as pesticides
	Hexachlorocyclohexanes (HCHs)	Banned	NL	Not listed
	Heptachlor	Banned	O	Obsolete or discontinued for use as pesticides
	Isazofos	Restricted	Ib	Highly hazardous
	Isoxathion	Banned	Ib	Highly hazardous
	Malathion	Permitted	III	Slightly hazardous
	Methamidophos	Banned	Ib	Highly hazardous
	Methyl parathion	Banned	Ia	Extremely hazardous
	o,p′-DDT	Banned	II	Moderately hazardous
	Parathion	Banned	Ia	Extremely hazardous

is frequently detected in Cambodia, the Philippines, and Vietnam (Whittle 2010). Residual metalaxyl was also detected in vegetable samples from Nepalese local farms with a mean concentration of 30.5 ng/g in conventional farming and 21.4 ng/g in IPM farming (Bhandari et al. 2019). Cabbage samples from farms located in Kinnaur district, India, were reported to be contaminated with methyl parathion and malathion with mean concentrations of 6.5 and 7.3 ng/g, respectively (Kumari and John 2019).

In Cambodia, a study conducted regarding pesticide use in 93 households in Boeung Cheung Ek Lake, Phnom Penh, Cambodia, revealed that the top 5 most applied compounds were cypermethrin (93.5%), dichlorvos (71.9%), mevinphos (11.8%), monocrotophos (11.8%), and methamidophos (5.4%) (Jensen et al.

Table 39.2 Detected pesticide compounds in vegetables from provinces around TSL

Provinces	Region	Commodity	Detected compounds		
			Fungicides	Herbicides	Insecticides
Battambang	Sangker River	Cabbage	Flutolanil Pyroquilon[a] (109.8 ng/g)	Diclofop-methyl Dimethametryn	Carbaryl Methoxychlor Mevinphos
		Chili	Phenylphenol (OPP)	Terbcarb (MBPMC)	
		Pepper		Dimethametryn	
		Spinach	Metalaxyl[a] (67.6 ng/g)		Buprofezin Chlorpyrifos
		Yardlong bean	Captan Fenpropimorph Fluquinconazole Pyroquilon[a] (91.4 ng/g)	Dimethametryn	Bendiocarb Chloropropylate Mevinphos
Kampong Chhnang	Chhnok Tru	Eggplant	Pyroquilon[a] (92.8 ng/g)		Disulfoton
		Chili	Diphenylamine	Desmedipham	
		Mung bean			Mevinphos
		Pea eggplant	Metalaxyl[a] (38.7 ng/g)		Chlorfenapyr
		Pepper		Diclofop-methyl	Disulfoton
		Thai eggplant		Bromacil	
		Tomato		Butylate Dimethametryn	
Kampong Thom	Stoung River	Eggplant	Fluquinconazole Phenylphenol (OPP)	Dimethametryn	
		Thai eggplant	Hexachlorobenzene Phenylphenol (OPP)	Dimethametryn	Disulfoton
Pursat	Dountri River	Cucumber			Chlorobenzilate
		Melon	Flusilazole		Thiometon
		Spinach	Captan		

					Cyfluthrin Cypermethrin Methoxychlor Mevinphos
	Pursat River	Bitter melon			Propoxur
		Corn	Fluquinconazole		Chlorfenapyr
		Sesame	Fluquinconazole		
		Yardlong bean			Mevinphos
Siem Reap	Chong Kneas	Thai eggplant	Fluquinconazole		
		Water spinach	Fluquinconazole		
	Kampong Phluk	Thai eggplant		Dimethametryn	
		Water spinach	Captafol	Diclofop-methyl	
		Yardlong bean	Diphenylamine Tebuconazole Trifloxystrobin		

[a]Target pesticide compounds for quantification

2011). In a similar survey by Whittle (2010), it was found that the most commonly used active compounds in Prey Veng, Cambodia, were cypermethrin (109 reports), permethrin (61), chlorfluazuron (57), monocrotophos (39), nereistoxin (34), and chlorpyrifos (29).

In our survey, metalaxyl was found in spinach (67.6 ng/g) from the Sangker River and in pea eggplant from Chhnok Tru (38.7 ng/g). The detected concentrations of metalaxyl were lower than the MRLs, which were set to be 0.5 mg/kg for cabbage and 2 mg/kg for spinach (FAO/WHO 2015). Pyroquilon was found in three samples (cabbage and yardlong bean from the Sangker River and eggplant from Chhnok Tru) with concentrations of 109.8, 91.4, and 92.8 ng/g, respectively. Other studies on pesticide contamination in vegetables from the Cambodian market also reported that some of the fresh produce was found to contain residues of organochlorines, organophosphate, and carbamates, exceeding the established MRLs (Swain et al. 2011; Wang et al. 2011). Wang et al. (2011) reported that the levels of DDTs (dichloro-diphenyl-trichloroethane) in vegetables from Kampong Cham were 1.09 ng/g in cabbage, 2.14 ng/g in cucumber, and 1.35 ng/g in the yardlong bean.

A separate study by Kimkhuy and Chhay (2014) reported that Cambodia ranks first among 13 countries in the region with the highest pesticide residue in vegetables, particularly the leafy vegetables from Kandal. Ramadan et al. (2020) also reported that metalaxyl residue was detected in tomato, potato, cucumber, and chili pepper in concentrations ranging from 0.007 to 0.419 mg/kg; the values that exceeded the MRLs were those found in tomato and potato. The concentrations of residual pesticides detected in most cases do not exceed the constitutionally determined safe levels. Nonetheless, these "safe levels" might overlook the actual health risk, as in the case of simultaneous exposure to two or more chemical contaminants occurring under real-life conditions and having synergistic effects (Nicolopoulou-Stamati et al. 2016).

Another concern about the application of pesticides in vegetable farms around TSL is that TSL is one of the main freshwater sources in Cambodia, and the surface runoff from the farms might contain pesticides that can pollute the soil and water in TSL. Additional problems arise from the common practice of some farmers to dispose of their waste including used pesticide containers directly into the river. Empty pesticide containers were reportedly thrown in the fields and irrigation canals or rivers. Usually, some of the pesticides are still left in empty pesticide containers, and when released to the waterways, these chemicals further contribute to the problem on water pollution. A survey on the practices of the use of pesticides among small-scale vegetable farmers in Ethiopia showed that most farmers mix their pesticides close to a river or water source (Mengistie et al. 2017). Similar practices were also noticed in Cambodia. Therefore, enforcement of policies, regulations, and legislation relating to environmental protection and the responsible use of pesticides should be prioritized to reduce the risks and impacts of agrochemicals on both humans and other living organisms and the environment of the TSL watershed.

Key Points

- Vegetable farms around TSL are affected by the problem of malpractices in pesticide application. The indiscriminate use of pesticides for pest control and lack of knowledge of the correct use and handling of pesticides has resulted in high levels of pesticide residues in locally produced vegetables.
- The vegetables harvested in the farms around TSL are contaminated by residual pesticides. The samples that we tested were below the MRLs, but other studies showed instances where levels of residual pesticides in the harvested vegetables from the farms around TSL exceeded the MRLs.
- Fluquinconazole (fungicide), dimethametryn (herbicide), and mevinphos (insecticide) were the most frequently detected pesticides in vegetables. They were found in vegetables such as cabbage, mung bean, spinach, and yardlong bean from Kampong Thom, Kampong Chhnang, Battambang, Pursat, and Siem Reap provinces.
- Co-occurrence of multiple pesticide residues is common among the obtained vegetable samples. Eight pesticide residues were detected in yardlong bean, whereas seven were detected in cabbage; both vegetables were produced in Battambang province.

References

Abang AF, Kouamé CM, Abang M, Hanna R, Fotso AK. Assessing vegetable farmer knowledge of diseases and insect pests of vegetable and management practices under tropical conditions. Int J Veg Sci. 2014;20:240–53. https://doi.org/10.1080/19315260.2013.800625.

Adjrah Y, Dovlo A, Karou SD, Eklu-Gadegbeku K, Agbonon A, de Souza C, Gbeassor M. Survey of pesticide application on vegetables in the Littoral area of Togo. Ann Agric Environ Med. 2013;20:715–20.

Bhandari G, Zomer P, Atreya K, Mol HGJ, Yang X, Geissen V. Pesticide residues in Nepalese vegetables and potential health risks. Environ Res. 2019;172:511–21. https://doi.org/10.1016/j.envres.2019.03.002.

de Bon H, Huat J, Parrot L, Sinzogan A, Martin T, Malézieux E, Vayssières J-F. Pesticide risks from fruit and vegetable pest management by small farmers in sub-Saharan Africa. A review. Agron Sustain Dev. 2014;34:723–36. https://doi.org/10.1007/s13593-014-0216-7.

Fan L, Niu H, Yang X, Qin W, Bento CPM, Ritsema CJ, Geissen V. Factors affecting farmers' behaviour in pesticide use: Insights from a field study in Northern China. Sci Total Environ. 2015;537:360–8. https://doi.org/10.1016/j.scitotenv.2015.07.150.

FAO/WHO. Codex committee on pesticide residues Codex Alimentarius Commission, Joint FAO and WHO Food Standard Programme. Beijing, China; 2015. http://www.fao.org/fao-who-codexalimentarius/codex-texts/dbs/pestres/commodities-detail/en/?c_id=310. Accessed 4 Aug 2020.

Jankowska M, Łozowicka B, Kaczyński P. Comprehensive toxicological study over 160 processing factors of pesticides in selected fruit and vegetables after water, mechanical and thermal processing treatments and their application to human health risk assessment. Sci Total Environ. 2019;652:1156–67. https://doi.org/10.1016/j.scitotenv.2018.10.324.

Jensen HK, Konradsen F, Jørs E, Petersen JH, Dalsgaard A. Pesticide use and self-reported symptoms of acute pesticide poisoning among aquatic farmers in Phnom Penh, Cambodia. J Toxicol. 2011;2011:1–8. https://doi.org/10.1155/2011/639814.

Kimkhuy K, Chhay N. Does Cambodia need integrated pest management? Past experience, present knowledge and future prospects; 2014.

Kumari D, John S. Health risk assessment of pesticide residues in fruits and vegetables from farms and markets of Western Indian Himalayan region. Chemosphere. 2019;224:162–7. https://doi.org/10.1016/j.chemosphere.2019.02.091.

Leong W-H, Teh S-Y, Hossain MM, Nadarajaw T, Zabidi-Hussin Z, Chin S-Y, Lai K-S, Lim S-HE. Application, monitoring and adverse effects in pesticide use: the importance of reinforcement of Good Agricultural Practices (GAPs). J Environ Manag. 2020;260:109987. https://doi.org/10.1016/j.jenvman.2019.109987.

Matsukawa M, Ito K, Kawakita K, Tanaka T. Current status of pesticide use among rice farmers in Cambodia. Appl Entomol Zool. 2016;51:571–9. https://doi.org/10.1007/s13355-016-0432-5.

Mengistie BT, Mol APJ, Oosterveer P. Pesticide use practices among smallholder vegetable farmers in Ethiopian Central Rift Valley. Environ Dev Sustain. 2017;19:301–24. https://doi.org/10.1007/s10668-015-9728-9.

Ministry of Agriculture F and F. Prakas on list of pesticides in Cambodia.pdf. Ministry of Agriculture, Forestry and Fisheries, Phnom Penh; 2019.

Nicolopoulou-Stamati P, Maipas S, Kotampasi C, Stamatis P, Hens L. Chemical pesticides and human health: the urgent need for a new concept in agriculture. Front Public Heal. 2016;4:1–8. https://doi.org/10.3389/fpubh.2016.00148.

Pang G, Chang Q, Bai R, Fan C, Zhang Z, Yan H, Wu X. Simultaneous screening of 733 pesticide residues in fruits and vegetables by a GC/LC-Q-TOFMS combination technique. Engineering. 2020;6:432–41. https://doi.org/10.1016/j.eng.2019.08.008.

Preap V, Sareth K. Current use of pesticides in the agricultura; products of Cambodia. In: Submitted for the FFTC-KU International Workshop on Risk Management on Agrochemicals through Novel Technologies for Food Safety in Asia, November 10–14; 2015. pp. 1–9.

Rajski Ł, Gómez-Ramos MM, Fernández-Alba AR. Application of LC-time-of-flight and orbitrap-MS/MS for pesticide residues in fruits and vegetables. In: Comprehensive analytical chemistry. Copyright © 2016 Elsevier B.V.; 2016. p. 119–54.

Ramadan MFA, Abdel-Hamid MMA, Altorgoman MMF, AlGaramah HA, Alawi MA, Shati AA, Shweeta HA, Awwad NS. Evaluation of pesticide residues in vegetables from the ASIR region, Saudi Arabia. Molecules. 2020;25:205. https://doi.org/10.3390/molecules25010205.

Schreinemachers P, Schad I, Tipraqsa P, Williams PM, Neef A, Riwthong S, Sangchan W, Grovermann C. Can public GAP standards reduce agricultural pesticide use? The case of fruit and vegetable farming in northern Thailand. Agric Human Values. 2012;29:519–29. https://doi.org/10.1007/s10460-012-9378-6.

Swain MR, Ray RC, Patra JK. Effects of pesticides on Cambodia farming and food production: Alternatives to regulatory policies; 2011. pp. 1–22.

Wang H-S, Sthiannopkao S, Du J, Chen Z-J, Kim K-W, Mohamed Yasin MS, Hashim JH, Wong CK-C, Wong M-H. Daily intake and human risk assessment of organochlorine pesticides (OCPs) based on Cambodian market basket data. J Hazard Mater. 2011;192:1441–9. https://doi.org/10.1016/j.jhazmat.2011.06.062.

Whittle B. Communities in Peril: Asian regional report on community monitoring of highly hazardous pesticide use; 2010. p. 156.

Part IX
Sanitation and Health Risk

Chapter 40
Water Use, Sanitation, and Health Conditions in Villages On/Around the Lake

Toru Watanabe, Sokneang In, Jian Pu, Hengsim Phuong, Sivmey Hor, Sengly Sroy, and Masateru Nishiyama

40.1 Survey on Water Use and Sanitation

Communities on/around Tonle Sap Lake (TSL), with approximately one million people, have adopted their living system based on specific hydrological conditions, forming three different human settlements: land-based (LB), water-based (WB), and water–land-based (WLB) villages (Chap. 2). The LB villagers are engaged more in farming and less in fishing because of the long distance from the lake. The WB villages are floating on the lake, where fishing is the primary occupation of the villagers. The WLB villages are located on the lake for 6 months during the rainy season and on the land for the remaining 6 months. Both agriculture and fishing are the main activities of the WLB villagers (ADB 2005; Sithirith 2014).

As the primary data on water, sanitation, and health risks were not available so far, we collected these data through cross-sectional questionnaire surveys and face-to-face interviews at LB, WLB, and WB villages around TSL from October 2016 to April 2018. According to Shukla (2008), the villages studied were purposely selected from three provinces (Kampong Chhnang, Kampong Thom, and Battambang) around TSL, involving a total of 27 villages: 13 LB, 6 WLB, and 8 WB villages.

In each province, the households and participants were selected randomly for interviews, giving all participants an equal chance to be selected independently of each other (Shukla 2008; Teddlie and Yu 2009). This technique allows the

T. Watanabe (✉) · M. Nishiyama
Yamagata University, Tsuruoka, Japan
e-mail: to-ru@tds1.tr.yamagata-u.ac.jp

S. In · H. Phuong · S. Hor · S. Sroy
Institute of Technology of Cambodia, Phnom Penh, Cambodia

J. Pu
The University of Tokyo, Tokyo, Japan

C. Yoshimura et al. (eds.), *Water and Life in Tonle Sap Lake*,
https://doi.org/10.1007/978-981-16-6632-2_40

identification of a representative group without further sub-categorization. The final set of samples is thus representative of the whole population in each village. Finally, 202, 132, and 208 households participated in this study from LB, WLB, and WB villages, respectively. The interviews were completed at the households considering the age groups (0–5, 6–17 years, and 18 years and older) and sex of correspondents, asking about water use (source of drinking water and its treatment), sanitation (using soap for hand washing and availability of toilets), and health conditions (the disease and its treatment) as well as general information of the interviewee (their family and economic situation). The interviews were conducted by researchers with similar experience and students trained by supervisors from the Institute of Technology of Cambodia. The interviews of the 0–5-year age group were answered by their parents or grandparents.

40.2 Drinking Water

The survey revealed the main source of the drinking water and water treatment for this purpose in the LB, WLB, and WB villages (Table 40.1). The main source in LB villages was well water (71.8%), whereas it was lake water in WLB (49.2%) and WB (52.9%) villages. Rainwater was the second source of drinking water for populations living in LB and WLB villages. On the other hand, bottled water was used for drinking in WB (47.1%) villages much more frequently than in LB (5.5%) and WLB (15.9%) villages. For water treatment before drinking, filtration using a variety of devices was applied commonly in LB (53.5%), WLB (34.9%), and WB (22.6%) villages. Boiling was the second popular treatment for drinking water, which was observed in 37.1%, 20.5%, and 32.7% households in LB, WLB, and WB villages,

Table 40.1 Main sources and treatments of drinking water–land-based (LB), water–land-based (WLB), and water-based (WB) villages

Water source (%)	*LB*	*WLB*	*WB*
Tap water	10.9	2.3	1.9
Rainwater	44.6	44.7	8.7
Bottled water	5.5	15.9	47.1
Well water	71.8	36.4	0.5
Lake water	5.5	49.2	52.9
Others	1.5	0.0	1.4
Water treatment (%)	*LB*	*WLB*	*WB*
Boiling	37.1	20.5	32.7
Filtration	53.5	34.9	22.6
Alum and boil	2.5	6.82	8.7
Alum and no boil	3.0	14.4	14.4
Filtration and boil	2.0	0.8	1.0

The values in the table mean the percentages of households using each water source and applying each water treatment among those belonging to LB, WLB, or WB villages. The multiple answers from each household were allowed respectively. The result was similar to other studies reporting that populations in LB areas could afford the price of a family-scale filtration system, because of their better socioeconomic status (Joffre and de Silva 2015).

40.3 Hygiene and Sanitation

Our direct observation found that people living in floating villages of TSL normally use the lake water for all their needs: washing, cooking, swimming, and drinking; they also use the lake as an open bathroom. Moreover, no affordable sanitation options seem to be available for the floating villages, indicating their poor sanitation environments. In our questionnaire surveys, 58.2% of households in WB villages, 28.8% in WLB villages, and 75.7% in LB villages had toilets. However, the toilets used in WB villages were often an open type, allowing direct discharge into the lake. This type of toilet is, of course, inadequate, as reported previously by Brown et al. (2010).

The target of the Millennium Development Goals of Cambodia was to halve the proportion of people without access to sanitation by 2015. Cambodia's national policy on rural water and sanitation envisions that every person in rural communities has sustained access to safe water supply and sanitation services and lives in a hygienic environment (Lake and Bukauskas 2009). However, studies found that hygiene awareness in communities was still low, as less than 15% of mothers wash their hands with soap after defecation, before preparing food, before feeding their child, before eating, and before cleaning the child's bottom (Oeur et al. 2015). Our survey, on the other hand, found a frequent use of soap, which was generally more in WB villages (80.3%) than in WLB (68.9%) and LB (69.8%) villages (Fig. 40.1).

The awareness of sanitation and hygiene practices of children who live near and on TSL was extremely low. In particular, the children in WLB villages have lower hygiene practice than those in WB and LB villages. The percentages of children aged less than 5 years who wash their hands with soap before eating and after defecation were similar (40–45%) in these three different types (Table 40.2). However, the percentage (56.7%) of children aged 6–17 years who wash their hands with soap was lower in WLB villages than those in WB (69.6%) and LB (75.6%) villages. This low awareness of hygiene practice among children in WLB villages might be because of poor education and lack of support from relevant ministries, government, and nongovernmental organizations (NGOs). Most NGOs and local governments have been focusing on people in WB villages for their supports such as improving water and sanitation and providing hygiene education programs. On the other hand, only a few NGOs support WLB villages because of the limited access to the villages located in remote areas. Not only filling this gap but also developing an effective education program for these villages will increase their awareness of sanitation and hygiene practices. For example, the risk assessment based on 3D simulation (Chap. 41), which can demonstrate graphically the risk of waterborne infectious

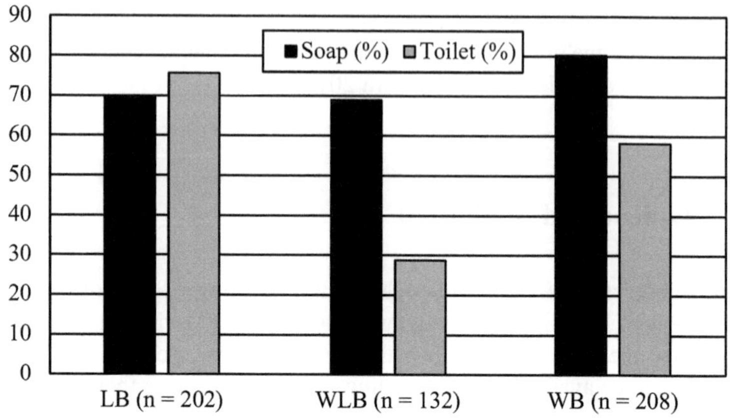

Fig. 40.1 Behavior of using soap for hand washing and availability of toilets in land-based (LB), water–land-based (WLB), and water-based (WB) villages. In the horizontal axis, n means the number of households investigated in LB, WLB, and WB villages. The vertical axis shows the percentage of adults using soap and the percentage of households equipped with toilets in these villages

Table 40.2 Percentage of children who wash hands with soap before eating and after defecation in the three types of villages

	LB		WLB		WB	
%	0–5 year	6–17 year	0–5 year	6–17 year	0–5 year	6–17 year
Yes	45	76	40	57	43	70
No	55	34	60	43	57	30

diseases in the villages and its reduction by possible countermeasures, is expected to make a great contribution to such a program.

40.4 Health Conditions

Figure 40.2 illustrates the occurrence frequency of normal diarrhea reported by LB, WLB, and WB villagers. Almost all the villagers experienced normal diarrhea at least once a year. The expectations of its annual frequency for adults in LB, WLB, and WB villages were 21.6, 24.4, and 24.8 times, respectively. The frequency of diarrhea in children in these villages (22.8 in LB, 26.6 in WLB, and 45.2 times in WB villages) was higher than that for adults, although larger proportions of children reported a frequency of less than once a year.

The LB villagers also experienced severe diarrhea, indicated by bloody diarrhea, vomiting, high fever, persistent diarrhea for 14 days or longer, and/or dehydration, less frequently than the WLB and WB villagers, regardless of age groups (Table 40.3). Diarrheal diseases are common waterborne diseases in Cambodia,

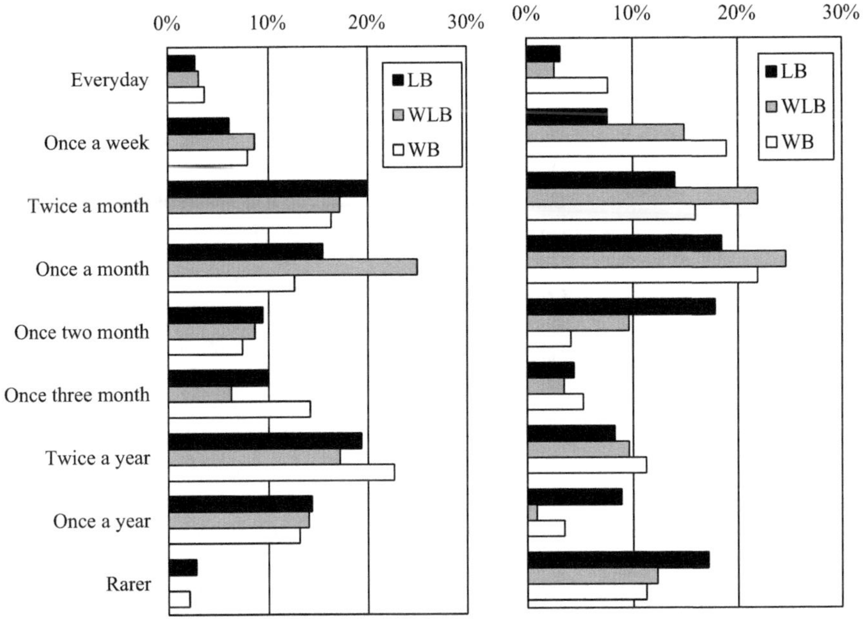

Fig. 40.2 Occurrence frequency of normal diarrhea reported by land-based (LB), water–land-based (WLB), and water-based (WB) villagers

Table 40.3 Experience of severe diarrhea, indicated by bloody diarrhea, vomiting, high fever, persistent diarrhea for 14 days or longer, and/or dehydration, in LB, water–land-based (WLB), and water-based (WB) villagers

%	LB	WLB	WB
Adults	8.4	22.0	18.8
Children	9.4	15.9	17.8

resulting from frequent exposures to the water and food contaminated by pathogenic microorganisms, such as *Escherichia coli*, *Shigella*, and rotavirus (McIver et al. 2014). The higher frequency of diarrheal diseases in WLB and WB villages is probably due to the poorer sanitary environment affected by the lake water. Although this result was based on the memory of villagers, the clear trend observed here was not affected by the recall errors. The on-site treatment of drinking water using poly-aluminum chloride and calcium hypochlorite, which is discussed in Chap. 42, is capable of reducing the frequency of diarrheal diseases in these villages drastically.

For the treatment of severe diarrhea, more than half the adult interviewees commonly used medicines in the three types of villages (Table 40.4). The second choice of disease treatment for them was to consult a doctor (18.6–24.2%). In line with this study, Yanagisawa et al. (2004) reported that self-medication and home

Table 40.4 Treatments of severe diarrhea selected by LB, water–land-based (WLB), and water-based (WB) villagers

Adults (%)	*LB*	*WLB*	*WB*
Do nothing	8.1	2.8	3.6
Use medicine	53.7	54.1	58.8
Consult a doctor	24.2	23.9	18.6
Go to a hospital/clinic	6.7	14.7	16.0
Others	7.4	4.6	3.1
Children (%)	*LB*	*WLB*	*WB*
Do nothing	4.9	1.0	1.7
Use medicine	52.0	56.4	61.7
Consult a doctor	24.4	20.8	26.9
Go to a hospital/clinic	13.0	19.8	8.6
Others	5.7	2.0	1.1

remedies are the first choices, followed by consulting a doctor in the other region of Cambodia. Self-medication should have saved lots of lives in Cambodia. On the other hand, it is alarming as this is one of the biggest contributors to the occurrence of antibiotic-resistant bacteria in the country (Chaps. 29 and 30; Vlieghe et al. 2013), indicating the necessity of its effective control. Children with severe diarrhea in LB, WLB, and WB villages were treated in the same manner as their parents. The most popular treatment was taking medicine (52–61.7%), followed by consulting a doctor (20.8–26.9%). The percentage of parents who did not give their children any treatment was rare (1–4.9%) in all types of villages, and the range was less than that of adults (2.8–8.1%) probably because severe diarrhea is sometimes fatal for children. Even so, the percentage of parents who took their children to a hospital was not so high (8.6–19.8%) because of their poor livelihood and/or limited access to healthcare facilities on/around TSL, especially in the WLB and WB villages.

Key Points

- The lake water is still widely used for drinking by people in WB and WLB villages, although most of them apply treatments of filtration, boiling, and alum. This fact, together with the low availability of toilets, indicates the considerable risk of waterborne infectious diseases in these villages.
- Adults in all three types of villages washed hands using soap before eating and after defecation, whereas children, especially those aged less than 5 years, sometimes did not do so. In addition to the proper management of lake water quality, education on better hygiene and sanitation should contribute to the effective reduction of health risks in this area.
- Almost all the interviewed adult villagers experienced normal diarrhea at least once a year with the expected frequency (21.6–24.8 times a year). Its occurrence in WLB and WB villages was more frequent than that in LB villages, which is consistent with the occurrence of severe diarrhea. This is probably because of the poorer sanitary environment in these villages.
- For the treatment of severe diarrhea, more than half of the adult interviewees selected were using medicines, followed by consulting a doctor. Similarly, the parents answered that they would select these two treatment options for their

children with severe diarrhea. The selection of disease treatment was common to all types of villages.

References

Asian Development Bank (ADB). The Tonle Sap Bassin Strategy. www.adb.org, Publication Stock No. 050105; 2005.

Brown M, Sodaneath H, Smith J, Hagan J. Sanitation in Floating Communities in Cambodia. October 28. Faith-based Education on Water, Sanitation and Hygiene (WASH) A toolkit for Christian schools and communities in Indonesia, n.d.; 2010.

Joffre OM, de Silva S. Community Water Access, Availability and Management Survey in the Tonle Sap Region, Cambodia, 2015. pp. 1–31.

Lake TS, Bukauskas K. Emerging Technologies for Sanitation and Human Waste Disposal in Developing Communities; 2009.

McIver LJ, Chan VS, Bowen KJ, Iddings SN, Hero K, Raingsey PP. Review of climate change and water-related diseases in Cambodia and findings from stakeholder knowledge assessments. Asia-Pac J Public Heal. 2014;28:49S–58S.

Oeur I, Dane S, Sopheakdey S. Community visioning and action plans: Tonle Sap hub. Penang, Malaysia: CGIAR Research Program on Aquatic Agricultural Systems. Program Report: AAS-2015-01. Research; 2015. Retrieved May 03, 2016, from http://aquaticcommons.org/17290/

Shukla P. Essentials of marketing research. 1st ed. New York: McGraw-Hill; 2008. ISBN:978-87-7681-411-3

Sithirith M. The patron–client system and its effect on resources management in Cambodia: a case in the tonle sap lake. Asian Polit Policy. 2014;6(4):595–609.

Teddlie C, Yu F. Mixed methods sampling: a typology with examples. J Mixed Methods Res. 2009;1(77):77–100.

Vlieghe E, Sary S, Lim K, Sivuthy C, Phe T, Parry C, De Smet B, Monidarin C, Baron E, Moore CE, Mfuko W, Asgari N, Chhorvoin O, Steenkeste N, Leyer C, Van Griensven J, Thai S, Jacobs J. First national workshop on antibiotic resistance in Cambodia: Phnom Penh, Cambodia, 16-18 November 2011. J Glob Antimicrob Resist. 2013;1:31–4.

Yanagisawa S, Mey V, Wakai S. Comparison of health-seeking behaviour between poor and better-off people health sector reform in Cambodia. Public Health. 2004;118:21–30.

Chapter 41
Health Risk Assessment of a Floating Village Based on a Three-Dimensional Hydraulic Model

Takashi Nakamura, Toru Watanabe, and Akino Kuwagaki

41.1 Hydraulic Model for Health Risk Assessment

As introduced in Chaps. 14–16, numerical hydraulic models can provide detailed information on the water current. The utilization of such information enables us to grasp the transport properties of various substances dissolved/suspended in a water body. Especially, as the hydraulic simulation describes the spatial profile of the current speed and direction, it has the advantage of the quantitative projection of temporal and spatial distributions of substances transported from nonuniform sources (e.g., toxic chemical compounds from factory sewage drains, pathogens from outfalls of wastewater, and heavy metal seepage from waste dumpsites). Among the various applications, a risk assessment of waterborne infectious diseases seems to be one of the most important utilizations of the hydraulic simulation. In tropical Asia, particularly in rural areas, drainage and sewage systems are still under planning or construction, and insufficient sanitation is commonly found. Such insufficient sanitation and extreme precipitation associated with the monsoon cause pathogen contamination of ambient water and consequently increase the risk of waterborne infectious diseases, such as diarrhea. The mortality rate among children under 5 years of age has a strong correlation with the insufficiency of the sanitation system (Cheng et al. 2012; Headey and Palloni 2019). Although the rate has been declining significantly since 2010, it is still relatively high in some

Supplementary Information The online version of this chapter (https://doi.org/10.1007/978-981-16-6632-2_41) contains supplementary material, which is available to authorized users.

T. Nakamura (✉) · A. Kuwagaki
Tokyo Institute of Technology, Tokyo, Japan
e-mail: tnakamur@tse.ens.titech.ac.jp

T. Watanabe
Yamagata University, Tsuruoka, Japan

C. Yoshimura et al. (eds.), *Water and Life in Tonle Sap Lake*,
https://doi.org/10.1007/978-981-16-6632-2_41

countries of tropical Asia, for example, Cambodia 2.8%, Bangladesh 3.0%, and Lao 4.7% (WHO 2020).

Hashimoto et al. (2014) simulated the flooding at Dhaka City in Bangladesh with a two-dimensional (2D) hydraulic model. By comparing their results with questionnaires on children's diarrhea, they concluded that morbidity was probably higher in areas with a maximum flooding depth of 0.7–1.2 m than in areas with a maximum flooding depth of 0.1–0.2 m. Kazama et al. (2012) estimated a spatial distribution of disease risk at Phnom Penh City in Cambodia with a 2D integrated hydraulic model. From a spatial distribution of coliform bacteria advected with a hydraulic model, the probability of waterborne diseases was evaluated quantitatively by using a dose–response model. As a result, it was unveiled that the location of the high-risk area is changed by the flooding magnitude. As stated above, 2D hydraulic simulations are useful in assessing the disease risk by shallow floodwater, in which the pathogen concentration tends to be uniformly distributed in the vertical direction.

On the other hand, three-dimensional (3D) hydraulic simulation enables us to conduct risk assessments in a water environment where the vertical distribution of pathogens is crucial. Hoyer et al. (2015) applied a 3D hydraulic model to a large alpine lake in the USA and simulated the transport of human waterborne pathogens from a recreational beach to drinking water intakes. Their model enabled the simulation of a vertical distribution of pathogens explicitly considering the vertical dispersion caused by a thermal stratification and selective degradation of pathogens by ultraviolet (UV) irradiation in the surface layer. The floating village tends to be located where the topography changes in a small area, such as a small meandering channel and the river mouth of tributaries. Therefore, a 3D hydraulic model is expected to be suitable for the simulation of flow that must have a complex 3D profile and change locally in the village.

41.2 Sanitation in Floating Villages

Tonle Sap Lake (TSL) contains approximately 90 floating villages. At most houses in the villages, residents use the lake water around the house for various purposes, drinking, cooking, bathing, washing, cleaning, and so on, as described in Chap. 40. Generally, a toilet in these villages is just a hole in the floor without a septic tank, although more than half of the households in water-based villages have a toilet, according to the data reported in Chap. 40. As excreta are dumped directly into lake water underneath the house without any treatment, the floating villages face a situation in which the residents can be infected easily by waterborne diseases by drinking/touching the unsafe lake water contaminated with fecal pathogens. According to Chap. 40, almost all the adults living in the floating village have experienced diarrhea at least once a year, and 18.8% of them have had severe diarrhea. Besides diarrhea, it has been suggested that a few people suffer from other waterborne diseases, such as parasitic infections and infectious gastroenteritis that spread through the lake water (Merali et al. 2014). To reduce the infection of waterborne diseases, it is important that the people gain awareness of these risks and change their behavior. A map of the disease risk in the village promises to be one of the most impressive materials to raise the people's awareness, because the map clearly shows the location of the unsafe water and how risky it is. In some previous studies, these kinds of disease risk maps were generated by using 2D hydraulic simulation models (Amano et al. 2012; Kazama et al. 2012). However, the maps were made to show the spatial distribution of the risk at a regional scale (up to 100 km); no past study has assessed a risk map at the village scale by solving a local structure of water flow with a 3D hydraulic model. Therefore, this chapter describes the application of the quantitative risk assessment of waterborne infectious diseases to one of the major floating villages, Chhnok Trou, to demonstrate the usefulness of a 3D hydraulic model and a risk map.

41.3 Simulation of *E. Coli* Transport

As feces is the main source of waterborne pathogens, *Escherichia coli* bacterium, which are found commonly in the intestines of people and animals, is normally used as an indicator to represent the extent of fecal pollution. In the quantitative microbial risk assessment (QMRA), the disease probability is evaluated by a dose–response model with the concentration of *E. coli* (Haas et al. 1999; WHO 2016). In the present work, the temporal and spatial distributions of the *E. coli* concentration, which is assumed to be dumped from each floating house, are assessed by solving the following advection–diffusion equation:

$$\frac{\partial C}{\partial t}+u\frac{\partial C}{\partial x}+v\frac{\partial C}{\partial y}+w\frac{\partial C}{\partial z}=$$

$$\frac{\partial}{\partial x}\left(\frac{K_H}{\sigma_C}\frac{\partial C}{\partial x}\right)+\frac{\partial}{\partial y}\left(\frac{K_H}{\sigma_C}\frac{\partial C}{\partial y}\right)+\frac{\partial}{\partial z}\left(\frac{K_V}{\sigma_C}\frac{\partial}{\partial z}\right)-\frac{C}{\tau_C} \tag{41.1}$$

where $C(t,x,y,z)$ is the concentration of *E. coli*; K_H and K_V are the kinematic viscosity in the horizontal and vertical directions, respectively; and σ_C is a Schmidt number for the diffusion process of *E. coli*. In this work, we set $\sigma_C = 1$. τ_C as the lifetime in which *E. coli* is reduced to e^{-1} times of the initial population. According to the field experiment conducted in TSL (Tan et al. 2017), τ_C was set at 10 h.

The locations of all houses (totally 1362 houses) in the Chhnok Trou floating village were obtained from a satellite image in Google Earth (captured on January 1, 2017). The emission of *E. coli* from houses to lake water was then modeled as follows: *E. coli* is excreted from each house every morning; the time of excretion is distributed randomly from 4 to 8 AM; the weight of feces excreted by each person is about 105 g (Tanaka et al. 1975); the *E. coli* concentration in feces is 10^9 colony-forming units (CFU)/g (Ervin et al. 2013); and there were five family members in each house (NIS 2002). As the emission of *E. coli* and water withdrawal are both

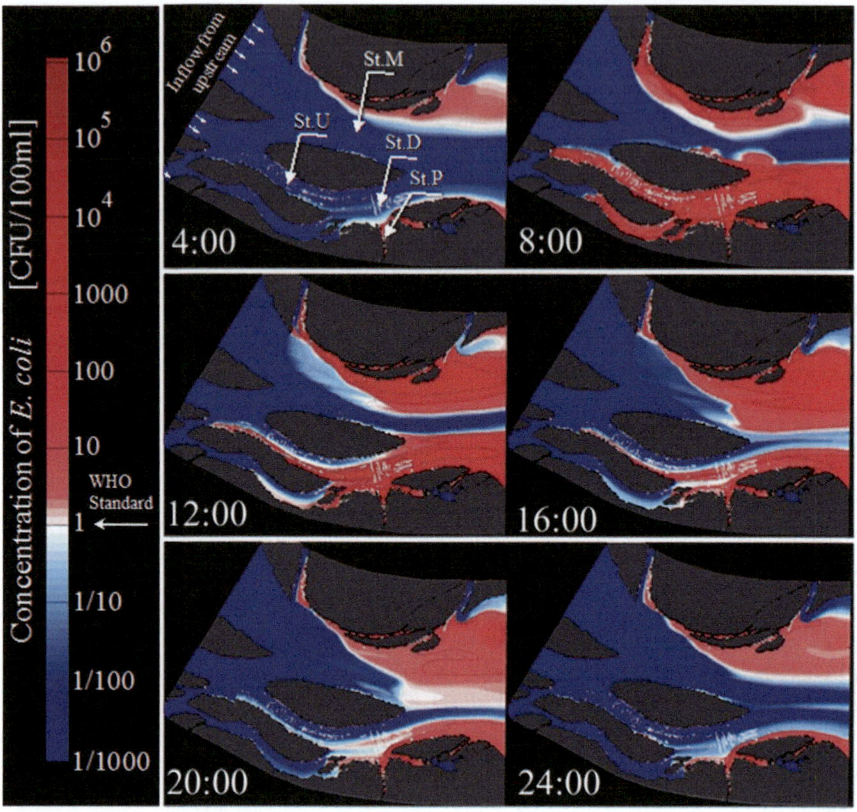

Fig. 41.1 Diurnal change of the *E coli*. Concentration calculated by the 3D transport simulation on March 10, 2018. The concentration at the water surface is shown, whereas a 3D profile of the concentration was calculated in the simulation. The movies are also available in the Supplementary Videos 41.1 and 41.2.

done at the water surface, the *E. coli* concentration is expected to have a nonuniform vertical distribution. Therefore, a 3D model, the Tokyo Institute of Technology–WAter Reservoir Model (TITech-WARM), was applied. The temporal–spatial change of *E. coli* concentration was projected by solving Eq. (41.1) along with hydraulic equations (see Chap. 17 for details).

The 3D simulation was conducted for 1 week in the dry season (from March 3 to 10, 2018). Figure 41.1 shows the calculated daily change in the *E. coli* concentration at the water surface on March 10, 2018. The calculated result showed that the reduction rate of the *E. coli* concentration depends strongly on the topographic condition. A spatial distribution of the flow speed is shown in Fig. 17.4 in Chap. 16. As shown in Fig. 41.2, in a channel where many houses are located and the water flow is relatively rapid (St. U), the *E. coli* concentration reduces to the WHO standard for the drinking water ($C < 1$ CFU/100 ml) before noon. On the other hand, in a stagnant water area, such as a clogged canal (St. P), the concentration

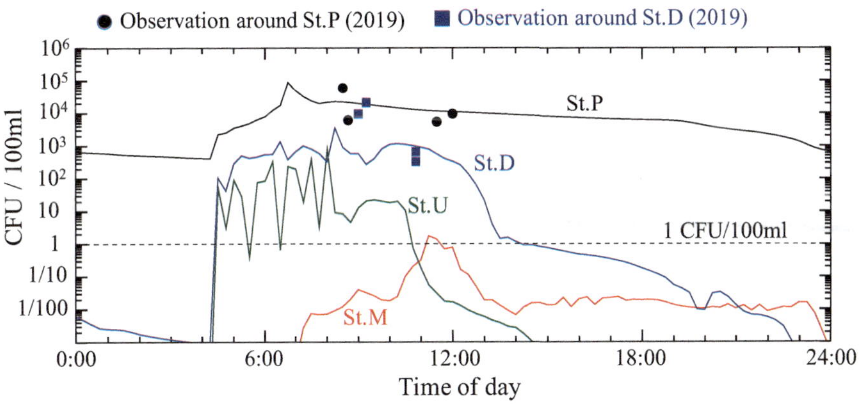

Fig. 41.2 Temporal change of the calculated *E. coli* concentration at the water surface on March 10, 2018. The locations of observational points (St. P, St. D, St. U, and St. M) are shown in Fig. 41.1. *Black circles* and *blue squares* represent data observed on May 14 and 15, 2019, respectively

remains 1000 times higher than the WHO standard throughout the day. This difference in the reduction rate suggests that the disease risk of pathogenic *E. coli* is possibly reduced by selecting an appropriate time and location to collect the lake water for drinking (e.g., in the evening, upstream of a central area where the houses are not crowded, and the water flow is relatively fast).

As exemplified in this simulation, in general, the simulation of the *E. coli* with the hydraulic model can produce persuasive materials to support environmental awareness activities and encourage the residents to change their behavior to access safe water even though the floating villages cause fecal pollution. The field observations of *E. coli* were made on March 14 and 15, 2019. In Fig. 41.2, the observational data are shown in black and blue symbols. Although it is not an accurate verification because the year is different from the simulation, the observed data agree with the calculation and the simulation seems to calculate reasonable *E. coli* transport to a certain extent.

41.4 Quantitative Microbial Risk Assessment (QMRA)

Based on the simulated 3D flow field, QMRA was applied to the calculated spatial distribution of *E. coli*. To estimate the probability of waterborne diseases (diarrhea, fever, gastroenteritis) caused by pathogenic *E. coli*, the following dose–response model of the beta-Poisson type was used (Hoang et al. 2011):

$$P(D) = 1 - \left[1 + \frac{D}{N_{50}}\left(2^{1/\alpha} - 1\right)\right]^{-\alpha} \tag{41.2}$$

where *P*(*D*) is the probability of disease after a person ingests *D* cells of *E. coli* in a day. As parameters, we used $\alpha = 0.1778$ and $N_{50} = 8.6 \times 10^7$. The model utilized is designed to evaluate the disease risk of pathogenic *E. coli.* As a non-pathogenic strain is counted in the *E. coli* concentration calculated in the transport simulation, the risk evaluated using the above model is expected to be larger than the actual probability of disease. From the *P*(*D*), we can estimate the probability of disease per *d* days P_d:

$$P_d = 1 - [1 - P(D)]^d \tag{41.3}$$

In this work, we assumed that the people ingest *E. coli* by drinking water, by accidental ingestion, and through polluted vegetables:

$$D = D_{\mathrm{drink}} + D_{\mathrm{accident}} + D_{\mathrm{vegetable}}\ [\mathrm{CFU/day}] \tag{41.4}$$

These three assumed ingestions were quantified as follows:

41.4.1 Ingestion by Drinking Water ($\mathbf{D}_{drink}$)

Although some people in the floating village use bottled water for drinking, the questionnaire survey (Chap. 40) showed that half of the people in the village use the lake water for drinking purposes, and about 30% of the people drink it without any water treatment. To assess the risk faced by the people drinking such untreated water, we estimated the dose by drinking (D_{drink}) based on the *E. coli* concentration calculated using the 3D hydraulic model:

$$D_{\mathrm{drink}} = L_{\mathrm{drink}} \times \overline{C}_{\mathrm{day}} \tag{41.5}$$

where L_{drink} is the volume of water that an adult drinks in a day, and we assumed $L_{\mathrm{drink}} = 2.6$ L/day (NHMRC and NZ MoH 2016). Although the *E. coli* concentration changes over time, as shown in Fig. 41.1, we used the average daily *E. coli* concentration at the water surface $\overline{C}_{\mathrm{day}}$ (Fig. 41.3a) because of the lack of detailed information on when the people withdraw the lake water.

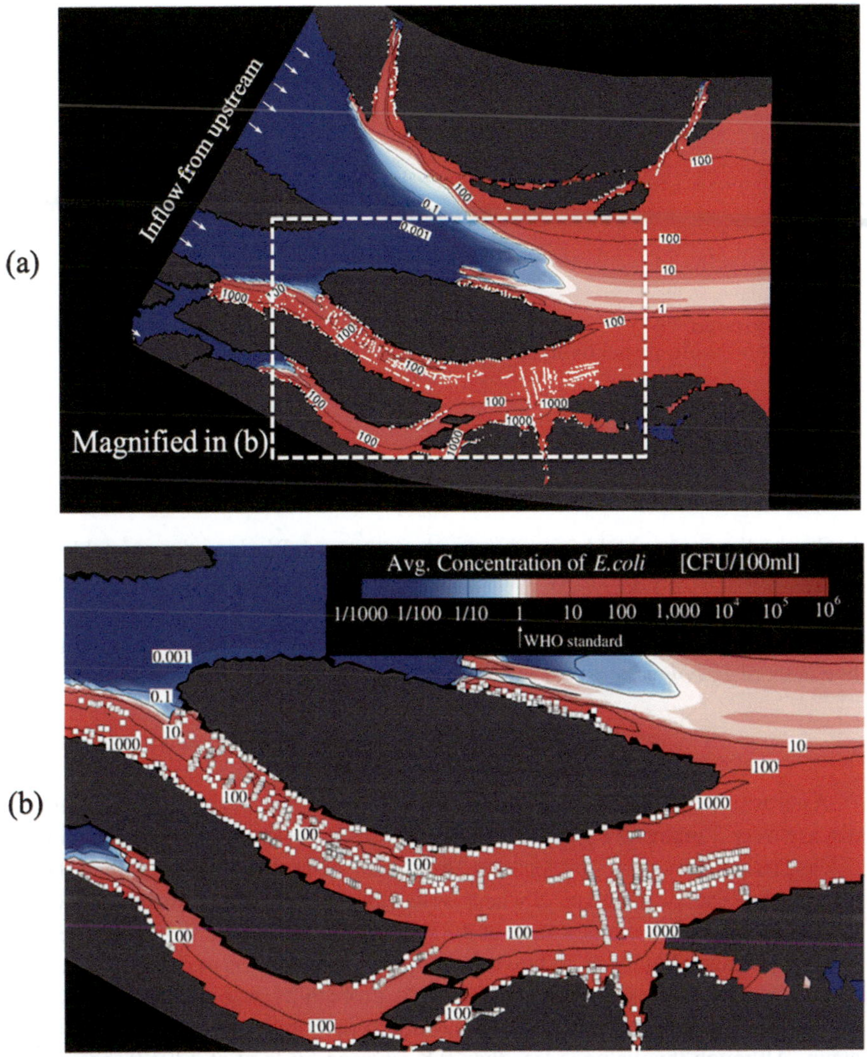

Fig. 41.3 Distribution of the daily average *E. coli* concentration at the water surface in (**a**) a whole simulated area and (**b**) the central area of Chhnok Trou village. *White squares* represent locations of houses

41.4.2 *Ingestion by Accident* ($D_{accident}$)

As to causes of accidental ingestion of the lake water, we assumed they both occurred from swimming and bathing in the lake:

$$D_{\text{accident}} = L_{\text{swimming}} \times \overline{C}_{\text{day}} + L_{\text{bathing}} \times \overline{C}_{\text{day}} \tag{41.6}$$

By assuming that the active time for swimming and bathing are $T_{\text{swimming}} = 10$ min and $T_{\text{bathing}} = 17$ min, respectively, the volumes of water ingestion were evaluated as follows:

$$L_{\text{swimming}} = T_{\text{swimming}} \times U_{\text{swimming}} \tag{41.7}$$

$$L_{\text{bathing}} = T_{\text{bathing}} \times U_{\text{bathing}} \tag{41.8}$$

where U_{swimming} and U_{bathing} are the volume of ingestion per unit time and were set to 10 and 50 ml/h, respectively, following Hoang et al. (2011).

41.4.3 Adhesion to Vegetables ($\mathbf{D}_{\textit{vegetable}}$)

In Cambodia, raw vegetables are commonly consumed. Washing the vegetables with contaminated water probably causes adhesion of the waterborne pathogens to vegetables. By referring to the QMRA for Hue City in Vietnam (Hoang et al. 2011), we assumed that an adult person eats 4.1 g of raw vegetables per day and five cells of *E. coli* adhere to 1 g of vegetables:

$$D_{\text{vegetable}} = 4.1[\text{g/day}] \times 5[\text{CFU/g}] = 20.5[\text{CFU/day}] \tag{41.9}$$

As a result of QMRA, the spatial distribution of the evaluated probability of disease over 3 months (P_{93}) is illustrated in Fig. 41.4. As shown in Fig. 41.4b, the disease risk clearly varies by location. Although the probability is less than 0.05 at the center of each water channel where water flows rapidly, the probability reaches a maximum of about ten times more at the shallow area along the shoreline where the water is stagnant. Figure 41.5 shows the frequency distribution of disease probability of the floating houses. The average disease probability was 0.235. This average probability was converted to 0.46 of a 7-month probability by applying Eq. (41.3) again. Although this value is less than 0.8 reported by In et al. (2017), based on a questionaries' survey, the order of magnitude of the probability is agreeable. As QMRA generally estimates the probabilities under many assumptions, we cannot expect that the estimated probabilities match the actual situation exactly. Therefore, instead of using the probabilities calculated by QMRA as they are, we should utilize QMRA to assess the effectiveness and benefits of the policies (WHO 2016) by comparing the possibilities, as explained below.

Even in the same village, the probability ranged widely from 0 to 0.997. Among all the houses in Chhnok Trou (1362 houses), 43.5% of houses (592 houses) are located where the disease risk is lower than 0.1. On the other hand, 32.7% of houses (446 houses) are located where the disease probability is higher than 0.2, meaning that at least one of five family members is infected in 3 months. This result suggests

(a)

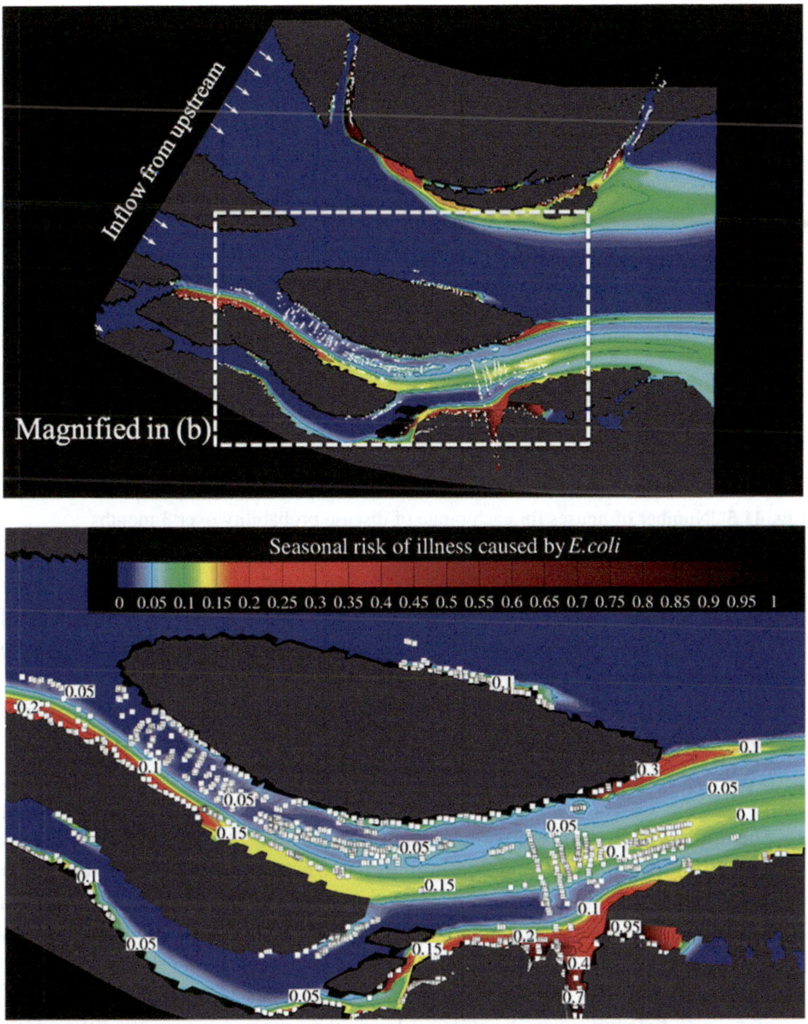

(b)

Fig. 41.4 Distribution of disease probability over 3 months in (**a**) a whole simulated area and (**b**) the central area of Chhnok Trou village. White squares represent locations of houses

that the possibility of the disease risk can be reduced to less than 0.2 by simply relocating 32.7% houses from the high-risk area to the low-risk area. According to our trial calculation on the assumption that the disease risk of the relocated houses is reduced to $P_{93} = 0.2$, the average disease probability was reduced from 0.235 to 0.121. Assuming that there are five family members per house, this reduction means that the number of patients is reduced from 1600 to 824 persons in 3 months.

As shown in this case study, the approach integrating the hydraulic model and QMRA visualizes the distribution of healthy risk in a floating village. Besides the floating village, the approach can be applied to various water areas and helps us

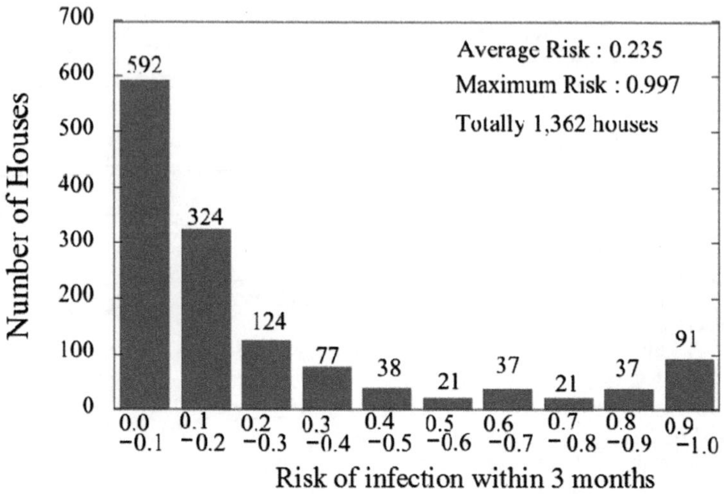

Fig. 41.5 Number of houses for each range of disease probability over 3 months

determine appropriate countermeasures based on the quantitative comparison of efficacy of possible options (e.g., encouraging hand washing with soap and simple purification of water before drinking, such as coagulation sedimentation). Moreover, the impacts of global warming and dam construction on health risks could be assessed if hydraulic conditions in the future could be assumed as computational conditions used in the hydraulic model.

Key Points

- The spread of *E. coli* that is dumped into the lake water from floating houses was simulated for Chhnok Trou floating village. The transport of *E. coli* because of the water flow was modeled with an advection–diffusion equation using a 3D hydraulic model, the Tokyo Institute of Technology–WAter Reservoir Model (TITech-WARM).
- The simulation of Chhnok Trou clearly showed the temporal and spatial change of *E. coli* distribution in the village. From the results, we can read the spatial and temporal distributions of the unsafe contaminated water. The simulation provides rather impressive results, which help with environmental awareness activities and encourage the residents to change their behavior to prevent waterborne diseases.
- From the simulated distribution of *E. coli*, the disease probability was evaluated using QMRA. Besides using QMRA to visualize the current health risks, it was confirmed that QMRA can help us take appropriate measures by evaluating the efficacy of measures quantitatively.

References

Amano A, Sakuma T, et al. Evaluation of diarrhea disease risk attributed to inundation water use on a local scale in Cambodia using hydrological model simulations. River Syst. 2012;20:185–96.

Australian National Health and Medical Research Council and the New Zealand Ministry of Health (NHMRC and NZ MoH). Nutrient reference values. 2016. https://www.nrv.gov.au/. Accessed 4 Feb 2021.

Cheng JJ, Schuster-Wallace CJ, et al. An ecological quantification of the relationships between water, sanitation and infant, child, and maternal mortality. Environ Health. 2012;11:4. https://doi.org/10.1186/1476-069X-11-4.

Ervin JS, Russell TL, et al. Characterization of fecal concentrations in human and other animal sources by physical, culture-based, and quantitative real-time PCR methods. Water Res. 2013;47:6873–82.

Haas CN, Rose JB, et al. Quantitative microbial risk assessment. New York, NY: John Wiley & Sons, INC; 1999. ISBN: 978-1-118-14529-6

Hashimoto M, Suetsugi T, et al. Assessing the relationship between inundation and diarrhoeal cases by flood simulations in low-income communities of Dhaka City, Bangladesh. Hydrol Res Lett. 2014;8:96–102.

Headey D, Palloni G. Water, sanitation, and child health: evidence from subnational panel data in 59 countries. Demography. 2019;56:729–52.

Hoang TM, Watanabe T, et al. Quantitative risk assessment of infectious diseases caused by waterborne *Escherichia coli* during floods in cities of developing countries. J Jpn Soc Water Environ. 2011;34:153–6.

Hoyer AB, Schladow SG, et al. A hydrodynamics-based approach to evaluating the risk of waterborne pathogens entering drinking water intakes in an large, stratified lake. Water Res. 2015;83:227–36.

Kazama S, Aizawa T, et al. A quantitative risk assessment of waterborne infectious disease in the inundation area of a tropical monsoon region. Sustain Sci. 2012;7:45–54.

Merali HS, Morgan JF, et al. The lake clinic - providing primary care to isolated floating villages on the Tonle Sap Lake, Cambodia. Rural Remote Health. 2014;14:2612.

National Institute of Statistics (NIS). General population census of Cambodia 1998, final census results. 2nd ed. Phnom Penh: Ministry of Planning; 2002. p. 88.

Tan R, Chanto MCT. et al. Survival of Escherichia coli K12 in Tonle Sap Lake and Tonle Sap water by using dialysis membrane as a supporter, Proceeding of the 2nd International symposium on conservation and management of tropical lakes; 2017. p. 133–9.

Tanaka Y, Hayashida S, et al. The relationship of the feces protein particles to rice protein bodiest. Agric Biol Chem. 1975;39:515–8.

WHO. Quantitative microbial risk assessment: application for water safety management. 2016. https://apps.who.int/iris/handle/10665/246195. Accessed 17 Jun 2021.

WHO. World Health Statistics 2020, 2020. https://apps.who.int/iris/bitstream/handle/10665/332070/9789240005105-eng.pdf. Accessed 27 Dec 2020).

Chapter 42
Virus Removal by Poly Aluminum Chloride and Calcium Hypochlorite

Porsry Ung, Reasmey Tan, Chanthol Peng, Sopheap Chheng, Sopheap Suon, Yasunori Tanji, and Kazuhiko Miyanaga

42.1 Life and Sanitation in Floating Villages

Water is essential for life and the environment; however, the amount of freshwater on the planet is limited. Water sources are threatened by anthropogenic pollutants, especially in developing countries, because of economic development without environmental protection (Chea et al. 2016; Kov et al. 2008). The lack of access to clean drinking water remains one of the major problems facing humanity in the twenty-first century. Waterborne diseases are usually derived from poor fecal disposal, which leads to the spread of pathogenic microorganisms in the water (Chaps. 40 and 41, Pichel et al. 2019).

More than 1.7 million people live and depend on Tonle Sap Lake, which is the largest flood-pulse freshwater lake in Southeast Asia. Many floating villages exist around the lake, classified into three types: water-based, land-based, and water–land-based villages. As a general observation, the kitchen and toilet are installed near each other at the rear of the floating houses. The villagers usually defecate, urinate, and discard household waste directly into the lake water. Although the lake water is contaminated continuously from fecal and household waste, villagers still use the lake water for their daily activities, such as bathing, washing, cooking, and drinking. In practice, villagers collect the lake water near or far from their home and settle the sediment by using a coagulant. After one-day settling, the supernatant becomes transparent, and it is used by the villagers with/without further treatment, such as

P. Ung
Ministry of Industry, Science, Technology and Innovation, Phnom Penh, Cambodia

R. Tan · C. Peng · S. Chheng · S. Suon
Institute of Technology of Cambodia, Phnom Penh, Cambodia

Y. Tanji · K. Miyanaga (✉)
Tokyo Institute of Technology, Tokyo, Japan
e-mail: miyanaga.kazuhiko@jichi.ac.jp

C. Yoshimura et al. (eds.), *Water and Life in Tonle Sap Lake*,
https://doi.org/10.1007/978-981-16-6632-2_42

filtration by cloth with coarse grids or boiling. Some villagers use the lake water directly without any treatment, whereas others can afford to buy bottled water for drinking.

Water usage becomes more challenging for villagers in the dry season, when the floating villages gather close to each other and the water quality worsens, with a bad odor. Notably, the concentration of *Escherichia coli*, used as an index of fecal contamination, in the surrounding water of floating village in Tonle Sap Lake is above the safety level recommended by the World Health Organization (WHO) (Chap. 41). Consequently, villagers, particularly children, are exposed to many waterborne diseases, and most of their income is spent on their healthcare (Chap. 40). The villagers understand the risks of contaminated water to their health, and they demand safe water and affordable sanitation (Brown et al. 2010). Therefore, the lake water treatment in place is critically important for the health of the villagers.

Chlorination and different physicochemical processes potentially eliminate microbes, including virus, from the water. Considering Stokes' law, viruses do not settle under static conditions owing to their size (mostly less than 0.1 μm). After being enmeshed into flocs that are produced in the coagulation process, waterborne virus particles settle down by gravity (Matsui et al. 2003). For such coagulation, poly aluminum chloride (PAC) has shown greater benefits than commercialized alum (Al $(SO_4)_2 \cdot 12H_2O$), as PAC has a wider suitable range of pH (6–9) for water treatment. A single water treatment with a coagulant could remove suspended solids and may partially remove microbes from the water. Thus, in general, an additional treatment is required to efficiently remove the microbes from water before consumption. To inactivate bacteria, hypochlorite is a well-known economical disinfectant that is used routinely in water and wastewater treatments. Therefore, this chapter describes the efficiency of PAC and calcium hypochlorite ($Ca(OCl)_2$) doses as chlorination for the removal and deactivation of viruses in water.

42.2 Virus Removal Experiment

In our experiment, the T4 phage was used as a model virus to evaluate the treatment efficiency of PAC and $Ca(OCl)_2$ in the suspension. The stock solution of the T4 phage was propagated in *E. coli* K12 host cells using the procedure mentioned in a previous study (Clokie and Kropinski 2009). The T4 phage culture was concentrated using 10% (w/v) polyethylene glycol 6000 and 4% (w/v) sodium chloride for overnight incubation at 4 °C. The phage solution was then centrifuged at 10,000 × *g* for 90 min, and the supernatant was discarded. Sodium–magnesium (SM) buffer and chloroform were added, and the phages were suspended in the solution. The solution was rested for 15 min and then centrifuged at 10,000 × *g* for 5 min; next, the water phase was transferred to a new tube, and SM buffer was added to adjust the phage concentration.

The experiment was conducted using 200 mL of distilled water, with 2 mL of T4 phages at a concentration of about 10^9 plaque-forming units (PFU)/mL, and adjusted with various doses of PAC: 0, 30, 50, and 100 mg/L. The experiments were conducted in duplicate, with a settling time of 0, 15, 60, 120, and 240 min. The aqueous samples were prepared by mixing the solution using a magnetic stirrer at a speed of 120 rpm for 5 min, and then they were left to settle by gravitation for 15, 60, 120, and 240 min. The phages in the supernatant were quantified using the plaque assay method.

The tests with $Ca(OCl)_2$ were performed with 100 mL of distilled water, with the addition of 1 mL T4 phage stock solution and modified doses of $Ca(OCl)_2$ at 0, 3, 5, and 10 mg/L. The solutions were homogenized using a magnetic stirrer at a speed of 120 rpm for 2 min for a contact time of 0, 1, 10, 20, and 40 min. The test sample was taken from the aqueous solution and used to quantify the T4 phage concentration using the plaque assay method.

The pH in the samples was measured using a WM-32EP dual-channel portable electric conductivity/pH meter. Turbidity was measured using the HANNA Instruments model HI 98703. This model measures water turbidity in a range between 0.00 and 1000 Nephelometric Turbidity Units (NTU), with $\pm 0.2\%$ accuracy. In addition, 10 mL of the sample was used to determine the concentration of free chlorine using the EUTECH Instruments model colorimeter C201. This instrument measures the concentration of free and total chlorine in the range of 0–6 mg/L, with a resolution of 0.01–0.1 mg/L and an accuracy of ± 0.02 to 0.2 mg/L (OAKION 2012).

42.3 Effect of Poly Aluminum Chloride (PAC)

42.3.1 PAC Dose on Water Quality and T4 Phages

The treatment efficiency of T4 phages using PAC was determined in distilled water at room temperature, without adjusting the pH or turbidity. The pH of each water sample changed depending on the PAC doses (Table 42.1). The higher dose of PAC, 100 mg/L, reduced the pH value of the water sample from 7.06 to 4.38. A high PAC concentration resulted in the high turbidity of the solution. The turbidity of each PAC dose after 240 min was decreased owing to the flocculation process by the coagulant. The neutralized interaction of negative and positive charges between the

Table 42.1 Effects of the initial poly aluminum chloride (PAC) concentration on pH and turbidity of water in the coagulation experiment

		Turbidity (NTU)	
PAC dose (mg/L)	pH	Before treatment (0 min)	After treatment (240 min)
0	7.06	1.08	0.32
30	6.67	3.23	0.36
50	4.50	4.82	0.47
100	4.38	12.8	0.45

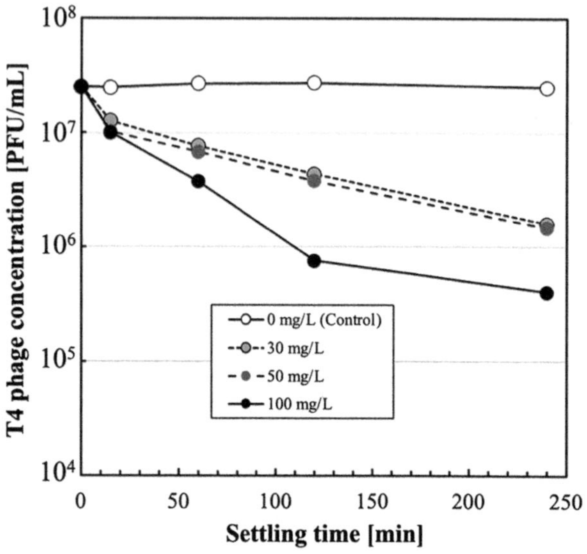

Fig. 42.1 Effect of poly aluminum chloride (PAC) dose on the T4 phage concentration in distilled water after a settling time of 15, 60, 120, and 240 min

particle and the coagulant is explained as a mechanism of precipitation (Mirzaiy et al. 2012).

There was no decrease in the T4 phage concentration in the control sample (without PAC) during the time course of the experiment; however, its concentration was reduced when PAC was applied at different concentrations (Fig. 42.1). Compared to the control condition, the T4 phage concentration decreased from 15 min when 30 mg/Land 50 mg/L PAC were applied. The T4 phage concentration showed a greater reduction when 100 mg/L PAC was introduced to the sample. There are two possible mechanisms of coagulation for this removal of T4 phages. The virions may aggregate as one infectious agent and act as a single phage to produce one plaque. If this was the case, one plaque would be produced by such an aggregate consisting of multiple virions. The other possibility is that the virus is inactivated after exposure to aluminum-hydrolyzing compounds during the coagulation process (Matsushita et al. 2011). The reduction of T4 phages in the study was caused by coagulation and floc formation, mainly because of the inactivating effect of dissolved PAC. Nevertheless, this mechanism needs further investigation, because even though coagulation of the highly alkaline aluminum cation of dissolved PAC occurs immediately after its addition to water, it needs more time to interact with the T4 phages (Kreißel et al. 2014).

The PAC dose of 100 mg/L was better than the other doses to remove T4 phages. However, it would not meet the guidelines for residual aluminum in drinking water (WHO 2004). The large amount of residual aluminum coagulant remaining in treated water may affect human health and may be associated with the increased turbidity and color in the distribution system, which might result in a concentration higher than the safe level, that is, 0.1–0.2 mg/L of aluminum (WHO 2013). The control of pH is also important to minimize the aluminum concentration, and the optimal pH to minimize the residual aluminum coagulant is around 6.5 (Kimura et al. 2013).

42.4 Effect of Calcium Hypochlorite ($Ca(OCl)_2$) on T4 Phages

In this study, the pH value of the control sample was 6.8, whereas that of each suspension sample was raised marginally throughout each experiment in the range from 7.0 to 7.5 (Table 42.2), which produced hypocaloric acid (HOCl) and hypochlorite anion (OCl^-) with similar proportions. The chlorine in water is in the form of HOCl at a pH of around 5, whereas it is OCl^- at a pH of 8.5. In this manner, pH plays a significant role in the efficiency of water treatment processes and in the formation of disinfection byproduct (DBP) in the suspension.

In the suspension, $Ca(OCl)_2$ behaves and functions in a manner similar to chlorine (Cl_2). Hypochlorite in the water is dissociated to produce OCl^- under the equilibrium state between different species of active chlorine, mainly, Cl_2, HClO, and OCl^-. The amount of each species depends strongly on the solution's physico-chemical parameters, such as pH, temperature, and ionic force. Among them, pH is the major factor influencing the hydro-chemical distribution of the various chlorine species (Escudero-Oñate 2014).

The concentration of free chlorine increased as the $Ca(OCl)_2$ doses increased, whereas the concentration of chlorine decreased over the time of the reaction (Fig. 42.2). On comparing the different doses of $Ca(ClO)_2$ from 1 to 10 mg/L, the higher doses of $Ca(ClO)_2$ in the suspension evidently showed a faster reduction in its concentration than the lower doses. This reduction was caused by volatilization and reaction with the phages.

The result of the experiment with the 1 mg/L of $Ca(OCl)_2$ dose showed that the viricidal effectiveness of chlorine diminished at a low dose (Fig. 42.3). For instance, 1 mg/L of $Ca(OCl)_2$ reduced T4 phages by 8.96% in the first minute and by 91.68% within 40 min. Similarly, the inactivation of T4 phages by 3 mg/L of $Ca(OCl)_2$ was achieved in 1 min of treatment and reduced T4 phages in suspension by 91.67% after 10 min of treatment. This dose showed a final overall reduction to 98.40% after 40 min of treatment. A dose of 5 mg/L of $Ca(OCl)_2$ removed the T4 phages faster than 1 and 3 mg/L of $Ca(OCl)_2$. The dose of 5 mg/L of $Ca(OCl)_2$ reduced T4 phages by 99.54% in 1 min and by 99.97% in 40 min. The dose of 10 mg/L of $Ca(OCl)_2$

Table 42.2 Dose of calcium hypochlorite ($Ca(OCl)_2$) and the final pH values

$Ca(OCl)_2$ dose (mg/L)	pH
0	6.81
1	7.05
3	7.29
5	7.35
10	7.42

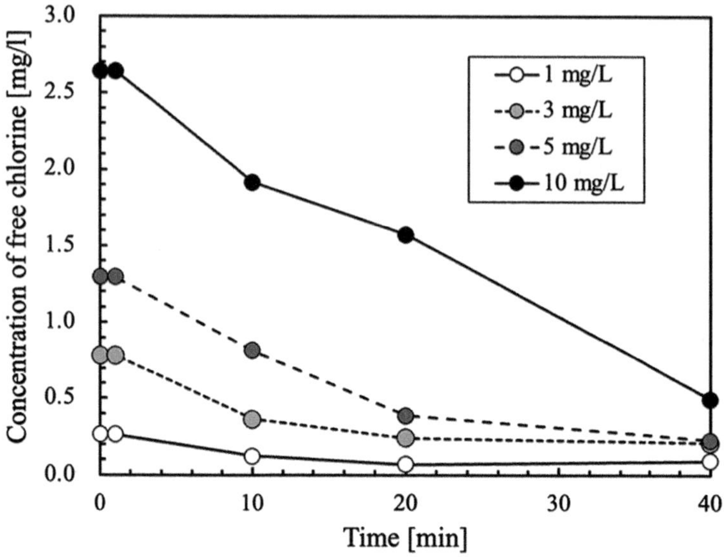

Fig. 42.2 Temporal shifts of free chlorine concentration in T4 suspension after introducing different initial concentrations of calcium hypochlorite ($Ca(OCl)_2$) of 1, 3, 5, and 10 mg/L

showed the highest effectiveness in removing T4 phages (99.81% within 1 min and 99.997% within 40 min) in the suspension. However, the free chlorine derived from 10 mg/L of $Ca(OCl)_2$ was higher than the free chlorine dose in WHO guidelines for drinking water standards (Spira and Grimbleby 1943). Hence, 5 mg/L of $Ca(OCl)_2$ is optimal for virus removal in the water treatment.

Overall, our experimental results imply that coagulant treatment might not be enough to remove the virus from the lake water, and thus, we recommend applying additional chlorination to ensure its safety. For further confirmation and development, we would recommend applying a similar experimental approach to the actual lake water and other types of viruses.

Key Points

- The experiment confirmed the removal of T4 phages in the coagulation process using PAC. The concentration of 100 mg/L of PAC showed the highest efficiency removal of T4 phages in the suspension among the doses from 0 to 100 mg/L. However, the concentration of T4 phages remained in the suspension after 240 min of treatment with PAC.
- The disinfection process in the suspension using calcium hypochlorite doses of 5 and 10 mg/L showed high removal rates of T4 phages of 99.99%.
- Based on the experimental results, coagulant treatment might not be enough to remove the virus from the lake water, and thus, we recommend applying additional chlorination to ensure its safety.

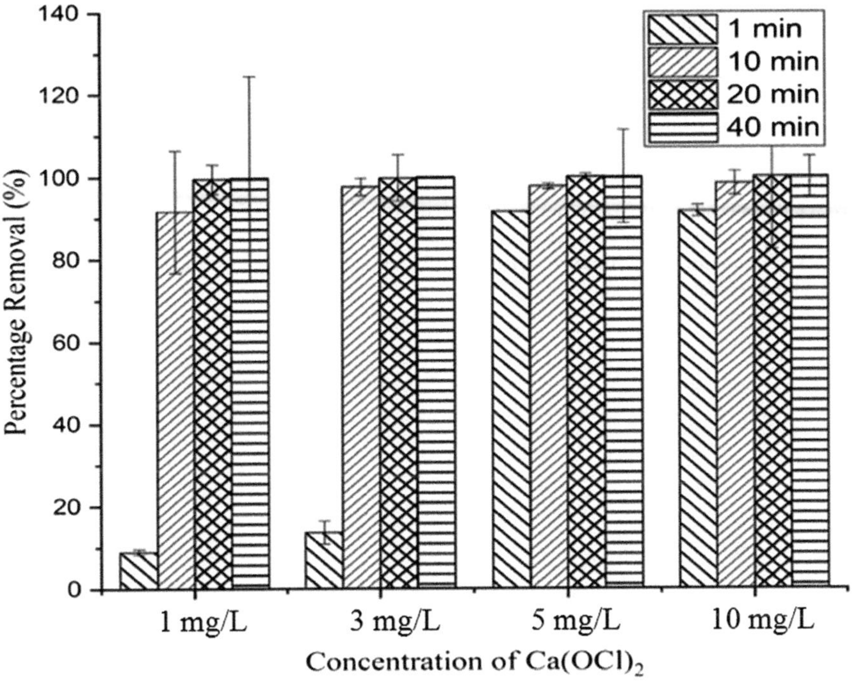

Fig. 42.3 Effects of calcium hypochlorite ($Ca(OCl)_2$) concentration and contact time on T4 phage removal

References

Brown M, Sodaneath H, Smith J, Hagan J. Sanitation in floating communities in Cambodia. Ministry of Rural Development. London: Water Aid, 2010.

Chea R, Grenouillet G, Lek S. Evidence of water quality degradation in lower Mekong basin revealed by self-organizing map. PLoS One. 2016;11(1)

Clokie MRJ, Kropinski AM. Bacteriophages: methods and protocols. Methods Mol Biol. 2009; xxii:307.

Escudero-Oñate C. *Survey of sodium and calcium hypochlorite. Skin (Los. Angeles)*. Copenhagen: The Danish Environmental Protection Agency; 2014. p. 7681.

Kimura M, Matsui Y, Kondo K, Ishikawa TB, Matsushita T, Shirasaki N, Al M. Minimizing residual aluminum concentration in treated water by tailoring properties of polyaluminum coagulants. Water Res. 2013;47(6):2075–84. Elsevier Ltd.

Kov P, Sok H, Roth S, Chhoeun K, Hutton G. Economic impacts of sanitation in Cambodia. *World Bank*, (February). 2008. http://www.wsp.org/library

Kreißel K, Bösl M, Hügler M, Lipp P, Franzreb M, Hambsch B. Inactivation of F-specific bacteriophages during flocculation with polyaluminum chloride - a mechanistic study. Water Res. 2014;51:144–51.

Matsui Y, Matsushita T, Sakuma S, Gojo T, Mamiya T, Suzuoki H, Inoue T. Virus inactivation in aluminum and polyaluminum coagulation. Environ Sci Technol. 2003;37(22):5175–80.

Matsushita T, Shirasaki N, Matsui Y, Ohno K. Virus inactivation during coagulation with aluminum coagulant. Chemosphere. 2011;85(4):571–6. Elsevier Ltd

Mirzaiy A, Takdastan A, Alavi N, Mohamadian H. Removal of turbidity, organic matter, coliform and heterotrophic bacteria by coagulants poly aluminum chloride from karoon river water in Iran. Asian J Chem. 2012;24(6):2389–93.

OAKION. Multiparameter colorimeters. C201, C301 C401 Multiparam Color. 2012;35645.

Pichel N, Vivar M, Fuentes M. The problem of drinking water access: a review of disinfection technologies with an emphasis on solar treatment methods. Chemosphere. 2019;218:1014–30. Elsevier B.V.

Spira L, Grimbleby FH. Chlorine in drinking water. J Hyg (Lond). 1943;43(2):142–5.

WHO. Chemicals from water treatment and distribution. 2004;2:55–60. https://www.who.int/water_sanitation_health/dwq/cmp130704chap8.pdf

WHO. Aluminium in drinking water. 2013. https://www.who.int/water_sanitation_health/publications/aluminium/en/

Part X
Environmental Shifts and Management

Chapter 43
Impact of Climate Change on the Hydrological Regime of Tonle Sap Lake

Hideto Fujii, Ichiro Yoneda, Yoichi Fujihara, Keisuke Hoshikawa, and Takashi Nakamura

43.1 Climate Change and Tonle Sap Lake

According to the fifth report of the Intergovernmental Panel on Climate Change (IPCC), the global average temperature rose to 0.85 °C between 1880 and 2012, and the sea level rose to 0.19 m between 1901 and 2010. The temperature and sea level would rise to 2.6–4.8 °C and 0.45–0.82 m at the end of this century, respectively, under the RCP8.5 scenario in which no strict measures are taken for global warming, and they would rise to 0.3–1.7 °C and 0.26–0.55 m under the RCP2.6 scenario in which severe measures are taken (IPCC 2014).

The effects of climate change on Tonle Sap Lake (TSL) have been studied, and future projections have been made by Eastham et al. (2008), Västilä et al. (2010), and Keskinen et al. (2015). Västilä et al. (2010) have predicted that the mean lake levels will rise by around 0.2 m and the inundation periods of TSL will be prolonged by 6–9% in 2050 compared to the baseline (BL) period in 1997–2000.

Arias et al. (2012) compared dry, average, and wet hydrological years and reported that the dry year had the largest impacts and the wet year had the least impacts by climate change; they also reported that the lake water level in 2030 would remain almost the same during the low-water months in April and May but increase by 0.2–0.75 m during the remaining months.

H. Fujii (✉) · I. Yoneda
Yamagata University, Tsuruoka, Japan
e-mail: fhideto@tds1.tr.yamagata-u.ac.jp

Y. Fujihara
Ishikawa Prefectural University, Nonoichi, Japan

K. Hoshikawa
Toyama Prefectural University, Imizu, Japan

T. Nakamura
Tokyo Institute of Technology, Tokyo, Japan

C. Yoshimura et al. (eds.), *Water and Life in Tonle Sap Lake*,
https://doi.org/10.1007/978-981-16-6632-2_43

On the other hand, Oeurng et al. (2019) assessed the impacts of climate change on river flows in 11 sub-basins of TSL using the Soil and Water Assessment Tool model, and the projected river flows from three GCMs (GFDL-CM3, GISS-E2-R-CC, and IPSL-CM5A-MR) indicated a likely decrease in both wet- and dry-season flows. The mean annual projected flow reductions were 9–29%, 10–35%, and 7–41% for the 2030s, 2060s, and 2090s. Their study, however, did not consider the reverse inflow from the Mekong River, which accounts for the majority of the TSL inflow, so their study did not conclude that the total inflow of TSL will decrease.

Furthermore, Keskinen et al. (2015) evaluated the impact of climate change on TSL, summarizing the outputs of the VMod hydrological model and EIA 3D floodplain model. They showed that climate change will cause changes to the rainfall and temperature in the area, impacting the runoff and water levels in the Mekong mainstream and the Tonle Sap system. However, their analysis indicated that the exact impact of climate change remains unclear, mainly because of differences among the GCMs applied to the Mekong basin. Even the direction of flow changes caused by climate change differed depending on the emission scenario and GCM used. Consequently, it is impossible to even say whether climate change will increase or decrease the flood-season water level in the Tonle Sap. Within the time frame used in their model, that is, by year 2042, climate change alone does not significantly impact the dry-season water level in the lake (Keskinen et al. 2015).

Hoang et al. (2016) modeled and analyzed changes in river flow regimes and hydrological extremes (high flow and low flow) to assess the hydrological impact of the Mekong basin using the CMIP5 climate projection. Their study found that the water cycle of the Mekong River intensifies under climate change and the annual river flow increases by 5–16% based on the ensemble average of future scenarios. Although individual scenarios of climate prediction and hydrological impacts have shown significant differences in flow changes, the ensemble of scenarios showed less uncertainty than CMIP3-based assessments.

Hoang et al. (2019) focused on the dynamics and mechanistic understanding of future hydrological changes under multiple drivers of climate change, hydropower development, and irrigation expansion factors in the Mekong River basin by comparing the impacts of individual and multiple drivers. They projected that the hydrological impact of climate change alone indicated that annual flow rates varied from −7% to +11% in the most downstream Kratie, with only one scenario member (i.e., HadGEM-RCP85) showing reduced flow rates. The discharge in the dry season increases significantly in the range of +15% to +20% from January to May and increases more in the upstream reaches, such as Vientiane and Mukdahan. The discharge in the rainy season also shows an increasing tendency, but the relative change is smaller than that in the dry season. Changes in the monthly discharge showed contrasting changes between the early rainy season (June and July) and the remaining rainy season. Thus, the scenario ensemble average showed more than −7% of decrease at Kratie in June and July, whereas it showed 3–8% of the increase in the high-flow period (August–September) at Kratie.

Hoang et al. (2016, 2019) analyzed the impacts of the mainstream of the Mekong River alone and did not analyze the impacts to TSL in their studies. However, considering that the annual flow rate and ensemble average discharge in the latter half of the rainy season at Kratie were projected to increase, the annual inflow and water level in the latter half of the rainy season of TSL are also estimated to rise.

To summarize the above studies, some studies predicted that the water level in the wet season will increase, especially in the dry hydrological year, whereas others indicated that inflow from TSL sub-basins is projected to decrease or that exact impacts of climate change remain unclear mainly because of uncertainties among GCMs.

43.2 Model and Boundary Conditions

The hydrological Nedbor-Afstromnings Model (NAM; refer to Chap. 10 for details) was applied to the TSL basin, and the Geomorphology-Based Hydrological Model (GBHM) was applied to Mekong River, the upstream part of Kratie (Yang et al. 2001). A hydrodynamic model, MIKE11, was used to simulate the area from Kratie of the Mekong mainstream and the TSL basin to the Vietnamese border at Tanchau and Chaudoc. Figure 43.1 shows the river network, cross sections, and boundary build-up by MIKE11. The one-dimensional hydraulic model MIKE11 has been used for several rivers in Southeast Asia (Tebakari et al. 2006; Kura and Tebakari 2012). MIKE11 solves the continuity and Saint-Venant equations. Finite differences are calculated using a staggered grid that alternately arranges the calculation points of the water level and discharge and solved using a six-point Abbott scheme of the implicit method.

43.3 Climate-Change Scenarios

Nine scenarios including the BL were set up to investigate the impacts of climate change on the hydrological regime of TSL and its inundation area (Table 43.1). The model calculations were performed by using downscaled (to 0.5°) and bias-corrected datasets provided by the Inter-Sectional Impact Model Intercomparison Project (ISI-MIP), which used the data from five GCMs from the Coupled Model Intercomparison Project Phase 5 (CMIP5) archives as the input.

The ISI-MIP is a project designed to synthesize impact projections in the agriculture, water, biome, health, and infrastructure sectors at different levels of global warming. Bias correction was conducted by calculating the cumulative frequency distribution of daily values of weather elements from WATCH and GCM weather data from 1960 to 1999 and determining a conversion function that matches the distribution of WATCH weather data (Hempel et al. 2013).

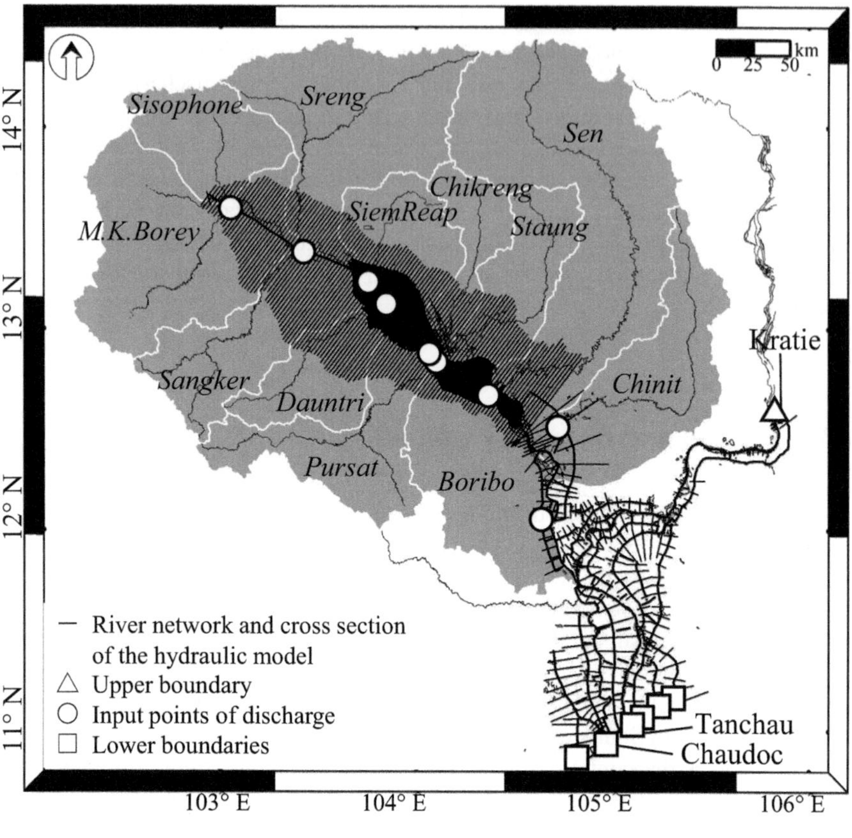

Fig. 43.1 The river network, river cross sections, and boundary conditions of the one-dimensional hydraulic model (MIKE11)

Table 43.1 Scenarios and GCMs for climate-change impact assessments on TSL

Scenario	Period	RCP	GCM
BL	1991–2000	–	–
CN2.6HG	2041–2050	RCP2.6	HadGEM2
CN2.6MIR	2041–2050	RCP2.6	MIROC
CN8.5HG	2041–2050	RCP8.5	HadGEM2
CN8.5MIR	2041–2050	RCP8.5	MIROC
CF2.6HG	2090–2099	RCP2.6	HadGEM2
CF2.6MIR	2090–2099	RCP2.6	MIROC
CF8.5HG	2090–2099	RCP2.6	HadGEM2
CF8.5MIR	2090–2099	RCP8.5	MIROC

HadGEM2-ES and MIROC-ESM-CHEM (hereinafter referred to as HadGEM2 and MIROC, respectively) were selected as GCMs. This is because the previous studies confirmed that MIROC and HadGEM2 can reproduce rainfall in the twentieth-century well in Southeast Asia and have been used to assess the impacts of climate change in the Mekong River basin (Eastham et al. 2008).

The Representative Concentration Pathway (RCP) was introduced by the IPCC Fifth Assessment Report, and four scenarios at the end of the twenty-first century were defined with a radiative forcing of 2.6, 4.5, 6.0, and 8.5 W/m^2 (Moss et al. 2010). Among them, the RCP2.6 and RCP8.5 scenarios were selected in this study.

We selected 1991–2000 as the BL period and set up two projection periods of 2041–2050 and 2090–2099 as the near future (CN) and far future (CF), respectively. The dry, average, and wet years for the BL period were determined based on the yearly maximum water level observed at Kg. Luong. Years 1998, 1999, and 2000 were determined as the dry, average, and wet years, respectively (Västilä et al. 2010; Arias et al. 2014). The dry, average, and wet years for the projection period were determined based on the discharge at Kratie, which was calculated from GBHM in its application to the Mekong River basin (Suif et al. 2017).

43.4 Impacts of Climate Change on Hydrological Regime

The projected water level at Kg. Luong in the near future and far future differed depending on the climate-change scenario (Fig. 43.2). The monthly mean water level in October, when the lake water level is the highest in the year, in the near and far future was projected to rise compared to the BL period. Their increments were +1.13 and +0.92 m in the dry year, +0.38 and +0.84 m in the average year, and +0.46 and +0.80 m in the wet year, respectively. The monthly mean water level in the flood season was found to rise because of climate change, regardless of the hydrological year, which is consistent with previous studies (Västilä et al. 2010; Arias et al. 2014).

Table 43.2 shows the annual minimum and maximum water levels, inundation areas, and storage volumes of TSL in the near and far future in three hydrological years. The peak inundation area of the lake was projected to increase, regardless of hydrological years, in both the near and far future. The increased inundation areas of the mean of four scenarios in the near and far future are as follows: +1337 and +993 km^2 in a dry year, +567 and +1040 km^2 in an average year, and +797 and +968 km^2 in a wet year. Dry years show the largest expansion of the inundation area in both the near and far future. Figure 43.3 shows the projected shifts of the flood period in the near and far future in three hydrological years from the BL flood duration. In the dry year, it is remarkable that the inundation starts earlier and ends later. The starting date of inundation in the average and wet years does not change much, but the ending date is delayed in both the near and far future. Here, flood refers to the water level higher than 2.44 m above the mean sea level, which is consistent with the value in Kummu and Sarkkula's (2008) study.

Key Points

- The impacts of climate change on the hydrological regime of TSL, specifically, the water level and inundation area, were evaluated using numerical models.

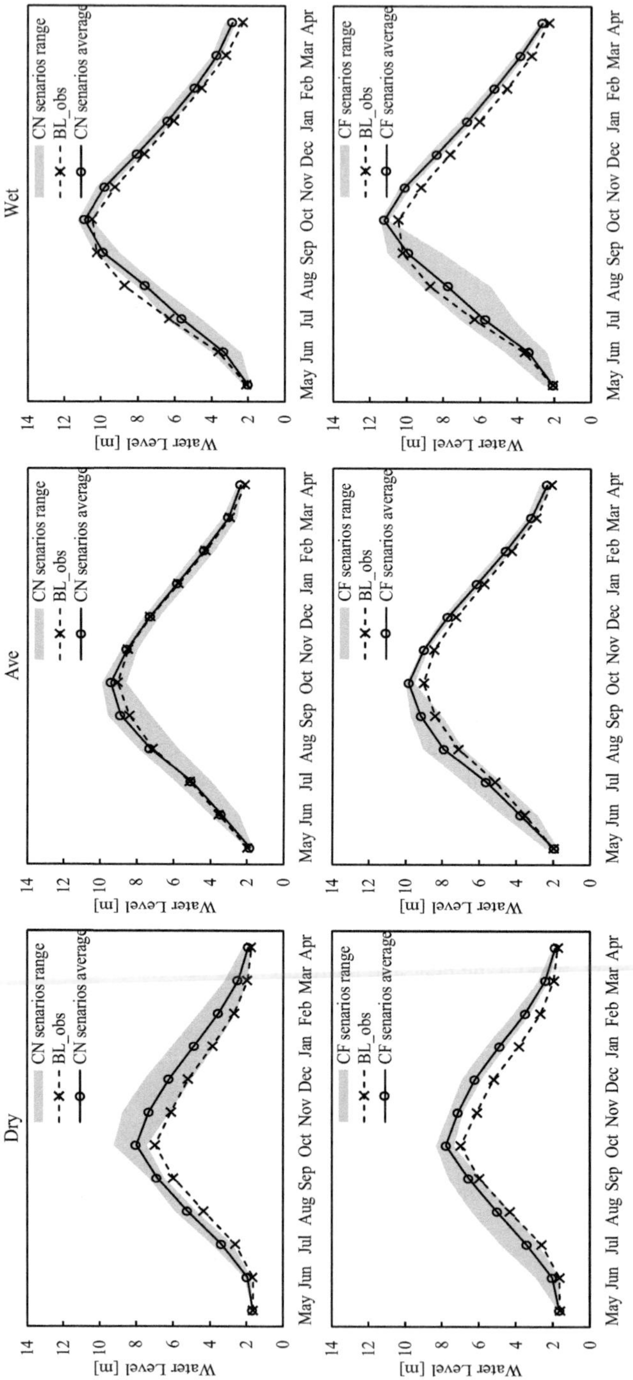

Fig. 43.2 The monthly water level at Kg. Luong in the near and far future in the dry, average, and wet hydrological years

Table 43.2 Annual minimum and maximum water level, inundation area, and storage volume of TSL in the near and far future in three hydrological years. The number in brackets shows the difference from the BL

Year		Annual minimum						Annual maximum					
		WL (m)		Area (km^2)		Volume (km^3)		WL (m)		Area (km^2)		Volume (km^3)	
Dry	BL	1.60		3,134		2.2		7.11		9,728		35.3	
	CN	1.62	(0.02)	3,160	(25)	2.3	(0.0)	8.22	(1.12)	11,065	(1,337)	47.4	(12.1)
	CF	1.60	(0.0)	3,134	(0.0)	2.2	(0.0)	7.94	(0.83)	10,721	(993)	44.1	(8.8)
Ave	BL	1.74		3,303		2.5		9.20		12,231		59.5	
	CN	1.65	(–0.09)	3,196	(–107)	2.3	(–0.2)	9.67	(0.47)	12,798	(567)	65.8	(6.4)
	CF	1.67	(–0.08)	3,212	(–91)	2.4	(–0.2)	10.07	(0.87)	13,271	(1,040)	71.4	(11.9)
Wet	BL	1.90		3,488		2.9		10.74		14,078		81.4	
	CN	1.82	(–0.07)	3,401	(–87)	2.7	(–0.2)	11.40	(0.67)	14,875	(797)	91.9	(10.5)
	CF	1.83	(–0.07)	3,404	(–83)	2.7	(–0.2)	11.55	(0.81)	15,046	(968)	94.2	(12.9)

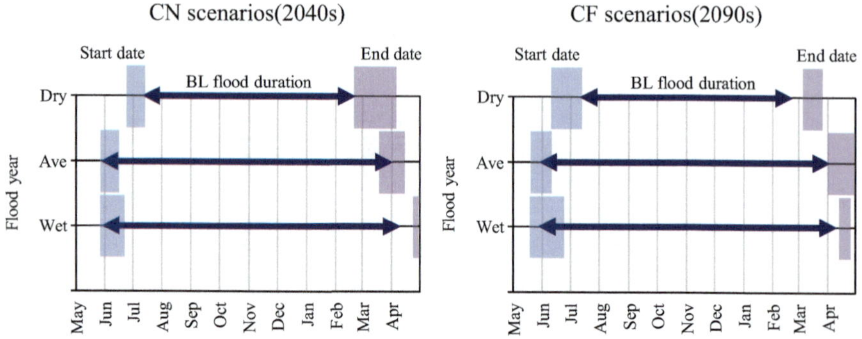

Fig. 43.3 The projected shift of the flood period due to climate change. *Arrows* show the flood duration in the BL year, and shaded areas are the range of four scenarios by two CGMs and two RCPs. The flood duration means the days when the water level is higher than 2.44 m above the mean sea level

- As a result of climate change, the mean water level in October, when the lake water level is the highest, rises regardless of hydrological years, and it was shown that the highest level in the dry year rises by 1.0 m in the 2040s. However, the minimum annual water level drops by 7–9 cm except in the dry year. Therefore, annual fluctuations in lake water levels are expected to increase.
- The inundation period was shown to be longer in both the near and far future compared to the BL period, especially in the dry year. The inundation period starts early and ends late, and the prolonged period was estimated to be around 30 days.
- As the projections in this chapter are based on two GCMs (i.e., HadGEM2 and MIROC-ESM-CHM), further studies are necessary to investigate the uncertainty of GCM models.

References

Arias ME, Cochrane TA, Kummu M, Lauri H, Holtgrieve GW, Koponen J, Piman T. Impacts of hydropower and climate change on drivers of ecological productivity of Southeast Asia's most important wetland. Ecol Model. 2014;272:252–63.

Arias ME, Cochrane TA, Piman T, Kummu M, Caruso BS, Killeen TJ. Quantifying changes in flooding and habitats in the Tonle Sap Lake (Cambodia) caused by water infrastructure development and climate change in the Mekong Basin. J Environ Manag. 2012;112:53–66.

Eastham J, Mpelasoka F, Mainuddin M, Ticehurst C, Dyce P, Hodgson G, Ali R, Kirby M. Mekong River basin water resources assessment: impacts of climate change. Water for a Healthy Country National Research Flagship: CSIRO; 2008.

Hempel S, Frieler K, Warszawski L, Schewe J, Piontek F. A trend-preserving bias correction - the ISI-MIP approach. Earth Syst Dynam. 2013;4:219–36.

Hoang LP, Lauri H, Kummu M, Koponen J, van Vliet MTH, Supit I, Leemans R, Kabat P, Ludwig F. Mekong river flow and hydrological extremes under climate change. Hydrol Earth Syst Sci. 2016;20(7):3027–41.

Hoang LP, van Vliet MTH, Kummu M, Lauri H, Koponen J, Supit I, Leemans R, Kabat P, Ludwig F. The Mekong's future flows under multiple drivers: how climate change, hydropower developments and irrigation expansions drive hydrological changes. Sci Total Environ. 2019;649:601–9.

IPCC. In: Core writing team, Pachauri RK, Meyer LA, editors. Climate change 2014: synthesis report. Contribution of working groups I, II and III to the fifth assessment report of the intergovernmental panel on climate change. Geneva, Switzerland: IPCC; 2014.

Keskinen M, Someth P, Salmivaara A, Kummu M. Water energy-food nexus in a trans-boundary River Basin: the case of Tonle Sap Lake, Mekong River Basin. Water. 2015;7:5416–36.

Kummu M, Sarkkula J. Impact of the Mekong river flow alteration on the Tonle Sap flood pulse. Ambio. 2008;7:85–92.

Kura S, Tebakari T. Hydrological impact of regional climate change in the Chao Phraya River, Basin, Thailand. Hydrol Res Lett. 2012;6:53–8.

Moss R, Edmonds J, Hibbard K, Manning M, Rose S, Vuuren D, Carter T, Emori S, Kainuma M, Kram T, Meehl G, Mitchell J, Nakicenovic N, Riahi K, Smith S, Ronald S, Thomson A, Weyant J, Wilbanks T. The next generation of scenarios for climate change research and assessment. Nature. 2010;463(7282):747–56.

Oeurng C, Cochrane TA, Chung S, Kondolf MG, Piman T, Arias ME. Assessing climate change impacts on river flows in the Tonle Sap Lake basin, Cambodia. Water. 2019;11(3):618. https://doi.org/10.3390/w11030618.

Suif Z, Yoshimura C, Saavedra O, Ahmad N, Hul S. Suspended sediment dynamics changes in Mekong river basin: possible impacts of dams and climate change. Int J GEOMATE. 2017;12 (34):140–5.

Tebakari T, Yoshitani J, Suvanpimol C, Miyamoto M, Yamada T. Development and validation of hydrological circulation model using MIKE11 in the Chao Phraya River basin, kingdom of Thailand. J Jpn Soc Hydrol Water Res. 2006;19:212–20.

Västilä K, Kummu M, Sangmanee C, Chinvanno S. Modelling climate change impacts on the flood pulse in the lower Mekong floodplains. J Water Clim Chang. 2010;1:67–86.

Yang D, Herath S, Oki T, Musiake K. Application of distributed hydrological model in the Asian monsoon tropic region with a perspective of coupling with atmospheric models. J Meteorol Soc Jpn. 2001;79(1B):373–85.

Chapter 44
Projection of Land Use and Land Cover in the Lake Basin

Kong Chhuon, Sreykeo Puok, Kim Lengthong, Ratino Sith, Ratha Doung, Khy Eam Eang, Rajendra Khanal, and Sytharith Pen

44.1 Land Use and Land Cover in the Lake Basin

Land use is a central component of ecological, social, and economic systems acting across various spatial scales. Land-use change is a complex and dynamic process that integrates natural and human systems. Land-use change has direct impacts on soil, water, and atmosphere and is thus related directly to many environmental issues (Lambin et al. 2003). Change in land use and land cover (LULC) is an important factor in the study of environmental change and its management (Van Asselen and Verburg 2013).

Tonle Sap Lake (TSL) is influenced by the surrounding basin where land-use change impacts hydrological processes, such as surface runoff generation, groundwater recharge, and soil erosion. Some studies have assessed the past forest cover (Potapov et al. 2019). Regarding the future LULC changes in the lake basin, for instance, in Cambodia, several studies have addressed the impact of land-use change on water supply for the Angkor temple and the surrounding population and the impact of land-use and climate changes on water resources in Srepok River Basin (Chim et al. 2019; Van Ty et al. 2012). Ruiter et al. (2017) studied the land-use change scenarios for hydrological impact assessment in a river of the TSL basin. All these studies highlighted the importance of forest conservation, which is strongly connected with water resources and hydrology. Therefore, this chapter provides a methodological framework for a systematic study of land-use simulation modeling and the projected results of LULC from different changing scenarios in the TSL

K. Chhuon (✉) · S. Puok · K. Lengthong · R. Sith · R. Doung · K. E. Eang · S. Pen
Institute of Technology of Cambodia, Phnom Penh, Cambodia
e-mail: chhuon.k@itc.edu.kh

R. Khanal
Tokyo Institute of Technology, Tokyo, Japan

Policy Research Institute, Kathmandu, Nepal

C. Yoshimura et al. (eds.), *Water and Life in Tonle Sap Lake*,
https://doi.org/10.1007/978-981-16-6632-2_44

basin using the CLUMondo model. This research integrated ArcMap and the CLUMondo model to identify the changes in land-use patterns in this basin from 2015 to 2045.

44.2 Model Application for the Projection

To project LULC changes in the TSL basin for the period until 2045, the CLUMondo model was used with a spatial resolution of 2 × 2 km to simulate future scenarios. CLUMondo is a dynamic and spatially explicit land-use model with the latest addition in the series of models for the conversion of land use and its effects (Verburg et al. 1999). CLUMondo has been designed specifically to simulate changes in land cover as well as changes in land-use intensity. At the core, the simulation of land-use change is based on an empirical analysis of location suitability combined with the dynamic simulation of competition and interactions between the spatial and temporal dynamics of land-use systems (Vliet 2015). This means that the CLUMondo model simulates changes relative to the initial land-use pattern. Thus, at least one land-use map is required, representing the situation at the beginning of the simulation period. LULC changes are related to a large number of biophysical and socioeconomic factors. These drivers are common variables that describe the demography, soil, geomorphology, climate, and infrastructural situation, based on which an allocation process is estimated. The driving factors are divided into four categories of climate, socioeconomic, soil type, and topography. The layers of different variables provide the model with regression parameters that are calculated in the regression analysis.

To capture LULC changes under different levels of socioeconomic development and environmental conservation, four scenarios for the period from 2015 to 2045 were developed according to the historical trend of LULC change, land-use planning, and biodiversity conservation targets. The differences among these demands reflected the importance given to land system functions corresponding to each scenario.

The land-use map of 2015, obtained from the Cambodian National Mekong Committee under the Ministry of Water Resources and Meteorology of Cambodia, was used as the baseline. It contains LULC of nine categories: water, urban area, annual crop, flooded forest, wood plantation, grassland, forest, agricultural, and shifting cultivation. The climate and topographic data were obtained from the Mekong River Commission, whereas socioeconomic and soil data were obtained from the National Institute of Statistics under the Ministry of Planning (Table 44.1). All these factors were input into the model to calculate the areal change of LULC, which depends on the social demands with reference to the existing land cover in 2015.

The first scenario, economic land concessions (ELC), is based only on social demand, population, and economics. It is the scenario that allows the development of industrial-scale agriculture and various activities. Second, moderate conservation (MCO) is a scenario based on social demand, the increasing population and

Table 44.1 Driving factors of LULC change

Data	Driving factor
Climate	Available water storage capacity in each grid
	Maximum temperature
	Minimum temperature
	Month of minimum precipitation
	Annual mean temperature
	River flood hazard
	Annual precipitation
Socioeconomic	General accessibility
	Population density
	Accessibility to domestic market
	Accessibility to international market
Soil	Topsoil gravel
	Topsoil sand
	Topsoil clay
	Topsoil silt
Topographic	Slope
	Elevation
	Very poor drainage
	Moderately well drained
	Terrain ruggedness index

economics, with moderate conservation of forests. MCO balances demand and supply. Third, strong conservation (SCO) is a scenario in which all the forest areas existing in the baseline land-use map are protected and preserved. High priority is given to preserve biodiversity or to change the land to a conservation area. The last scenario, Sustainable Development Goals 2030 (SDG2030), is based on restoration to achieve 60% of forest cover in Cambodia by 2030.

The restrictions of land system conversion at a specific location were introduced based on the policies, natural conditions, and history of land use. We also considered the demands of cash crop, wood production, subsistence crop, and built-up area. All the services were defined as the productive capacity of each cell (i.e., area) of a LULC type. The conversion order was estimated based on the realities of land-use services corresponding to LULC types. The conversion resistance factor for each land system was estimated based on the costs of conversion, information on local policies on land use, and scenario setup. In the model, each land-use type was assigned a dimensionless factor that represents the relative elasticity to conversion, ranging from 0 (easy conversion) to 1 (irreversible change) (Table 44.2).

Table 44.2 Conversion of each LULC in four scenarios

Conversion resistance	ELC	MCO	SCO	SDG2030
Urban area	0.9	0.9	0.9	0.9
Annual crop	0.7	0.5	0.5	0.4
Wood plantation	0.7	0.5	0.4	0.4
Grassland	0.6	0.5	0.5	0.28
Forest cover	0.4	0.4	0.9	1.0
Agricultural	0.7	0.5	0.5	0.4
Shifting cultivation	0.4	0.4	0.4	0.36
Water	1.0	1.0	1.0	1.0
Flooded forest	0.9	0.9	1.0	0.9

Table 44.3 LULC changes from 2015 to 2045 under the different scenarios

		Area in 2045 (ha)			
LULC	Area in 2015 (ha)	ELC	MCO	SCO	SDG2030
Urban area	254,800	713,200 (+179%)	714,000 (+180%)	713,200 (+179%)	712,000 (+179%)
Annual crop	1383,200	2,343,600 (+69%)	2,332,800 (+68%)	2,358,800 (+70%)	2,346,000 (+69%)
Wood plantation	189,200	253,600 (+34%)	254,000 (+34%)	256,800 (+35%)	252,800 (+33%)
Grassland	1,108,000	337,600 (−69%)	191,600 (−82%)	0 (−100%)	0 (−100%)
Forest cover	2,171,200	1,378,400 (−36%)	1,524,800 (−29%)	1,692,400 (−22%)	1,711,600 (−21%)
Agricultural	2,233,600	2,480,400 (+11%)	2,489,600 (+11%)	2,485,200 (+11%)	2,468,000 (+10%)
Shifting cultivation	166,800	0 (−100%)	0 (−100%)	400 (−99%)	16,400 (−90%)
Water	353,200	353,200 (±0%)	353,200 (±0%)	353,200 (±0%)	353,200 (±0%)
Flooded forest	277,200	277,200 (±0%)	277,200 (±0%)	277,200 (±0%)	277,200 (±0%)

44.3 Projected LULC

The results showed that areas of grassland, forest cover, and shifting cultivation have strong reductions, even though the demands for most land system commodities and services are similar among all scenarios (Table 44.3, Fig. 44.1). The spatial changes of LULC in the TSL basin from 2015 to 2045 changed significantly in urban areas located close to the TSL floodplain. Marked changes also occurred spatially at the upland area where the land cover is dominantly forest. The model showed that the forest cover will be lost by 36% in 2045 if we conduct business as usual or continue having ELCs. In this case of ELC scenario, the annual loss rate will be 1.2% over 30 years, starting from 2015 as the baseline to 2045. This average annual loss rate is comparable to the annual deforestation rate in Cambodia, which was around 1.2%

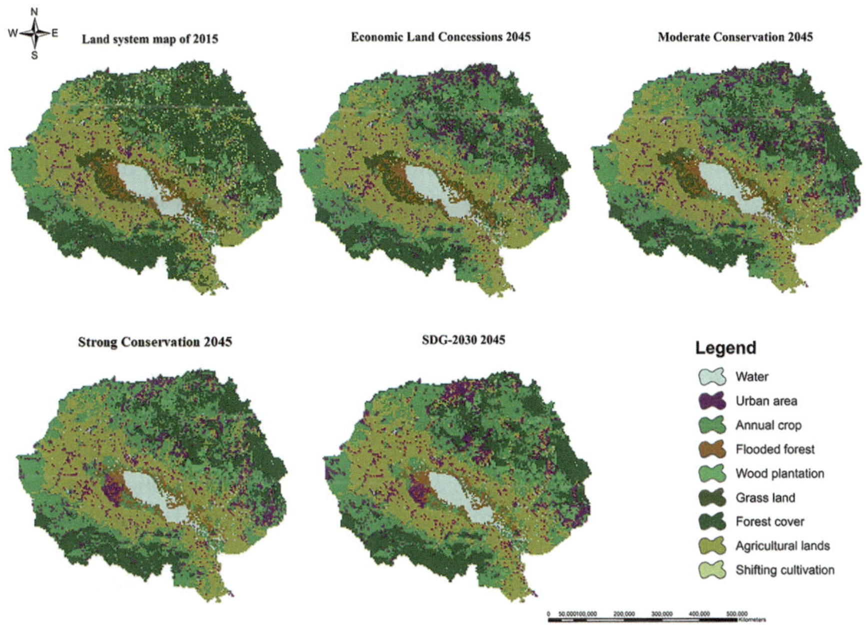

Fig. 44.1 LULC in 2015 and 2045 under the four different scenarios

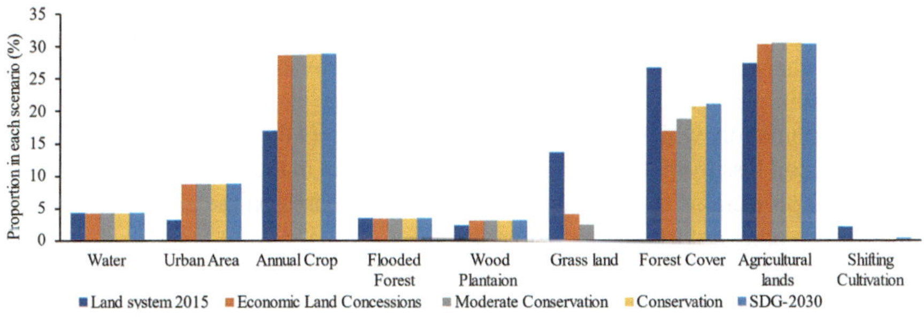

Fig. 44.2 The proportions of LULCs in 2015 and 2045 under the four scenarios

from 2000 to 2017, analyzed by Potapov et al. (2019). This annual loss rate is exactly the same as the steady high rates of forest loss within flooded forests around TSL, at 1.2% within the 25-year period from the past (Sapana Lohani et al. 2020).

LULC changes from 2015 to 2045 showed that the area of intensive annual crop and agricultural land increase in all scenarios, whereas grassland, forest cover, and shifting cultivation are markedly decreased, as shown in Table 44.3. The water and flooded forests were confirmed to be unchanged for all the scenarios. The other LULCs increased and decreased significantly for each scenario. The total areas of agricultural land, urban area, annual crop land, and wood plantation increase to approximately 30.5%, 8.8%, 28.9%, and 3.1%, respectively, in 2045 (Fig. 44.2).

However, those types of LULC would be taken over in grassland, forest land, and shifting cultivation. The grassland decreases to 0%, and the shifting cultivation decreases to nearly 0% in SCO and SDG2030 scenarios, whereas the forest cover remains at about 21%, even though the conversion resistance for forest cover was not permitted or was very small in those scenarios. This result indicates that as a result of the conversion matrix, forest cover was allowed to be converted to agricultural and shifting cultivation in these scenarios. Until 2045, the forest cover in the basin was assumed to shift differently among the four scenarios, which are −10%, −8%, −6%, and − 6% in the proportion of forest cover for ELC, moderate conservation, strong conservation, and SDG2030, respectively. For both ELC and MCO, the shifting cultivation land will decrease to 0% from 2030, whereas grassland will remain at more than 4.5% and 19% in 2045.

All scenarios showed that the cash crop, wood product, subsistence crop, and built-up area can be converted by both smallholders and large-scale land acquisition. In the ELC and MCO scenarios, most of the increase of the urban built-up and annual crop areas are converted from forest cover and shifting cultivation. However, the urban area and annual crop area are converted from grassland and shifting cultivation in the SCO and SDG2030 scenarios. Given these results, the area of each LULC changed over the simulation period at an almost constant rate, except for the areas of open water and flooded forest, which remained unchanged. The main reason for such results is that all the data of scenarios, including cash crop production, wood production, subsistence crop production, and built-up area, were defined at a constant rate of annual increase.

Therefore, the model simulation presented the land system change that is a major driver of change in the spatial pattern and overall provision of ecosystem service. These results can guide and support future land-use planning, management, and policy formulation, which are the main factors that affect the water quality, sediment transport, soil loss from the upland area, and ecosystem in the TSL basin. In addition, the LULC projections for the future are useful for research on the effects of land-use changes on other environmental components.

To enrich the projection, further studies could use satellite data to classify each land-use type by using image classification techniques and including climate data to clarify the suitability of land-use change for the demands and services of LULC. Nevertheless, this analysis addressed several main forces that drive LULC changes and to make projection on the future change of LULC.

Key Points

- The four scenarios represent the different pathways of managing the land resources in the TSL basin, considering multiple demands for commodities and services. They were then used to project LULC for the period 2015–2045 using the CLUMondo model.
- The model simulation showed that the grassland, forest cover, and shifting cultivation would be converted to other land-use types, such as annual crops, wood plantations, urban areas, and agricultural land, for all scenarios.

- The result also indicated that the forest would be restored in the SDG2030 scenario and preserved in the conservation scenario. However, the forest cover will disappear after 2030 for all scenarios if there is no resource supply from outside to support the increasing local need. Many parts of forest cover in the lowland would be replaced with wood plantations, urban areas, and annual crop areas.
- This projection of LULC is important to determine the variations of the ecosystem in the TSL basin. For further research, our results can be used as input data to assess the water quality and hydrological balance in this basin.

References

Chim K, Tunnicliffe J, Shamseldin A, Ota T. Land use change detection and prediction in upper Siem Reap River, Cambodia. Hydrology. 2019;6(3):64. https://doi.org/10.3390/hydrology6030064.

Lambin EF, Geist HJ, Lepers E. Dynamics of L and -U se and L and -C over C Hange in T Ropical R Egions. Annu Rev Environ Resour. 2003;28(1):205–41. https://doi.org/10.1146/annurev.energy.28.050302.105459.

Lohani S, Dilts TE, Weisberg PJ, Null SE, Hogan ZS. Rapidly accelerating deforestation in Cambodia's Mekong River basin: a comparative analysis of spatial patterns and drivers. Water. 2020;12:2191. https://doi.org/10.3390/w12082191.

Potapov P, Tyukavina A, Turubanova S, Talero Y, Hernandez-Serna A, Hansen MC, Saah D, Tenneson K, Poortinga A, Aekakkararungroj A, et al. Annual continuous fields of woody vegetation structure in the Lower Mekong region from 2000–2017 Landsat time-series. Remote Sens Environ. 2019;232:111278.

Ruiter J, Chhuon K, van Vliet J. Land use change scenarios for hydrological impact assessment in a tropical river of Tonle Sap basin. In: The 2nd International Symposium on Conservation and Management of Tropical Lakes. San Francisco: Academia; 2017.

Van Asselen S, Verburg PH. Land cover change or land-use in-tensification: simulating land system change with a globa-scale land change model. Glob Change Biol. 2013;19(12):3648–67. https://doi.org/10.1111/gcb.12331.

Van Ty T, Sunada K, Ichikawa Y, Oishi S. Scenario-based impact assessment of land use/cover and climate changes on water resources and demand: a case study in the Srepok River basin, Vietnam-Cambodia. Water Resour Manag. 2012;26(5):1387–407. https://doi.org/10.1007/s11269-011-9964-1.

Verburg PH, De Koning GHJ, Kok K, Veldkamp A, Bouma J. A spatial explicit allocation procedure for modelling the pattern of land use change based upon actual land use. Ecol Model. 1999;116:45–61.

Vliet JV. Improving land change simulation capacity to reduce conflict from competing land demands. Amsterdam, The Netherlands: Institute for Environmental Studies, VU University Amsterdam; 2015.

Chapter 45
Permissible Phosphorus Load

Kaing Vinhteang, Theng Vouchlay, Uk Sovannara, and Chihiro Yoshimura

45.1 The Concept of Total Maximum Daily Load

The total maximum daily load (TMDL) refers to the maximum mass of pollutants from all contributing sources that a water body is capable of assimilating while meeting the existing water-quality standards (EPA 2007). This concept has been widely applied in the management of surface water around the world (Iwanyshyn et al. 2008; Luo and Zhang 2017). It helps policymakers enforce constraints on the allowable level of pollutant loads to a water body. Even the TMDL is expressed daily; however, the TMDL can be expressed as a unit mass per month or a unit mass per annum.

The concentration of a pollutant has been used as an indicator to assess water quality in the water environment and at sources of the pollution (e.g., factory and industry) to protect human and ecosystem health (Cahn and Hartz 2014). However, the use of concentration has some shortcomings for closed and semi-closed water bodies. This is because these water bodies tend to accumulate pollutants depending on their hydrodynamics and internal processes. For instance, the environmental degradation of closed water bodies, such as lakes and bays, might occur even though all the relevant producers of the pollutants meet the effluent standards. In addition, the concentration does not necessarily determine the effect of individual producers of pollutants on the water-quality impairments in the water bodies. Such shortcomings can be overcome by controlling the pollutant load (i.e., mass flux). The pollutant load also allows us to allocate allowable loads of the individual pollution sectors

K. Vinhteang (✉) · T. Vouchlay
Tokyo Institute of Technology, Tokyo, Japan

Institute of Technology of Cambodia, Phnom Penh, Cambodia
e-mail: kaing.v.aa@m.titech.ac.jp

U. Sovannara · C. Yoshimura
Tokyo Institute of Technology, Tokyo, Japan

C. Yoshimura et al. (eds.), *Water and Life in Tonle Sap Lake*,
https://doi.org/10.1007/978-981-16-6632-2_45

(e.g., agriculture, industry, water treatment plant) to avoid environmental degradation of a concerned water body.

Establishing the relationship between the lake water-quality target and pollutant loading is a critically important process of TMDL development. It enables environmental managers and decision-makers to propose management options and to decide desirable load reductions. The TMDL is estimated using various approaches, from simple mass balance calculations to complex water-quality modeling. The TMDL estimation varies based on various factors, including the waterbody type, the complexity of flow conditions, and the pollutant type causing the impairment (EPA 2007). Then, the integration of monitored data and the water-quality model allows the TMDL developer to associate responses of a certain waterbody to flow and loading conditions.

To develop the TMDL, water-quality indicators and numeric water-quality targets of the concerned pollutant must be specified. For eutrophication, typically, four parameters are used to indicate the relationship between water quality and the trophic status: total phosphorus (TP), chlorophyll-a, Secchi depth, and hypolimnion oxygen (Nürnberg 1996). In the case of Tonle Sap Lake (TSL), the TP concentration is used as the water-quality indicator because phosphorus (P) is considered a limiting nutrient and a significant driver of the eutrophication process in TSL (see Chap. 24). Moreover, the seasonal flood pulse in this chapter is categorized into four periods: low-water period (March–May), rising-water period (June–August), high-water period (September–October), and receding-water period (November–February), to benefit the practice of lake management, to cope with vegetation and fish life histories in the lake and its floodplain, and to cope with agricultural practices that are synchronous with the seasonality of the flood pulse (Arias et al. 2014). Arias et al. (2014) also divided these periods based on the changing water level in the lake.

Various sources contribute P to TSL, and the majority are nonpoint sources (e.g., tributaries; see Chaps. 24 and 25 for details). In TSL, the TMDL can be defined as the permissible phosphorus load (PPL) from the lake tributaries. This chapter presents the PPL in TSL, estimated using a static model (Vollenweider model) and a dynamic model (P dynamic model). We applied the same values of the parameters related to atmospheric deposition and internal loading, such as settling velocity and resuspension of P, in both models. Then, we established the relationship between the TP load and eutrophication in TSL.

45.2 PPL Estimated by Vollenweider Model

The Vollenweider model has been applied to link the in-lake TP concentration targets and the TP input load to a lake. This model has been developed based on mass balance to link the occurrence of eutrophication to its P balance, assuming the concentration in a lake (Vollenweider 1975). In this model, the change in the P mass over time is equal to the input P mass minus the total of its output mass and removal (i.e., sedimentation including the primary production process) in the lake. In

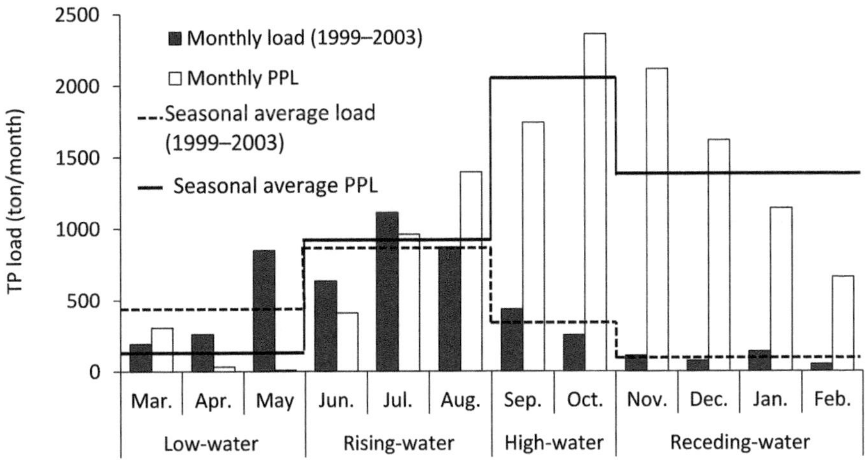

Fig. 45.1 The monthly and seasonal PPL to meet the target TP concentration (30 μg/L) and the actual TP load from tributaries from 1999 to 2003

addition, the P input from sediment resuspension and atmospheric deposition were included in this study. The mass balance of the Vollenweider model is written as.

$$\frac{dM}{dt} = Q_0C_0 - QC - \frac{w_0}{H_0}M + AR + Ak_a \tag{45.1}$$

where M is the total mass of TP in lake, t is the time, C_0 is the TP concentration of inflow, C is the TP concentration in the lake, Q_0 is the inflow discharge, Q is the outflow discharge, w_0 is the settling velocity of TP (12 m/month), H_0 is the lake depth, A is the lake surface area, R is the internal loading rate of TP, k_a is the P atmospheric deposit rate, and $R + k_a = 0.24$ g/m^2/month. The settling velocity, internal loading rate, and atmospheric deposit rate of P were obtained from the calibrated P dynamic model (Chap. 25).

The TP concentration of 30 μg/L is set as the target concentration to determine the PPL for the lake as this value has been used as the threshold corresponding to the eutrophic state used to represent the eutrophic status of many freshwater lakes and reservoirs (Carlson 1977). Based on the Vollenweider model, the PPL of TSL was calculated on a monthly basis. Thus, the PPL represents the maximum allowable load that prevents the whole lake area from being eutrophic. The hydrological data, settling velocity of P, sum of the internal loading rate of TP, and atmospheric deposition were described in Chap. 25. Under the assumption of a steady state ($dM/dt = 0$), the PPL is expressed by output and storage terms as concentration times volume. We assumed here that the outflow TP concentration is equal to that of the lake water.

The results showed that the estimated monthly average PPL of TSL ranged from 8.5 to 2357.3 tons/month (the annual load was 12,735.5 tons/year), with the highest and lowest PPL in October and May, respectively (Fig. 45.1). Seasonally, the

Vollenweider model showed the highest and lowest PPL in the high-water (September–October) and low-water (March–May) periods of approx. 2048.4 and 116.1 tons/month, respectively. Compared to the TP actual load from the lake tributaries in 1999–2003, this result suggests that the lake can accept more load from the late rising-water to the early low-water periods (August–March). However, from April to July, the lake tributaries' load needs to be reduced to meet the target water-quality goal (Fig. 45.1). The permissible additional load to the baseline would be 49.1; 1704.1; and 1289.3 tons/month in the rising-water, high-water, and receding-water periods. In this case, the load should be reduced by 318.1 tons/month, on average, in the low-water period.

The difference of the PPL of TSL among periods should be the consequence of the high variation of inflow and outflow, resulting in the significant variation of the lake volume seasonally. The lake volume appeared to differ by 20 times between the dry season (covering receding-water and low-water periods) and the rainy season (covering rising-water and high-water periods), with the lowest volume around 2.5 km^3 and the highest of 70–85 km^3 in the dry and rainy seasons, respectively (Chap. 16). The lake volume for each period (from 1999 to 2003) was approx. 7.2, 30.6, 70.6, and 32.2 km^3 in low-water, rising-water, high-water, and receding-water periods, respectively.

Given the highly dynamic nature of TSL in terms of hydrodynamics, the P-related processes in the lake, such as P release from sediment, P sedimentation, and transport, are possibly highly active. Thus, the steady-state assumption might not provide an accurate estimation of the PPL for TSL. We checked this issue by introducing the dynamic model application for the PPL estimation.

45.3 PPL Estimated by the Dynamic Model

We also applied the P dynamic model based on the process-based hydrodynamic model (2D-LIE) (see Chap. 25 for more details of the model). The daily TP load and surface area of the lake during 1999–2003 were calculated from monthly and seasonal averages using the result from the 2D-LIE model before the simulation in the P dynamic model. Likewise, the daily eutrophic area was calculated from the monthly and seasonal average eutrophic area. Then, the monthly and seasonal percentages of the eutrophic area (%EAs) were determined as the ratio of corresponding monthly and seasonal average eutrophic area to the monthly and seasonal average surface area of the lake.

The %EA values from model simulation in response to actual loads from lake tributaries (during 1999–2003) are shown in Fig. 45.2. In general, the TP load and % EA fluctuated seasonally. The low-water period (March–May) and rising-water period (June–August) were the hydrological periods when the TP load from the tributaries was high and when the eutrophic area was relatively large. The %EAs in the low- and rising-water periods were 6.1% and 6.6%, respectively. The TP input load from the tributaries was 434.2 tons/month in the low-water period and

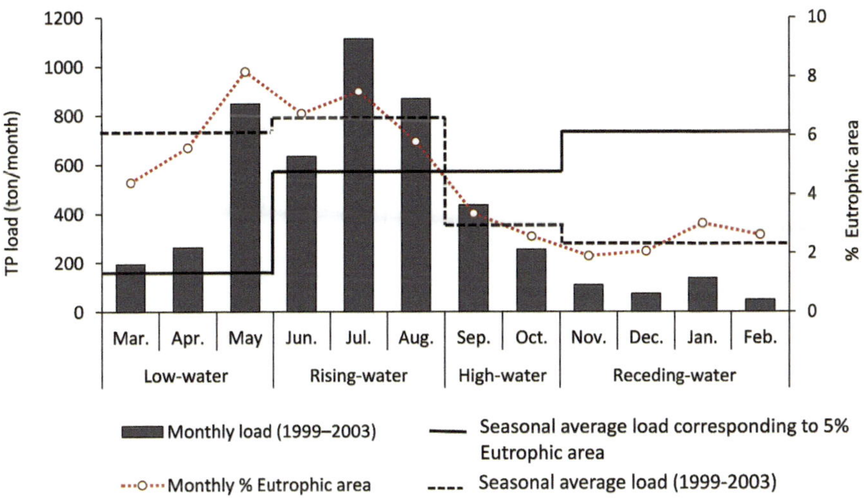

Fig. 45.2 The actual monthly and seasonal TP input load from lake tributaries and the proportion of the eutrophic area of the lake (>30 μg/L) during 1999–2003

872.0 tons/month in the rising-water period. These two periods were the most alarming because eutrophication occurred. Meanwhile, the %EA was only 3.0% and 2.3% in the high- and receding-water periods, respectively.

To determine the PPLs of the tributaries from September 1999 to December 2003, the TP concentration in tributaries was varied, whereas the load from villages and the model parameters (settling velocity and internal and atmospheric loading rate) were kept unchanged (i.e., baseline). There is a constraint in estimating the PPL within the designated proportion (%EA) for every month using the P dynamic model. When trying to minimize or maximize the PPL in the early months to meet the target %EA, it will affect the following month's value. As a result of this constraint, the PPL results from the P dynamic model were presented seasonally. We calculated the seasonal %EA corresponding to the TP input load response to various TP input loads from lake tributaries to clarify the importance of seasonality in-lake management practice.

As a result, to control the lake with the limitation of the eutrophic area within 5%, the maximum permissible loads were found to be 158.0, 571.7, 572.8, and 733.6 tons/month for the low-, rising-, high-, and receding-water periods, respectively (Figs. 45.2 and 45.3). In such case, the annual permissible load was 6269.4 tons/year. To meet the target water-quality standards with the limitation of the eutrophic area within 5%, the TP input from the lake tributaries needs to be reduced by 64% and 34% in the low- and rising-water periods, whereas it can still be increased by 66% and 695% in the high- and receding-water periods, respectively, from the simulation from 1999 to 2003. The PPL reduction and increase can provide a scientific basis for the authorities to make decisions on water pollution management.

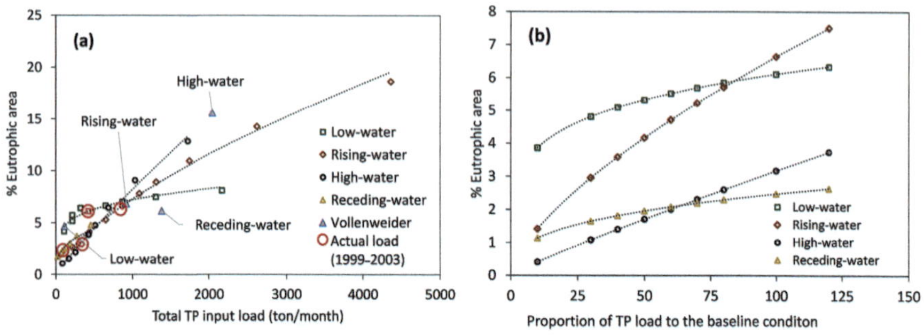

Fig. 45.3 The seasonal %EA of the inundated area of lake response to (**a**) the TP input load from lake tributaries and (**b**) the proportion of the TP load to the baseline condition. The results were obtained from varying the TP concentration in tributaries, whereas the load from villages and the model parameters were kept unchanged

45.4 PPL: Characteristics of TSL

The Vollenweider and P dynamic models showed the highest and lowest PPLs in the high- and low-water periods, respectively. These highest and lowest PPLs were synchronous with the highest and lowest water volumes of the lake, which were approx. 70.6 and 7.2 km^3 in high- and low-water periods, whereas they were 30.6 and 32.2 km^3 in rising- and receding-water periods, respectively.

The Vollenweider model estimated the seasonal average PPL ranging from 116.1 to 2048.4 tons/month to prevent eutrophication. In the simulation using the dynamic model, however, such estimates of the PPL resulted in 4.7–15.6% of the eutrophic area of the inundated area (Fig. 45.3a). As the Vollenweider model was built based on the concept of a "completely mixed reactor" (Vollenweider 1975), the difference in results of both models can be explained by the weak mixing process and the dynamic distribution of the P concentration in TSL.

In the P dynamic model, with the limitation of the eutrophic area within 5%, the PPL in the receding-water period (733.6 tons/month) was higher than that in the high-water period (572.8 tons/month). In contrast, in the case of the Vollenweider model, the PPL in the high-water period was higher than that in the receding-water period. The results of the Vollenweider model are based on the water volume and the hydraulic retention time of the lake, which is the ratio of the lake volume to the total inflow. The water volume in the lake in the receding-water period was less than that in the high-water period, on average. In addition, the hydraulic retention time in the high-water period was longer than that in the receding-water period (Chap. 15). A longer retention time generally tends to promote eutrophication. Therefore, the PPL in the receding-water period was lower than that in the high-water period.

It was also noted that the rising-water period is the reverse flow period from the Mekong River to the lake and the receding-water periods is the time when the water drains out of the lake. Moreover, from the water balance calculation of Kummu and Sarkkula (2008), the lake inflow is derived mainly from the Mekong main stem (57%

Table 45.1 PPLs in TSL and other lakes

Lake	Country	Average volume (km^3)	Average water surface area (km^2)	Mean depth (m)	Average PPL per water surface (tons/year/km^2)
TSL	Cambodia	65	9812	2.6	0.64 (0.19–0.90)
Taihu[a]	China	4.66	2339	1.9	0.20
Okeechobee[b]	USA	5.2	1720	2.7	0.08
Onondaga[c]	USA	0.13	12	4.5	1.92
Simcoe[d]	Canada	11	7222	16	0.01

[a]Hu et al. (2020)
[b]Effler et al. (2002)
[c]Effler et al. (2002)
[d]Winter et al. (2007)

of total inflow), at which 52% is from the TSR and 5% from overland flooding. From our calculation, on average, the Mekong River contributes about 57.8% and 41.4% of the TP load from the tributary system during the rising-water period (June–August) and reverse flow period (May–September), respectively. Therefore, the retention time and the PPL of TSL were considerably controlled by the Mekong River flood pulse.

In terms of the PPL per lake surface area, TSL is comparable to Taihu Lake in China and is lower than Onondaga Lake in the USA, whereas it is higher than others in North America (Table 45.1). The differences are associated with the P-related process and flow characteristics in the lake. TSL is located in the tropical monsoon climate, which possibly explains the relatively high settling velocity of P in TSL (12 m/month, including the process of primary production) compared to temperate lakes (e.g., 10 m/year (Vollenweider 1975); refer to Chap. 31 for primary production). In addition, the hydraulic retention time of TSL is about 9 months and 25 days (Chap. 15), which is longer than that of Taihu Lake (4 months and 5 days (Yang and Liu 2010). Such a long hydraulic retention further promotes P removal in the lake, which possibly results in the high PPL of TSL.

The P source in TSL is dominated by the internal loading for all the periods of the TP load (Table 45.2), which is a major reason for the weak response of the %EA to the external loading (Fig. 45.3b). This means that the reduction of internal loading seems to be an effective measure to control the eutrophication of the lake. The importance of internal loading is related mainly to the flushing rate, loading history, and chemical characteristics of the sediment (Marsden 1989). P can be released from active sediment within the top 20 cm. There are some methods to significantly reduce internal loading, such as removing P-rich surface sediment layers or adding iron or alum to increase the sediment sorption capacity (Canfield Jr et al. 2020; James and Pollman 2011; Sondergaard et al. 2001). However, sediment dredging is costly and complex; it is conducted only when controlling the external loading still does not meet the target. In addition, adding chemical substances to the lake might lead to secondary pollution. Therefore, the control of the external loading or the P

Table 45.2 The seasonal TP load contribution from the primary sources (estimates are based on the actual input load and the P dynamic model for 2000–2003)

	Periods	Local tributaries	Mekong River	Village	Internal and atmospheric	Total
TP load (tons/month)	Low water	420.8	13.4	30.4	1465.0	1929.6
	Rising Water	351.9	520.0	42.7	2546.0	3460.6
	High water	291.9	63.1	59.8	3593.6	3997.8
	Receding water	96.4	0.0*	11.6	2375.0	2478.9
% contribution to TP load	Low water	21.8	0.7	1.6	76.0	100.0
	Rising Water	10.2	15.0	1.2	73.6	100.0
	High water	7.0	1.6	1.5	90.0	100.0
	Receding water	3.7	0.00	0.5	95.8	100.0

Note: The TP load from Mekong River is the TP input through TSR. The TP load from local tributaries is the TP input through lake tributaries excluding TSR
- * No input load as there is no reverse flow during this period.

input from lake tributaries would be the first countermeasure option against eutrophication of TSL. Notably, the low- and rising-water periods are the important periods for which countermeasures are required for load reduction.

This chapter provides the first estimation of the PPL of P for eutrophication control for TSL. Some knowledge gaps and recommendations for future work from this study are listed below:

(i) The margin of safety should be included in the PPL estimation, taking into account any uncertainty in data, model, and future hydrological conditions.
(ii) The unique hydrodynamic process of TSL induced by the Mekong River's flood pulse is critical in changing the sediment-nutrient distribution by inducing sediment-P release. Even though this chapter discussed some relationship of PPL to the Mekong River's flood pulse, additional analyses and discussion should be conducted.
(iii) As the TSL is in a tropical region with a high temperature, with an average of 30–38 °C, the biological process in the lake is highly productive, which impacts the number of P species in the lake. The investigation of the effects of temperature on the P and PPL is also recommended in future studies.
(iv) Based on the estimated PPL, it is essential to conduct TP load allocation for each of the major sources in the basin and propose good management practices for TSL. Currently, the TP loads from each land use type within each tributary basin are not available.

Key Points

- The Vollenweider model estimated the lowest and highest seasonal average PPLs in TSL as about 116.1 tons/month in the low-water period (i.e., March–May) and 2048.4 tons/month in the high-water period (September–October), respectively.
- Based on the P dynamic model to control the lake with the limitation of the eutrophic area within 5% for all seasons, the maximum PPLs were 158.0, 571.7, 572.8, and 733.6 tons/month for the low-, rising-, high-, and receding-water periods, respectively.
- The low- and rising-water periods are the important periods for which countermeasures are required to mitigate eutrophication.
- The differences in the PPL of the Vollenweider and P dynamic models indicate the weak mixing process and the dynamic distribution of the TP concentration in TSL. Thus, we recommend the results from the P dynamic model for PPL application to the TSL basin.

Acknowledgments The authors thank the Mekong River Commission and Water Utilization Program of Finland Component Project for the provision of data for this work.

References

Arias ME, Cochrane TA, Elliott V. Modelling future changes of habitat and fauna in the Tonle Sap wetland of the Mekong. Environ Conserv. 2014;41(2):165–75.

Cahn M, Hartz T. Load vs. concentration: implications for reaching water quality goals. University of California Cooperative Extension Monterey County; 2014

Canfield DE Jr, Bachmann RW, Hoyer MV. Restoration of Lake Okeechobee, Florida: mission impossible? Lake Reserv Manag. 2020;37(1):1–17.

Carlson RE. A trophic state index for lakes 1. Limnol Oceanogr. 1977;22(2):361–9.

Effler SW, O'Donnell SM, Matthews DA, Matthews CM, O'Donnell DM, Auer MT, Owens EM. Limnological and loading information and a phosphorus total maximum daily load (TMDL) analysis for Onondaga Lake. Lake Reserv Manag. 2002;18(2):87–108.

EPA, U. Options for expression of daily loads in TMDLs. Washington, DC: US Environmental Protection Agency; 2007.

Hu K, Wang Y, Feng B, Wu D, Tong Y, Zhang X. Calculation of water environmental capacity of large shallow lakes–a case study of Taihu Lake. Water Policy. 2020;22(2):223–36.

Iwanyshyn M, Ryan M, Chu A. Separation of physical loading from photosynthesis/respiration processes in rivers by mass balance. Sci Total Environ. 2008;390(1):205–14.

James RT, Pollman CD. Sediment and nutrient management solutions to improve the water quality of Lake Okeechobee. Lake Reserv Manag. 2011;27(1):28–40.

Kummu M, Sarkkula J. Impact of the Mekong River flow alteration on the Tonle Sap flood pulse. AMBIO J Hum Environ. 2008;37(3):185–92.

Luo Y, Zhang J. Application of a load duration curve for establishing TMDL programs upstream of the Tiaoxi River within the Taihu watershed, China. J Coast Res. 2017;80:80–5.

Marsden MW. Lake restoration by reducing external phosphorus loading: the influence of sediment phosphorus release. Freshw Biol. 1989;21(2):139–62.

Nürnberg GK. Trophic state of clear and colored, soft-and hardwater lakes with special consideration of nutrients, anoxia, phytoplankton and fish. Lake Reserv Manag. 1996;12(4):432–47.

Sondergaard M, Jensen PJ, Jeppesen E. Retention and internal loading of phosphorus in shallow, eutrophic lakes. TheScientificWorldJOURNAL. 2001;1:427–42.
Vollenweider RA. Input-output models. Schweiz Z Hydrol. 1975;37(1):53–84.
Winter JG, Eimers MC, Dillon PJ, Scott LD, Scheider WA, Willox CC. Phosphorus inputs to Lake Simcoe from 1990 to 2003: declines in tributary loads and observations on lake water quality. J Great Lakes Res. 2007;33(2):381–96.
Yang S-Q, Liu P-W. Strategy of water pollution prevention in Taihu Lake and its effects analysis. J Great Lakes Res. 2010;36(1):150–8.

Chapter 46
Effects of Environmental Factors on Eutrophication

Theng Vouchlay, Uk Sovannara, Vinhteang Kaing, Tomohiro Tanaka, Hidekazu Yoshioka, and Chihiro Yoshimura

46.1 Major Factors for Eutrophication in TSL

Eutrophication from the over-enrichment of nutrients is among the serious threats to aquatic ecosystems around the world, causing various negative ecological impacts (i.e., algal bloom and fish kills; Hupfer and Hilt 2008; Smith and Schindler 2009). The rate and extent of the eutrophication process have been accelerated markedly by the increase in anthropogenic stressors, such as the soaring human population, rapid urbanization, expanding use of agricultural fertilizers, and untreated municipal wastewater. In Tonle Sap Lake (TSL) and its adjacent systems, major changes have also been occurring as a result of anthropogenic activities over the past several decades (see a detailed review by Uk et al. 2018). In this ecosystem, similar to most other freshwater lakes around the world, phosphorus (P) seems to be a major limiting nutrient (see Chap. 24). With this background, this chapter aims to elucidate the effects of P in the external sources and internal processes on P dynamics and eutrophication in TSL as well as on the concentration of total phosphorus (TP) in the water outflow to Tonle Sap River (TSR), based on spatial sensitivity analyses. Scenario analyses were also conducted to assess the effects of population growth and

T. Vouchlay (✉) · V. Kaing
Tokyo Institute of Technology, Tokyo, Japan

Institute of Technology of Cambodia, Phnom Penh, Cambodia
e-mail: theng.v.aa@m.titech.ac.jp

U. Sovannara · C. Yoshimura
Tokyo Institute of Technology, Tokyo, Japan

T. Tanaka
Kyoto University, Kyoto, Japan

H. Yoshioka
Shimane University, Shimane, Japan

C. Yoshimura et al. (eds.), *Water and Life in Tonle Sap Lake*,
https://doi.org/10.1007/978-981-16-6632-2_46

the impact of the TP concentration in the local tributaries on an intake point of the water supply in Siem Reap province.

46.2 Spatial Sensitivity Analyses

P dynamics as well as eutrophication of TSL were assessed using the P dynamics model described in Chap. 25. This model was governed by horizontal advection–dispersion, sedimentation, internal loading, and inputs of TP from local tributaries, TSR, atmospheric deposition, and villages on TSL in two dimensions. The model was calibrated and validated from September 1999 to December 2003. The root mean square error of the TP concentration in the calibration was 15 μg/L (12 μg/L in the validation), whereas the observed TP concentration ranged from 1 to 70 μg/L. The calibrated settling rate was 0.4 m/day, and the internal and atmospheric loading rate was 8.0 mg/m^2/day. These calibration results and the boundary condition in Chap. 25 were set as the baseline conditions.

In the model framework, the effects of the model boundary condition (i.e., the TP concentration in the local tributaries and TSR and the village loading) and the model parameters (i.e., the settling rate and the internal and atmospheric loading rate) on (1) the TP concentration in and the eutrophic area of TSL and (2) the TP concentration in the outflow from TSL to TSR were analyzed using spatial sensitivity analyses. The spatial sensitivity analyses of the model boundary conditions illustrated how the changes in the TP concentration in the local tributaries, TSR, and village loading affect the distribution of the TP concentration in TSL. Although the model parameters are hardly controlled in environmental management, the spatial sensitivity analyses elucidated the dynamics of the lake and their importance compared to the boundary conditions. The spatial sensitivity index (*SSI*) is expressed in Eq. (46.1) (Nelson et al. 2018):

$$SSI_{x,y} = \frac{\left(C_{\max,x,y} - C_{\min,x,y}\right)}{\left(I_{\max,x,y} - I_{\min,x,y}\right)} \times \frac{\overline{I}_{x,y}}{\overline{C}_{x,y}} \tag{46.1}$$

where subscripts x and y are coordination points; $I_{\max}$ and $I_{\min}$ are the maximum and minimum and of each parameter or boundary condition in its variation from the baseline, respectively; $C_{\max}$ and $C_{\min}$ are the maximum and minimum TP concentrations corresponding to the variation of each parameter or boundary condition, respectively; and $\overline{C}$ and $\overline{I}$ are average values of the variations of the TP concentration and of each parameter or boundary condition, respectively. To determine the *SSI* of a parameter or a boundary condition, that parameter or boundary condition was varied from the baseline condition by factors from 0.0 to 2.0 with intervals of 0.25 (i.e., from −100% to +100% of the baseline), whereas other parameters and boundary conditions were kept unchanged (i.e., the baseline condition). The *SSI* was arranged for each of four hydrological periods based on TSR inflow/outflow to elucidate the effects of the boundary flow exchanges on the TP concentration in TSL, namely, the low-water period (April–May), inflow period (June–September), early outflow period (October–December), and late outflow period (January–March). A higher *SSI* value indicates a stronger impact of a concerned parameter or boundary condition on the corresponding variable (i.e., the TP concentration in TSL and in the outflow from TSL to TSR).

46.2.1 TP Concentration and Eutrophic Area in TSL

Regarding the seasonal *SSI*s of the TP concentration in TSL under the effect of each model parameter and boundary condition (Fig. 46.1), the settling rate showed the highest *SSI*. The second highest *SSI* was shown by the internal and atmospheric loading rate; their impact was almost uniform for all periods.

The settling rate affected the TP concentration in the whole lake in all periods. In the low-water period, when the water flowed out from TSL to TSR, the effect of the settling rate was high at the middle of the lake and weak in the floodplain. In the inflow period, when the water reversed from TSR to TSL, the settling rate highly affected the TP concentration almost everywhere in the lake ($SSI > 3.0$) except in some northwestern and northeastern parts of the floodplain ($SSI < 3.0$). Then, in the early outflow period, it showed low impact ($SSI < 2.5$) in the most northwestern part

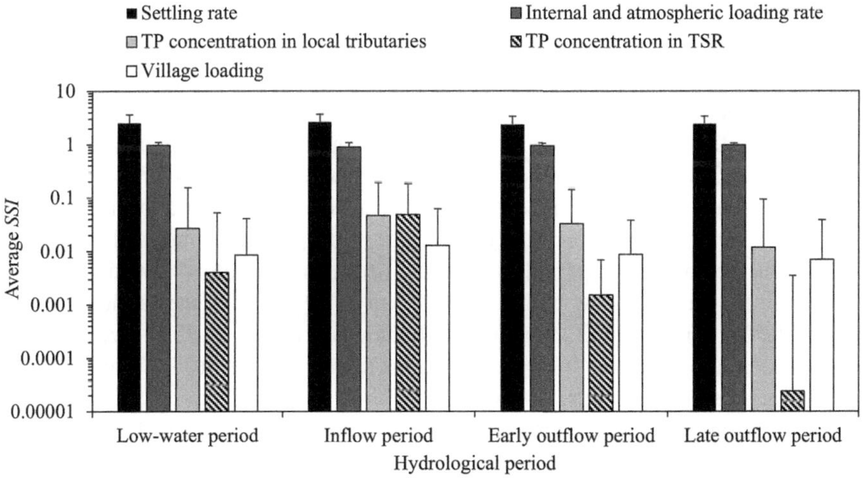

Fig. 46.1 Average *SSI* of the TP concentration showing the effect of each parameter and boundary condition in four hydrological periods: low-water period (April–May), inflow period (June–September), early outflow period (October–December), and late outflow period (January–March). The error bar indicates a standard deviation

Table 46.1 Averages of the baseline eutrophic area and TSL surface area in the four hydrological periods

	Unit	Low-water period (Apr.–May)	Inflow period (Jun.–Sep.)	Early outflow period (Oct.–Dec.)	Late outflow period (Jan.–Mar.)
Lake area	km^2	6004	11,416	12,822	7201
Eutrophic area	km^2	418	635	282	236
%EA	%	7.0	5.6	2.2	3.3

of TSL and high impact along the water flow from TSL to TSR ($SSI > 2.5$). In the late outflow period, the water flowed out continuously from the lake to TSR, resulting in the low sensitivity ($SSI < 2.5$) of the TP concentration to the settling rate in the middle of the lake. The high impact ($SSI > 2.5$) was confirmed around the floodplain where the TP mass did not flow out.

When the settling rate increased to 150% of the baseline, the proportion of the eutrophic area to the lake area (%EA) decreased to 3.5%, 2.7%, 1.0%, and 1.4% in the low-water, inflow, early outflow, and late outflow periods, respectively (refer to Table 46.1 for the baseline condition). When the settling rate decreased to 25–75% of the baseline, it was noticeable that the 50% decrease of the settling rate was the critical point of the lake eutrophication because its corresponding %EA increased drastically compared to the 25% decrease of the settling rate. With the 50% decrease of the settling rate, %EA increased to 80%, 78%, 86%, and 88% in those four periods, respectively.

Regarding the TP concentration in the local tributaries, the *SSI* of the TP concentration in TSL was relatively high around the outlet of the local tributary flows in all periods (Fig. 46.2). The strong effect of the TP concentration in the local tributaries was confirmed in the inflow and early outflow periods. In contrast, in the low-water and late outflow periods, it showed a weak effect because of the low water inflows and as some tributaries were dry. When the TP concentration in the local tributaries increased by 75% of the baseline, the eutrophic area increased by 49 km^2 (0.8%) in the low-water period, 181 km^2 (1.6%) in the inflow period, 120 km^2 (0.9%) in the early outflow period, and 27 km^2 (0.4%) in the late outflow period. When the TP concentration in the local tributaries decreased by 75%, it reduced the eutrophic area by 135 km^2 (2.2%), 199 km^2 (1.7%), 113 km^2 (0.9%), and 58 km^2 (0.8%) in those four periods, respectively.

In addition, the *SSI* of the TP concentration in TSL indicated that the TP concentration in TSR highly affected the area of TP concentration in TSL, particularly in the inflow period, followed by the early outflow period. In the low-water and late outflow periods, there was almost no impact ($SSI \sim 0.0$). This is because TSR mostly flowed into TSL only in the inflow period. The change of the TP concentration in TSR in the inflow period also affected the TP concentration in the lake in the early outflow period. Such an impact was shown along the water flow from the inlet of TSR (Prek Kdam) to the middle of the lake. In the inflow period, when the TP

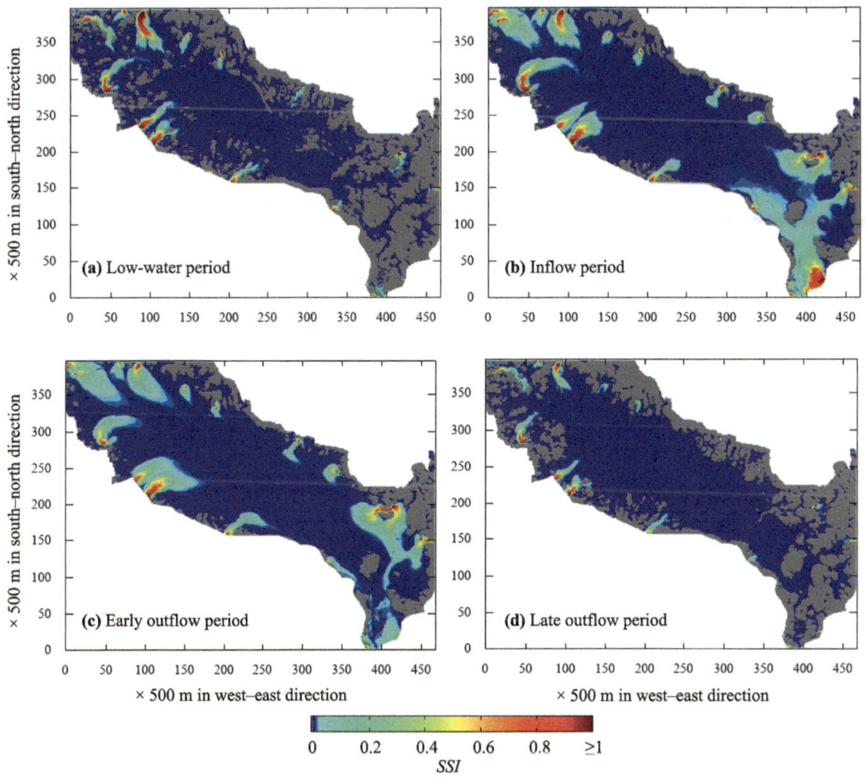

Fig. 46.2 The *SSI* of the TP concentration in TSL showing the degree of the effect of the TP concentration on the local tributaries in four hydrological periods

concentration in TSR increased or decreased by 75%, %EA increased by 1.9% or decreased by 1.7%, respectively. Similar to the effect of the local tributaries, the sensitivity to village loading was high in the areas around the villages.

46.2.2 TP Concentration of the Outflow from TSL to TSR

We also assessed the daily outflow to TSR under the effects of the model parameters and boundary conditions from September 1999 to December 2003. In the baseline condition, the outflow TP concentration was 0.7, 15, and 2.7 μg/L in the low-water, early outflow, and late outflow periods, respectively. The result confirmed that the model parameters were more influential on the outflow TP concentration than the boundary conditions (Fig. 46.3).

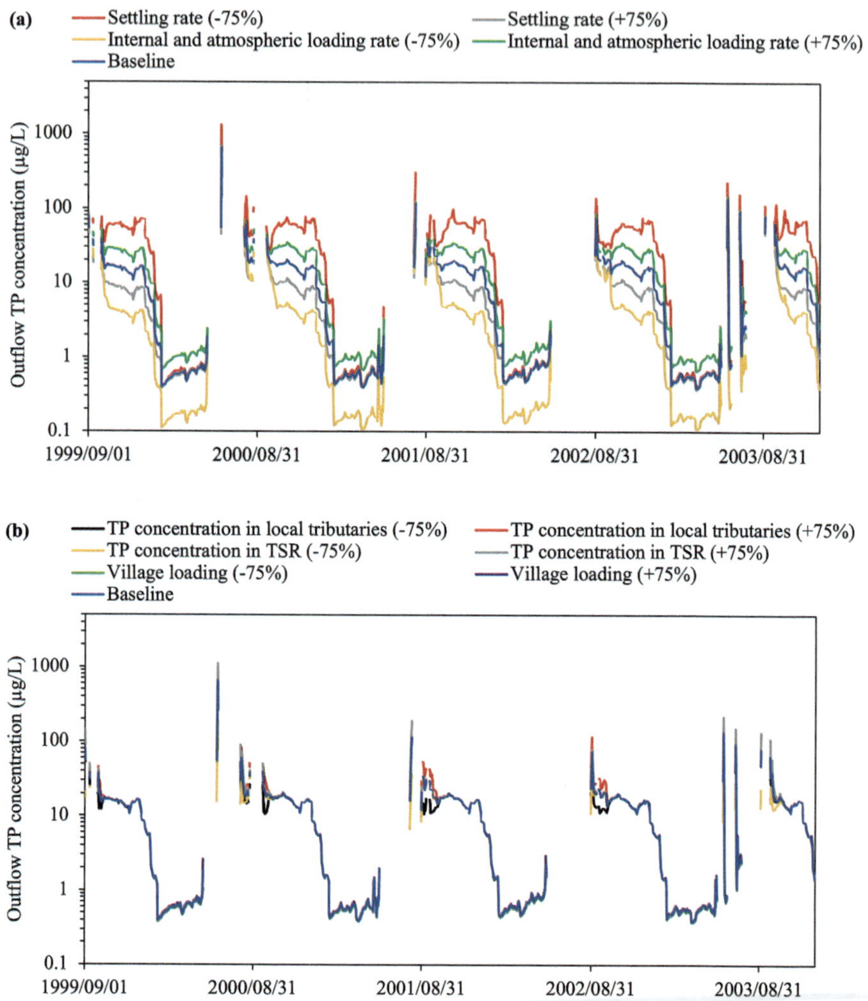

Fig. 46.3 Variations of the outflow TP concentration to TSR under the effect of (**a**) the model parameters (the settling rate and internal and atmospheric loading rate) and (**b**) the model boundary conditions (the TP concentration in the local tributaries, TSR, and the village loading) in 1999–2003. No result is shown for the period of the reverse flow from TSR to TSL

The settling rate showed the highest impact to the outflow TP concentration among the model parameters and the boundary conditions (Fig. 46.3). When it decreased by 75%, the outflow TP concentration increased by 7.1%, 269%, and 298% of the baseline in the low-water, early outflow, and late outflow periods, respectively. The increase in the settling rate by 75% reduced the outflow TP concentration by 4.1%, 45%, and 40% in the low-water, early outflow, and late outflow periods, respectively. This indicates the significant impact of the settling rate in the early outflow and late outflow periods. Regarding the internal and atmospheric loading, when it increased or decreased by 75%, the outflow TP concentration increased or decreased by 71% in the low-water and early outflow periods, respectively; such a shift was 74% in the late outflow period.

Regarding the boundary conditions, when the TP concentration in the local tributaries increased or decreased by 75%, the outflow TP concentration remained almost same as the baseline in all the periods, except for the early outflow when the outflow TP concentration increased or decreased by 1.9% from the baseline. TSR and the villages showed a slight impact on the outflow TP concentration. The highest impact of the TP concentration in TSR (i.e., inflow) was confirmed in the early outflow period. When the TP concentration in TSR increased or decreased by 75%, it increased or decreased the outflow TP concentration by 1.3% from the baseline in the early outflow period, respectively. When the village loading decreased by 75% of the baseline, the outflow TP concentration decreased by 4.7%, 0.4%, and 1.0% in the low-water, early outflow, and late outflow periods, respectively.

46.3 Scenario Analyses

The scenario analyses considered two factors: (1) the impact of population growth on eutrophication in TSL and (2) the impact of TP concentration in the local tributaries on the water intake of a water treatment plant in Siem Reap.

46.3.1 Population Growth

Population in Cambodia is increasing rapidly. Sewage from the floating villages is discharged into the lake directly without any treatment because of the lack of a wastewater treatment facility (Brown et al. 2010; Takahashi et al. 2002). Thus, the TP load from the floating villages to TSL may increases every year. However, if the people in the floating villages move to the land and adequate wastewater treatment facilities are developed, the TP loading from the villages will be reduced.

By assuming that the population growth rate remains constant at 1.54% annually (National Institute of Statistics 2010), the impacts of population growth in 2020, 2040, and 2060 on the eutrophication in TSL were assessed in comparison to the baseline in 1999–2003. The total populations in the study domain are 1,874,812 in 2000; 2,545,106 in 2020; 3,455,033 in 2040; and 4,690,254 in 2060. As the impact of the villages on the eutrophic area was small compared to that on the whole lake area, the results of the eutrophic area were compared to the baseline of the eutrophic area in TSL.

The result showed that the eutrophic area in 2020, 2040, and 2060 was increased from the baseline eutrophic area from June to September because of the higher number of the villages in the inundated area (Fig. 46.4). In September, the eutrophic area increased by 2.3%, 5.7%, and 10.5% of the baseline eutrophic area (490 km^2) in

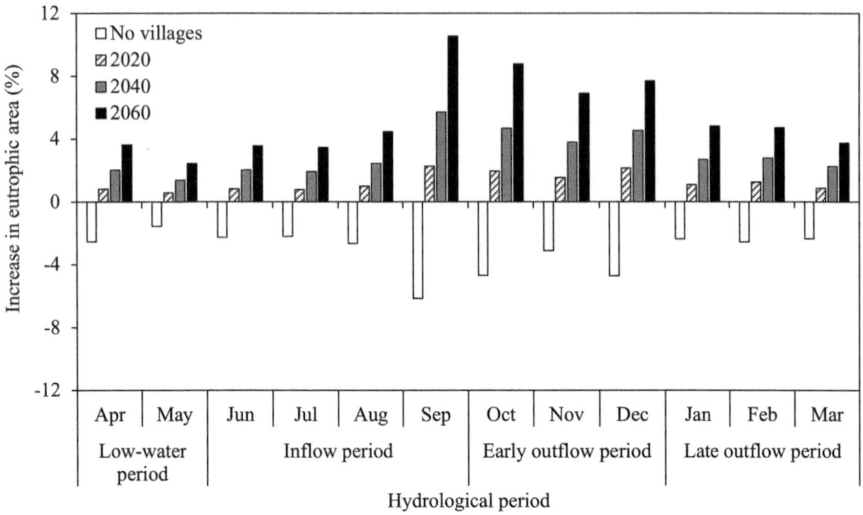

Fig. 46.4 Effect of population growth on the eutrophic area in TSL

2020, 2040, and 2060, respectively. The lowest impact was confirmed in May. If there was no TP loading from the villages, the monthly eutrophic area would be reduced in the range from 1.6% to 6.2% of the baseline eutrophic area.

46.3.2 *Water Supply in Siem Reap*

In the National Strategic Plan for the Rural Water Supply, Sanitation and Hygiene (RWSSH), 2014–2025, a target was set to reach 100% of water supply and sanitation in both rural and urban areas (World Bank Group and WSP 2015). In 2020, the Asian Development Bank was planning to build a new water supply system in Siem Reap, which is the largest tourism province in Cambodia. Its water source will be the lake water, and the intake point will be located at 13.2273°N and 103.8893°E. The activities around TSL (e.g., agriculture, urbanization, population growth) might impact the water quality of the water intake. Blue-green algae (or cyanobacteria) containing toxins (Chap. 28) might cause serious problems in water supply processes (Davis and Glen 2009). Therefore, the impact of the TP concentration in the local tributaries on the TP concentration at the water intake point was assessed using the phosphorus dynamics model.

In this assessment, we obtained the results of the TP concentration at the intake point or the nearest wet location when the intake point was dry from the scenario analyses. They showed that if the TP concentration in the local tributaries in the range of −75% to +75% of the baseline, there was almost no impact (<1.0% in change) on the TP concentration of the intake water. However, the TP concentration of all scenarios in the water intake ranged from 11 to 30 μg/L (20 μg/L in average). resulting in eutrophication. The concentration is generally high under the low-water condition, which is caused by the internal loading. Therefore, it is recommended that an additional treatment process should be installed in the treatment plant for a stable supply of safe water.

46.4 Effective Countermeasures

To control the eutrophication in TSL is challenging and needs substantial management efforts, as TSL is unique because of its seasonal hydrological shift and involves many environmental factors, as described in Sect. 46.1. Therefore, an effective countermeasure against eutrophication must be developed for the lake's conservation. The results of the spatial sensitivity analyses indicated that the internal processes of sedimentation and internal loading are more influential on eutrophication in TSL than the external inputs from the tributaries, TSR, and villages. However, the internal processes are difficult to control in environmental management. Thus, in this section, only external processes are discussed as the target for feasible options.

Among the external processes, the TP concentration in the local tributaries was the most influential on the TP concentration in TSL. Interestingly, during the inflow period, the 75% reduction of TP concentration in each of the local tributaries and TSR showed the same percentage reduction of the eutrophic area from the baseline condition in TSL (i.e., by 31%). Thus, the effect of those external loading on TP concentration in TSL should be controlled during the inflow period. In the case of village loading, even though the TP loading from the villages from 1999 to 2003 slightly affected the eutrophication in TSL, it could cause concern in the near future, especially in the inflow and early outflow periods, if no proper wastewater management is implemented.

Therefore, an effective countermeasure against eutrophication in TSL is to control the TP concentration in the local tributaries. In addition, the wastewater from the villages on TSL should be treated well before being discharged into TSL. As for further investigations, climate change, dam construction, and floodplain activities (e.g., land use change, agriculture, and aquaculture) might also be important factors affecting the eutrophication in TSL.

Key Points

- The internal processes of phosphorus (sedimentation and internal loading) were more influential than the external loading from the local tributaries, TSR, and villages on the TP concentration in TSL as well as the outflow TP concentration from TSL to TSR.
- On the assumption of the steady population growth, the village loading could increase the average eutrophic area by 1.2% in 2020, 2.8% in 2040, and 5.1% in 2060 from the baseline eutrophic area in 1999–2003.

- The effective countermeasure against eutrophication in TSL is to control the TP concentration in the local tributaries.
- The effects of climate change, dam construction, and floodplain activities might be important factors affecting the eutrophication, which require further investigation.

Acknowledgments The authors thank MRC and WUP-FIN for the provision of the observation data for the presented modeling work.

References

Brown M, Sodaneath H, Smith J, Hagan J. Sanitation in floating communities in Cambodia. Ministry of Rural Development, October, 28; 2010.

Davis JL, Glen S. Impacts of eutrophication on the safety of drinking and recreational. Water. In: Encyclopedia of Life Support Systems (EOLSS); 2009. Vol. 2.

Hupfer M, Hilt S. Lake restoration. In: Encyclopedia of ecology, five-volume set. Oxford: Elsevier Inc.; 2008. p. 2080–93.

National Insitute of Statistics. Census map layer and databases in 2008. National Institute of Statistics: Phnom Penh, Cambodia; 2010.

Nelson NG, Munoz-Carpena R, Phlips EJ, Kaplan D, Sucsy P, Hendrickson J. Revealing biotic and abiotic controls of harmful algal blooms in a shallow subtropical Lake through statistical machine learning. Environ Sci Technol. 2018;52(6):3527–35.

Smith VH, Schindler DW. Eutrophication science: where do we go from here? Trends Ecol Evol. 2009;24(4) February:201–7.

Takahashi Y, Doi R, Enomoto H. The Kingdom of Cambodia: From Reconstruction to Sustainable Development. Part II Chapter 2 Section 7. The Environment. Country Study for Japan's ODA to the Kingdom of Cambodia; 2002. p. 280–315.

Uk S, Yoshimura C, Siev S, Try S, Chantha Y, Shangshang O, Seingheng L, Asia E. Tonle Sap Lake: current status and important research directions for environmental management. Lakes & Reservoirs: Science, Policy and Management for Sustainable Use. 2018;23:177–89.

World Bank Group & WSP. Water supply and sanitation in Cambodia, turning finance into services for the future. International Bank for Reconstruction and Development/The World Bank; 2015. p. 1–70.

Chapter 47
Application of a Sorbent Derived from Lake Sediment and Bivalve Shells for Phosphorus Removal

Chompey Den, Eden M. Andrews, Winarto Kurniawan, and Hirofumi Hinode

47.1 Phosphorus Loading in the Lake Basin

The water quality of Tonle Sap Lake (TSL) is influenced mainly by human activities and the reverse inflow from the Mekong River (Oyagi et al. 2017; see also Chaps. 22 and 23). The water quality of the lake is strongly affected by the reverse inflow during the rainy season and by the discharge from tributaries around the lake and floating villages during the dry season. Human activities have strongly impacted the water quality of the lake, and domestic wastewater is one of the major sources of phosphorus (P) effluent to the water body (Grzmil and Wronkowski 2006; Oyagi et al. 2017). Phosphorus is considered one of the indispensable nutrients of life; however, a high concentration of phosphorus contributes to the eutrophication process (Yu et al. 2010; Kuwahara and Yamashita 2017; see also Chap. 24).

On the TSL surface during the dry season, the green algal blooms caused by the high phosphorus concentration have noticeably appeared (Chap. 31, Bonheur and Lane 2002). Phosphorus was detected in both northern and southern parts of the lake during the dry season (Chap. 24, Chea et al. 2016). The concentration of total phosphorus (TP) of 0.013–0.56 mg/L was detected in the upstream northern part lake; this range was mostly higher than the water quality standard in public water areas for biodiversity conservation in Cambodia, which ranges from 0.005 to 0.05 mg/L. Field sampling during 2016–2017 (Chap. 24) also showed that the average TP concentration in the surface water was higher in the dry season (March, June) than in the rainy season (September, December). The increase in TP

C. Den (✉)
Tokyo Institute of Technology, Tokyo, Japan

Institute of Technology of Cambodia, Phnom Penh, Cambodia
e-mail: mariquit.e.ab@m.titech.ac.jp

E. M. Andrews · W. Kurniawan · H. Hinode
Tokyo Institute of Technology, Tokyo, Japan

C. Yoshimura et al. (eds.), *Water and Life in Tonle Sap Lake*,
https://doi.org/10.1007/978-981-16-6632-2_47

during the low-water period could be a result of both internal loading (sediment resuspension releasing P) and external loading (wastewater from floating villages discharging directly into the lake). There are approximately 170 floating villages in Cambodia, with a population over one million (Thuok et al. 1996), and the TP contribution from the floating villages into the lake was estimated from a unit load of 1.1 g-TP/day/capita (Takeuchi et al. 2005). Therefore, it is essential to decrease the amount of P discharge from floating villages as well as inflow rivers into the lake for the control of eutrophication and sustainable lake management.

Nowadays, various techniques have been developed to remove P (in the form of phosphate) from water and wastewater, such as physical, chemical, and biological treatment (Clark et al. 1997; Krishna and Haridas 2008; Yu et al. 2010; Luo et al. 2016). However, these methods are either expensive or inefficient. Recently, adsorption has become an attractive technology and has increasingly garnered the interest of researchers because of its high efficiency, simplicity, ease in operability, and easy sludge handling (Yu et al. 2010; Okano et al. 2013). Furthermore, the development of adsorbents from low-cost precursors, such as waste or abundantly available material, is possible, thus reducing the overall cost. Various adsorbents, such as calcium silicate hydrate (CSH), zeolite, fly ash, and modified coir pith, have been applied for phosphate removal (Krishna and Haridas 2008; Yu et al. 2010).

CSH has been recognized as a promising adsorbent to remove phosphate from wastewater (Zhang et al. 2019). CSH can be synthesized from the reaction between calcium oxide (CaO) and silica (SiO_2) through a hydrothermal method. These precursors are available abundantly in TSL in the form of lake sediment and bivalve shells. TSL sediment is composed mainly of silica (Chap. 21), and this composition makes it favorable as a potential material for CSH synthesis (Den et al. 2020). Moreover, some of TSL's channels cannot be navigated during the dry season because of sedimentation, and this would threaten transport between the capital and regional centers and inhibit the migration of fish into the lake (Penny et al. 2005; Kummu et al. 2008). By dredging and utilizing the sediment for CSH synthesis, two of TSL's problems (sedimentation and high phosphate concentration) can be solved simultaneously. On the other hand, bivalve shells are a rich source of calcium carbonate, which can be converted easily into CaO. Furthermore, the CSH product after phosphate removal can be separated easily from wastewater and could be used as a fertilizer in agricultural fields (Okano et al. 2015), which makes it a solution to solve the environmental problem, ensuring the sustainability of the lake. Here, the possibility of CSH synthesis from TSL sediment and bivalve shells is introduced. The method and results have been reported in a separate paper (Den et al. 2020), and the discussion in this chapter focuses mainly on the conditions relevant to TSL.

Table 47.1 Chemical compositions (wt.%) of TSL sediment and bivalve shells from X-ray fluorescence spectrometry

Sample	SiO_2	CaO	MgO	Al_2O_3	K_2O	TiO_2	Fe_2O_3
Sediment	80.65	0.40	1.33	15.01	1.38	0.41	1.5
Bivalve Shell	1.53	98.46	n.d.	n.d.	n.d.	n.d.	n.d.

n.d. not detected
The composition is presented in oxides of elements (Den et al. 2020)

47.2 CSH Synthesized from Sediment and Bivalve Shells

The sediment was collected from the bottom of TSL during the dry season. As the sediment composition did not vary much across TSL (Chap. 21), the sediment from sampling point P1 shown in Fig. 21.1 was used as the representative material. The bivalve shells mixed with the sediment were separated from the sediment before treatment. Both the sediment and bivalve shells samples were dried separately in an oven at 105 °C for 24 h to remove any adsorbed moisture content, then crushed manually in a mortar, and sieved through a 75-μm aperture.

The first preparation step is the extraction of silicate from the lake sediment and the formation of CaO from bivalve shells. The chemical compositions of the sediment and bivalve shells are shown in Table 47.1. The fine fraction of the sediment was mixed with sodium hydroxide powder (Waco; 97 wt.% NaOH) at the weight ratio of 1:1.2, respectively (Fotovat et al. 2009). The mixture was then fused at 600 °C for 90 min. The fused mass was cooled to room temperature and subsequently mixed with deionized water at the weight ratio of 1:5, respectively. The mixture was then stirred at 300 rpm for 2 h at 25 °C. The suspension was filtered to obtain a clear supernatant (sodium silicate solution) using filter paper. In the practical synthesis of CSH, the separation of the supernatant can be omitted to reduce waste generated during synthesis process. The omission of filtration (by also using undissolved sediment) will decrease the adsorption capacity of the sorbent; however, the sorbent will still be able to remove phosphate. To obtain CaO, ground bivalve shells were calcined at 800 °C for 1 h (Yu et al. 2010). The formation of CaO was confirmed using X-ray diffraction (XRD).

Then, CSH was synthesized from sodium silicate solution from the sediment and calcined bivalve shell (CaO) powder at a Ca/Si molar ratio of 0.83 to 1.5, respectively. The mixture was then stirred at 300 rpm for 1 h, followed by hydrothermal treatment in a polypropylene bottle at 110 °C for 12 h. Subsequently, the solid product was recovered by filtration and thorough washing with deionized water to remove free $Ca(OH)_2$. The precipitated solid was finally dried at 90 °C for 3 h and then ground and sieved through a 75-μm aperture. For identification, the samples prepared with Ca/Si molar ratios of 0.83, 1, and 1.5 were denoted as CSH Ca/Si: 0.83, CSH Ca/Si: 1, and CSH Ca/Si: 1.5, respectively (refer to Den et al. (2020) for further details).

47.3 Phosphorus Adsorption by CSH

The adsorption experiments were carried out using simulated wastewater containing 150 mg/L of P, prepared by dissolving KH_2PO_4 in deionized water (phosphorus exists as phosphate in the solution). The adsorption tests were performed with an initial pH of 4.82.

The XRD patterns of CSH before and after the contact with the phosphate solution were compared (Fig. 47.1). A peak of CSH (tobermorite) was observed in CSH Ca/Si: 1 and CSH Ca/Si: 1.5 at 6.6 °C, indicating a successful CSH synthesis. Regarding the diffraction intensity and crystalline structure of the solid, the synthetic CSH was considered to be a low crystalline CSH sample. Furthermore, no peak of CSH appeared in CSH Ca/Si: 0.83, and this result is attributed to the amorphous phase (no crystalline form).

After 5 h of contact, the peak of CSH obviously disappeared, and a weak peak of calcium hydrogen phosphate hydrate was observed at 16.82 °C in CSH Ca/Si: 1.5. Moreover, no new peak was observed in CSH Ca/Si: 1 and CSH Ca/Si: 0.83 after contact with the phosphate solution as no crystalline structure formed. Elemental analysis showed that fresh CSH is composed of Si, Ca, and other elements, as shown in Table 47.2. After 5 h of contact, an atomic percentage of 18.23% of P content was detected in CSH, which means that P was removed successfully by CSH. The elemental mapping images revealed that after contact with the P solution, the CSH also contained a substantial amount of P (Fig. 47.2).

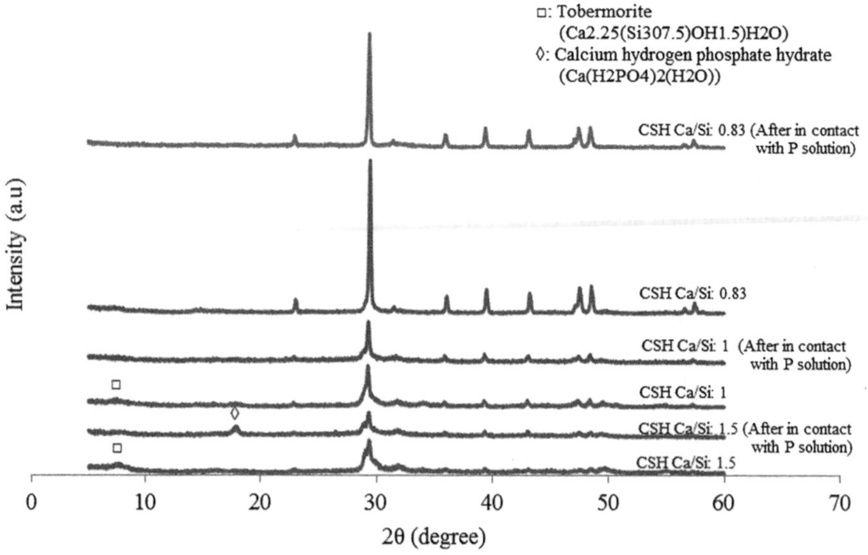

Fig. 47.1 XRD patterns of CSH before and after contact with the phosphate solution. Conditions of adsorption: dosage 2 g/L, conc. 150 mg/L, pH 4.8, contact time 300 min, and temperature 25 °C (Den et al. 2020)

Table 47.2 Chemical composition in atomic percentage (%) of CSH before and after contact with the phosphate solution

Samples	Si	Ca	Al	P	Fe	Ca/Si
Before contact with P solution						
CSH Ca/Si: 1	33.58	64.90	1.88	n.d.	0.41	1.96
After contact with P solution						
CSH Ca/Si: 1	19.04	61.30	n.d.	18.23	0.35	

n.d. not detected

Fig. 47.2 Scanning electron microscope (SEM) and elemental mapping (P) images of CSH Ca/Si: 1 after contact with the phosphate solution. Conditions of adsorption: dosage 2 g/L, conc. 150 mg/L, pH 4.8, contact time 300 min, and temperature 25 °C

In 5 h of contact, the P removal capacity and amount of Si released changed from 32 to 65 mg/g and from 28 to 34 mg/L, respectively, when the Ca/Si molar ratio increased from 0.83 to 1 (Fig. 47.3). An increase to a molar ratio of 1.5 did not increase the P removal capacity further, although the amount of Si ions released increased slightly.

The effect of the initial pH after contact with the phosphate solution on CSH was investigated at pH 3.0–11.0. After contact with the phosphate solution, all the pH values at equilibrium were observed to be higher than the initial pH. This result could be attributed to the release of OH^- ions into the solution (Kuwahara and Yamashita 2017). The high P removal capacity was confirmed in acidic conditions ($pH < 7$), and the maximum P removal of 67 mg/g was observed at the initial solution pH of 3 (Fig. 47.4). The P removal capacity decreased with the further increase of the pH of the solution. For application in TSL, as the pH of lake water is in the range of 4–7 (Chap. 23), the removal result is expected to be lower than our experimental value.

The effect of pH on the P removal capacity could be attributed to the P speciation in the solution (Liu et al. 2012; Xie et al. 2014; Zhang et al. 2019). P ions are found mainly in the form of $H_2PO_4^-$ and HPO_4^{2-} in the pH range between 5 and 10 (Karageorgiou et al. 2007; Liu et al. 2012). When the pH is lower than 7, $H_2PO_4^-$ ions are the predominant form. In the pH range of 7–10, HPO_4^{2-} species is present in the solution. At pH values between 10 and 12, HPO_4^{2-} and PO_4^{3-} ions

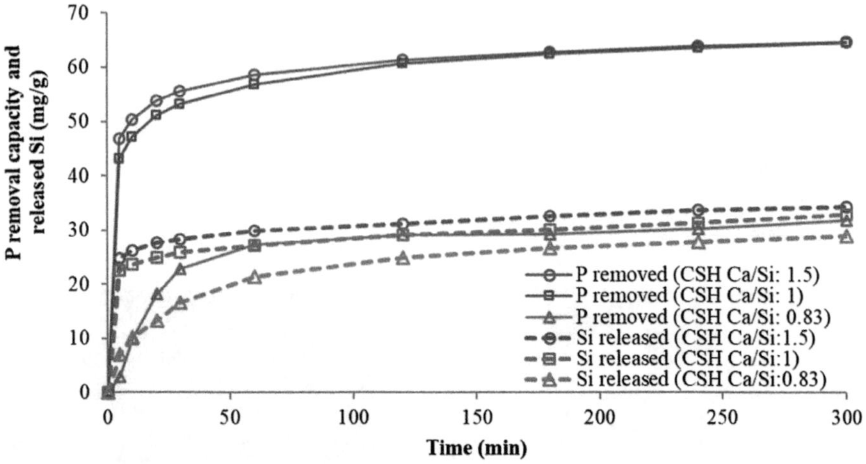

Fig. 47.3 P removal capacity and amount of Si released. Conditions of adsorption: dosage 2 g/L, conc. 150 mg/L, pH 4.8, contact time 300 min, and temperature 25 °C

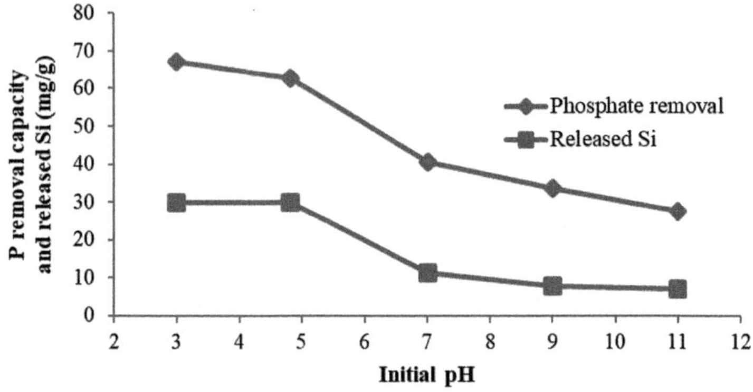

Fig. 47.4 Effect of initial pH on the P removal capacity and Si released. Conditions of adsorption: dosage 2 g/L, conc. 150 mg/L, contact time 300 min, and temperature 25 °C

are the main forms. The active sites of CSH show affinity toward $H_2PO_4^-$ ions, as shown in Fig. 47.5.

47.4 Adsorption Isotherms

Langmuir and Freundlich adsorption isotherm models were adopted to fit the phosphate adsorption data. The equations are as follows:

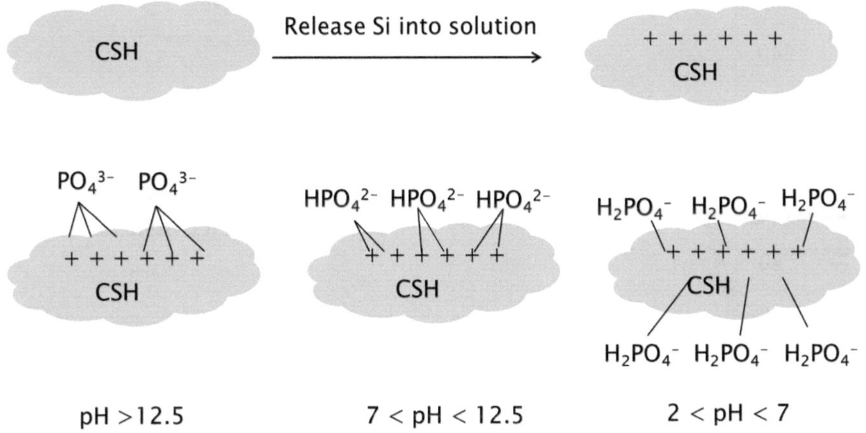

Fig. 47.5 Mechanism of P removal by CSH at different pH values

$$\text{Langmuir isotherm}: q_e = \frac{q_m b C_e}{1 + bC_e} \tag{47.1}$$

$$\text{Freundlich isotherm}: q_e = K_f C_e^{1/n} \tag{47.2}$$

where q_e is the equilibrium uptake, C_e is the equilibrium concentration, q_m is the maximum uptake capacity, b is the Langmuir constant, and K_f and n are the Freundlich constants.

The adsorption experiment was performed using different initial P concentrations ranging from 150 to 2000 mg/L with CSH Ca/Si: 1 for 180 min at a temperature of 25 °C. The root mean square error obtained from the Langmuir model (12.3) was lower than that from the Freundlich model (18.5), indicating that the experimental data fit the Langmuir model better than the Freundlich model in the adsorption of equilibrium data (Fig. 47.6). This implies that the adsorption behavior is a monolayer adsorption (only one adsorbent can occupy one adsorption site).

The maximum adsorption capacity of phosphate onto CSH Ca/Si: 1 obtained from the nonlinear Langmuir isotherm model was 280.83 mg/g, which was higher than that of other adsorbents reported in previous studies (Table 47.3). A comparison of the Langmuir constant showed that to obtain the same level of P removal, a higher concentration of the CSH developed in this work is required. However, this should not be considered a drawback, because CSH can be applied in the area where the P concentration is high in TSL, such as near toilets of the floating village. In addition, the CSH product after contact with the P solution contained not only P but also other important nutrients for vegetation, such as Si, Ca, and Fe, which can be used as a fertilizer in farming areas around TSL. However, apart from P, which existed as adsorbed species, and Si, which was found to be easily leached to the solution, the leachability of other minerals is unknown, and the effectiveness of the CSH product

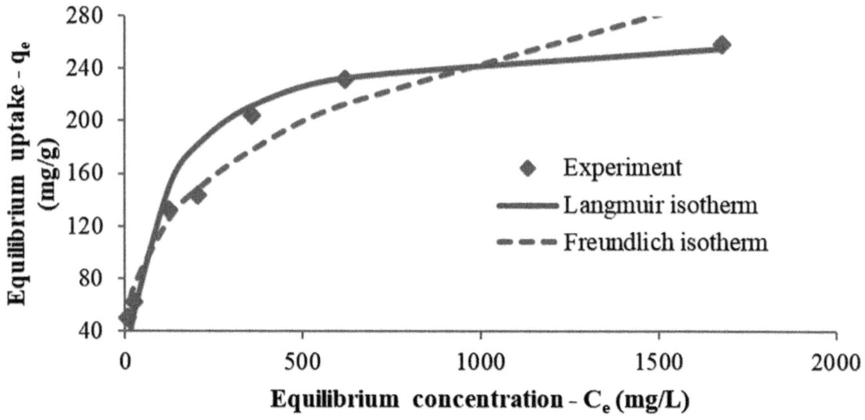

Fig. 47.6 Adsorption isotherm and fitted models

Table 47.3 Comparison of adsorption capacity of CSH Ca/Si: 1 with other adsorbents for phosphate removal

Adsorbent	q_m (mg/g)	*b* (L/mg)	References
CSH Ca/Si: 1	280.83	0.0068	This work
CP-Fe-I	70.92	0.10	Krishna and Haridas (2008)
CSH (oyster shell)	0.72	7.79	Chen et al. (2013)
Magnetic CSH	55.84	20.08	Peng et al. (2018)

as a fertilizer has not been proved yet. Further studies are needed to elucidate the leachability and bioavailability of the elements to confirm CSH's effectiveness as a fertilizer.

Key Points

- TSL sediment and bivalve shells were used successfully as precursors of CSH synthesis using a hydrothermal method.
- The adsorption isotherm showed that P adsorption was well fitted to the Langmuir equation and the maximum P removal capacity onto CSH was 280.83 mg/g, which was higher than that of other adsorbents reported in previous studies.
- The CSH product after P removal contained not only P but also other important nutrients, such as Si, Ca, and Al, and thus can potentially be used as a fertilizer. However, studies on the leachability and bioavailability of those elements need to be conducted to confirm CSH's effectiveness as a fertilizer.

References

Bonheur N, Lane BD. Natural resources management for human security in Cambodia's Tonle Sap Biosphere Reserve. Environ Sci Policy. 2002;5:33–41.

Chea R, Grenouillet G, Lek S. Evidence of water quality degradation in lower Mekong basin revealed by self-organizing map. PLoS One. 2016;11(1):e01145527. https://doi.org/10.1371/journal.pone.0145527.

Chen J, Cai Y, Clark M, Yu Y. Equilibrium and kinetic studies of phosphate removal from solution onto a hydrothermally modified oyster shell material. PLoS One. 2013;8:e60243.

Clark T, Stephenson T, Pearce PA. Phosphorus removal by chemical precipitation in a biological aerated filter. Wat Res. 1997;31(10):2557–63.

Den C, Marquit EG, Kurniawan W, Hinode H. Phosphate removal from wastewater using calcium silicate hydrate synthesized from lake sediment and bivalve shell. J Chem Eng Jpn. 2020;53(7): 287–95.

Fotovat F, Kazemian H, Kazemeini M. Synthesis of Na-A and faujasitic zeolites from high silicon fly ash. Mater Res Bull. 2009;44:913–7. https://doi.org/10.1016/j.materresbull.2008.08.008.

Grzmil B, Wronkowski J. Removal of phosphates and fluorides from industrial wastewater. Desalination. 2006;189:261–8.

Karageorgiou K, Paschalis M, Anastassakis GN. Removal of phosphate species from solution by adsorption onto calcite used as natural adsorbent. J Hazard Mater. 2007;A139:447–52. https://doi.org/10.1016/j.jhazmat.2006.02.038.

Krishna KA, Haridas A. Removal of phosphate from aqueous solutions and sewage using natural coir pith. J Hazard Mater. 2008;152:527–35. https://doi.org/10.1016/j.jhazmat.2007.07.015.

Kummu M, Penny D, Sarkkula J, Koponen J. Sediment: curse or blessing for Tonle Sap Lake? Ambio. 2008;37:158–63.

Kuwahara Y, Yamashita H. Phosphate removal from aqueous solution using calcium silicate hydrate prepared from blast furnace slag. ISIJ Int. 2017;57(9):1657–64. https://doi.org/10.2355/isijinternational.ISIJINT-2017-123.

Liu Y, Sheng X, Dong Y, Ma Y. Removal of high- concentration phosphate by calcite: effect of sulfate and pH. Desalination. 2012;289:66–71. https://doi.org/10.1016/j.desal.2012.01.011.

Luo W, Hai FI, Price WE, Guo W, Ngo HH, Yamamoto K, Nghiem LD. Phosphorus and water recovery by a novel osmotic membrane bioreactor-reverse osmosis system. Bioresour Technol. 2016;200:297–304. https://doi.org/10.1016/j.biotech.2005.10.029.

Okano K, Miyamaru S, Kitao A, Takano H, Aketo T, Toda M, Honda K, Ohtake H. Amorphous calcium silicate hydrates and their possible mechanism for recovering phosphate from wastewater. Sep Purif Technol. 2015;144:63–9. https://doi.org/10.1016/j.seppur.2005.01.043.

Okano K, Uemoto M, Kagami J, Miura K, Aketo T, Toda M, Honda K, Ohtake H. Novel Technique for phosphorous recovery from aqueous solutions using amorphous calcium silicate hydrates (A-CSHs). Wat Res. 2013;47:2251–9. https://doi.org/10.1016/j.watres.2013.01.052.

Oyagi H, Endoh S, Ishikawa T, Okumura Y, Tsukawaki K. Seasonal changes in water quality affected by water level fluctuations in lake Tonle Sap, Cambodia. Geogr Rev Jpn B. 2017;90(2): 53–65.

Peng L, Dai H, Wu Y, Dai Z, Li X, Lu X, Wu Y, Dai Z, Li X, Lu X. Performance and adsorption mechanism of a magnetic calcium silicate hydrate composite for phosphate removal and recovery. Wat Sci Technol. 2018;2017:578–91.

Penny D, Cook G, Im SS. Long-term rates of sediment accumulation in the Tonle Sap, Cambodia: a threat to ecosystem health? J Paleolimnol. 2005;33:95–103.

Takeuchi T, Takahashi Y, Sina C. Sewage water quality of Phnom Penh city. J Wat Environ Technol. 2005;3(1):133–43.

Thuok N, Ahmed M, Nuov S. Cambodia's great lake: How to sustain its ecological and economic diversity, Berkeley, USA, (June), 5–8. Bloomington, IN: Indiana University; 1996.

Xie Y, Li Q, Zhao X, Luo Y, Wang Y, Peng X, Wang Q, Su J, Lu Y (2014) Removing and recovery phosphate from poultry wastewater using amorphous ceramics. J Chem 132,582. https://doi.org/10.1155/2014/132582 Hindawi Publishing Corporation.

Yu Y, Wu R, Clark M. Phosphate removal by hydrothermally modified fumed silica and pulverized oyster shell. J Colloid Interface Sci. 2010;350:538–43. https://doi.org/10.1016/j.jcis.2010.06.033.

Zhang Z, Wang X, Zhao J. Phosphate recovery from wastewater using calcium silicate hydrate (C-S-H): sonochemical synthesis and properties. Environ Sci Wat Res Technol. 2019;5:131–9. https://doi.org/10.1039/c8ew00643a.

Chapter 48
Management of Flooded Forests and Fish Resources

Lim Puy

48.1 Major Frameworks for Environmental Conservation

Tonle Sap Lake (TSL) is socially, economically, and culturally important for Cambodians and is a site of global ecological and conservation significance (see also Chaps. 1–5 and 31–34). In 1997, the three core areas of the lake were designated as the UNESCO international biosphere reserves (Royal Decree, dated April 10, 2001). Nevertheless, this productive ecosystem is vulnerable to various stressors (e.g., urbanization and overexploitation of natural resources), adversely affecting the welfare of the lake ecosystem, environment, and biodiversity.

Recognizing the need to conserve the lake's environment and natural resources, the Royal Government of Cambodia has made considerable efforts via various legislative instruments, one of which is the establishment of the Tonle Sap Authority (TSA) in 2009. The TSA's mission is to coordinate the management, conservation, and development in the TSL areas. Since 2009, the TSA has been evaluating ecosystems, focusing on the main human activities on natural resources, such as hydrology, flooded forest exploitation, agriculture, fishery, degradation of water quality, and lifestyle in floating villages. The TSA aims to propose a road map promoting the sustainable development of the ecosystem of TSL, under the National Strategic Development Plan 2019–2023 and Cambodia Sustainable Development Goals 2016–2030. This approach is to compare the past activities with the current situation to elucidate the positive and negative impacts on the natural environment. With this background, this chapter presents the state of the land use and land cover (LULC) of TSL, with a focus on flooded forests and the status of fish resources, including the potential threats to and recommendations for them.

L. Puy (✉)
Tonle Sap Authority, Phnom Penh, Cambodia
e-mail: puy.lim@toulouse-inp.fr

C. Yoshimura et al. (eds.), *Water and Life in Tonle Sap Lake*,
https://doi.org/10.1007/978-981-16-6632-2_48

48.2 The Floodplain and Flooded Forests

The TSL area and the floodplain, surrounded by national road numbers 5 and 6, have been divided into three zones, namely, Zone 1 (395,578 ha), Zone 2 (369,865 ha), and Zone 3 (647,406 ha) (Fig. 48.1a). Zone 1 is to be used for housing and the production of rainy-season rice; Zone 2 (the buffer zone of the TSL-protected area) for the development and production of floating rice (rice is transplanted in inundated rice fields and grown in the existing water level), recession rice (the growth starts when the water level decreases), and dry-season rice (rice crops depend on the irrigation system); and Zone 3 as the strictly protected flooded forests. The definition of Zone 3 was issued by government sub-decree no. 197 (dated September 29, 2011) to protect the flooded forests, including the three core areas of the biosphere reserve, namely, Prek Toal, Boeung Tonle Chhmar, and Stung Sen core zones.

The flooded forest area along with a type of shrub area (resulting from the degradation of the flooded forest) predominates the landscapes of the protected area along the shores of the lake. In the protected area of TSL, particularly in Zone 3, the built-up area is characterized mainly by floating villages. These villages move seasonally, according to the water level of the lake. As their source of income is mainly fishing-related activities, villagers take up residence close to the shores of the lake, which is generally good for fishing. The floating houses are simple structures, constructed of wood or other natural materials. Moreover, many public facilities, such as schools, health centers, temples/churches, and shops, are also available, floating on the lake water, in each major village. In the area of seasonal inundation, typical farming stilt houses are wooden and have high floors because they are subjected to seasonal inundation and frequent and short flood events in the rainy season. The rest of the rural area, located further away from the lake, is not markedly different from other rural areas of the country, relying mostly on rainwater for field cultivation of rice and other crops. Human settlements are observed along the national roads that are connected to the urban area.

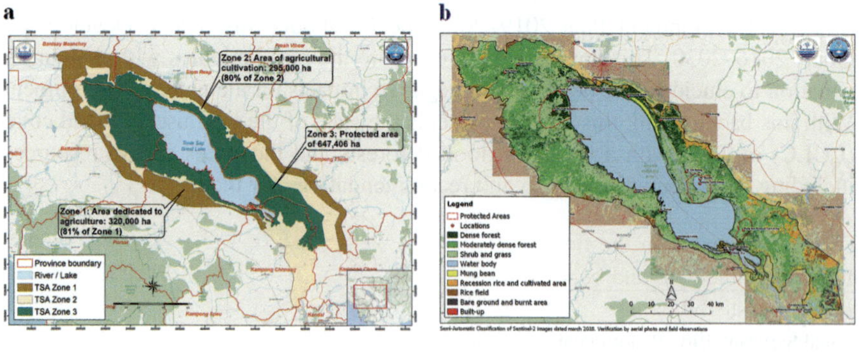

Fig. 48.1 (**a**) Map showing Zone 1, Zone 2, and Zone 3 and (**b**) LULC of the Zone 3 protected area based on the image classification of the Sentinel-2 satellite, dated March 2018

Table 48.1 Definition of each land cover classification (class typology) observed in the protected area of TSL

Class name	Description	Proportion to the total area of Zone 3 (%)
Dense forest	Forest with tall trees	12.2
Moderately dense forest	Mixed forest of gallery forests and shrubs	32.3
Shrub and grass	Shrubland and grassland	39.5
Mung bean	Mung bean seasonally cultivated along the shoreline	0.4
Rice field	Rainfed rice fields (not cultivated at the date of the image)	9.6
Recession rice and cultures	Dry-season rice fields or other crops (incl. vegetables)	6.9
Bare soils	Mineral soils or burned soils	1.9
Built-up	All the urban areas and roads	0.1

The spatial analysis of LULC in the TSL's floodplain based on Sentinel-2 satellite imagery (Fig. 48.1b), conducted by TSA in 2019 focusing on Zone 3 (Table 48.1), showed that dense forests cover 12.2% of the total area of Zone 3. Most of these forests are located in the less accessible parts of Zone 3 and along the shoreline of the lake and have been preserved. Moderately dense forests (or mixed forests) cover 32.3% of Zone 3, whereas shrubland and grassland, characterizing an extreme state of forest degradation, represent 39.5% of Zone 3. The classes corresponding to the rice fields have been identified as comprising 9.6% of Zone 3 (recession rice and other cultivated areas, 6.9%, and harvested rice fields, 2.7%). These classes are distributed mainly in the external boundary of Zone 3, near the villages and the lakeshore (e.g., the mung bean fields located between Kampong Phluk and Kampong Khleang represent 0.4% of Zone 3). Bare ground and burned area and built- up area represent 1.9% and 0.1% of Zone 3, respectively.

Based on Fig. 48.1b, the LULC classification shows clearly that Zone 3 is characterized by TSL and the surrounding forests and shrubland, which comprise a wide area of the predominant lowland rice fields. This rice-field area reaches as far as the outer boundary of the protected area at the limit of Zone 2.

48.2.1 Threat to Flooded Forests

The flooded forest in the floodplain is the core of the lake's ecosystem, providing numerous functions and benefits and offering heterogeneous habitats for fauna and flora (refer also to Chaps. 31–34). However, as the population and economy of the country grow, the demand for the resources from the lake (e.g., food and energy) has intensified, resulting in the degradation of the lake's natural resource and ecological system. Deforestation and LULC changes in the TSL basin and its inundation area are substantial. Originally, flooded forests covered approximately one million hectares around the lake, which was subsequently reduced to approximately 614,000 ha by the 1960s, 362,000 ha by 1991, and 350,000 ha by 1997 (ADB 2005). Given the importance of the flooded forests discussed in Chap. 32, such a significant loss of the flooded forests would lead to ecological degradation that could negatively affect the fauna and flora communities and local communities. Thus, the protection of the flooded forests is inevitable to ensuring the sustainability of this nature-based livelihood of the local people.

The main causes of the loss of flooded forests include enlarging land and rice fields and settling illegal water reservoirs for rice fields at an industrial scale. In addition, forest degradation is due to harvesting firewood to make smoked fish, catching wild animals by setting fires (at the household scale), exporting precious trees to neighboring countries, implementing various fish trapping methods, and harvesting of big trees by fishermen. According to the International Union for Conservation of Nature (ICUN 2016), two main factors drive forest fires in the floodplain: accidental and intentional factors. The former is caused by the accidental spread of fire from the utilization of smoke to harvest honey, burn firewood, and leave cooking fires unattended. For the latter, the farmers intentionally burn the flooded forests to convert the land to rice fields, to catch wild animals, and to set long fishing nets (Sach Daiy) across river channels.

48.2.2 Recommendation

As mentioned earlier, a sub-decree was issued in 2011 to designate Zone 3 as the protected flooded forests. As of now, this legal framework needs enforcement. For substantial protection, we need the recolonization of flooded forests in the cleared areas and a good, comprehensive functioning of the flooded forests and aquatic ecosystems, linked to seasons and hydrological balance.

It is important to identify and remap flooded forest areas to observe the ecological consequences of the forest loss as well as to identify the potential zones to reforest. To achieve the aforementioned objectives, both the geographical information system and remote sensing should be used. The flooded forests have been mapped in 2005, 2010, 2013, and 2018; the latest mapping was based on satellite images by Sentinel-2A and Sentinel-2B and aerial photography (Fig. 48.1b).

Ecological research studies on the flooded forests are also needed. These studies should focus on the inventory of species hosted by the flooded forests (i.e., trees, other vegetation, and fauna), reproduction and natural colonization, reforestation, and the role of flooded forests as physical and biological habitats over time and space. Specific research should also be conducted on the roles and benefits of the flooded forests along the river and stream located around TSL.

It is important to enforce the legal framework and define appropriate scrutiny to cover the surface area efficiently to protect and delegate the responsibility of the flooded forest management to the local authorities. The required action is to create a network of observation stations focusing notably on forest fires during the dry season, illegal forest exploitation, and illegal hunts and fishing.

48.3 Fish Resources

The TSL ecosystem provides heterogeneous physical habitats and food for fish (refer to Chaps. 31 and 33). Of the 167 fish species surveyed by the TSA from 2010 to 2019, around 20 species have been exploited at an industrial scale for processing and market-fresh supply to consumers. Other aquatic animals, such as water snakes, mollusks, and turtles, are also exploited for market-fresh and exported.

During the 5-year period (2015–2020), the TSA conducted surveys on 11 fish-landing ports around TSL (Table 48.2) and on-site surveys focused on three types of fishing gear: the fishing lot fence system (*Bor*), gillnet (*Mong*), and horizontal cylinder traps (*Lob*). The average catch estimation was around 37,513 tons/year. The total catch was divided into 28% for fresh market, 32% for processing, and 40% for fish feed. Fish used for processing can be divided into six categories by order of importance in terms of the proportion used for processing: (1) fish fermentation (in Khmer: *Prahok, Pa'ak, Mam*), (2) dried fish, (3) salted fish, (4) fish smoked on the skewer, (5) fish ball, and (6) fish sauce.

Table 48.2 Fish-landing ports around TSL during the 5-year period (2015–2020)

		UTM coordinates (m)		
No.	Fish-landing port	Latitude	Longitude	Scale
1	Psar Chhnang	465,388	1,356,247	Large
2	Chhnock Trou	440,803	1,382,955	Large
3	Kampong Luong	413,990	1,389,895	Large
4	Reang Til	393,029	1,414,184	Medium
5	Kanchor	398,454	1,396,981	Medium
6	Kampong Khleang	401,632	1,444,731	Large
7	Kampong Phluk	388,138	1,459,161	Medium
8	Chong Khneas	373,090	1,463,425	Large
9	Me Chrey	357,243	1,469,266	Medium
10	Prek Toal	356,216	1,462,967	Medium
11	Sla Ket	306,504	1,451,246	Large

Small scale = family (subsistence) fishing; medium scale = artisanal fishing; large scale = industrial fishing. UTM coordinate (m) = Universal Transverse Mercator coordinate in meters

48.3.1 Fish Catch and Its Shift

A large variety of fish species in TSL are in disequilibrium regarding fish stock and fish catches because of illegal fishing and overfishing. Our estimated fish stock in TSL was 540,000 tons in 2008–2009, and the permissible fish catch was 180 kg/ha/year. However, the actual total fish catch in that period from the lake was 537,000 tons, revealing the obvious overfishing and the depleted fish stock. Baran et al. (2001) also indicated a high average fish catch of 230 kg/ha/year from the TSL floodplain system, which was higher than the fish yield estimates from other floodplains, such as 51–130 kg/ha/year from the Bangladesh natural floodplain (MRAG 1997), 26–41 kg/ha/year from the Brazilian Amazon flooded forests (WWF 2002), and 40–73 kg/ha/year from the Nam Ngum Reservoir in Lao (Mattson et al. 2001).

Regardless of the opening or closing of the fishery period, problems of overfishing and the operation of modern and illegal fishing gear have notoriously depleted the fish stocks. During 2010–2011, after an alert on fish stock at an unprecedented low level, 36 fishing lots were cancelled to protect the fish resources.

Large Stationery Bagnet (*Lot Dai*) fishery (Deap 1999) is located in a section of approximately 37 km in the Tonle Sap River (TSR) (Fig. 48.2) in Kandal Province and Phnom Penh City. Within this *Lot Dai* section, there were 15 ranks (rows) and 64 *Lot Dai* units in 2011–2012 and only 13 ranks (rows) and 43 *Lot Dai* units in 2019–2020. During 2011–2019, the TSA had followed up the catch of *Lot Dai* along the TSR in the season from November to February. During the 2019–2020 season, the TSA identified 109 fish species in 74,233 samples.

After year 2012–2013, our survey on the fishing *Lot Dai* in Kandal Province and Phnom Penh City showed a slight increase in fish catch and fish diversity, especially endangered species (Table 48.3, Fig. 48.3). However, since 2014, the total fish catch on these *Lot Dai* has decreased gradually to approximately 5–11% of those of 2011–2012. Protection measures must be implemented and continued in the future years to increase fish biomass and catch, particularly against the illegal fishing gear and fishing during the closed season.

48.3.2 Recommendations

The fish from the TSL is a major source of animal protein (~60%) for both local and urban consumption (Nam and Song 2011), and it has been exploited by many economic sectors. Research to estimate the fish stock and productivity of the lake must be carried out, and regulations against illegal and overfishing are necessary to prevent the depletion of the fish stock in TSL. In addition, a decrease up to 25% of fish catch in TSL was reported because of its high vulnerability to climate and land-

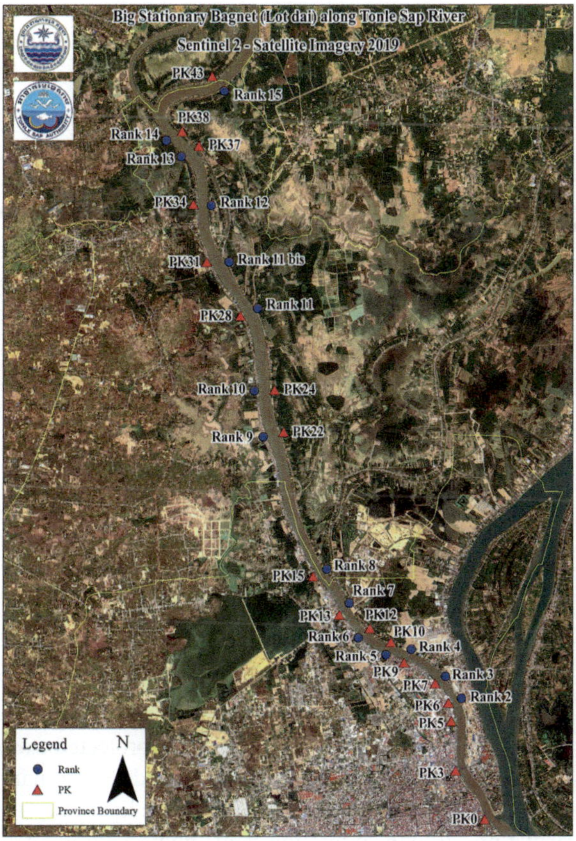

Fig. 48.2 *Lot Dai* rank (*blue points*) along the Tonle Sap River in 2019 (the imagery of Sentinel-2)

Table 48.3 Fish catch by *Lot Dai* from the 2011–2012 to the 2019–2020 seasons

Season	Catch (ton)	No. of species	No. of sample
2011–2012	41,559	83	64,700
2012–2013	5326	83	38,660
2013–2014	13,400	91	18,945
2014–2015	8543	86	41,747
2015–2016	4512	88	50,996
2016–2017	8223	92	59,671
2017–2018	5625	101	52,103
2018–2019	5385	99	48,616
2019–2020	2140	109	74,233
Total	94,713	109	449,671

use changes (Kao et al. 2020). As fish habitats have been degraded gradually, with expanded rice cultivation (Gray 2008; Mahood et al. 2020), the likelihood of decreasing fish catch is possible. To that end, the objectives and actions required are recommended as follows:

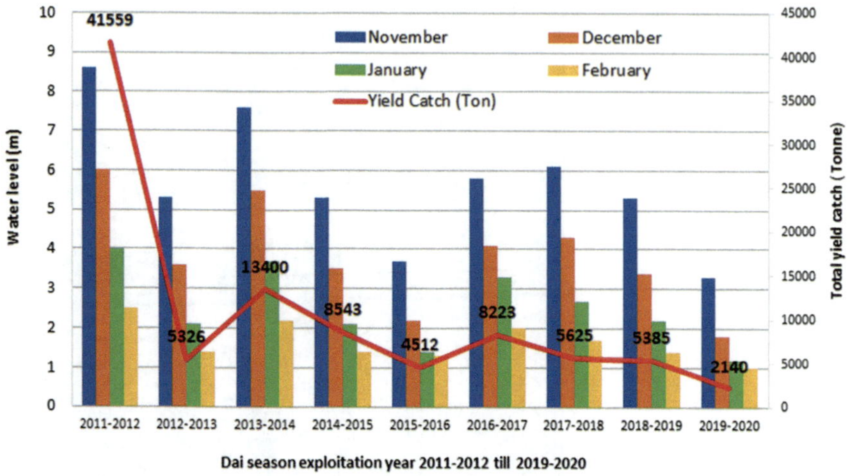

Fig. 48.3 Annual catch statistics of *Lot Dai* in Kandal/Phnom Penh and monthly water level in Tonle Sap River at Prek Kdam station from November to February between 2011 and 2020

Law enforcement needs to be strictly in place against illegal activities so that laws are implemented effectively. Incentive policies should be created to encourage the implementing agency to carry out the work.

It is important to conduct scientific research to define a rational regulation of the fishery, including (1) sanctuary zones (protected zone), (2) opening and closing periods for fishing (fishing calendar), and (3) annual quotas for fish catches. The actions required include analyzing the fish life cycle—particularly reproduction, egg spawning, growth, migration, and food location—with a focus on major species, endangered species, and high-commercial-value species. Understanding the migration patterns of fish species of TSL, MR, the Bassac River, and other related regions is also needed. Based on such research (e.g., fish stock and lake productivity), the annual quotas for fish catch also need to be identified.

It is also important to review the local management of fish resources by revisiting the organization of community fisheries and establishing an aquaculture center to release fingerlings to restock fish. The required actions are (1) to (re)define the organization, budget, and technical skills necessary for the functioning of community fisheries and (2) to select fingerlings to produce and create new stock.

For the long term, we also should improve fishery revenue, propose fish-processing infrastructure, and reformulate fish market infrastructure to comply with average standards. The actions required are to assess existing processing activities and the most profitable ones, to recommend processing activities at community and medium scales, and to propose a relocation (if necessary) of fish markets and infrastructure improvements.

Taking climate and land-use changes into account, relevant government agencies are expected to scrutinize the actual water use and management within the TSL basin and floodplain. Such actions may increase the water use efficiency, thereby lowering the water extraction from TSL. These government agencies also need to pay more attention to protect the water quality within TSL, because access to clean water (as measured by the proportion of a population using drinking water and sanitation services in the basin) is suggested to be a characteristic of the lake where fish catches are less vulnerable to climate and land-use changes (Kao et al. 2020). Investments in sanitation indicated improved fish habitats, thereby potentially contributing to healthier fish stocks.

Key Points

- The flooded forest area along with a type of shrub area predominates the landscapes of the protected area along the shores of the lake. Deforestation and LULC changes in the TSL basin and its inundation area are substantial.
- A large variety of fish species have been threatened by illegal fishing and overfishing in TSL and have caused a disequilibrium in fish stock and fish catches, resulting in the shrinkage of the in-lake fish stock available for fishery.
- It is necessary to monitor the dynamics of the flooded forest areas to capture their ecological shifts and identify the potential zones to reforest using selected species from the local tree nursery in the TSL region.
- It is also important to redefine a rational regulation for the fishery, including sanctuary zones (protected zone) and opening and closing periods for fishing (fishing calendar), and to determine the annual quotas for fish catches.
- Relevant government agencies should pay more attention to ensure proper water quantity and quality within the TSL basin, such that fish catches in TSL can be potentially less vulnerable to climate and land-use changes.

Acknowledgments The author would like to thank Tonle Sap Authority (TSA) for providing the data for this work.

References

ADB. The Tonle Sap Basin strategy. Manila, Philippines: ADB; 2005.

Baran E, Van Zalinge N, Bun NP, Baird I, Coates D. Fish resource and hydrobiological modelling approaches in the Mekong Basin. In: ICLARM, Penang, Malaysia and the Mekong River Commission Secretariat. Phnom Penh: RePEc; 2001. p. 60.

Deap L. The Bagnet (Dai) Fishery in the Tonle Sap River. 1999. http://www.mekonginfo.org/assets/midocs/0002656-biota-the-bagnet-dai-fishery-in-the-tonle-sap-river.pdf. Accessed 14 Jun 2021.

Gray TNE. The conservation and ecology of the Bengal Florican Houbaropsis bengalensis in Cambodia: grasslands, people and management. University of East Anglia: PhD thesis, BUSTARDS.ORG; 2008.

ICUN. Flooded forest fires: a major threat to the Tonle Sap. 2016. https://www.iucn.org/news/cambodia/201607/flooded-forest-fires-major-threat-tonlesap#:~:text=Two%20fators%20drive%20foest%20fire,Sach%20Daiy)%20across%20river%20channels. Accessed 11 Mar 2021.

Kao YC, Rogers MW, Bunnell DB, et al. Effects of climate and land-use changes on fish catches across lakes at a global scale. Nat Commun. 2020;11:2526.

Mahood SP, Poole CM, Watson JEM, Sharma S, Garnett ST, Mackenzie RA. Agricultural intensification is causing rapid habitat change in the Tonle Sap floodplain, Cambodia. Wetl Ecol Manag. 2020;28(5):713–26.

Mattson NS, Balavong V, Nilsson H, Phounsavath S, Hartmann WD. Changes in fisheries yield and catch composition at the Nam Ngum Reservoir, Lao PDR. In: De Silva SS, editor. Reservoir and Culture-based Fisheries: Biology and Management, Proceedings of an International Workshop, Bangkok, Thailand, 15–18 February 2000, vol. 98. Canberra, Australia: ACIAR Proceedings; 2001. p. 48–55.

MRAG. Fisheries dynamics of modified floodplains in southern Asia. In: Final Technical Report. London: ODA; 1997.

Nam S, Song SL. Fisheries management and development in Tonle Sap great Lake, Cambodia. Paper presented to the consultation on development trends in fisheries and aquaculture in Asian Lakes and reservoirs, 20–23 September 2011, Wuhan, China; 2011

Royal Decree (dated 10 April 2001) Tonle Sap Biosphere Reserve.

Sub-decree No.197 (dated 08/09/2011) on marking of the boundaries of flooded forest sites for 647.406 hectares in 6 provinces around Tonle sap lake (Kampong Chhnang, Pursat, Battambang, Banteay Meanchey, Siemreap and Kampong Thom Province).

WWF. Sustainably managed Amazon fisheries 60% more productive. 2002. https://wwf.panda.org/?4961/Sustainably-managed-Amazon-fisheries-60-more-productive. Accessed 14 Jun 2021.

Part XI
Outlook for Sustainability

Chapter 49
Transdisciplinary Research Collaboration for Environmental Conservation

Rajendra Khanal, Uk Sovannara, Ly Sophanna, Ratino Sith, Kong Chhuon, Binaya Raj Shivakoti, Pham Ngoc Bao, Chihiro Yoshimura, Hideto Fujii, Winarto Kurniawan, Kazuhiko Miyanaga, Toru Watanabe, Sambo Lun, Chantha Oeurng, Chanvorleak Phat, Reasmey Tan, Sokneang In, Kimleang Kheurn, and Aiko Yamashita

49.1 TDRC Project for the Lake Science and Management

Several publications are available on review of TDRC in sustainability science (Stokols 2006; Mobjörk 2010; Brandt et al. 2013). One of the most widely used quality criteria for TDRC has been well categorized by Bergmann et al. (2005), which includes (i) members and project formulation, (ii) project execution and methodology, and (iii) results, products, and publications.

Complex environmental systems can be better understood by examining the following key areas: (i) instrumentation and method development, (ii) monitoring and statistical analysis, (iii) treatment methods, (iv) toxicity assessment, (v) risk

R. Khanal (✉)
Tokyo Institute of Technology, Tokyo, Japan

Policy Research Institute, Kathmandu, Nepal

U. Sovannara · C. Yoshimura · W. Kurniawan · K. Miyanaga
Tokyo Institute of Technology, Tokyo, Japan
e-mail: miyanaga.kazuhiko@jichi.ac.jp

L. Sophanna
Ministry of Environment, Phnom Penh, Cambodia

Tokyo Institute of Technology, Tokyo, Japan

R. Sith · K. Chhuon · S. Lun · C. Oeurng · C. Phat · R. Tan · S. In · K. Kheurn
Institute of Technology of Cambodia, Phnom Penh, Cambodia

B. R. Shivakoti · P. N. Bao
Institute for Global Environmental Strategies, Hayama, Japan

H. Fujii · T. Watanabe
Yamagata University, Tsuruoka, Japan

A. Yamashita
Japan International Cooperation Agency, Tokyo, Japan

C. Yoshimura et al. (eds.), *Water and Life in Tonle Sap Lake*,
https://doi.org/10.1007/978-981-16-6632-2_49

categorization, (vi) science communication, (vii) policy formulation, and (viii) capacity development (Khanal 2021). Policy questions that encompass complex environmental systems can rarely be answered by a single study or even by a single discipline. There is a prerequisite of evidence synthesis by researchers to aid decision-makers in evidence-based policy synthesis. One of the approaches undertaken by the government of Japan in unwinding this complexity of environmental systems is through the Science and Technology Research Partnership for Sustainable Development (SATREPS).

SATREPS is funded by the Government of Japan through the Japan Science and Technology Agency, Japan Agency for Medical Research and Development, and Japan International Cooperation Agency. The projects under this framework are transdisciplinary research collaboration between Japanese and foreign institutions in developing countries. One of the targets of SATREPS is to enhance environmental sustainability by researching global issues including but not limited to energy, environment, bioresources, food security, disaster prevention and mitigation, and infectious disease control through fundamental, applied, and innovative policy-based research, evidence-based policies, and capacity development in partner countries (https://www.jst.go.jp/global/english/). As of June 2020, a total of 157 SATREPS projects have been commenced in 52 countries, of which 84, 41, 23, and 9 are in Asian, African, Latin American, Caribbean regions, respectively.

SATREPS project—Establishment of Environmental Conservation Platform of Tonle Sap Lake (henceforth called TSL platform)—was implemented in Cambodia from April 2016 to March 2021 (later extended to March 2022 because of the covid-19 pandemic). This TDRC was designed to meet long-term objective to establish a framework to realize the sustainable environmental conservation of TSL. Within the project period, the scientific and administrative capacity for water environmental management of TSL is expected to be strengthened on the basis of the following outputs: (1) environmental database of TSL, (2) Water Environment Analytical Tool, and (3) Tonle Sap Water Environmental Platform. To meet these targets, the focus was made especially on (i) establishing and enhancing state-of-the-art research facilities for data generation in Cambodia, promoting evidence-based policies through primary, secondary, and modeled data, and (ii) capacity development at the individual and institutional levels both at the academic and policy levels. The TDRC team for the TSL platform includes the Tokyo Institute of Technology, Yamagata University, and Institute for Global Environment Strategies from Japan and the Institute of Technology of Cambodia (ITC), Royal University of Phnom Penh, Ministry of Water Resource and Meteorology, Ministry of Environment, and Tonle Sap Authority from Cambodia. Figure 49.1 shows the framework for SATREPS implementation and the expected impact after its implementation.

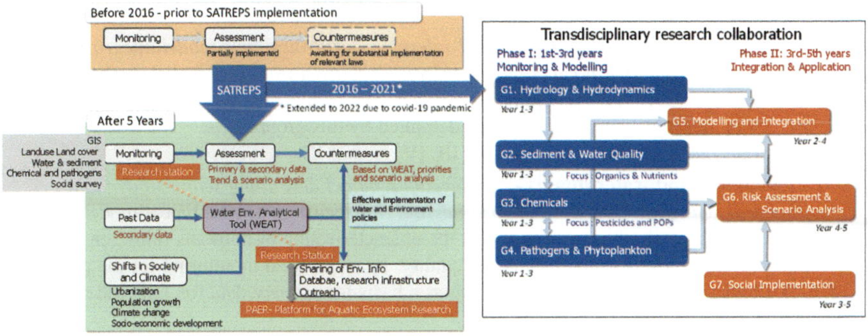

Fig. 49.1 Impact assessment of SATREPS implementation

49.2 Implementation and Framework of SATREPS

The team of TDRC of TSL platform was divided into seven working groups (WGs), comprising WG1, hydrology and hydrodynamics; WG2, sediment and water quality; WG3, chemicals; WG4, pathogens and phytoplankton; WG5, modeling and integration; WG6, risk assessment and scenario analysis; and WG7, social implementation. Each group comprised project members both in Cambodia and Japan. There were approximately 20 and 15 PhD holders along with the undergraduate and graduate students, policymakers, and bureaucrats in Japan and Cambodia, respectively. Seven WGs were formed by focusing on each research task for monitoring, assessment, evaluation, policy discussion, and social implementation (Fig. 49.1 and Table 49.1). For the social and policy implementation, one of the distinct features of TDRC, the inputs from all the WGs were processed and analyzed for the pragmatic implementation in policy formulation and environmental management for the decision-makers.

49.3 Major Outcomes from SATREPS

The activities and achievements from the TDRC of the SATREPS project are summarized into three main points: (i) capacity development, (ii) technology transfer and knowledge sharing, and (iii) scientific policy impacts (Fig. 49.2).

The chapters from each WG, as summarized in Table 49.1, significantly contributed to understanding the mechanism of hydrology and hydrodynamics, sediment discharge water quality (basic, chemical, and microbial), modeling application, and risk assessment as a baseline to understand .cognition of water and life in TSL.

Capacity development at the individual level (students, researchers, and bureaucrats) and institution level (instruments for environmental surveillance and monitoring at government and academic institutions) have been achieved through the TSL platform TDRC implementation. More specifically, capacity development at one of

Table 49.1 Scope of work and responsibilities of each working group (WG)

WG 1. Hydrology and hydrodynamics—*To develop a hydrodynamic model integrated with modules for describing hydrological and meteorological processes.* This model works as a basis for water quality and risk assessment in this project. The bathymetry survey and integrated 1D, 2D, and 3D models have been developed and applied by this group Chapters contributed to this book—Parts II, III, IV, X, and XI
WG 2. Sediment and water quality—*To develop a model describing sediment dynamics including sediment yield, transport, deposition, and resuspension.* This sediment model has been integrated with the hydrodynamic model for the characterization of nutrients (nitrogen and phosphorous), groundwater, and lake water quality. Intensive field monitoring for the period 2016–2019 was also performed by this group as a basis for primary data for TSL water quality assessment Chapters contributed to this book—Parts IV, V, and XI
WG 3. Chemicals—*To develop a model describing pollutant dynamics mainly focusing on organic matter, nutrients, trace metals, and chemicals from agriculture.* The task involves spatiotemporal monitoring of those pollutants, source identification, and modeling of their fate and transport in the target regions Chapters contributed to this book—Parts IV, VIII, X, and XI
WG 4. Pathogens and phytoplankton—*To develop a model describing distributions of pathogenic bacteria and virus as well as phytoplankton.* The task involves spatiotemporal monitoring of those microbes, source identification, and modeling of their fate and transport in the target regions Chapters contributed to this book—Parts VI, VII, IX, and XI
WG 5. Model integration—*To integrate and validate the overall model, which will include hydrodynamics (surface water and groundwater), sediment dynamics, and pollutants as an integrated water environment analytical tool (WEAT).* This WEAT forms a basis for the TSL platform for further analysis and application for water quality conservation. The WEAT application and outputs can be used by any individuals or institutions, including researchers, civil society, environmental managers, and policymakers Chapters contributed to this book—Parts II, V, X, and XI
WG 6. Risk assessment—*To quantify the health risks of people living on floating houses and the lakeshore and to propose several options of environmental management based on water flow and various scenarios.* A primary survey based on site visit and questionnaire survey and available epidemiological data was also performed to assess the reliability of the method quantitatively Chapters contributed to this book—Parts VI, IX, and XI
WG 7. Social implementation—*To apply the developed model and outcomes from scenario analyses from group 6 for environmental conservation involving the Cambodian government and local communities.* These tasks are being combined with the capacity building of environmental managers, officers, and local people. This group work arranges an effective plan combining training courses, workshops, symposiums, and other activities and conducts such plan with a help of other core members in this project Chapters contributed to this book—Parts I, VII, X, and XI

the partner institutes, ITC, is demonstrated in Fig. 49.2. This helix of capacity development includes the identification of key questions for fundamental and applied research, sharing the research outcomes through a series of workshops and seminars, publications, social contribution, and communication outreach through social networking services. Overall, capacity development has contributed toward self-realization and contribution in education as a fundamental base for the Cambodian hub for the world. Briefly, the achievements of excellent potential of academic

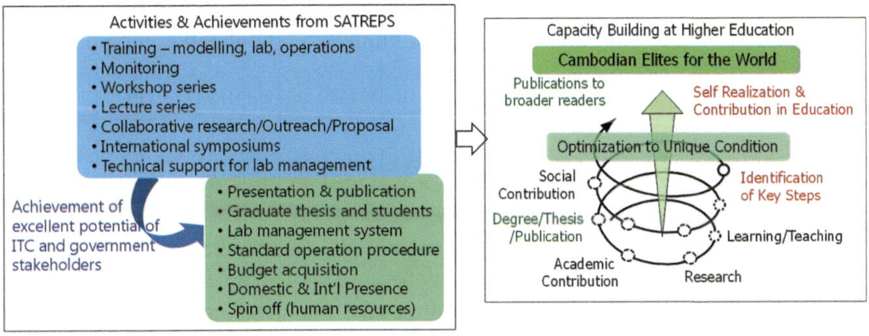

Fig. 49.2 Activities, achievements, and capacity development in the SATREPS project

and government officials have been demonstrated through multiple peer-reviewed journal publications, international conference presentations, graduate theses, panel discussions, and policy outputs.

Technology transfer has been done in setting up of state-of-the-art research facilities at ITC. An advanced environmental laboratory has been set up consisting of but not limited to water quality analysis, microbial analysis, and environmental modeling. The water quality analysis equipment consists of a multiparameter real-time sensor to determine the location, temperature, pH, conductivity, salinity, turbidity, total dissolved solids, blue-green algae, chlorophyll a, oxidation–reduction potential, and depth. The researchers have been trained in quality analysis such as the calibration of analytical equipment, statistical data analysis, cleaning after usage, and operation, care, and maintenance of the equipment. Other advanced equipment includes organic carbon analyzer, particle size analyzer, ion chromatography, gas chromatography–mass spectrometry, ultrapure water purification system, and basics for microbiological analysis including next-generation gene sequencing, autoclaves, and clean bench. The modeling tools include 1D, 2D, and 3D hydrodynamic modeling. Additionally, facilities for the sample storage and the standard operating procedures for the disposal of laboratory wastes have also been partly set. Several workshops, seminar, annual international symposium, and policy discussion have also been conducted.

This technology transfer and knowledge sharing mainly aim to emphasize knowledge synthesis, organizing knowledge, sustainable technology transfer, and multiple stakeholders (academic, bureaucrats, students, and civil society) for collective discussion, feedback analysis, and decision-making (Kloprogge and Van Der Sluijs 2006; Pohl 2008).

The most significant scientific impacts of SATREPS have been on theses at bachelor, master, and PhD degrees, nearly 100 conference proceedings, approximately 30 peer-reviewed international journal publications, policy recommendation book, and this technical book comprising water and life in TSL. Additionally, project members delivered keynote speeches and discussed as panelists in an international forum for sustainable lake basin management and organized awareness trainings at the local level including but not limited to school and communities.

49.4 Challenges and Hurdles

Many challenges occur in TDRC, which mainly include the integration of scientific knowledge, local knowledge, and jointly generated knowledge (Stokols 2006; Wiek 2007; Mobjörk 2010). Table 49.2 summarizes the main challenges and hurdles for the international TDRC in the format of the strength, weakness, opportunities, and threat (SWOT) analysis. The most significant strength is the involvement of international experts and sharing based on the evidence-based policies between multiple institutions and stakeholders in international scenarios. However, sometimes, the involvement of multiple stakeholders can be a weakness as it might lead to a division of tasks and hence responsibilities—such that the transparency in decision-making and accountability of the decision are in discord. In the case of TSL, for instance, the communication reached out to several additional stakeholders of the lake management, for instance, agricultural, fishery, energy, and transportation sectors. Moreover, TSL locates within a transboundary river, which is the Mekong River, and hence, the upstream and downstream countries are also important stakeholders, which make social implementation more complicated and exhausting. Therefore, the challenge for all stakeholders is the continuous accumulation of environmental

Table 49.2 Results from the SWOT analysis of the international TDRC

Strength	***Weakness***
• International cooperation • Experts • Institutional framework • Multiple stakeholder—division of task • Technology and knowledge transfer for coordination and action-oriented goals	• Limited experience of collaboration under large international research projects • Multiple stakeholders—responsibilities • Transparency in decision-making and accountability • Social implementation as a project outcome
Opportunities	*Threat*
• Implication of scientific knowledge to political and societal decision making • Practical solution • Sustainability • Platform for conservation	• Quick turnover of and dependency on project members • Operation of sophisticated equipment and model • Tradeoff between economic development and environmental conservation

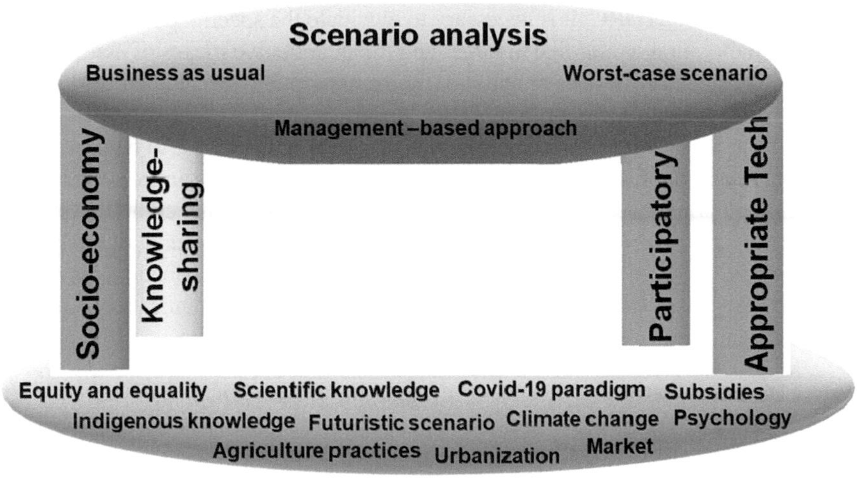

Fig. 49.3 Framework for transdisciplinary research collaboration

data, knowledge, and tools, which establish a basis for collaboration and codesign of the environment.

Based on the experiences of the SATREPS project, a continuous supply of research resources to institutions in developing countries for gaining research outcomes, which is known as project sustainability, must be secured. We also recommend taking into account at the initial stage of projects major potential hurdles such as the breakdown of sophisticated equipment, continuity of long-term monitoring, and the swift turnover of project members. The basis for any evidence policies lies on equity and equality, knowledge sharing based on science and indigenous technology, climate change, urbanization, and subsidy to name a few. The major pillars include socioeconomy, knowledge sharing, participatory approach, and appropriate technology. Finally, scenario analysis must be done for decision-making considering business as usual, worst-case scenarios, and management-based approach (Fig. 49.3).

The key criteria for the successful TDRC—SATREPS on the TSL platform being one of them—include the competency of the project members, problem formulation, criteria for the objective evaluation, financing, workflow, methodology and integration, results and communication outreach, and social implementation of the project outcomes and impacts on the society. This TDRC for the TSL platform has been found to be contributing to solving problems of environmental degradation of TSL through evidence-based policies with a special focus on long-term TSL sustainability.

Key Points

- TDRC formed by the inclusion of the academicians, bureaucrats, policymakers, local people, and business entity helps attain pragmatic solutions for environmental conservation and management.

- In international TDRC, it is necessary to elucidate the specific policy impact that the collaboration would like to target and the quantitative means of assessing the same, during TDRC implementation and after policy formulation sustainably even after the completion of the TDRC project.
- It is necessary to clarify whether the outcomes from TDRC are to be used solely for research analysis, policy formulation, or sustainable management, consequently sharing the experiences with the broader international community.
- The key criteria for the success of TDRC include the competency of the project members, problem formulation, criteria for the objective evaluation, financing, workflow, methodology and integration, results and communication outreach, and social implementation of the project outcomes and impacts.

References

Bergmann M, Brohmann B, Hoffmann E, Loibl MC, Rehaag R, Schramm E, Voß JP. Quality criteria of transdisciplinary research: A guide for the formative evaluation of research projects. Frankfurt am Main, Germany: ISOE-Studientext; 2005. p. 13.

Brandt P, Ernst A, Gralla F, Luederitz C, Lang DJ, Newig J, Newig J, Reinert F, Abson D, Von Wehrden H. A review of transdisciplinary research in sustainability science. Ecol Econ. 2013;92:1–15.

Khanal R. Challenges of science-policy interlinkages. JJpn Soc Hydrol Wat Res. 2021;34(3): 144–5.

Kloprogge P, Van Der Sluijs JP. The inclusion of stakeholder knowledge and perspectives in integrated assessment of climate change. Clim Chang. 2006;75(3):359–89.

Mobjörk M. Consulting versus participatory transdisciplinarity: a refined classification of transdisciplinary research. Futures. 2010;42(8):866–73.

Pohl C. From science to policy through transdisciplinary research. Environ Sci Pol. 2008;11(1): 46–53.

Stokols D. Toward a science of transdisciplinary action research. Am J Community Psychol. 2006;38(1–2):63–77.

Wiek A. Challenges of transdisciplinary research as interactive knowledge generation–experiences from transdisciplinary case study research. GAIA-Ecol Perspect Sci Soc. 2007;16(1):52–7.

Chapter 50
Recommendations for Further Research and Environmental Management

Chihiro Yoshimura, Pham Ngoc Bao, Hideto Fujii, Tomohiro Tanaka, Sokly Siev, Rajendra Khanal, Kazuhiko Miyanaga, Ly Sophanna, Eden M. Andrews, Toru Watanabe, and Uk Sovannara

50.1 This Book and Steps Forward

This book compiles all the relevant available information, knowledge, and techniques to delineate the uniqueness and importance of the water and life of the Tonle Sap Lake (TSL). However, similar to other themes in environmental science, we have identified several knowledge gaps as described in each chapter. Thus, we sincerely hope that this book will be used as a solid and concise basis to further develop science on TSL in the general framework of tropical limnology and to

C. Yoshimura (✉) · K. Miyanaga · E. M. Andrews · U. Sovannara
Tokyo Institute of Technology, Tokyo, Japan
e-mail: yoshimura.c.aa@m.titech.ac.jp; miyanaga.kazuhiko@jichi.ac.jp

P. N. Bao
Institute for Global Environmental Strategies, Hayama, Japan

H. Fujii · T. Watanabe
Yamagata University, Tsuruoka, Japan

T. Tanaka
Kyoto University, Kyoto, Japan

S. Siev
Institute of Technology of Cambodia, Phnom Penh, Cambodia

Ministry of Industry, Science, Technology and Innovation, Phnom Penh, Cambodia

R. Khanal
Tokyo Institute of Technology, Tokyo, Japan

Policy Research Institute, Kathmandu, Nepal

L. Sophanna
Ministry of Environment, Phnom Penh, Cambodia

Tokyo Institute of Technology, Tokyo, Japan

C. Yoshimura et al. (eds.), *Water and Life in Tonle Sap Lake*,
https://doi.org/10.1007/978-981-16-6632-2_50

develop substantial management policy and practices for wise use of this precious water body in domestic and international perspectives.

From this standpoint, we, all the part editors, carefully selected and concisely summarized the important remaining knowledge gaps and directions for lake environmental management in this chapter as our recommendation. We believe that this chapter will help you pick up the focuses of your research and policy development together with a good overview. In case some detailed explanation is required on each topic, please refer to the corresponding parts or chapters.

50.2 Recommendations for Further Research

50.2.1 Socioeconomics and Governance (Part I)

The livelihoods and well-being of the local communities in the TSL region are linked with the environmental conditions of the lake-floodplain system, which is one of the excellent research themes in cultural anthropology. To enhance the resilience of the local communities, it is important to further strengthen our understanding of the potential impacts of environmental and climate stresses on them. To achieve this goal, there is a strong and urgent need for further understanding of the relationship among the livelihoods, governance, and lake ecosystem based on accurate and reliable data of the lake ecosystem, particularly the fish community.

50.2.2 Climate and Hydrology (Part II)

It is necessary to enhance the network of climatic and hydrological observations in the lake basin as the current network is obviously insufficient in this large and dynamic basin. Such an observation network needs to be capable of monitoring climatic conditions and river discharge based on updated rating curves. Given the scale of the lake, we additionally recommend research and development to fully utilize satellite-based remote sensing techniques for monitoring hydrological processes together with a cloud-based platform for data sharing and modeling integrating ground observation. This approach for environmental monitoring together with the use of archived data will also help us understand the shifts in local climate, environment, and ecosystem covering the past few decades.

50.2.3 Hydrodynamics (Part III)

We should continue the use of observation-based approaches, such as a 1-year measurement of water velocity profile and time series analysis of lake water levels, for elucidating the unique hydrodynamics in TSL. With data integration, the 2D hydrodynamic model should be used for longer periods and future climate change scenarios to project possible changes of hydrodynamics over the lake. At the village scale, the 3D hydrodynamic model can be combined with environmental assessments, in which a vertical profile of flow velocity is important (e.g., sediment dynamics and fate and transport of pollutants and pathogens). The integration of groundwater exchange and wind effect in hydrodynamic models would widen their applicability for broader issues related to hydrology and geochemistry.

50.2.4 Sediment Dynamics (Part IV)

Further research should be conducted to fully understand the sediment balance of TSL, the sedimentation–resuspension process, and those related to the Mekong River basin by integrating field observation, remote sensing, and modeling techniques. In particular, the relationship among wind conditions, turbulence, and sediment resuspension in the lake needs to be elucidated. Furthermore, the shift in sediment load and dynamics should be quantified and modeled in relation to forest cover change and water resource management in local basins and the Mekong River basin. This is because these factors will possibly cause an imbalance in the sediment dynamics in TSL and its floodplain, resulting in geomorphological shifts and environmental and ecological deteriorations in the lake in short- and long-term periods.

50.2.5 Physicochemical Water Quality (Part V)

The assessment of the historical and current trends of spatiotemporal variation in basic water quality, nutrients, and chlorophyll *a* indicated some hotspots in TSL where the water quality matrix significantly varies. In particular, areas of floating villages, hypoxia, and eutrophication are predominated. Nevertheless, further research is required to gain a full understanding of groundwater exchange, photochemical reactions, and biogeochemistry at the water–sediment interface in relation to the flow pattern, sediment dynamics, and primary production. We also need to cover the emission and absorption of greenhouse gases in this lake. These remaining knowledge gaps are possibly elucidated by the integration of isotopic analysis and process-based modeling of those water quality components.

50.2.6 Microbial Community (Part VI)

In general, physicochemical (e.g., aerobic/anaerobic, pH, temperature, and nutrients) and biological (e.g., prey–predator relationship) factors affect the microbial consortia. Thus, it is important to investigate not only microbial community composition but also their interactions with the factors in the lake. Furthermore, the productivity and functional roles of bacterial and viral communities in the lake ecosystem are our next challenge, which will elucidate the importance of the microbial loop in biogeochemistry and the food web. For those challenges, culture- and gene-based approaches are available. Gene-based analysis can reveal the whole microbial consortia, including unculturable microbes. Thus, both approaches should be properly used and combined to gain a comprehensive understanding of the microbial community.

50.2.7 Flora and Fauna (Part VII)

Flooded forests and aquatic vegetation are ideal and important habitats for both aquatic and terrestrial biota. They are critically important to maintain intact ecosystems and ecosystem functions/services. However, their ecological role has been poorly investigated, and they are under imminent threat from anthropogenic pressures. Therefore, further research should be conducted to understand their spatio-temporal distribution, the ecological traits of key species including migration, and the multispecies interactions with the explicit linkage to physical and chemical conditions in the floodplain. For this purpose, it would be beneficial to combine models of hydrodynamics, water quality, primary production, and fish migration and production, together with other techniques, for example, using isotopes and environmental DNA.

50.2.8 Chemical Pollution (Part VIII)

Chemical pollution in TSL is caused mainly by heavy metals, pesticides, and antibiotics. These chemicals come from direct disposal from the floating villages and runoff via the tributaries. Persistent organic pollutants are evidently seeping into the food chain. In addition, plastic pollution is now apparent in TSL and is still critically understudied. The current status of the pollution highlights the importance of the establishment of standard analytical methods and strategic monitoring of the pollutants in TSL. Research on the understanding of their fates (e.g., distribution, bioaccumulation, biomagnification, biodegradation, photodegradation, and transport) in relation to the unique flood pulse is also required, which can help us develop and recommend pollution remediation techniques.

50.2.9 Sanitation and Health Risk (Part IX)

Almost all of the people living on TSL have frequently experienced diarrheal diseases probably due to the poor sanitary environment, which is characterized by the use of lake water as a source of drinking water and the absence of adequate toilet. The environmental sanitation can be improved by onsite treatments of lake water to provide safer drinking water as well as fecal matter to prevent lake water pollution. Therefore, the development of low-cost technologies is highly demanded for the safe use of lake water and the proper treatment of fecal matter. In addition, to gain public acceptance of the developed technologies, quantitative microbial risk assessment targeting not only *Escherichia coli* but also more emerging pathogens, such as norovirus, should be conducted with practical scenarios of implementing countermeasures.

50.2.10 Environmental Shifts and Management (Part X)

Environmental management should be implemented based on a sufficient understanding of the interrelations of environmental processes and human dimensions (e.g., socioeconomics, ecosystem service, climate change, land use, and waste management). In the case of TSL, we are still at the initial stage of the investigation to gain a comprehensive overview of such a complex system. Therefore, while knowledge and techniques are being updated as aforementioned, they need to be actively used for scenario analysis concerning climate change, land use/land cover, health risk, agriculture, and aquaculture for policy integration for realizing sustainable development. For this purpose, we highly recommend transdisciplinary research and science communication as described in Chap. 49.

50.3 Recommendations for Environmental Management

50.3.1 Socioeconomics and Governance (Part I)

The identified environmental and climate change impacts on the livelihoods and well-being of the local communities in the TSL region must be well addressed through collective actions among actors, including researchers, policymakers, environmentalists, relevant governmental agencies, and private sectors. Fair and open dialogs based on scientific evidence among local and international stakeholders must be considered, which could facilitate the scientific evidence-based policymaking process. The existing governance system is still complex and cannot guarantee sound environmental management and good governance of the lake ecosystem.

Therefore, government interventions need to be implemented to improve the performance of the current governance system.

50.3.2 Climate and Hydrology (Part II)

The development of the many hydropower dams on the upper reach of the Mekong River, together with the impacts of climate change, has greatly caused concerns over the potential threat to the biological diversity resulting from a significant change to the hydrological regimes in TSL. The development and integration of tools, management plans, and measures and policies about water resources management are required to rigorously evaluate and mitigate the abovementioned impacts on the TSL ecosystem for a sustainable regional development. Capacity building programs are also needed to secure sufficient human resources for the use of those tools. In addition, it is important to strengthen the cross-cutting collaboration among all stakeholders for the sake of the ease of data sharing.

50.3.3 Hydrodynamics (Part III)

The long-term monitoring of hydrodynamic processes is essential for determining hydrodynamic changes possibly due to climate change and the shifted flow regime in the Mekong River. With regard to the environmental management, the 2D-LIE model helps identify the hydrodynamic condition required for fishery, agriculture, and the unique ecosystem. This model suggested that the average retention time is approximately 4 months, which may be considered as the turnover time of the lake. In addition, local hydrodynamics at floating villages can be analyzed using the 3D non-hydrostatic hydraulic model as demonstrated at the floating village of Chhnok-Trou. The simulated flow field serves as a basis for understanding the water quality conditions and habitat conditions of aquatic flora and fauna.

50.3.4 Sediment Dynamics (Part IV)

TSL is extremely sensitive to sediment dynamics, while the available sediment data and knowledge are still limited. Therefore, we recommend the development of a system for monitoring sediment concentration, load, and quality in the scales of tributary basins and the whole lake. The monitored data are useful for further development of sediment transport models for the TSL basin, which are required for understanding sediment balance in TSL and implementing various scenario analyses concerning the shifting factors of land use/land cover, water resources, and climate change. Furthermore, the high concentration of lead in the sediment and the lake water emphasized its potential risk to humans and the ecosystem. Thus, continuous monitoring of lead and the development of methods for mitigating its effect are necessary.

50.3.5 Physicochemical Water Quality (Part V)

For the effective management of the lake, continuous long-term monitoring of the water quality is deemed necessary. Considering the size of TSL, it is laborious and capital-intensive to continue frequent ground-based monitoring. Thus, it is highly recommended to integrate continuous monitoring by fixing multi-sensor probes at some points in the lake (ideally, real-time online system) and remote sensing techniques (i.e., satellite imagery) to minimize the cost of the ground survey. In addition, given the serious pollution at floating villages, the collection and treatment systems of solid waste and wastewater are urgently required for major floating villages. It is also important to restrict the number or density of the floating houses for maintaining a healthy environment and the unique ecosystem in TSL and its floodplain.

50.3.6 Microbial Community (Part VI)

The concentration of *E. coli* in the floating villages is higher than that in other areas of the lake. Moreover, microcystin that is produced by blue-green algae is strongly influenced by eutrophication and metal availability. To reduce the risk of waterborne disease, relatively simple treatment methods (e.g., sand filtration, flocculation, and chlorination) must be introduced to remove bacteria, algae, and viruses from the lake water for daily use at the village or household level. In addition, integrated lake basin management is highly recommended for controlling pollution and reducing the health risk of waterborne diseases. To this end, the key options are the introduction of wastewater treatment facilities in major communities in the lake basin and the reduction of loads of nutrients, pesticides, and antibiotics to the lake.

50.3.7 Flora and Fauna (Part VII)

Continuous monitoring of endemic species and their habitats in the flooded forests is necessary to understand the status of the floodplain ecosystems and develop preventive and adaptive measures against anthropogenic threats and climate change. Based on such understanding and development, a practical guideline for conservation and wise use of the floodplain ecosystem should be formulated and put into practice. Law enforcement needs to be strict against illegal activities so that the guidelines and countermeasures can be effectively implemented. This will need strong and close collaborative work among relevant governmental agencies and institutions. Furthermore, such enforcement might lead to the identification of some potential wetlands for the conservation of flora and fauna, in particular, waterbirds.

50.3.8 Chemical Pollution (Part VIII)

The use of agrochemicals in Cambodia should be regulated. Based on their current distribution and risk, the current environmental legislation needs to expand the list of chemicals and establish standard analytical protocols for monitoring those chemicals. Urgently, farmers should be trained in good agricultural practices and proper use and handling of such chemicals. Capacity building is also vital for all the TSL stakeholders. In parallel, scientists, engineers, environmental managers, and policymakers should work together to address pollution problems through policy development, environmental monitoring, and remediation for conservation and sustainable use of the lake, avoiding serious degradation of the TSL ecosystem. Pollution prevention is fundamentally more desirable than pollution mitigation.

50.3.9 Sanitation and Health Risk (Part IX)

Hygienic behaviors are not commonly practiced among people living in the TSL region. For instance, washing hands with soap before eating and after defecation is easy and effective to avoid infection with pathogens originating from the lake. Hygienic behaviors eventually help people avoid lake water pollution. For this purpose, we strongly recommended the development of an educational program to promote hygienic behaviors. The educational program may use visualized materials based on 2D and 3D simulations of health risks under various scenarios, which can effectively help people understand the importance of hygienic practices. Such simulations tell us where and when people can obtain safe water and which countermeasure and treatment facility should be introduced to villages on TSL.

50.3.10 Environmental Shifts and Management (Part X)

To realize sustainable development of the TSL region, research and management strategies need to be strengthened to (1) bridge the gap between researchers and policymakers via a series of evidence-based policy dialogs; (2) promote appropriate technologies that are economic, effective, environmentally friendly, and locally implementable for water and wastewater treatment; (3) assess health and ecological risks in the areas of floating villages; (4) remap the flooded forest, conservation area, and ecologically sensitive zones based on fundamental ecological understanding; and (5) develop and implement the policy framework for integrated lake basin management. On top of these, climate change needs to be additionally considered to develop mitigation and adaptation measures.

50.4 Final Remarks

At the same time, even after 5 years of intensive research combining literature survey, this book clearly highlights the limitation of our understanding of the TSL ecosystem. To take the next challenges, we humbly hope that the book will serve as a reference and encourage further investigations to improve our understanding of the ecosystem and our capacity for efficient evidence-based environmental management. Thus, we welcome any constructive feedback and comments, discussion, and collaboration on any topics related to the TSL ecosystem. Throughout the project, we have also confirmed again the importance of transdisciplinary research collaboration among various institutions and a network of scientists and researchers in various fields of expertise. We believe that transdisciplinary networks would advance our limnological knowledge, which would allow researchers, scientists, policymakers, and stakeholders to develop management strategies toward enhanced lake environmental sustainability.

Before closing, all the book editors and the part editors express sincere gratitude to all the authors who spent their substantial amount of time on writing and editing each corresponding chapter. We also appreciate the internal and external reviewers who provided us with their valuable expert comments to help us refine the quality of each chapter.

GPSR Compliance

The European Union's (EU) General Product Safety Regulation (GPSR) is a set of rules that requires consumer products to be safe and our obligations to ensure this.

If you have any concerns about our products, you can contact us on ProductSafety@springernature.com

In case Publisher is established outside the EU, the EU authorized representative is:

Springer Nature Customer Service Center GmbH
Europaplatz 3
69115 Heidelberg, Germany

Batch number: 10755214

Printed by Printforce, the Netherlands